Nonparametric Inference

This book provides a comprehensive and balanced treatment of both classical and modern methods in nonparametric inference. It begins with foundational topics such as order statistics, ranks, and confidence intervals for medians and percentiles before progressing to distribution-free tests, robust estimators, regression quantiles and U-statistics. Advanced topics include nonparametric density and regression estimation, model diagnostics, empirical likelihood, and survival analysis, including nonparametric Bayesian and maximum likelihood estimators. The book uniquely integrates these topics into a single resource, making it distinct from other texts in the field.

Key Features:

- A balanced blend of classical methods (e.g., rank and sign tests) and modern techniques (e.g., bootstrap, empirical likelihood, and nonparametric regression).
- Comprehensive coverage of nonparametric density and regression estimation, model diagnostics, and survival analysis, including Bayesian and maximum likelihood approaches.
- Unique inclusion of empirical likelihood inference, a broadly applicable and essential methodology for contemporary graduate courses.
- Numerous exercises and notes at the end of chapters to reinforce concepts and provide historical context.
- Designed for both teaching and reference, offering up-to-date techniques in nonparametric inference.

This text is ideal for a two-semester course on nonparametric inference for graduate students in statistics, applied mathematics, machine learning, and computer science. It also serves as a valuable reference for researchers and practitioners interested in nonparametric methods. Its comprehensive scope, including empirical likelihood, nonparametric Bayes, and bootstrap methodologies, makes it a unique resource. Notes at the end of each chapter provide insights into the chronological development of the field, while numerous exercises help reinforce the concepts and methodologies presented.

CHAPMAN & HALL/CRC

Texts in Statistical Science Series

Recently Published Titles

Statistics in Survey Sampling
Jae Kwang Kim

Time Series: A Data Analysis Approach Using R
Robert H. Shumway and David S. Stoffer

Bayesian Statistical Methods: With Applications to Machine Learning
By Brian J. Reich, Sujit K. Ghosh

Experimental Design for Data Science and Engineering
By V. Roshan Joseph

Survival Analysis: Principles and Applications in Clinical Trials and Beyond
Song Yang

Foundations of Bayesian Statistics for Data Scientists: With R and Python
Alan Agresti, Maria Kateri, Ranjini Grove, and Antonietta Mira

A Course in Regression and Smoothing Methods
Zhiqiang Tan

A Course in Large-sample and High-dimensional Theory
Zhiqiang Tan

Markov Chain Monte Carlo
Stochastic Simulation for Bayesian Inference
Dani Gamerman, Hedibert Lopes, and Flávio Bambirra Gonçalves

Introduction to Modern Randomization-Based Design and Analysis for Causal Inference
Tirthankar Dasgupta and Donald B. Rubin

The Analysis of Time Series
An Introduction with R, Eighth Edition
Haipeng Xing and Chris Chatfield

Regressions in Covariances, Dependencies and Graphs
Mohsen Pourahmadi and Aramayis Dallakyan

Nonparametric Inference
Hira L. Koul, Anton Schick, and Palaniappan Vellaisamy

For more information about this series, please visit: https://www.routledge.com/Chapman--HallCRC-Texts-in-Statistical-Science/book-series/CHTEXSTASCI

Nonparametric Inference

Hira L. Koul
Michigan State University

Anton Schick
Binghamton University, New York

Palaniappan Vellaisamy
University of California, Santa Barbara

CRC Press
Taylor & Francis Group
Boca Raton London New York

CRC Press is an imprint of the
Taylor & Francis Group, an **informa** business
A CHAPMAN & HALL BOOK

First edition published 2027
by CRC Press
2385 NW Executive Center Drive, Suite 320, Boca Raton FL 33431

and by CRC Press
4 Park Square, Milton Park, Abingdon, Oxon, OX14 4RN

CRC Press is an imprint of Taylor & Francis Group, LLC

ISBN: 978-1-032-95613-8 (hbk)
ISBN: 978-1-032-95620-6 (pbk)
ISBN: 978-1-003-58577-0 (ebk)

DOI: 10.1201/9781003585770

Typeset in LMRoman
by KnowledgeWorks Global Ltd.

Publisher's note: This book has been prepared from camera-ready copy provided by the authors.

Dedication

In the memory of Excellent and Inspiring Professors

Sharashchandra Raghunath Adke and Vasant Prabhakar Bhapkar
and Shyam Pyari Koul for her continued support

Hira L. Koul

Jeanette for her unwavering support

Anton Schick

Kolammal Nageswari for her unconditional support

Palaniappan Vellaisamy

Contents

Preface

In nonparametric statistical inference the population distribution or density function is not assumed to be of any parametric form. Instead they are assumed to belong to a given class of distributions or density functions and the inference is made about the entire distribution or density function. Nonparametric inference procedures are developed under minimal assumptions that are broadly applicable.

Historically, the development of nonparametric inference took place at an accelerated rate in the second half of the twentieth century. Initially, it focussed on the development of distribution free tests mostly in additive models. These tests are typically based on signs and ranks of data and empirical distribution functions. The associated confidence intervals and various robust estimators were developed in the 1960s followed by the development of the empirical likelihood based inference procedures in the 1990s. Simultaneously, nonparametric estimation of density and regression functions took place in the 1960's. The application of nonparametric reliability theory to health sciences saw an accelerated development soon after 1972.

This book begins with a review of some basic notions of probability theory followed by a chapter on order statistics and ranks. This chapter also discusses universally applicable confidence intervals for population median and percentiles. The next chapter discusses the notion of testing of hypotheses, distribution free tests and confidence intervals for location parameters in the one and two sample location models. It also contains some rank tests for the one way classification problem and in the linear regression setup. The chapter on estimation includes some R, M and minimum distance estimators in location models including the linear regression model. It also contains a description of regression quantiles and regression rank scores. The book also contains chapters on U statistics, nonparametric density function estimation and nonparametric regression function estimation. The chapter on model diagnostics includes asymptotically distribution free tests for fitting an error distribution in location-scale models including the linear regression model. It also includes some tests of lack-of-fit of a parametric regression model and for comparing two regression functions. Other chapters are devoted to the topics like empirical likelihood inference and nonparametric maximum likelihood and nonparametric Bayes estimators of survival and cumulative hazard functions for randomly right-censored data.

There are many texts on various aspects of nonparametric inference with emphasis on either classical methods like rank and sign rank tests or with emphasis on some modern techniques like nonparametric density or regression functions estimation. Our aim is to provide a balanced mixture of both classical and modern methods in nonparametric inference in a single book. The book can serve as a text for a two semester course on nonparametric inference at the M.Sc/Ph.D. level in statistics and applied mathematics. It contains several up-to-date techniques in nonparametric inference. Some key modern topics such as the bootstrap, nonparametric regression and density functions estimation, model diagnostics and nonparametric and Bayesian survival analysis and empirical likelihood makes this book different from many other books at this level. No other currently available single book meant for M.A./M.Sc. in Statistics contains topics on empirical likelihood, nonparametric Bayes and maximum likelihood estimator of survival function or bootstrap methodology. We believe that this book will also serve as a useful reference for anyone interested in the theories and methods of nonparametric inference.

Some chapters of this book were used for teaching a masters' level one semester (about 33 hours) course for statistics students at the Indian Institute of Technology Bombay and some were used for teaching an M.Sc / Ph.D. level course at the Michigan State University (MSU). The chapters on density and regression function estimation, empirical likelihood and parts of chapter on U statistics were also taught at the Binghamton University, New York and MSU.

The lead author visited the Department of Mathematics of the Indian Institute of Technology Bombay, over several years since 2013, while the other two authors visited the Department of Statistics and Probability, the Michigan State University, several times. The authors are grateful to these departments for facilitating these visits. They are also grateful to Dr. Vijay Kumar of Thapar Institute of Engineering and Technology, Patiala, Panjab, India and to Mr. Shashi Kumar of KnowledgeWorks Global Ltd., India, for providing help with some LaTex issues.

Hira L. Koul
Anton Schick
Palaniappan Vellaisamy

June 8, 2026

Author Bios

Hira L. Koul secured his doctorate in statistics from the University of California, Berkeley, in 1967. He joined the Department of Statistics and Probability, Michigan State University (MSU), on January 1, 1968. Since January 1, 2018, he has been Professor Emeritus at the MSU after serving there as a faculty member for 50 years. His areas of research include nonparametric inference, inference on short- and long-memory processes, time series analysis and survival analysis. He has published around 150 papers, several monographs and books and guided 35 doctoral theses. He is a Fellow of the American Statistical Association and of the Institute of Mathematical Statistics and Past President of the International Indian Statistical Association. He was a recipient of a *Humboldt Research Award* for senior scientists in October 1995 and a *Distinguished Faculty Award* at the MSU, 2005.

Anton Schick earned his doctorate in statistics from Michigan State University in 1983. He spent one year at Tufts University before joining the Department of Mathematical Sciences at Binghamton University in the fall of 1984. He retired as full professor on September 1, 2024, after forty years of service that included two terms as chair. His research has focused on the characterization and construction of efficient statistical inference procedures in nonparametric and semiparametric models with an emphasis on regression and time series models, on curve estimation with parametric rates, on inference with incomplete data, and on the empirical likelihood approach. He has published one hundred twenty research papers and guided ten doctoral theses.

Palaniappan Vellaisamy is currently a Visiting Professor in the Department of Statistics and Applied Probability, University of California, Santa Barbara, USA. He completed his Ph.D. degree in statistics from Indian Institute of Technology Kanpur in 1989. He worked as a Research Associate from July 1989 to December 1990 at the Indian Statistical Institute, New Delhi. Then he joined in 1991 as an Assistant Professor in the Department of Mathematics at Indian Institute of Technology Bombay, India. He became a full professor in 2003 and retired in June 2024. His research areas include statistical inference, applied probability, and fractional stochastic processes. He has published more than 120 research papers in various journals of statistics and probability and has guided 11 Ph.D. dissertations. He is currently an Associate Editor for *Statistics and Probability Letters* and *The Journal of Indian Statistical Association*.

1

Introduction

In parametric statistical inference, one often assumes the knowledge of the sampling population up to an unknown finite-dimensional parameter, and the inference is made about this unknown parameter. In nonparametric inference, functions such as distribution, density or regression functions are the parameter of interest. It is assumed that the given function belongs to a certain class of functions and is not of a parametric form. Such classes are often described by some smoothing assumptions. The given data is used to make inferences about the functions or functionals thereof. The nonparametric inference procedures are developed under minimal assumptions, which make them broadly applicable.

Chapter 2 discusses some basic notions of probability theory, some distributions, expectation of a random variable and some key results of probability theory needed in this book. Order statistics play a fundamental role in nonparametric inference for a distribution on the real line. The confidence intervals based on order statistics for population percentiles achieve their confidence levels for all sample sizes, two or larger, and for all distributions, continuous or not. Chapter 3 discusses these inference procedures based on order statistics.

In statistics one is often interested in comparing two or more treatments. In a sociological study comparing the two treatments, Wilcoxon (1945) observed that the assumption of normality of the populations was not tenable. Instead of using the classical two sample t test, he ordered the combined two samples and ranked them from the smallest to the largest, and then replaced the two sample means in the two sample t test statistic by the averages of the ranks from the first and second sample. This statistic is called the Wilcoxon two sample rank sum statistic. The null distribution of this statistic does not depend on the underlying population as long as it is continuous. A test statistic is said to be distribution free if its null distribution does not depend on the underlying population. The Wilcoxon two sample rank sum statistic is a prime example of distribution free statistics. Some general notions of testing of hypotheses, the interval estimation of the location parameters and tests of hypotheses pertaining to the location parameters in the one and two sample location models using the Wilcoxon signed rank and the Wilcoxon two sample rank sum statistics, respectively, are discussed in Chapter 4. This chapter also discusses the normal scores signed rank and rank sum tests for the one and two sample location models, the Kruskal-Wallis rank test for the k-sample location model and an analog of the Wilcoxon rank sum test in the linear regression model.

A class of statistics that is often found useful in everyday use of statistical notions is known as the class of U statistics. Hoeffding (1948) provided a unified way of deriving various statistical properties of such statistics. Hájek (1968) crystalized Hoeffding's method of the proof of the asymptotic normality of U statistics. Chapter 5 discusses Hájek's projection lemma and exhibits its usefulness in proving the asymptotic normality of several statistics in the one and two sample settings. Hall (1984) proved a very useful lemma about the asymptotic normality of a class of degenerate U statistics. This chapter includes an extension of this result to a class of weighted degenerate U statistics.

DOI: 10.1201/9781003585770-1

These results are found useful in proving the asymptotic normality of some goodness-of-fit and lack-of-fit test statistics later in Chapter 9.

Pitman's asymptotic relative efficiency (ARE) of any two given tests is measured by comparing the asymptotic sample sizes needed to achieve the given power at the given level. Based on this notion, when testing the hypothesis about one sample population mean, the Wilcoxon signed rank test needs roughly 100 observations to achieve the same precision as the one sample t test based on 96 observations, provided the error distribution is Gaussian. This is a remarkable property. The Wilcoxon signed rank test is valid under broader assumptions than the t test, and needing extra 4 observations is not much of a price to pay for this. This property led researchers in the 1950s and 1960s to the discovery of broader classes of signed rank and rank sum tests. These classes include the so-called normal scores signed rank and rank sum tests, which are as good as the one and two sample t tests, respectively, at the normal error distribution and much better than the t test at non–normal error distributions in the ARE sense. Chapter 5 also contains the definition of ARE and its derivation for some tests.

The decade of the 1960s saw the development of robust estimators based on ranks and signed ranks. These are the so-called R estimators. They were developed by Hodges and Lehmann (1963) and Sen (1963a) for the one and two sample location models and by Jurečková (1969, 1971) for the linear regression model. Huber (1964, 1981) developed another class of estimators, the so-called M estimators. These estimators are obtained by using a modification of the likelihood score. In the location models that include regression models, all of these estimators are robust against contaminated error distributions and some of them are also asymptotically efficient at some error distributions. These estimation methods and a minimum distance estimation method in the one and two sample location models and in the linear regression model are described in Chapter 6 in some detail. This chapter also includes a short section on the regression quantiles and regression rank scores as developed by Koenker and Bassett (1978) and Gutenbrunner and Jurečková (1992).

The two classical problems of importance are that of making inference about a given population density and a general regression function. Both of these functions are assumed to belong to some classes of smooth functions without assuming them to be of parametric forms. Chapter 7 discusses nonparametric kernel density estimators, while Chapter 8 is devoted to Nadaraya-Watson and local polynomial regression function estimators. In particular, consistency and asymptotic normality of these estimators is derived in these chapters. In addition, Chapter 7 discusses some bandwidth issues, plug-in estimators of $\int f(x \pm y) f(y)\, dy$, where f is a density on $\mathbb{R}$ and some consistent estimators of $\int f^2(x)\, dx$ in the linear regression model, where f is an error density on $\mathbb{R}$. It also contains Beran's minimum Hellinger distance estimators of the one sample location parameter.

Chapter 9 contains some tests for the goodness-of-fit of a given error distribution function in the one sample location-scale model and in the linear regression model with random and non-random covariates. The classical Kolmogorov-Smirnov goodness-of-fit tests based on the residual empirical distribution function are often not asymptotically distribution free. This chapter discusses asymptotically distribution free tests based on the so-called Khmaladze (1981) transform of the residual empirical process. A class of goodness-of-fit tests based on integrated squared difference between a kernel density estimators and its null expectation are also discussed in this chapter. The above mentioned Hall's lemma about degenerate U statistics is found useful in proving the asymptotic normality of these test statistics.

Another problem of practical importance is that of checking the validity of a given parametric regression model. This problem is known as testing for the lack-of-fit of a given regression model. A class of lack-of-fit tests based on certain minimized integrated square distances between a nonparametric regression function estimate and the parametric model being fitted are also presented in this chapter. The Hall's lemma is also found useful in proving the asymptotic normality of these minimum distance test statistics. This chapter also contains some asymptotically optimal tests for testing the equality of the two regression functions against the one-sided alternatives in the two independent regression models setting. This problem arises when comparing the average effects of the two treatments over a range of predicting variables.

Wilks Theorem, see Wilks (1938), says that under some conditions on the given parametric model, when testing a parametric null hypothesis, the asymptotic null distribution of the likelihood ratio test statistic is a chi-square distribution. Owen (1988, 1990) derived a similar result using empirical likelihood. In principle, then, this result disposes of any parametric model assumptions and is broadly applicable. The inference based on empirical likelihood can be free of model assumptions. Chapter 10 discusses some basic notions of empirical likelihood and its usefulness in constructing confidence sets and tests in several important statistical problems. It also contains two proofs of the fact that the empirical d.f. is a nonparametric maximum likelihood estimator of the given population distribution function.

Survival analysis is concerned with analyzing the times of the occurrences of some events. For example, in medical science, the time it takes to cure a patient of a disease is of interest. In health-related studies, data observed is often censored. As an example consider a clinical experiment where patients are put on a treatment for a certain time period. If a patient is cured within the period of study, then we have a complete observation on the time to cure. But if a patient is not cured during the time period or leaves the study for some reason, then we have a randomly right-censored observation on the time to cure. Chapter 11 contains some nonparametric inference procedures for completely observed data and when data is randomly right-censored. In particular, it contains the Kaplan and Meier (1958) derivation of the nonparametric maximum likelihood estimators of the survival and cumulative hazard functions and a description of nonparametric Bayes estimator of the survival function due to Susarla and Van Ryzin (1976) for randomly right-censored data. It also contains a description of the estimators of parameters in the randomly right-censored linear regression model due to Buckley and James (1979) and Koul, Susarla and Van Ryzin (1981) and a brief description of Cox's proportional hazards model.

2

Some Preliminaries

2.1 Introduction

In this chapter we shall discuss some preliminary notions needed in the rest of the book, including the rudimentary definitions of probability distributions, random variables and random vectors, expectations, and some notions of convergence. We shall include the statements of important convergence results for integrals and limit theorems for averages and sums of random variables. This is done for the sake of completeness and easy reference.

Some notation and conventions. The following notation and conventions will be used in the book. By $y := x$ we mean y is defined to be x. For any positive integer p, the p-dimensional Euclidean space is denoted by $\mathbb{R}^p$ with the convention $\mathbb{R} = \mathbb{R}^1$. For any set D, the statement $x \in D$ stands for x belongs to D. For any $z \in \mathbb{R}^p$, z' denotes its transpose and $\|z\|$ its Euclidean norm. We also use M' to denote the transpose of the matrix M. Let f be a function from an open subset of $\mathbb{R}$ into $\mathbb{R}$. If f is differentiable, we denote its derivative by $\dot{f}$, and if f is twice differentiable, we denote its second derivative by $\ddot{f}$. For any $a, b \in \mathbb{R}$, $a \wedge b := \min(a, b)$, $a \vee b := \max(a, b)$. For any function f defined on the domain $\mathcal{D}$ to $\mathbb{R}^q$, q a positive integer, $\|f\|_\infty := \sup_{x \in \mathcal{D}} \|f(x)\|$.

Let f and g be real-valued functions with a common domain D. Then their sum $f + g$ is defined pointwise, i.e., $(f + g)(x) := f(x) + g(x)$, $x \in D$. Similarly, one defines their difference $f - g$, their product fg, and their ratio f/g, where the latter requires that g does not take the value 0. We write $f \le g$ if $f(x) \le g(x)$ for all $x \in D$. We also use the conventions

$$\{f < g\} := \{x \in D : f(x) < g(x)\}, \quad \{f \le g\} := \{x \in D : f(x) \le g(x)\},$$
$$\{f = g\} := \{x \in D : f(x) = g(x)\}, \quad \{f \ne g\} := \{x \in D : f(x) \ne g(x)\}.$$

For real-valued functions $f, f_1, f_2, \cdots$ with a common domain, the functions $\inf_{n \ge 1} f_n$, $\sup_{n \ge 1} f_n$, $\limsup_{n \to \infty} f_n$ and $\liminf_{n \to \infty} f_n$ and $\lim_{n \to \infty} f_n$ are defined pointwise. We write $f_n \uparrow f$ if for each x in the common domain, $f_1(x) \le f_2(x) \le f_3(x) \le \ldots$ and $\lim_{n \to \infty} f_n(x) = f(x)$.

2.2 Probability and Distributions

In this section, we discuss some basic probability notions and some distributions. A good reference for these notions is the book of Loève (1963). We start with the following definition.

Definition 2.2.1. *Sigma field.* Let Ω be a nonempty set. A collection $\mathscr{A}$ of subsets of Ω is called a *sigma field* or *σ-field* on Ω if it satisfies the following conditions. (i) The set Ω belongs to $\mathscr{A}$. (ii)

DOI: 10.1201/9781003585770-2

If $A \in \mathscr{A}$ then its complement $A^c := \{w \in \Omega : w \notin A\} \in \mathscr{A}$. (iii) If $A_i, i \geq 1$ is a countable collection of elements in $\mathscr{A}$, then $\cup_{i\geq 1} A_i \in \mathscr{A}$.

Let $\emptyset$ denote the empty set. Since it is the complement of Ω, $\emptyset$ belongs to $\mathscr{A}$. For any positive integer n, if we take $A_i = \emptyset$ for $i > n$ in (iii) of Definition 2.2.1, we see that the sigma field $\mathscr{A}$ contains all finite unions of its members. It follows from (ii), (iii) and De Morgan's law that the intersection of a countable collection of events is also an event. Indeed, $\cap_{i\geq 1} A_i = \left(\cup_{i\geq 1} A_i^c\right)^c$ belongs to $\mathscr{A}$ if $A_i, i \geq 1$ belong to $\mathscr{A}$.

The smallest sigma field is $\{\emptyset, \Omega\}$ and the largest sigma field is the power set of Ω which is the collection of all subsets of Ω. One can show that the intersection of a nonempty class of sigma fields on Ω is again a sigma field. This can be used to show that for any collection $\mathcal{C}$ of subsets of Ω there is a smallest sigma field containing $\mathcal{C}$. This sigma field is called the *sigma field generated by* $\mathcal{C}$.

On the real line $\mathbb{R}$, we work with the σ-field generated by the intervals $\{(-\infty, t]; t \in \mathbb{R}\}$. This σ-field is called the *Borel σ-field on* $\mathbb{R}$ and denoted by $\mathscr{B}$. It contains all open and hence all closed subsets of $\mathbb{R}$. On $\mathbb{R}^k$, k a positive integer, we work with the σ-field generated by the sets $\{(-\infty, t_1] \times \cdots \times (-\infty, t_k]; t_1, \cdots, t_k \in \mathbb{R}\}$. This σ-field is called the *Borel σ-field on* $\mathbb{R}^k$ and denoted by $\mathscr{B}_k$. It contains all open and hence all closed subsets of $\mathbb{R}^k$. Note that $\mathscr{B} = \mathscr{B}_1$.

Definition 2.2.2. *Probability space.* A triple $(\Omega, \mathscr{A}, P)$ is called a *probability space* if Ω is a nonempty set, $\mathscr{A}$ is a σ-field on Ω, and P is a *probability measure* on $\mathscr{A}$, i.e., a function from $\mathscr{A}$ to $[0, 1]$ satisfying (a) $P(\Omega) = 1$ and (b) for any countable collection $A_i, i \geq 1$ of disjoint members of $\mathscr{A}$, $P\left(\cup_{i\geq 1} A_i\right) = \sum_{i\geq 1} P(A_i)$. The property (b) is called sigma-additivity.

In applications, Ω represents the set of possible outcomes of an experiment and is called the sample space, $\mathscr{A}$ represents the class of subsets of Ω to which probabilities are assigned and P does this assignment. The elements of $\mathscr{A}$ are called events, and $P(A), A \in \mathscr{A}$ is read as the probability of the occurrence of the event A or, in short, the probability of A. The events or members of $\mathscr{A}$ are denoted by capital letters like A, B, etc.

Some properties of the probability measure P. We shall now describe some often used properties of the probability measure P. Let A, B and $A_1, A_2, \cdots$ be events. Then the following hold.

1. $P(\emptyset) = 0$.
2. $P(A^c) = 1 - P(A)$.
3. $A \subset B$ implies $P(A) \leq P(B)$.
4. $P(A \cup B) = P(A) + P(B) - P(A \cap B)$.
5. $P(A) = P(A \cap B) + P(A \cap B^c) \leq P(A \cap B) + P(B^c)$.
6. *Boole's inequality.* $P\left(\cup_i A_i\right) \leq \sum_i P(A_i)$.
7. *Continuity from below.* If $A_1 \subset A_2 \subset A_3 \subset \cdots$ and $A := \cup_{n\geq 1} A_n$, then $\lim_{n\to\infty} P(A_n) = P(A)$.
8. *Continuity from above.* If $A_1 \supset A_2 \supset A_3 \supset \cdots$ and $A := \cap_{n\geq 1} A_n$, then $\lim_{n\to\infty} P(A_n) = P(A)$.

Definition 2.2.3. A probability space is called *discrete* if Ω is a countable set, $\mathscr{A}$ is the collection of all subsets of Ω and for some nonnegative function π defined on Ω such that $\sum_{\omega\in\Omega} \pi(\omega) = 1$, the probability measure P is defined to be $P(A) = \sum_{\omega\in A} \pi(\omega)$, $A \subseteq \Omega$.

Let $(\Omega, \mathscr{A}, P)$ be a probability space. Recall that $\mathscr{B}$ denotes the Borel σ-field on $\mathbb{R}$. Let X be a function from Ω to $\mathbb{R}$. Then the following are equivalent.

1. For every $t \in \mathbb{R}$, $\{X \le t\} := \{\omega \in \Omega; X(\omega) \le t\}$ belongs to $\mathscr{A}$.
2. For every $B \in \mathscr{B}$, $\{X \in B\} := \{\omega \in \Omega; X(\omega) \in B\}$ belongs to $\mathscr{A}$.

This follows as the collection $\mathcal{C} := \{B \in \mathscr{B}; \{X \in B\} \in \mathscr{A}\}$ is a σ-field on $\mathbb{R}$ and hence equals $\mathscr{B}$ if it contains the intervals $(-\infty, t]$, $t \in \mathbb{R}$.

Definition 2.2.4. Random Variable. A function X from Ω to $\mathbb{R}$ that satisfies $\{X \le t\} \in \mathscr{A}$ for all $t \in \mathbb{R}$ is called a *random variable (r.v.)*.

With a r.v. X we associate the function F_X defined by

$$F_X(t) := P(X \le t), \quad t \in \mathbb{R},$$

and the probability measure P_X on $\mathscr{B}$ defined by

$$P_X(B) := P(X \in B) = P\{\omega \in \Omega; X(\omega) \in B\}, \quad B \in \mathscr{B}.$$

We call F_X the *distribution function (d.f.) of* X and P_X the *distribution of* X. The function F_X is right-continuous, nondecreasing, and has limits

$$F_X(-\infty) := \lim_{t \to -\infty} F_X(t) = 0 \quad \text{and} \quad F_X(\infty) := \lim_{t \to \infty} F_X(t) = 1.$$

These properties of F_X follow from those of the probability measure P. Moreover, the left-hand limit $F_X(x_-) := \lim_{t \to x_-} F(t)$ exists and equals $P(X < x)$, for each $x \in \mathbb{R}$. Because $\{X \le x\} = \{X < x\} \cup \{X = x\}$, we have the identity

$$\pi_X(x) := P(X = x) = F_X(x) - F_X(x_-), \quad x \in \mathbb{R}.$$

The set $J_X = \{x \in \mathbb{R}; \pi_X(x) > 0\}$ is countable. This follows from the identity $J_X = \bigcup_{n \ge 1} A_n$ and the fact that $A_n = \{x \in \mathbb{R}; \pi_X(x) \ge 1/n\}$ has at most n elements. Note that J_X is empty if and only if F_X is continuous.

A r.v. X is said to be *discrete* if $\sum_{x \in J_X} \pi_X(x) = 1$. In this case, for all $t \in \mathbb{R}$, $B \in \mathscr{B}$,

$$F_X(t) = \sum_{x \in J_X,\, x \le t} \pi_X(x); \qquad P_X(B) = \sum_{x \in J_X,\, x \in B} \pi_X(x).$$

We call F_X and P_X *discrete* d.f. and distribution, respectively, and π_X the *probability mass function*.

A r.v. X is said to be *continuous* if F_X is continuous. A r.v. X is said to have density f if f is a nonnegative function on $\mathbb{R}$ such that $\int_{-\infty}^{\infty} f(x)\,dx = 1$ and

$$F_X(t) = \int_{-\infty}^{t} f(x)\,dx, \quad t \in \mathbb{R}.$$

In this case F_X is said to be *absolutely continuous* with almost everywhere (a.e.) derivative f. If f is continuous, then F_X is differentiable with derivative $\dot{F}_X = f$.

Motivated by the above, a nondecreasing right-continuous function F from $\mathbb{R}$ to $[0,1]$ with $F(-\infty) = 0$ and $F(\infty) = 1$ is called a distribution function. We shall show in Theorem 2.9.1 below that for every d.f. F there is a r.v. X such that $F_X = F$.

Now consider two random variables (r.v.'s) X and Y with respective distribution functions (d.f.'s) F_X and F_Y. These d.f.'s are often called marginal d.f.'s to distinguish them from the joint d.f. $F_{X,Y}$ of X and Y defined as

$$F_{X,Y}(x,y) := P(X \le x, Y \le y), \quad x, y \in \mathbb{R},$$

where $(X \le x, Y \le y) := \big(\{\omega \in \Omega; X(\omega) \le x\} \cap \{\omega \in \Omega; Y(\omega) \le y\}\big)$. Since $F_{X,Y}$ is a function of two variables, it is often called a bivariate distribution function. The joint distribution of X and Y is the probability measure $P_{X,Y}$ on $\mathscr{B}_2$ defined by

$$P_{X,Y}(B) = P\big((X,Y) \in B\big), \quad B \in \mathscr{B}_2.$$

We say X and Y are identically distributed if $F_X = F_Y$ and write $X =_D Y$ to express this. The marginal d.f.'s of X and Y can be derived from their joint d.f. via the formulas

$$F_X(x) := F_{X,Y}(x,\infty) = P(X \le x), \quad F_Y(y) = F_{X,Y}(\infty, y) = P(Y \le y), \quad x, y \in \mathbb{R}.$$

We say X and Y have a joint density f if f is a nonnegative function on $\mathbb{R}^2$ such that

$$F_{X,Y}(s,t) = \iint I(x \le s, y \le t) f(x,y)\, dx\, dy, \quad s, t \in \mathbb{R}.$$

This implies that $\iint f(x,y)\, dx\, dy = 1$. Any nonnegative function defined on $\mathbb{R}^2$ with this property is called a *bivariate density*. If X and Y have a joint density f, then

$$P\big((X,Y) \in B\big) = \iint I\big((x,y) \in B\big) f(x,y)\, dx\, dy, \quad \forall B \in \mathscr{B}_2.$$

In this case X and Y have marginal densities

$$f_X(x) := \int_{\mathbb{R}} f(x,y)\, dy, \quad f_Y(y) := \int_{\mathbb{R}} f(x,y)\, dx.$$

We say X and Y have a joint probability mass function π if π is a nonnegative function on $\mathbb{R}^2$ with countable support $A = \{\pi > 0\}$ such that

$$F_{X,Y}(s,t) = \sum_{(x,y) \in A, x \le s, y \le t} \pi(x,y), \quad s, t \in \mathbb{R}.$$

This implies $\sum_{(x,y)\in A} \pi(x,y) = 1$. Any nonnegative function defined on $\mathbb{R}^2$ with countable support and having this property is called a *bivariate probability mass function*. If X and Y have a joint probability mass function π, then X and Y are discrete r.v.'s with marginal probability mass functions

$$\pi_X(x) = \sum_{y:(x,y)\in A} \pi(x,y), \qquad \pi_Y(y) = \sum_{x:(x,y)\in A} \pi(x,y).$$

The r.v.'s X and Y are said to be *independent*, if

$$P\big(X \in A, Y \in B\big) = P\big(X \in A\big)P\big(Y \in B\big), \quad \forall A, B \in \mathscr{B}.$$

This implies

$$F_{X,Y}(s,t) = F_X(s)F_Y(t), \quad \forall s, t \in \mathbb{R}.$$

It can be shown that the latter also implies the former. If X and Y have marginal densities f_X and f_Y, then X and Y are independent if the function f defined by

$$f(x,y) := f_X(x) f_Y(y), \quad x, y \in \mathbb{R},$$

is a joint density of X and Y. If X and Y have marginal probability mass functions π_X and π_Y, then X and Y are independent if the function π defined by

$$\pi(x,y) := \pi_X(x)\pi_Y(y), \quad x, y \in \mathbb{R},$$

is the joint probability mass function of X and Y. Independence of X and Y implies that the joint distribution $P_{X,Y}$ of (X,Y) is the product measure $P_X \otimes P_Y$ of the marginal distributions, i.e., $P_X \otimes P_Y$ is the unique probability measure on $\mathscr{B}_2$ that satisfies

$$P_X \otimes P_Y(A \times B) = P_X(A)P_Y(B), \quad \forall A, B \in \mathscr{B}_1.$$

2.3 Integration with a Probability Measure

Let $(\Omega, \mathscr{A}, P)$ be a probability space. For any function γ from Ω into the extended real line $[-\infty, \infty]$, let $\{\gamma \le t\} := \{\omega \in \Omega; \gamma(\omega) \le t\}$, $t \in \mathbb{R}$. Such a function γ is said to be measurable if $\{\gamma \le t\} \in \mathscr{A}$ for all $t \in \mathbb{R}$. When working with the extended real numbers, we use the conventions that $a + \infty = \infty$, for all $a > -\infty$ and $a \times \infty := -\infty, 0, \infty$, for $a < 0, a = 0, a > 0$, respectively.

For any event A, the indicator function $I_A(\cdot)$ of A is defined as follows.

$$I_A(x) = \begin{cases} 1, & \text{if } x \in A, \\ 0, & \text{if } x \in A^c. \end{cases} \tag{2.3.1}$$

For the sake of brevity we shall often write $I(A)$ for the indicator function and read that $I(A) = 1$, if A occurs and $I(A) = 0$, if A does not occur.

Let $\mathfrak{M}_+$ denote the set of all measurable functions from Ω to $[0, \infty]$. We shall first define the integral $\int \gamma \, dP \equiv \int_\Omega \gamma(\omega) \, dP(\omega)$, for γ in $\mathfrak{M}_+$. A γ in $\mathfrak{M}_+$ is said to be *simple* if the range R of γ consists of a finite number of nonnegative real values so that $\gamma = \sum_{y \in R} y I(\gamma = y)$. In this case we set

$$\int \gamma \, dP := \sum_{y \in R} y P(\gamma = y).$$

In particular, if γ is a constant function, say $\gamma(\omega) = c$, $\omega \in \Omega$, for some finite $c \ge 0$, then $R = \{c\}$ and $\int \gamma \, dP = cP(\Omega) = c$. If $\gamma = I_A$ is the indicator function of an event A different from $\emptyset$ and Ω, then $R = \{0, 1\}$ and $\int I_A \, dP = 0 \times (1 - P(A)) + 1 \times P(A) = P(A)$.

For $\gamma \in \mathfrak{M}_+$, let γ_n denote the simple function

$$\gamma_n(\omega) := \sum_{j=1}^{n2^n} \frac{j-1}{2^n} I\Big(\frac{j-1}{2^n} \le \gamma(\omega) < \frac{j}{2^n}\Big) + nI(\gamma(\omega) \ge n), \quad \omega \in \Omega. \tag{2.3.2}$$

Then

$$\int \gamma_n \, dP = \sum_{j=1}^{n2^n} \frac{j-1}{2^n} P\Big(\frac{j-1}{2^n} \le \gamma < \frac{j}{2^n}\Big) + nP(\gamma \ge n).$$

Moreover, the sequence $\int \gamma_n \, dP$ is nondecreasing and hence has a limit in $[0, \infty]$. We define

$$\int \gamma \, dP = \lim_{n \to \infty} \int \gamma_n \, dP.$$

Note that this limit may be ∞ even if γ is real-valued.

This integral has the following properties. The proofs of the many results stated below can be found in Loève (1963) or Athreya and Lahiri (2006).

$$\int (f+g)\,dP = \int f\,dP + \int g\,dP, \qquad f,g \in \mathfrak{M}_+,$$
$$\int cf\,dP = c\int f\,dP, \qquad c \in [0,\infty], f \in \mathfrak{M}_+,$$
$$\int f\,dP \le \int g\,dP, \qquad f,g \in \mathfrak{M}_+, f \le g,$$
$$\text{For } f \in \mathfrak{M}_+, \quad \int f\,dP = 0 \quad \text{implies} \quad P(f=0)=1.$$

Theorem 2.3.1. [Monotone Convergence]. *Let $\gamma, \gamma_1, \gamma_2, \dots$ be nonnegative measurable functions such that $\gamma_n \uparrow \gamma$. Then, $\int \gamma_n\,dP \to \int \gamma\,dP$.*

Lemma 2.3.1. [Fatou Lemma]. *Let $\gamma_1, \gamma_2, \dots$ be nonnegative measurable functions. Then,*

$$\liminf_{n\to\infty} \int \gamma_n\,dP \ge \int \liminf_{n\to\infty} \gamma_n\,dP.$$

For a measurable function γ from Ω into $[-\infty,\infty]$, the integral $\int \gamma\,dP$ is defined as follows. Write $\gamma = \gamma^+ - \gamma^-$, where $\gamma^+ := \max(\gamma,0)$ and $\gamma^- := \max(-\gamma,0)$. Note that γ^+ and γ^- belong to $\mathfrak{M}_+$. For example, the measurability of γ^+ follows from $\{\gamma^+ \le t\} = \emptyset$ if $t < 0$ and $\{\gamma^+ \le t\} = \{\gamma \le t\}$ for $t \ge 0$. This allows us to define

$$\int \gamma\,dP = \int \gamma^+\,dP - \int \gamma^-\,dP,$$

as long as at least one of the integrals on the right-hand side is finite. In this case, we say the integral of γ exists. If both integrals are finite, then $\int \gamma\,dP$ is finite, and we say γ is integrable. This happens if and only if $\int |\gamma|\,dP = \int \gamma^+\,dP + \int \gamma^-\,dP$ is finite.

The next fact shows that the integral of a function of a random variable can be transformed into an integral of the function with respect to the distribution of the random variable. Let X be a r.v. defined on $(\Omega, \mathscr{A}, P)$ with distribution P_X and d.f. F_X. Let h be a measurable function from $\mathbb{R}$ to $\mathbb{R}$ such that $\int |h(X)|\,dP$ is finite. Then,

$$\int h(X)\,dP = \int_\Omega h(X(\omega))\,dP(\omega) = \int_{\mathbb{R}} h(x)\,dP_X(x) = \int h dP_X. \tag{2.3.3}$$

We shall often write $\int h(x)\,dF_X(x)$ instead of the more precise $\int h(x)\,dP_X(x)$. If F_X has a density f_X (a probability mass function π_X), then

$$\int h(X)\,dP = \int h(x) f_X(x)\,dx, \qquad \left(\int h(X)\,dP = \sum_{x:\pi(x)>0} h(x)\pi(x)\right).$$

Product Integrals. Let $(\Omega_1, \mathscr{A}_1, P_1)$ and $(\Omega_2, \mathscr{A}_2, P_2)$ be two probability spaces. For any two sets A_1, A_2, define $A_1 \times A_2 := \{(\omega_1,\omega_2); \omega_1 \in A_1, \omega_2 \in A_2\}$. Let $\Omega := \Omega_1 \times \Omega_2$ and $\mathscr{A} := \mathscr{A}_1 \otimes \mathscr{A}_2$ be the product sigma-field generated by $\{A_1 \times A_2 : A_1 \in \mathscr{A}_1, A_2 \in \mathscr{A}_2\}$. Let $P := P_1 \otimes P_2$ denote the product probability measure defined on $\mathscr{A}$ by the relation $P(A_1 \times A_2) = P_1(A_1)P_2(A_2)$, $A_j \in \mathscr{A}_j, j = 1,2$. It is known that this product measure exists and is unique, see, e.g., Loève (1963) or Athreya and Lahiri (2006). Let γ be a real-valued function, and consider the condition

$$\int \Big(\int \big|\gamma(w_1, w_2)\big| \, dP_1(w_1) \Big) \, dP_2(w_2) < \infty. \tag{2.3.4}$$

Let

$$J_1(\gamma) := \int \Big(\int \gamma(w_1, w_2) \, dP_1(w_1) \Big) \, dP_2(w_2),$$
$$J_2(\gamma) := \int \Big(\int \gamma(w_1, w_2) \, dP_2(w_2) \Big) \, dP_1(w_1).$$

We are ready to state the Fubini Theorem.

Theorem 2.3.2. *If either γ is nonnegative or* (2.3.4) *holds, then* $J_1(\gamma) = J_2(\gamma) = \int \gamma \, dP$.

In other words, for an integrable γ or for nonnegative γ, integration with respect to the product probability measure $P = P_1 \otimes P_2$ can be done by interchanging the order of integration with respect to P_1 and P_2.

The above theorem can be generalized to several probability spaces $(\Omega_i, \mathscr{A}_i, P_i)$, $i = 1, \cdots, k$. Their product space is $(\Omega, \mathscr{A}, P)$, where $\Omega = \Omega_1 \times \cdots \times \Omega_k$, $\mathscr{A} = \mathscr{A}_1 \otimes \cdots \otimes \mathscr{A}_k$ is the sigma field generated by $\mathcal{C} = \{A_1 \times \cdots \times A_k : A_i \in \mathscr{A}_i, i = 1, \cdots, k\}$ and $P = P_1 \otimes \cdots \otimes P_k$. Let γ be a measurable function defined on Ω that is either nonnegative or satisfies $\int_\Omega |\gamma(\omega_1, \cdots, \omega_k)| \, dP(\omega_1, \cdots, \omega_k) < \infty$. The Fubini theorem then says that the integral with respect to the product probability measure P can be evaluated by repeated integrations with the probability measures P_i's in arbitrary order. For example, for $k = 3$, the formula becomes

$$\int \gamma \, dP = \int \Big(\int \Big(\int \gamma(\omega_1, \omega_2, \omega_3) \, dP_{i_1}(\omega_{i_1}) \Big) \, dP_{i_2}(\omega_{i_2}) \Big) \, dP_{i_3}(\omega_{i_3}),$$

where (i_1, i_2, i_3) is a permutation of $(1, 2, 3)$. We usually skip the parentheses and just write

$$\int \gamma \, dP = \iiint \gamma(\omega_1, \omega_2, \omega_3) \, dP_{i_1}(\omega_{i_1}) \, dP_{i_2}(\omega_{i_2}) \, dP_{i_3}(\omega_{i_3}).$$

General integration. In the above we described the integral with respect to a probability measure. This type of integral extends to more general measures. A measure m is a sigma-additive function from $\mathscr{A}$ to $[0, \infty]$ such that $m(\emptyset) = 0$. Recall sigma-additivity means $m(\cup_i A_i) = \sum_i m(A_i)$ for any sequence $A_1, A_2, \ldots$ of disjoint events. The measure m is called finite if $m(\Omega)$ is finite. Thus a probability measure P is a finite measure with $P(\Omega) = 1$. The measure m is called sigma-finite if there is a sequence A_n, $n \geq 1$ of events such that $m(A_n)$ is finite for each n and $\bigcup_{n \geq 1} A_n = \Omega$. If m is a measure, then we call $(\Omega, \mathscr{A}, m)$ a measure space. We call a measure space finite if the measure is finite and sigma-finite if the measure is sigma-finite.

Assume now that m is a measure. The integral with respect to m is then defined analogous to that with respect to a probability measure. For a simple γ in $\mathfrak{M}_+$ with range R, we set $\int \gamma \, dm = \sum_{y \in R} m(\gamma = y)$, for a general γ in $\mathfrak{M}_+$, we define $\int \gamma \, dm = \lim \int \gamma_n \, dm$, with γ_n as in (2.3.2), and for a measurable γ, we set $\int \gamma \, dm = \int \gamma^+ \, dm - \int \gamma^- \, dm$ as long as at least one of the integrals on the right-hand side is finite. A measurable function γ is called integrable if $\int \gamma \, dm$ is finite. This means that $\int \gamma^+ \, dm$ and $\int \gamma^- \, dm$ are finite and are implied if $\int |\gamma| \, dm$ is finite. With this definition all the results for a probability measures carry over to general measures if we replace

P by m, with the exception of Theorem 2.3.2 which requires the measures m_1 and m_2 replacing P_1 and P_2 to be sigma-finite and that $P(f = 0) = 1$ just prior to Theorem 2.3.1 needs to be replaced by $m(f \neq 0) = 0$.

Let G be a nondecreasing right continuous function on $\mathbb{R}$. Then there exists a unique sigma-finite measure m_G on the Borel sigma-field $\mathscr{B}$ such that $m_G((a, b]) = G(b) - G(a)$ for all real a and b with $a < b$. If $G(-\infty) = \lim_{t \to -\infty} G(t) = 0$ and $G(\infty) := \lim_{t \to \infty} G(t) = 1$, then G is a d.f. and m_G is a probability measure on $\mathscr{B}$. In an abuse of notation, we often write $\int \gamma \, dG$ to denote the integral $\int \gamma \, dm_G$. The measure m_G for the function $G(x) = x$, $x \in \mathbb{R}$, is called the Lebesgue measure on the real line and commonly denoted by λ. It assigns to each interval its length. The k fold product measure of λ with itself is denoted by λ_k and called the Lebesgue measure on $\mathscr{B}_k$. Note that λ_k can be recursively defined by $\lambda_1 = \lambda$ and $\lambda_k = \lambda_{k-1} \otimes \lambda$. The measure λ_2 assigns to each rectangle its area, while the measure λ_3 assigns to each box its volume.

In view of the above convention, we write $\int \gamma(x) \, dx$ for the integral $\int \gamma \, d\lambda$. For most functions there is no difference between the Lebesgue integral as defined above and the Riemann integral as introduced in calculus. A reader unfamiliar with abstract integration may pretend that the Lebesgue integral and the Riemann integral are the same. The Lebesgue integral provides better convergence results, which play a crucial role in some of the proofs.

Example 2.3.1. Consider the function $\gamma(x) = I(0 \leq x < 1)\sqrt{x}$. The Lebesgue integral is the limit

$$\begin{aligned} \int \gamma \, d\lambda := & \lim_{n \to \infty} \sum_{j=1}^{2^n} \frac{j-1}{2^n} \lambda\Big(\frac{j-1}{2^n} \leq \sqrt{x} < \frac{j}{2^n}\Big) \\ = & \lim_{n \to \infty} \sum_{j=1}^{2^n} \frac{j-1}{2^n} \frac{j^2 - (j-1)^2}{2^{2n}} \\ = & \lim_{n \to \infty} \sum_{j=1}^{2^n} \frac{(j-1)(2j-1)}{2^{3n}} \\ = & \lim_{n \to \infty} \sum_{j=1}^{2^n} \frac{2j^2 - 3j + 1}{2^{3n}}. \end{aligned}$$

Using the identities

$$\sum_{j=1}^{N} j = \frac{N(N+1)}{2}, \qquad \sum_{j=1}^{N} j^2 = \frac{(2N+1)(N+1)N}{6}$$

one verifies that the above limit equals $2/3$ which is the value of the Riemann integral

$$\int_0^1 \sqrt{x} \, dx = \frac{2}{3} x^{3/2} \Big|_0^1 = \frac{2}{3}.$$

2.4 Expectation

Let X be a r.v. If the integral $\int X \, dP$ makes sense, then we set $E(X) := \int X \, dP = \int x dF_X(x)$ and call $E(X)$ the expectation of X or the mean of X. We say X has finite expectation (or finite mean) if $E(X)$ is finite. This is equivalent to $E(|X|) < \infty$.

Similarly, if g is a measurable function from $\mathbb{R}$ to $\mathbb{R}$ such that $\int |g(x)|\,dF_X(x) < \infty$, then the r.v. $g(X)$ has finite expectation

$$E\big(g(X)\big) = \int g(X)\,dP = \int g(x)\,dF_X(x).$$

Example 2.4.1. We shall now give some examples of expectations.

(a) If A is an event and c a real number, then $E(I_A) = P(A)$ and $E(cI_A) = cE(I_A) = cP(A)$. In particular, $E(c) = E(cI_\Omega) = cP(\Omega) = c$.

(b) If X has finite range taking values $x_1, \cdots, x_m$, then X can be expressed as $\sum_{j=1}^m x_j I(X = x_j)$. In this case, X has finite expectation $E(X) = \sum_{j=1}^m x_j P(X = x_j)$. This follows from the result for simple functions and the definition of the integral. Furthermore, $g(X)$ has finite expectation $E\big(g(X)\big) = \sum_{j=1}^m g(x_j)P(X = x_j)$, for every function g.

(c) If the range of X is contained in the set of nonnegative integers, then X can be expressed as $X = \sum_{j=0}^\infty jI(X = j)$ and thus has expected value $E(X) = \sum_{j=1}^\infty jP(X = j)$.

(d) Using Fubini's theorem, the expected value of a nonnegative r.v. X can also be expressed as

$$\begin{aligned} E(X) &= \int X\,dP = \iint I(0 < x < X)\,dx\,dP = \iint I(0 < x < X)\,dP\,dx \qquad (2.4.1)\\ &= \int_0^\infty P(X > x)\,dx. \end{aligned}$$

A similar argument used with $x < X$ replaced by $x \le X$ gives the formula

$$E(X) = \int_0^\infty P(X \ge x)\,dx.$$

(e) The above facts applied to $X^+ = \max(0, X)$ and $X^- = \max(0, -X)$ yield the identities

$$\begin{aligned} E(X^+) &= \int_0^\infty P(X^+ > x)\,dx = \int_0^\infty P(X > x)\,dx = \int_0^\infty \big(1 - F_X(x)\big)\,dx,\\ E(X^-) &= \int_0^\infty P(-X \ge x)\,dx = \int_0^\infty P(X \le -x)\,dx = \int_{-\infty}^0 F_X(x)\,dx. \end{aligned}$$

Thus, if X has finite expectation, then the definition $E(X) := E(X^+) - E(X^-)$ yields the identity

$$E(X) = \int_0^\infty \big(1 - F_X(x)\big)\,dx - \int_{-\infty}^0 F_X(x)\,dx.$$

This formula remains valid as long as one of the integrals on the right-hand side is finite.

(f) The formula (2.4.1) and the substitution $x = t^m$ give the identity

$$\begin{aligned} E\big(|X|^m\big) &= \int_0^\infty P\big(|X|^m > x\big)\,dx = \int_0^\infty P\big(|X|^m > t^m\big)m\,t^{m-1}\,dt\\ &= \int_0^\infty P(|X| > t)m\,t^{m-1}\,dt. \end{aligned}$$

for each positive integer m.

Let $X_1, X_2, \cdots, X_n$ be r.v.'s and define the *empirical d.f.*

$$F_n(x) := \frac{1}{n}\sum_{i=1}^n I(X_i \le x), \quad x \in \mathbb{R}. \qquad (2.4.2)$$

For each $n \geq 1$, $F_n(x)$ is a d.f. in $x \in \mathbb{R}$. It puts the mass $1/n$ at each X_i. Thus, by part (b) of the above example, we have the identity

$$\int g(x)dF_n(x) := \frac{1}{n}\sum_{i=1}^{n} g(X_i)$$

for any function g. The empirical d.f. F_n plays a central role in nonparametric inference as will be seen in later chapters of this book.

If X has a density f or probability mass function π and $g(X)$ has finite expectation, then,

$$E(g(X)) = \int g(x)f(x)\,dx, \quad E(g(X)) = \sum_{x:\pi(x)>0} g(x)\pi(x). \tag{2.4.3}$$

Let (X, Y) be two r.v.'s with the bivariate d.f. $F_{X,Y}$. For any measurable function $h(X, Y)$ of (X, Y), if $\int |h(x, y)|\,dF_{X,Y}(x, y) < \infty$, then analogous to (2.3.3) and (2.4.3), we define

$$\begin{aligned} E(h(X,Y)) &:= \int\int h(x,y)\,dF_{X,Y}(x,y), \\ &:= \int\int h(x,y)\,f_{X,Y}(x,y)\,dx\,dy, \quad \text{if } (X,Y) \text{ has density } f_{X,Y}, \\ &:= \sum_{(x,y)\in\mathbb{R}^2} h(x,y)\pi(x,y), \quad \text{if } (X,Y) \text{ is discrete with mass function } \pi. \end{aligned}$$

We now mention some properties of expectation and some useful inequalities. Their proofs can be found in Loève (1963) or Athreya and Lahiri (2006).

Let X and Y be two r.v.'s with finite expectations. Then the *linearity* property says

$$E(aX + bY) = aE(X) + bE(Y), \quad \forall\, a, b \in \mathbb{R},$$

and the *monotonicity* property states: If $P(X \leq Y) = 1$, then $E(X) \leq E(Y)$.

If $X_1, X_2, \ldots$ are nonnegative r.v.'s, then

$$E\Big(\sum_{j=1}^{\infty} X_j\Big) = \sum_{j=1}^{\infty} E(X_j).$$

Moments. Let X be a r.v. Then, for $0 < s < t < \infty$, we have the inequality $|X|^s \leq I(|X| \leq 1) + |X|^t I(|X| > 1) \leq 1 + |X|^t$. This inequality and the monotonicity property of expectation show that $E(|X|^t) < \infty$ implies $E(|X|^s) < \infty$. Now let k be a positive integer. If X^k has finite expectation, i.e., $E(|X|^k)$ is finite, then we call its expectation $E(X^k)$ the k-th *moment* of X and say X has a finite k-th moment. If X has a finite k-th moment, then it has a finite mean μ, and

$$E((X-\mu)^k) = E\Big(\sum_{j=0}^{k}\binom{k}{j}X^j(-\mu)^{k-j}\Big) = \sum_{j=0}^{k}\binom{k}{j}(-\mu)^{k-j}E(X^j)$$

is finite. In this case we call $E((X - \mu)^k)$ the k-th *central moment* of X. The *variance* $\mathrm{Var}(X)$ of X is the second central moment if X has a finite second moment and is set to be infinity otherwise. If X has a finite second moment, then we have the identity

$$\mathrm{Var}(X) = E((X-\mu)^2) = E(X^2 - 2\mu X + \mu^2) = E(X^2) - 2\mu E(X) + \mu^2 = E(X^2) - \mu^2.$$

This formula is useful in computing the variance.

The positive square root of $\text{Var}(X)$ is called the *standard deviation* of X. If X is a r.v. with finite mean μ and finite positive standard deviation σ, then the r.v. $Z := (X - \mu)/\sigma$ has mean $E(Z) = 0$ and variance $V(Z) = 1$ and is called the standardized X.

We now mention some important inequalities involving expectations.

Cauchy-Schwarz Inequality. For any two r.v.'s X and Y,

$$E(|XY|) \le \sqrt{E(X^2)E(Y^2)}. \tag{2.4.4}$$

Hölder Inequality. For any two r.v.'s X and Y and numbers p and q greater than 1 satisfying $\frac{1}{p} + \frac{1}{q} = 1$,

$$E(|XY|) \le \Big(E(|X|^p)\Big)^{1/p}\Big(E(|Y|^q)\Big)^{1/q}. \tag{2.4.5}$$

Analogously, suppose g_1, g_2 are measurable functions on $\mathbb{R}$ and p, q are as above. Then,

$$\int |g_1(t)g_2(t)|dt \le \Big(\int |g_1(t)|^p dt\Big)^{1/p}\Big(\int |g_2(t)|^q dt\Big)^{1/q}. \tag{2.4.6}$$

Moment Inequality. For any r.v. X and numbers $0 < r < s$,

$$E\Big(|X|^r\Big) \le \Big(E(|X|^s)\Big)^{r/s}. \tag{2.4.7}$$

The Cauchy–Schwarz inequality (2.4.4) follows from the Hölder inequality (2.4.5) upon taking $p = q = 2$. The moment inequality (2.4.7) follows from (2.4.5) upon applying it with $p = s/r$, $q = s/(s - r), Y = 1$, where the role of $|X|$ is played by $|X|^r$.

Markov Inequality. For any r.v. X and $p > 0$,

$$t^p P(|X| > t) \le E(|X|^p I(|X| > t)) \le E(|X|^p), \quad \forall t > 0. \tag{2.4.8}$$

This follows from the monotonicity property of expectation and because $t^p P(|X| > t) = E(t^p I(|X| > t)$ and $t^p I(|X| > t) \le |X|^p I(|X| > t)$.

Chebyshev inequality. If the r.v. Y has finite mean, then

$$P(|Y - E(Y)| > t) \le t^{-2}\text{Var}(Y), \qquad \forall 0 < t < \infty. \tag{2.4.9}$$

The inequality is trivially true if $\text{Var}(Y) = \infty$. It follows from (2.4.8) applied with $p = 2$ and $X = Y - E(Y)$. Needless to say, all of the above inequalities are of value only if their right-hand sides are finite.

Another important inequality is that of Jensen (1906). For this we first recall the notion of a convex function. A function g from a subinterval D of $\mathbb{R}$ into $\mathbb{R}$ is said to be *convex* if, for all $0 < \alpha < 1$,

$$g(\alpha x + (1 - \alpha)y) \le \alpha g(x) + (1 - \alpha)g(y), \quad \forall x, y \in D.$$

If the inequality is strict for all x, y in D, $x \ne y$, then g is said to be *strictly convex*. A sufficient condition for a twice differentiable function g to be convex is that its second derivative $\ddot{g}$ is positive. The function g is said to be *concave* if $-g$ is convex. Strict concavity is defined similarly.

Jensen's Inequality. Let g be function from a subinterval D of $\mathbb{R}$ into $\mathbb{R}$ and X be a r.v. with finite mean and range contained in D. Then

$$g(E(X)) \le E(g(X)), \qquad \text{if } g \text{ is convex}, \tag{2.4.10}$$

$$g(E(X)) \ge E(g(X)), \qquad \text{if } g \text{ is concave}. \tag{2.4.11}$$

These inequalities are strict if g is strictly convex or strictly concave.

Example 2.4.2. We give some applications of the Jensen inequality. Let X be a positive r.v. with $E(X) < \infty$. Since the function $g(x) = 1/x$, $x > 0$ is convex, by (2.4.10), $1/E(X) \le E(1/X)$. Since the function $g(x) = \log(x)$, $x > 0$, is concave, by (2.4.11), $\log(E(X)) \ge E(\log(X))$. Since the function $g(y) = e^{ty}$, $y \in \mathbb{R}$, is convex for every $t \in \mathbb{R}$, by (2.4.10), for any r.v. Y with finite mean, $e^{tE(Y)} \le E(e^{tY})$, for all $t \in \mathbb{R}$.

The following lemma gives a characterization of the independence of the two r.v.'s X and Y in terms of expectations.

Lemma 2.4.1. *The r.v.'s X and Y are independent if and only if*

$$E\big(h(X)g(Y)\big) = E\big(h(X)\big)E\big(g(Y)\big)$$

holds for all $h(X)$ and $g(Y)$ with finite expectations.

Proof. Taking $h(X) = I(X \le s)$ and $g(Y) = I(Y \le t)$ for real s and t gives the independence of X and Y. Independence of X and Y implies that the joint distribution $P_{X,Y}$ is the product measure $P_X \otimes P_Y$ of the marginal distributions. Hence,

$$\begin{aligned} E\big(h(X)g(Y)\big) &= \int h(x)g(y)\, dP_X(x)\, dP_Y(y) \\ &= \int h(x)\, dP_X(x) \int g(y)\, dP_Y(y) = E\big(h(X)\big)E\big(g(Y)\big) \end{aligned}$$

whenever $h(X)$ and $g(Y)$ have finite expectations. □

Covariance and Correlation. Let X and Y be r.v.'s having finite means $\mu = E(X)$ and $\nu = E(Y)$ and finite variances. Then, the covariance and correlation of X and Y, denoted by $\mathrm{Cov}(X,Y)$ and $\mathrm{Corr}(X,Y)$, respectively, are defined by

$$\begin{aligned} \mathrm{Cov}(X,Y) &:= E((X-\mu)(Y-\nu)) = E(XY) - E(X)E(Y), \\ \mathrm{Corr}(X,Y) &:= \frac{\mathrm{Cov}(X,Y)}{\sqrt{\mathrm{Var}(X)\mathrm{Var}(Y)}}, \qquad \text{if } \mathrm{Var}(X) > 0, \mathrm{Var}(Y) > 0, \\ &:= 0, \qquad \text{otherwise.} \end{aligned}$$

Use the inequality (2.4.4) to show that $|\mathrm{Corr}(X,Y)| \le 1$. Note that $\mathrm{Cov}(X,X) = \mathrm{Var}(X)$ and $\mathrm{Corr}(X,X) = 1$, if $\mathrm{Var}(X) > 0$.

We say that X and Y are uncorrelated if $\mathrm{Corr}(X,Y) = 0$ or equivalently, if $\mathrm{Cov}(X,Y) = 0$. If X and Y are independent, then $E(XY) = E(X)E(Y)$ holds and this implies that X and Y are uncorrelated. The converse need not be true. For example, let $X = -1, 0, 1$ with equal probability $1/3$ and let $Y = |X|$. Then $\mathrm{Cov}(X,Y) = 0$ while clearly X and Y are not independent.

2.5 Random Vectors

In this section we shall extend results for two random variables to several random variables. For this it is convenient to introduce the notion of a random vector. We assume throughout that $(\Omega, \mathscr{A}, P)$ is the probability space. Also recall that $\mathscr{B}_k$, the Borel sigma field on $\mathbb{R}^k$, is generated by the sets $\{\mathcal{I}_t, t \in \mathbb{R}^k\}$, where

$$\mathcal{I}_t = (-\infty, t_1] \times \cdots \times (-\infty, t_k], \quad t = (t_1, \cdots, t_k)' \in \mathbb{R}^k.$$

Definition 2.5.1. A function X from Ω into $\mathbb{R}^k$ is called a *random vector* if $\{X \in B\}$ is an event for all $B \in \mathscr{B}_k$. In this case we also call X a k-dimensional random vector or a random vector of dimension k.

Let $X = (X_1, \cdots, X_k)'$ be a random vector. If we take $B = (-\infty, t] \times \mathbb{R}^{k-1}$, then $\{X \in B\} = \{X_1 \leq t\}$. From this we conclude that X_1 is a random variable. Similarly, one shows that $X_2, \cdots, X_k$ are r.v.'s. The converse is also true. Indeed, if $X_1, \cdots, X_k$ are r.v.'s, then, for every $t = (t_1, \cdots, t_k)' \in \mathbb{R}^k$, the set $\{X_1 \leq t_1, \cdots, X_k \leq t_k\}$ is an event which equals $\{X \in \mathcal{I}_t)$. Since the collection of sets $\{B \in \mathscr{B}_k, \{X \in B\} \in \mathscr{A}\}$ is a sigma field and contains the sets $\mathcal{I}_t$, $t \in \mathbb{R}^k$, it equals $\mathscr{B}_k$. This shows that $\{X \in B\}$ is an event for each $B \in \mathscr{B}_k$. We summarize this in the following lemma.

Lemma 2.5.1. *A function $X = (X_1, \cdots, X_k)'$ from Ω into $\mathbb{R}^k$ is a random vector if and only if its coordinates $X_1, \cdots, X_k$ are random variables.*

With a random vector $X = (X_1, \cdots, X_k)$ we associate the probability measure P_X defined by

$$P_X(B) = P(X \in B), \quad B \in \mathscr{B}_k.$$

We call P_X the distribution of X or the joint distribution of $X_1, \cdots, X_k$ in which case we also write $P_{X_1,\cdots,X_k}$ instead of P_X. The distribution function (d.f.) of X is the function F_X defined by

$$F_X(t) = P(X \in I_t), \quad t \in \mathbb{R}^k.$$

We refer to F_X also as the joint d.f. of $X_1, \cdots, X_k$ and then write $F_{X_1,\cdots,X_k}$ for F_X. Since F_X is a function of k variables, it is often called a k-variate distribution function.

Let g be a function from $\mathbb{R}^k$ into $\mathbb{R}$ that is integrable with respect to P_X. Then, the r.v. $g(X)$ has finite expectation and analogous to (2.3.3), we have

$$E\big(g(X)\big) = \int g(x)\, dF_X(x),$$

where the integral is a k-tuple integral.

The random vector X is said to have a density f if f is a nonnegative function defined on $\mathbb{R}^k$ satisfying

$$P(X \in B) = \int I(x \in B) f(x)\, dx = \int I\big((x_1, \cdots, x_k)' \in B\big) f(x_1, \cdots, x_k)\, dx_1, \cdots, x_k,$$

for all B in $\mathscr{B}_k$. In particular, $\int f(x)\, dx = 1$.

If X has density f, then each coordinate X_j has a marginal density f_{X_j} which is obtained from f by integrating over all variables other than the one associated with X_j:

$$f_{X_1}(x_1) = \int f(x_1, \cdots, x_k)\, dx_2 \cdots dx_k,$$

$$f_{X_j}(x_j) = \int f(x_1, \cdots, x_k)\, dx_1 \cdots dx_{j-1}, dx_{j+1}, \cdots d_{x_k}, \quad j = 2, \cdots, k-1,$$

$$f_{X_k}(x_k) = \int f(x_1, \cdots, x_k)\, dx_1 \cdots dx_{k-1}.$$

Similarly, one finds joint densities of subvectors $\big(X_{i_1}, \cdots, X_{i_m}\big)$. For example, if $k = 5$, then X_2 and X_5 have joint density

$$f_{X_2,X_5}(x_2, x_5) = \int f(x_1, x_2, x_3, x_4, x_5)\, dx_1\, dx_3\, dx_4.$$

If each coordinate of X has finite expectation, then $E(X) := \big(E(X_1), \cdots, E(X_k)\big)'$ is called the vector of the expectation of the random vector X. This implies $E(a'X) = a'E(X)$, for all $a \in \mathbb{R}^k$. Suppose $\|X\|^2 = X_1^2 + \cdots + X_k^2$ has finite expectation. Then each of the r.v.'s $X_1, \cdots, X_k$ has a finite second moment. Thus their variances are finite and $\mu_{ij} := E\big(X_i X_j\big)$ and the covariances $\text{Cov}(X_i, X_j)$, $i, j = 1, \cdots, k$ are well defined. In this case, we call the matrix V with entries $V_{ij} = \text{Cov}(X_i, X_j)$, $i, j = 1, \cdots, p$, the dispersion matrix or the variance-covariance matrix of X. For $a = (a_1, \cdots, a_k)'$ and $b = (b_1, \cdots, b_k)'$ in $\mathbb{R}^k$, the linearity of the expectation is used to derive

$$\text{Cov}(a'X, b'X) = \Big(\sum_{i=1}^{n} a_i X_i, \sum_{i=1}^{n} b_i X_i\Big) = \sum_{i=1}^{n}\sum_{j=1}^{n} a_i b_j \text{Cov}(X_i, X_j) = a'Vb,$$

$$\text{Var}(a'X) = \text{Var}\Big(\sum_{i=1}^{n} a_i X_i\Big) = \sum_{i=1}^{n} a_i^2 \text{Var}(X_i) + 2 \sum_{1 \le i < j \le n} a_i a_j \text{Cov}(X_i, X_j) = a'Va.$$

If X_i's are uncorrelated, i.e., if $\text{Cov}(X_i, X_j) = 0$, $1 \le i < j \le n$, then V is a diagonal matrix and

$$\text{Var}(a'X) = \text{Var}\Big(\sum_{i=1}^{n} a_i X_i\Big) = \sum_{i=1}^{n} a_i^2 \text{Var}(X_i).$$

Since $\text{Cov}(X_i, X_j) = \text{Cov}(X_j, X_i)$ and $a'Va = \text{Var}(a'X) \ge 0$, for all $a \in \mathbb{R}^k$, the matrix V is symmetric and nonnegative definite.

Remark 2.5.1. A random matrix M is a matrix whose entries are random variables. If these random variables have finite expectations, then we call the matrix of these expected values the expected value of M and denote it by $E(M)$. In this case we say M has finite expectation. If $X = (X_1, \cdots, X_k)'$ is a random vector of dimension k, then XX' is a random matrix of dimension $k \times k$ whose i, j entry equals $X_i X_j$. If each component of X has a finite second moment, then X has mean vector μ and XX' and $(X - \mu)(X - \mu)'$ have finite expectations. Moreover, we have the identity $E(X - \mu)(X - \mu)') = E(XX') - \mu\mu'$ as the i, j entry of each matrix equals the covariance of X_i and X_j. Thus the variance-covariance matrix V of X can be written as

$$V = E\big((X - \mu)(X - \mu)'\big) = E\big(XX'\big) - \mu\mu'.$$

We now discuss independence for random vectors.

Definition 2.5.2. Let $Y_1, \cdots, Y_m$ be random vectors with respective dimensions $k_1, \cdots, k_m$. Then $Y_1, \cdots, Y_m$ are said to be independent if

$$P\big(Y_1 \in B_1, \cdots, Y_m \in B_m\big) = \prod_{j=1}^{m} P\big(Y_j \in B_j\big), \quad \forall B_1 \in B_{k_1}, \cdots, B_m \in \mathscr{B}_{k_m}.$$

If we take $k_1 = \cdots = k_m = 1$, this gives the definition of independence for random variables. If we also take $B_i = (-\infty, t_i]$ for $i = 1, \cdots, m$, then we obtain

$$F_{Y_1, \cdots, Y_m}(t_1, \cdots, t_m) = \prod_{j=1}^{m} F_{Y_j}(t_j), \quad t_1, \cdots, t_m \in \mathbb{R}.$$

The latter also implies independence of the r.v.'s $Y_1, \cdots, Y_m$. This remains true even for random vectors $Y_1, \cdots, Y_m$, i.e., when $k_j \ge 1, j = 1, \cdots, m$.

Analogous to Lemma 2.4.1, we have the following result.

Lemma 2.5.2. *The random vectors $Y_1, \cdots, Y_m$ are independent if and only if*

$$E\Big(\prod_{j=1}^{m} h_j(Y_j)\Big) = \prod_{j=1}^{m} E(h_j(Y_j))$$

holds for all r.v.'s $h_1(Y_1), \cdots, h_m(Y_m)$ with finite expectations.

If each random vector Y_j has a density f_j, then $Y_1, \cdots, Y_m$ are independent if the function f defined by

$$f(y) := \prod_{j=1}^{m} f_j(y_j), \quad y = (y_1', \cdots, y_m')' \in \mathbb{R}^{k_1+\cdots+k_m}$$

is a density of the random vector $Y = (Y_1', \cdots, Y_m')'$.

Definition 2.5.3. A sequence $Y_1, Y_2, \cdots$ of random vectors is said to be independent if $Y_1, \cdots, Y_n$ are independent for each integer n greater than one. We say the k-dimensional random vectors $Z_1, \cdots, Z_n$ are independent and identically distributed (i.i.d.) if $Z_1, \cdots, Z_n$ are independent and $Z_1, \cdots, Z_n$ have the same distribution. A set of r.v.'s or random vectors $X_1, \cdots, X_n$ is called a *random sample from a d.f. F* if $X_1, \cdots, X_n$ are i.i.d. r.v.'s each having d.f. F.

2.6 Transforms

In this section we introduce two probability transforms, the moment generating function and the characteristic function. We first introduce them for random variables and then discuss their extension to random vectors.

Let X be a r.v. and set

$$M_X(t) = E(e^{tX}), \quad t \in \mathbb{R}.$$

Note that $M_X(0) = 1$, but that M_X may be infinite for all other t. We call M_X a *moment generating function* (m.g.f.) if $M_X(t)$ is finite for all t in an open interval containing 0. The name stems from the following fact. If M_X is a m.g.f., then M_X is infinitely often differentiable in this open interval and its k-th derivative at $t = 0$ is the k-th moment of X:

$$E(X^k) = \frac{d^k M_X(t)}{dt}\Big|_{t=0}.$$

The m.g.f. uniquely determines the distribution. More precisely if X and Y are r.v.'s with respective moment generating functions M_X and M_Y and d.f.'s F_X and F_Y, then $F_X = F_Y$ is equivalent to $M_X = M_Y$.

Let $X_1, \cdots, X_n$ be independent r.v.'s, with X_i having m.g.f. M_{X_i}, $i = 1, \cdots, n$. Set $S = X_1 + \cdots + X_n$. Then, for $t \in \mathbb{R}$, we obtain

$$M_S(t) = E(e^{tS}) = E(e^{t(X_1+\cdots+X_n)}) = E\Big(\prod_{i=1}^{n} e^{tX_i}\Big) = \prod_{i=1}^{n} E(e^{tX_i}) = \prod_{i=1}^{n} M_{X_i}(t).$$

In particular, if $X_1, \cdots, X_n$ are i.i.d. with common m.g.f. M, then

$$M_S(t) = \big(M(t)\big)^n, \quad t \in \mathbb{R}.$$

Now consider a random vector $X = (X_1, \cdots, X_k)$ and set

$$M_X(t) = E(e^{t'X}) = E(e^{t_1X_1+\cdots+t_kX_k}), \quad t = (t_1, \cdots, t_k)' \in \mathbb{R}^k.$$

We call M_X a moment generating function if $M_X(t)$ is finite for all t with $\|t\| < \delta$ for some $\delta > 0$. Again the moment generating function uniquely determines the distribution of the random vector. More precisely, if X and Y are k-dimensional random vectors with respective moment generating functions M_X and M_Y and d.f.'s F_X and F_Y, then $F_X = F_Y$ is equivalent to $M_X = M_Y$.

Suppose M_X is a moment generating function. Then, the m.g.f. of X_j is given by

$$M_{X_j}(s) = M_X(se_j), \quad s \in \mathbb{R}, \ j = 1, \cdots, k,$$

where e_j is the standard unit vector in $\mathbb{R}^k$ whose jth coordinate is 1 and all other coordinates are 0. For example, for $k = 3$, we have $M_{X_1}(s) = M_X((s, 0, 0)')$, $M_{X_2}(s) = M_X((0, s, 0)')$ and $M_{X_3}(s) = M_X((0, 0, s)')$.

If the coordinates $X_1, \cdots, X_k$ of X are independent, then we have

$$M_X(t) = \prod_{j=1}^{n} M_{X_j}(t_j) = \prod_{j=1}^{k} M_X(t_j e_j), \quad t = (t_1, \cdots, t_k)' \in \mathbb{R}^k.$$

Moreover, if

$$M_X(t) = \prod_{j=1}^{k} M_X(t_j e_j), \quad t = (t_1, \cdots, t_k)' \in \mathbb{R}^k, \ \|t\| < \delta,$$

for some $\delta > 0$, then $X_1, \cdots, X_k$ are independent.

Let $b \in \mathbb{R}^m$ and A be an $m \times k$ matrix. Then we calculate

$$E\Big(e^{s'(b+AX)}\Big) = E\Big(e^{s'b}e^{(s'AX)}\Big) = e^{s'b}E\Big(e^{(A's)'X}\Big).$$

From this we conclude that if X has a moment generating function M_X, then $b + AX$ has moment generating function

$$M_{b+AX}(s) = e^{s'b}M_X(A's), \quad s \in \mathbb{R}^m. \tag{2.6.1}$$

Since moment generations function do not always exist, their use is limited. A related transform that always exists and has some of the features of the moment generating function is the characteristic function. To introduce this function we need to work with complex numbers.

Let ι denote the imaginary number that solves $\iota^2 = -1$. Recall that $e^{\iota tx} = \cos(tx) + \iota\sin(tx)$ and $|e^{\iota tx}|^2 = e^{\iota tX}e^{-\iota tX} = \cos^2(tx) + \sin^2(tx) = 1$, for all t, x. Let X be a random variable. Thus, as a function of X, $e^{\iota tX}$ is integrable for every $t \in \mathbb{R}$.

The *characteristic function of* X is the function ψ_X defined by

$$\psi_X(t) = E\Big(e^{\iota tX}\Big), \quad t \in \mathbb{R}.$$

Note that $\psi(0) = 1$ and $|\psi(t)| \leq E\Big(\big|e^{\iota tx}\big|\Big) = 1$, for all $t \in \mathbb{R}$. The characteristic function uniquely determines the distribution. In other words, if X and Y are random variables with respective characteristic functions ψ_X and ψ_Y and d.f.'s F_X and F_Y, then $\psi_X = \psi_Y$ if and only if $F_X = F_Y$.

While the existence of m.g.f. implies finite moments of all orders, but this is no longer true for the characteristic function. However, we have the following result. If X has a finite k-th moment, then ψ_X is k times differentiable with k-th derivative $\psi^{(k)}(t) = E\Big((\iota X)^k e^{\iota tX}\Big) = \iota^k E\Big(X^k e^{\iota tX}\Big)$. Thus $\psi^{(k)}(0) = \iota^k E(X^k)$ and $E(X^k) = \psi_X^{(k)}(0)\iota^k$.

Now let $X = (X_1, \cdots, X_k)'$ be a k-dimensional random vector. Its characteristic function ψ_X is given by

$$\psi_X(t) = E\Big(e^{\iota t'X}\Big) = E\Big(e^{\iota(t_1X_1+\cdots+t_kX_k)}\Big), \quad t = (t_1, \cdots, t_k)' \in \mathbb{R}^k.$$

Again, the characteristic function determines the distribution uniquely. Moreover, $X_1, \cdots, X_k$ are independent if and only if

$$\psi_X(t) = \prod_{j=1}^{k} \psi(t_j e_j), \quad \forall t = (t_1, \cdots, t_k)' \in \mathbb{R}^k.$$

2.7 Conditional Expectation

Here we shall discuss the conditional expectation of the random variables with respect to a sub sigma-field $\mathscr{A}_0$ of $\mathscr{A}$ and mention some of its properties. A sub sigma field $\mathscr{A}_0$ is a sigma-field on Ω such that every A in $\mathscr{A}_0$ belongs to $\mathscr{A}$. In what follows X is a random variable with $E\big(|X|\big) < \infty$.

Definition 2.7.1. A conditional expectation of X, given $\mathscr{A}_0$, is a r.v. X^* with the following properties:

(a) The r.v. X^* is $\mathscr{A}_0$ measurable: $\{X^* \le t\} \in \mathscr{A}_0$ for all $t \in \mathbb{R}$.
(b) For all $A \in \mathscr{A}_0$, $E(I_A X) = E(I_A X^*)$.

We denote this conditional expectation by $E(X|\mathscr{A}_0)$.

Applying (b) with $A = \Omega$, yields the law of the iterated expectation:

$$E(X) = E\big(E(X|\mathscr{A}_0)\big). \tag{2.7.1}$$

Also, the requirement (b) can be shown to be equivalent to
(c) For every bounded $\mathscr{A}_0$ measurable r.v. Z, $E\big(ZX\big) = E\big(ZX^*\big)$.
From this we immediately derive $E(YX|\mathscr{A}_0) = YE(X|\mathscr{A}_0)$ for each bounded $\mathscr{A}_0$ measurable Y. This is even true if Y is unbounded and YX has finite expectation.

Example 2.7.1. Suppose $\mathscr{A} = \{\emptyset, \Omega\}$. Then, $X^* := E(X|\mathscr{A}_0) = E(X)$.

Example 2.7.2. Suppose $\mathscr{A}_0 := \{\emptyset, B, B^c, \Omega\}$ for some event B with $0 < P(B) < 1$. Then

$$X^* := E(X|\mathscr{A}_0) = \frac{E(XI_B)}{P(B)} I_B + \frac{E(XI_{B^c})}{1 - P(B)} I_{B^c}.$$

This follows as $E\big(X^* I_B\big) = E\big(XI_B\big)$ and $E\big(X^* I_{B^c}\big) = E\big(XI_{B^c}\big)$.

Example 2.7.3. Let B belong to $\mathscr{A}$ and set $\mathscr{A}_B := \{A \in \mathscr{A} : A \cap B \in \{\emptyset, B\}\}$. Thus $\mathscr{A}_B$ is the collection of all events that either contain B or whose complement contains B. The class $\mathscr{A}_B$ is a subsigma field of $\mathscr{A}$. If $P(B)$ is positive, then we obtain

$$E(X|\mathscr{A}_B) = \frac{E(XI_B)}{P(B)} I_B + XI_{B^c}.$$

and if $P(B)$ is zero, we obtain $E(X|\mathscr{A}_B) = XI_{B^c}$. In particular, applying this with XI_B in place of X yields

$$E\big(XI_B\big|\mathscr{A}_B\big) = I_B E\big(XI_B\big)/P(B), \quad \text{if } P(B) > 0, \qquad (2.7.2)$$
$$= 0, \quad \text{otherwise.}$$

If $\mathscr{A}_0$ is the sigma-field generated by the k-dimensional random vector Y, i.e., $\mathscr{A}_0 = \{A : A = \{Y \in B\}$ for some $B \in \mathscr{B}_k\}$, then we write $E(X|Y)$ instead of $E(X|\mathscr{A}_0)$. Note that $E(X|Y)$ is of the form $m(Y)$, for some function m. From now on we set $m(y) := E(X|Y = y)$. This function is a solution to the equations

$$E\big(I(Y \in B)X\big) = E\big(I(Y \in B)m(Y)\big), \quad \forall B \in \mathscr{B}.$$

Example 2.7.4. Let X be a r.v. with density f satisfying $\int |x| f(x)\,dx < \infty$. We wish to compute $E\big(X\big|\,|X|\big)$. For this we first need to find a density of $Y := |X|$. Since the d.f. F of X is continuous, we calculate

$$G(y) := P(|X| \le y) = P(-y \le X \le y) = F(y) - F(-y),$$
$$= \int_{-y}^{y} f(x)\,dx = \int_0^y \big(f(-x) + f(x)\big)\,dx, \quad y \ge 0.$$

Hence Y has density $g(y) = \big(f(y) + f(-y)\big)I(y > 0)$.

Next, for a $B \in \mathscr{B}$, let $B_+ = B \cap (0, \infty)$. Then,

$$\begin{aligned} E(I(|X| \in B)X) &= \int I(x \in B_+)xf(x)\,dx + \int I(-x \in B_+)xf(x)\,dx \\ &= \int I(x \in B_+)xf(x)\,dx + \int I(x \in B_+)(-x)f(-x)\,dx \\ &= \int I(x \in B_+)(xf(x) - xf(-x))\,dx \\ &= \int I(y \in B_+)\frac{yf(y) - yf(-y)}{f(y) + f(-y)}g(y)\,dy = E\big(I(Y \in B)m(Y)\big) \end{aligned}$$

with

$$m(y) := E\big(X\big|\,|X| = y\big) = \frac{yf(y) - yf(-y)}{f(y) + f(-y)}I(y > 0).$$

If $f(x) = f(-x), \forall x \in \mathbb{R}$, then $g(y) = 2f(y)I(y > 0)$ and $m(y) \equiv 0$.

Example 2.7.5. Let $f_{X,Y}$ be a joint density of X and Y. The conditional density of X, given $Y = y$, is

$$f_{X|Y}(x|y) := \frac{f_{X,Y}(x, y)}{f_Y(y)}, \qquad f_Y(y) > 0,$$
$$:= 0, \qquad f_Y(y) = 0.$$

Then, $E\big(X|Y = y\big) = 0$, if $f_Y(y) = 0$. And for those y with $f_Y(y) > 0$,

$$E(X|Y = y) = \int x f_{X|Y}(x|y)\,dx = \frac{\int x f_{X,Y}(x, y)\,dx}{f_Y(y)}.$$

Example 2.7.6. Suppose X and Y are discrete with joint probability mass function $\pi_{X,Y}$. Then, $E(X|Y=y) = 0$, if $\pi_Y(y) = 0$ and if $\pi_Y(y) > 0$, then

$$E(X|Y=y) = \frac{\sum_x x\,\pi_{X,Y}(x,y)}{\pi_Y(y)} = \frac{\sum_x x\,\pi_{X,Y}(x,y)}{\sum_x \pi_{X,Y}(x,y)}.$$

Example 2.7.7. Here, we treat the case of discrete X and continuous Y. More precisely, we assume that there is a nonnegative function f such that $\sum_x \int f(x,y)\,dy = 1$ and

$$P(X=x, Y \in B) = \int I(y \in B) f(x,y)\,dy, \qquad \forall x \in \mathbb{R},\ B \in \mathscr{B}.$$

Let $f_Y(y) := \sum_x f(x,y)$. Then the conditional probability mass function of X, given $Y=y$, is

$$\begin{aligned} \pi_{X|Y}(x|y) &= \frac{f(x,y)}{f_Y(y)}, && f_Y(y) > 0, \\ &= 0, && f_Y(y) = 0. \end{aligned}$$

Hence, the conditional expectation of X, given Y, is

$$\begin{aligned} E(X|Y=y) &= \frac{\sum_x x f(x,y)}{f_Y(y)}, && f_Y(y) > 0, \\ &= 0, && f_Y(y) = 0. \end{aligned} \tag{2.7.3}$$

Example 2.7.8. Suppose X is continuous and Y is discrete, i.e., there is a nonnegative function f defined on $\mathbb{R}^2$ such that $\sum_y \int f(x,y)\,dx = 1$ and

$$P(X \in B, Y=y) = \int I(x \in B) f(x,y)\,dx, \qquad \forall y \in \mathbb{R},\ B \in \mathscr{B}.$$

Let $\pi_Y(y) := \int f(x,y)\,dx$. Then

$$\begin{aligned} E(X|Y=y) &= \frac{\int x f(x,y)\,dx}{\pi_Y(y)}, && \pi_Y(y) > 0, \\ &= 0, && \pi_Y(y) = 0. \end{aligned} \tag{2.7.4}$$

Next, we define the *conditional variance*. For this, assume furthermore that X has a finite second moment. Then the conditional variance of X, given Y, is defined to be

$$\mathrm{Var}(X|Y) := E\Big((X - E(X|Y))^2 \big| Y\Big) = E(X^2|Y) - (E(X|Y))^2. \tag{2.7.5}$$

We shall now state some properties of the conditional expectation that are used in the later parts of the book. See Loève (1963) for their proofs. Let X, Y be r.v.'s with finite expectations and $g(X,Y)$ be a random variable with finite expectation. Then,

$$E(X) = E(E(X|Y)), \quad E\big(g(X,Y)|X=x\big) = E\big(g(x,Y)|X=x\big),\ \forall x \in \mathbb{R}. \tag{2.7.6}$$

$$E\big(g(X,Y)|X=x\big) = E\big(g(x,Y)\big),\ \forall x \in \mathbb{R}, \text{ if } X \text{ and } Y \text{ are independent.} \tag{2.7.7}$$

$$E(XY) = E\big(YE(X|Y)\big), \quad \text{if } E(|XY|) < \infty. \tag{2.7.8}$$

2.8 Some Special Distributions

In this section we shall describe some well-known distributions that often arise in practice. We begin with some discrete distributions, followed by distributions having densities. Recall if X is a discrete r.v., then its probability mass function is $\pi(x) = P(X = x)$.

Bernoulli distribution. Let $0 \le p \le 1$, $q := 1 - p$. A r.v. X is said to have Bernoulli distribution with parameter p, and we write $X \sim_D \text{Ber}(p)$, if X takes values $0, 1$ with respective probabilities q, p, i.e, if the probability mass function of X is

$$\begin{aligned} \pi(x) := p^x q^{1-x} &= q, \qquad && x = 0 \\ &= p, && x = 1. \end{aligned}$$

In this case, $E(X) = p$, $\text{Var}(X) = pq$. Moreover, the m.g.f. of X is $M_X(t) = E(e^{tX}) = pe^t + q$ and the characteristic function (ch.f.) of X is $\psi_X(t) = E\big(e^{\iota tX}\big) = pe^{\iota t} + q$, for all $t \in \mathbb{R}$.

Binomial distribution. For a given positive integer n and a $0 \le p \le 1$, a r.v. X is said to have binomial distribution with parameters n, p, and we write $X \sim_D \text{Bin}(n, p)$, if

$$\pi(x) := \binom{n}{x} p^x q^{n-x}, \quad x = 0, 1, \cdots, n.$$

If $X \sim_D \text{Bin}(n, p)$, then $E(X) = np$, $\text{Var}(X) = npq$, its m.g.f. is $M_X(t) = \big(pe^t + q\big)^n$ and its ch.f. is $\psi_X(t) = \big(pe^{\iota t} + q\big)^n$. From this fact we derive that if $X_1, \cdots, X_m$ are independent with $X_i \sim_D \text{Bin}(n_i, p)$, $i = 1, \cdots, m$, then $X_1 + \cdots + X_m \sim_D \text{Bin}(n, p)$ with $n = n_1 + \cdots + n_m$. In particular, if $X_1, \cdots, X_n$ are independent $\text{Ber}(p)$ r.v.'s, then their sum $X_1 + X_2 + \cdots + X_n \sim_D \text{Bin}(n, p)$. Binomial distribution can also arise from certain dependent and non-identical Bernoulli r.v.'s, see, e.g., Vellaisamy and Punnen (2006).

Poisson distribution. A r.v. X is said to have Poisson distribution with parameter λ, and we write $X \sim_D \text{Poiss}(\lambda)$, if λ is positive and X has probability mass function

$$\pi(x) := \frac{\lambda^x e^{-\lambda}}{x!}, \qquad x = 0, 1, \cdots.$$

If $X \sim_D \text{Poiss}(\lambda)$, then $E(X) = \lambda$ and $\text{Var}(X) = \lambda$. This is the unique discrete distribution for which $E(X) = \text{Var}(X)$. Its m.g.f. and ch.f. are $M_X(t) = e^{\lambda(e^t - 1)}$, $\psi_X(t) = e^{\lambda(e^{\iota t} - 1)}$, $t \in \mathbb{R}$. If $X_i, 1 \le i \le n$ are independent r.v.'s and $X_i \sim_D \text{Poiss}(\lambda_i)$, $1 \le i \le n$, then $X_1 + \cdots + X_n \sim_D \text{Poiss}(\lambda)$, with $\lambda := \lambda_1 + \cdots + \lambda_n$.

Uniform distribution. Let $-\infty < a < b < \infty$ be given. A r.v. X is said to have a uniform distribution on the interval (a, b), and we write $X \sim_D \text{Unif}(a, b)$, if its density and d.f. are

$$\begin{aligned} f(x) &:= \frac{1}{b-a} I(a < x < b), \\ F(x) &:= 0, && x \le a, \\ &:= \frac{x-a}{b-a}, && a < x < b, \\ &:= 1, && x \ge b. \end{aligned}$$

Of particular interest is the r.v. $U \sim_D \text{Unif}(0,1)$. Its d.f. and density are

$$\begin{aligned} F_U(u) = P(U \le u) &:= 0, \qquad u \le 0, \\ &:= u, \qquad 0 < u < 1, \\ &:= 1, \qquad u \ge 1, \\ f_U(u) &:= I(0 < u < 1). \end{aligned} \tag{2.8.1}$$

If $X \sim_D \text{Unif}(a,b)$, then $U = (X-a)/(b-a) \sim_D \text{Unif}(0,1)$. Moreover, $E(X) = (a+b)/2$, $\text{Var}(X) = (b-a)^2/12$.

Exponential distribution. A r.v. X is said to have an exponential distribution with parameter $\lambda > 0$, and we write $X \sim_D \text{Exp}(\lambda)$, if X has density

$$f(x) := \lambda e^{-\lambda x} I(x > 0).$$

Its m.g.f. is finite for $t < \lambda$ and is given for such t by

$$M_X(t) = \lambda \int_0^\infty e^{tx} e^{-\lambda x}\, dx = \lambda \int_0^\infty e^{-(\lambda - t)x}\, dx = \lambda(\lambda - t)^{-1}.$$

Its k-th order derivative is $M_X^{(k)}(t) = \lambda k!/(\lambda - t)^{k+1}$ yielding $E(X^k) = k!/\lambda^k$. In particular, $E(X) = 1/\lambda$ and $\text{Var}(X) = 1/\lambda^2$. The ch.f. is $\psi_X(t) = \lambda/(\lambda - \iota t)$.

Gamma distribution. The Gamma function is defined by $\Gamma(a) := \int_0^\infty x^{a-1} e^{-x} dx$, $a > 0$. Integration by parts shows that for any $a > 0$, $\Gamma(a+1) = a\Gamma(a)$. A r.v. X is said to have gamma distribution with parameters $a > 0, b > 0$, and we write $X \sim_D \text{Gamma}(a,b)$, if X has density

$$f(x) := \frac{b^a}{\Gamma(a)} x^{a-1} e^{-bx} I(x > 0).$$

Observe that the density of $\text{Gamma}(1,\lambda)$ is the same as that of $\text{Exp}(\lambda)$. If $X \sim_D \text{Gamma}(a,b)$, then $Y = bX \sim_D \text{Gamma}(a,1)$ and

$$E(Y^r) = \frac{1}{\Gamma(a)} \int_0^\infty y^{a+r-1} e^{-y} dy = \frac{\Gamma(a+r)}{\Gamma(a)}, \quad r + a > 0.$$

In particular, $E(Y) = \Gamma(a+1)/\Gamma(a) = a$, $E(Y^2) = \Gamma(a+2)/\Gamma(a) = (a+1)a$ and $\text{Var}(Y) = a(a+1) - a^2 = a$. This in turn implies $X = Y/b \sim_D \text{Gamma}(a,b)$ and

$$E(X) = \frac{a}{b}, \quad E(X^2) = \frac{a(a+1)}{b^2}, \quad \text{Var}(X) = \frac{a}{b^2}, \quad M_X(t) = \left(\frac{b}{b-t}\right)^a, \quad t < b.$$

If $X_1, \cdots, X_m$ are independent with $X_i \sim_D \text{Gamma}(a_i, b)$, $i = 1, \cdots, m$, then $X_1 + \cdots + X_m \sim_D \text{Gamma}(a,b)$ with $a = a_1 + \cdots + a_m$. In particular, if $X_1, \cdots, X_m$ are i.i.d. $\text{Exp}(\lambda)$, then $X_1 + \cdots + X_m \sim_D \text{Gamma}(m, \lambda)$.

Beta distribution. A r.v. X is said to have a beta distribution with parameters $a > 0, b > 0$, and we write $X \sim_D \text{Beta}(a,b)$, if X has density

$$f(x) := \frac{\Gamma(a+b)}{\Gamma(a)\Gamma(b)} x^{a-1}(1-x)^{b-1} I(0 < x < 1). \tag{2.8.2}$$

The kth moment of this X is

$$E(X^k) = \frac{\Gamma(a+k)}{\Gamma(a+b+k)} = \prod_{j=0}^{k-1} \frac{a+j}{a+b+j}, \quad k \ge 1.$$

Thus $E(X) = a/(a+b)$, $E(X^2) = a(a+1)/((a+b)(a+b+1))$ and $\text{Var}(X) = (ab)/((a+b)^2(a+b+1))$. The beta distribution with $a = 1, b = 1$ is the uniform distribution on $[0, 1]$.

Chi-squared distribution. Let q be a positive integer. A r.v. X is said to have a chi-squared distribution with q degrees of freedom, and we write $X \sim_D \chi^2_q$, if X has density

$$f(x; q) = \frac{1}{2^{q/2}\Gamma\left(\frac{q}{2}\right)} x^{\frac{q}{2}-1} e^{-x/2}, \qquad x \geq 0. \tag{2.8.3}$$

This is the density of the Gamma$(q/2, 1/2)$ distribution. Thus, the mean and variance of the χ^2_q distribution are $E(X) = q$, $\text{Var}(X) = 2q$, and the sum of independent chi-squared random variables is chi-squared again. More precisely, if $X_1, \cdots, X_m$ are independent with $X_i \sim_D \chi^2_{q_i}$, $i = 1, \cdots, m$, then $X_1 + \cdots + X_m \sim_D \chi^2_q$ with $q = q_1 + \cdots + q_m$.

Non-central chi-squared distribution. A r.v. X is said to have non-central chi-squared distribution with k degrees of freedom and with non-centrality parameter $\delta > 0$, and we write $X \sim_D \chi^2_k(\delta)$, if X has density

$$f(x; k, \delta) := \sum_{j=0}^{\infty} \frac{e^{-\delta/2}(\delta/2)^j}{j!} f(x; k+2j), \quad x \geq 0, \tag{2.8.4}$$

where $f(x; k+2j)$'s are as in (2.8.3) with $q = k + 2j$. Note that this is also a density of $X_{k+2N} := \sum_{j=0}^{\infty} I(N = j)X_{k+2j}$, where $N \sim_D$ Poisson$(\delta/2)$ is independent of the sequence $X_1, X_2, \cdots$ of i.i.d. χ^2_1 random variables. Thus, we can think of a non-central chi-squared distribution as a Poisson mixture of chi-squared random variables. Clearly, $f(x; k, 0)$ is the density of a χ^2_k r.v..

Symmetric distribution. A r.v. X is said to be symmetrically distributed about zero if and only if, $X =_D -X$. If X has a density f that is even, i.e., $f(-x) = f(x)$, for all $x \in \mathbb{R}$, then X is symmetrically distributed about zero.

Double exponential distribution. A r.v. Y is said to have standard double exponential or Laplace distribution if X has density

$$f_Y(y) := \frac{1}{2} e^{-|y|}, \qquad y \in \mathbb{R}.$$

The d.f. of this r.v. is

$$\begin{aligned} F_Y(y) = \int_{-\infty}^{y} f_Y(s)\, ds &= \frac{1}{2} e^{y}, & y < 0, \\ &= 1 - \frac{1}{2} e^{-y}, & y > 0. \end{aligned}$$

Because f is even, Y is symmetrically distributed around 0, $E(Y) = 0$, $E(Y^2) = 2$, $\text{Var}(Y) = 2$. Let $X = Y + \theta$, for some $\theta \in \mathbb{R}$. Then

$$F_X(x) = P(Y + \theta \leq x) = F_Y(x - \theta), \quad f_X(x) = f_Y(x - \theta) = \frac{1}{2} e^{-|x-\theta|}, \quad x \in \mathbb{R}.$$

Note that $X - \theta =_D -(X - \theta)$. Moreover, $E(X) = \theta$, $\text{Var}(X) = 2$.

Univariate normal distributions. We say a r.v. X has a normal distribution with parameters $\mu \in \mathbb{R}, \sigma > 0$ and write $X \sim_D \mathcal{N}(\mu, \sigma^2)$, if X has density

$$\varphi(x) := \frac{1}{\sigma\sqrt{2\pi}} e^{-(x-\mu)^2/(2\sigma^2)}, \quad x \in \mathbb{R}.$$

Let $X \sim_D \mathcal{N}(\mu, \sigma^2)$. Then $E(X) = \mu$, $\text{Var}(X) = \sigma^2$ and $Z = (X - \mu)/\sigma$ is standard normal, i.e., $Z \sim_D \mathcal{N}(0, 1)$. The m.g.f. and ch.f. of Z are $M_Z(t) = e^{t^2/2}$ and $\psi_Z(t) = e^{-t^2/2}$, while those of X are $M_X(t) = E\big(e^{t\sigma Z + t\mu}\big) = e^{t\mu + (t^2\sigma^2/2)}$, $\psi_X(t) = e^{\iota t\mu - (t^2\sigma^2/2)}$, $t \in \mathbb{R}$. If $Z \sim_D \mathcal{N}(0, 1)$, then $Z^2 \sim_D \chi_1^2$, and if $Z_1, \cdots, Z_n$ are i.i.d. $\mathcal{N}(0, 1)$, then $Z_1^2 + \cdots + Z_n^2 \sim \chi_n^2$.

Let $Z_1, \cdots, Z_k$ be independent $\mathcal{N}(0, 1)$ r.v.'s. Then the random vector $(Z_1, \cdots, Z_k)'$ has m.g.f.

$$M_Z(t) = E\Big(e^{t_1 Z_1 + \cdots + t_k Z_k}\Big) = \prod_{j=1}^{k} E(e^{t_j Z_j}) = e^{\sum_{j=1}^{k} t_j^2/2} = e^{\|t\|^2/2}, \quad t = (t_1, \cdots, t_k)' \in \mathbb{R}^k.$$

Now set $Y = b + AZ$ for some $b \in \mathbb{R}^m$ and some $m \times k$ matrix A. Then the formula (2.6.1) yields

$$M_Y(s) = e^{s'b} M_Z(A's) = e^{s'b + (1/2)\|A's\|^2} = e^{s'b + (1/2)s'AA's}, \quad s \in \mathbb{R}^p.$$

Note that the matrix $A'A$ is a dispersion matrix, i.e., $A'A$ is symmetric and nonnegative definite. This motivates the following definition.

Multivariate normal distribution. The p-dimensional random vector $X = (X_1, \cdots X_p)'$ is said to have a p-variate normal distribution with parameters μ and Σ, and write $X \sim_D \mathcal{N}_p(\mu, \Sigma)$, if $\mu \in \mathbb{R}^p$, Σ is a symmetric and nonnegative definite $p \times p$ matrix, and X has m.g.f.

$$M_X(t) = e^{t'\mu + (1/2)t'\Sigma t}, \quad t \in \mathbb{R}^p.$$

Let σ_{ij} denote the (i, j)th entry of the matrix Σ, $i, j = 1, \cdots, p$. The marginal m.g.f. of X_j is then

$$M_{X_j}(s) = M_X(se_j) = e^{s\mu_j + (s^2/2)\sigma_{jj}}, \quad s \in \mathbb{R}.$$

This shows that $X_j \sim_D \mathcal{N}(\mu_j, \sigma_{jj})$ and gives $E(X_j) = \mu_j$ and $\text{Var}(X_j) = \sigma_{jj}$. The m.g.f. of $X_i + X_j$ is given by

$$M_{X_i + X_j}(s) = M_X(s(e_i + e_j)) = e^{s(\mu_i + \mu_j) + (1/2)s^2(\sigma_{ii} + 2\sigma_{ij} + \sigma_{jj})}, \quad s \in \mathbb{R}.$$

This yields $\text{Var}(X_i + X_j) = \sigma_{ii} + 2\sigma_{ij} + \sigma_{jj}$, from which we conclude that σ_{ij} is the covariance of X_i and X_j. This shows that X has mean vector $E(X) = \mu$ and dispersion matrix Σ.

Next, we shall mention some additional properties of a p-variate normal random vector X.

(i) Let $b \in \mathbb{R}^m$ and A be a $m \times p$ matrix, then $b + AX$ has m.g.f.

$$M_{b+AX}(s) = e^{s'b} M_X(A's) = e^{s'b + s'A\mu + (1/2)A\Sigma A'}, \quad s \in \mathbb{R}^m.$$

This shows that $b + AX \sim \mathcal{N}_m(b + A\mu, A\Sigma A')$.

(ii) If Σ is a diagonal matrix, then

$$M_X(t) = e^{t'\mu + (1/2)\sum_{j=1}^{p} t_j^2 \sigma_{jj}} = \prod_{j=1}^{p} e^{t_j\mu_j + (1/2)t_j^2\sigma_{jj}} = \prod_{j=1}^{p} M_{X_j}(t_j), \quad t = (t_1, \cdots, t_j)' \in \mathbb{R}^p.$$

This shows that $X_1, \cdots, X_n$ are independent if and only if Σ is diagonal. In other words, if the components of a multivariate normal random vector are uncorrelated, then they are independent. The multivariate normal distribution is a rare distribution for which the components being uncorrelated implies their independence.

Suppose Σ is positive definite so that the determinant $\det(\Sigma)$ of Σ is positive and the inverse of the matrix Σ, denoted by Σ^{-1}, exists. Then, the random vector X has density

$$\varphi(x) := \frac{1}{\Big((2\pi)^p \det(\Sigma)\Big)^{1/2}} e^{-\frac{1}{2}(x-\mu)'\Sigma^{-1}(x-\mu)}, \quad x \in \mathbb{R}^p.$$

2.9 Distribution of a Function of Random Variables

Often we are interested in obtaining the distribution of a function of r.v.'s or the joint distribution of several functions of r.v.'s. We begin with a result that shows how to construct a random variable with a given d.f. from a uniform r.v. U. For this we need the left-inverse F^{-1} of a d.f. F, which is defined by

$$F^{-1}(u) := \inf\{x \in \mathbb{R}; F(x) \geq u\}, \quad 0 < u < 1. \tag{2.9.1}$$

Note the following facts.

$$F\big(F^{-1}(u)\big) \geq u, \qquad \forall 0 < u < 1, \tag{2.9.2}$$

with equality holding if and only if F is continuous. Also

$$F^{-1}\big(F(x)\big) \leq x, \qquad \forall x \in \mathbb{R}, \tag{2.9.3}$$

with equality holding if and only if F is strictly increasing on $\mathbb{R}$.

Since F^{-1} is nondecreasing, we can extend F^{-1} to a function on $[0,1]$ by setting

$$F^{-1}(0) := \lim_{u \to 0_+} F^{-1}(u), \qquad F^{-1}(1) := \lim_{u \to 1_-} F^{-1}(u).$$

Note that these limits may not be finite.

Theorem 2.9.1. *Let F be a d.f. on $\mathbb{R}$ and U be a* $\mathrm{Unif}(0,1)$ *r.v. Then* $X := F^{-1}(U) \sim_D F$.

Proof. Fix an $x \in \mathbb{R}$. By the right-continuity and monotonicity of F, the set $\{0 < u < 1 : u \leq F(x)\}$ equals the interval $\{0 < u < 1 : x \geq F^{-1}(u)\}$. Hence, $P\big(F^{-1}(U) \leq x\big) = P\big(U \leq F(x)\big) = F(x)$. □

Theorem 2.9.2. *Let X and U be independent r.v.'s with X having d.f. F and $U \sim$* $\mathrm{Unif}(0,1)$. *Then* $Y := F(X_-) + U(F(X) - F(X_-)) \sim_D \mathrm{Unif}(0,1)$. *In particular, if F is continuous, then* $F(X) \sim_D \mathrm{Unif}(0,1)$.

Proof. Let $0 < t < 1$. We need to show $P(Y \leq t) = t$. Let $q = \sup\{x : F(x_-) \leq t\}$. Since $x \to F(x_-)$ is left-continuous, we have $F(q_-) \leq t \leq F(q)$. The event $\{Y \leq t\}$ equals $\{X < q\} \cup \{X = q, F(q_-) + U(F(q) - F(q_-)) \leq t\}$. This implies $P(Y \leq t) = F(q_-) + (F(q) - F(q_-))P(U(F(q) - F(q_-)) \leq t - F(q_-)$. If $F(q_-) = F(q)$, then we obtain $P(Y \leq t) = F(q) = t$. If $F(q_-) < F(q)$, then we calculate $(F(q) - F(q_-))P(U(F(q) - F(q_-)) \leq t - F(q_-)) = (F(q) - F(q_-))(t - F(q_-))/F(q) - F(q_-) = (t - F(q_-)$ and obtain again $P(Y \leq t) = t$. □

Corollary 2.9.1. *Let X be a r.v. with d.f. F. Then $P(F(X) \leq t) \leq t \leq P(F(X_-) \leq t)$ holds for all $0 < t < 1$.*

Proof. Let Y be as in the previous theorem. Then we have $F(X_-) \leq Y \leq F(X)$ and this implies the inequality $P(F(X) \leq t) \leq P(Y \leq t) \leq P(F(X_-) \leq t)$ for all $t \in \mathbb{R}$. This is the desired inequality as $P(Y \leq t) = t$ for $0 < t < 1$. □

Suppose g is an increasing and differentiable function from the open interval D onto the open interval $\mathcal{I}$. Let g^{-1} denote the inverse function of g which maps $y \in \mathcal{I}$ to x in D if $g(x) = y$. If $\dot{g}(x)$ does not vanish for x in D, then g^{-1} is also differentiable with derivative $J(y) = 1/\dot{g}(g^{-1}(y))$ for $y \in \mathcal{I}$. Now let X be a random variable with $P(X \in D) = 1$. Then we have $P\big(g(X) \le t\big) = P\big(X \le g^{-1}(t)\big)$ for all $t \in \mathcal{I}$. Assume furthermore that X has a density f_X. Then, the substitution $x = g^{-1}(y)$ gives

$$\begin{aligned} P\big(g(X) \in B\big) &= P\big(g(X) \in B \cap \mathcal{I}\big) = P\big(X \in g^{-1}(B \cap \mathcal{I})\big) \\ &= \int_{g^{-1}(B\cap\mathcal{I})} f_X(x)\,dx = \int_{B\cap\mathcal{I}} f_X(g^{-1}(y))J(y)\,dy, \\ &= \int_B f_X(g^{-1}(y))J(y)I(y \in \mathcal{I})\,dy \end{aligned}$$

for all Borel sets B. This shows that $Y = g(X)$ has density

$$f_Y(y) = f_X(g^{-1}(y))J(y)I(y \in \mathcal{I}).$$

The above can be repeated if g is decreasing rather than increasing. In this case, we need to replace $J(y)$ by its absolute value $|J(y)|$.

Example 2.9.1. If X has density f_X, then $Y = \exp(X)$ has density $f_Y(y) = f_X(\log(y))/yI(y > 0)$ and $Z = aX + b$ has density $f_Z(z) = f_X((z-a)/b)/|b|$ for real numbers a and b, $b \neq 0$. If $U \sim_D \text{Unif}(0,1)$, then U has density $f(u) = I(0 < u < 1)$ and $Y = -\log(U)$ has density $f_Y(y) = I(0 < \exp(-y) < 1)|-\exp(-y)|I(y > 0) = \exp(-y)I(y > 0)$. This shows that $-\log(U) \sim_D \text{Exp}(1)$.

Example 2.9.2. Let X have density f. Then $|X|$ has density $f_{|X|}(x) = (f(x) + f(-x))I(x > 0)$. This was shown in Example 2.7.4. We shall now find a density of $Y := X^2$. Since Y equals $|X|^2$, $P(|X| > 0)$ equals one, and the increasing function $g(x) = x^2$, $x > 0$, has inverse $g^{-1}(y) = \sqrt{y}$, $y > 0$, we derive

$$f_Y(y) = \frac{f(-\sqrt{y}) + f(\sqrt{y})}{2\sqrt{y}} I(y > 0).$$

If f satisfies $f(x) = f(-x)$ for all x, then f_Y simplifies to

$$f_Y(y) = \frac{f(\sqrt{y})}{\sqrt{y}} I(y > 0).$$

Now let $g := (g_1, \cdots, g_p)'$ be a one-to-one function from an open subset D of $\mathbb{R}^p$ onto some open subset $\mathcal{C}$ of $\mathbb{R}^p$. Assume that the components $g_j, 1 \le j \le p$ are continuously differentiable. For any square matrix A, let $\det A$ denote its determinant. The *Jacobian* of the transformation g is

$$J_g(x) := \det \begin{bmatrix} \frac{\partial g_1(x)}{\partial x_1} & \cdots & \frac{\partial g_1(x)}{\partial x_p} \\ \vdots & & \vdots \\ \frac{\partial g_p(x)}{\partial x_1} & \cdots & \frac{\partial g_p(x)}{\partial x_p} \end{bmatrix}, \qquad x = (x_1, \cdots, x_p)' \in \mathbb{R}^p.$$

Let g^{-1} denote the inverse transformation, i.e., $g^{-1}(y) = x$ if and only if $y = g(x)$, for all $x \in D$ and $y \in \mathcal{C}$. There is a well-known relation between the Jacobian of g^{-1} and that of g in the case $J_g(x)$ does not vanish for any x in D. It is given by

$$J_{g^{-1}}(y) = \frac{1}{J_g\big(g^{-1}(y)\big)}.$$

Now, let $X := (X_1, \cdots, X_p)'$ be a random vector with $P(X \in D) = 1$ and density f_X. Suppose $J_g(x) \neq 0,\ x \in D$. Then a density of the random vector $Y := g(X)$ is given by

$$f_Y(y) = f_X(g^{-1}(y))\left|J_{g^{-1}}(y)\right| = \frac{f_X(g^{-1}(y))}{\left|J_g(g^{-1}(y))\right|}, \qquad y \in \mathcal{C}. \tag{2.9.4}$$

For the proof see Bickel and Doksum (1977; pp 485–487) and the references therein.

Example 2.9.3. Consider the case $p = 2$, $X = (X_1, X_2)'$ and the transformation $g(x) = (x_1 + x_2, x_2)'$. Clearly, this transformation satisfies the above conditions for (2.9.4). Also note that

$$J_g(x) = \det \begin{bmatrix} 1 & 1 \\ 0 & 1 \end{bmatrix} = 1 \neq 0, \qquad \forall x \in \mathbb{R}^2.$$

Hence, a density of $Y = (X_1 + X_2, X_2)'$ is

$$f_Y(y) = f_X(y_1 - y_2, y_2), \qquad y \in \mathbb{R}^2.$$

From here we obtain the density of the sum $Y_1 = X_1 + X_2$ to be

$$f_{Y_1}(y_1) = \int f_X(y_1 - y_2, y_2)\, dy_2, \quad y_1 \in \mathbb{R}.$$

If X_1 and X_2 are independent, then the density of the sum is

$$f_{Y_1}(y) = \int f_{X_1}(y_1 - y_2) f_{X_2}(y_2)\, dy_2.$$

Example 2.9.4. As a second example, consider the affine transformation $Y = AX + b$, where A is an invertible $p \times p$ matrix and $b \in \mathbb{R}^p$. Here, $g(x) = Ax + b$ and $g^{-1}(y) = A^{-1}(y - b)$ and the Jacobian is $J_{g^{-1}}(y) = \det A^{-1} = 1/\det A$. Thus $f_Y(y) = f_X(A^{-1}(y - b))/|A|$. As special case, take $p = 3$ and consider the permutation

$$g(x_1, x_2, x_3) = \begin{bmatrix} x_2 \\ x_3 \\ x_1 \end{bmatrix} = \begin{bmatrix} 0 & 1 & 0 \\ 0 & 0 & 1 \\ 1 & 0 & 0 \end{bmatrix} \begin{bmatrix} x_1 \\ x_2 \\ x_3 \end{bmatrix}.$$

Then $f_Y(y_1, y_2, y_3) = f_X(y_3, y_1, y_2)$.

Example 2.9.5. Let X_1, X_2 be independent random variables with $X_1 \sim_D \text{Gamma}(\alpha, \gamma)$ and $X_2 \sim_D \text{Gamma}(\beta, \gamma)$. Then the random vector $X = (X_1, X_2)'$ has density

$$\begin{aligned} f_X(x) &= \frac{\gamma^\alpha x_1^{\alpha-1} \exp(-\gamma x_1)}{\Gamma(\alpha)} I(x_1 > 0) \frac{\gamma^\beta x_2^{\beta-1} \exp(-\gamma x_2)}{\Gamma(\beta)} I(x_2 > 0). \\ &= \frac{\gamma^{\alpha+\beta} x_1^{\alpha-1} x_2^{\beta-1} \exp\big(-\gamma(x_1 + x_2)\big)}{\Gamma(\alpha)\Gamma(\beta)} I(x_1 > 0, x_2 > 0). \end{aligned}$$

Now consider the transformation $g(x) = (x_1/x_1 + x_2, x_1 + x_2)'$, $x_1 > 0, x_2 > 0$. It has Jacobian

$$J_g(x) = \det \begin{bmatrix} \dfrac{x_2}{(x_1 + x_2)^2} & -\dfrac{x_1}{(x_1 + x_2)^2} \\ 1 & 1 \end{bmatrix} = \frac{1}{x_1 + x_2}, \quad x_1 > 0, x_2 > 0,$$

and inverse $g^{-1}(y) = (y_1y_2, y_2 - y_1y_2)'$, $0 < y_1 < 1, y_2 > 0$. Thus $Y = g(X)$ has density

$$\begin{aligned} f_Y(y) &= \frac{\gamma^{\alpha+\beta}(y_1y_2)^{\alpha-1}(y_2 - y_1y_2)^{\beta-1}\exp(-\gamma y_2)}{\Gamma(\alpha)\Gamma(\beta)}|y_2|I(y_1y_2 > 0, y_2 - y_1y_2 > 0). \\ &= \frac{\Gamma(\alpha+\beta)}{\Gamma(\alpha)\Gamma(\beta)}y_1^{\alpha-1}(1-y_1)^{\beta-1}I(0 < y_1 < 1)\frac{\gamma^{\alpha+\beta}y_2^{\alpha+\beta-1}\exp(-y_2/\gamma)}{\Gamma(\alpha+\beta)}I(y_2 > 0). \end{aligned}$$

We conclude that $Y_1 = X_1/(X_1 + X_2)$ and $Y_2 = X_1 + X_2$ are independent, Y_1 has a Beta(α, β) distribution and $X_1 + X_2$ a Gamma$(\alpha + \beta, \gamma)$ distribution.

2.10 Modes of Convergence

In this section we review various notions of convergence and some important useful facts when dealing with sequences of random variables. For this we shall need the weak convergence of distribution functions as defined next. All limits are taken as $n \to \infty$, unless mentioned otherwise.

Definition 2.10.1. Let $F, F_1, F_2, \cdots$ be d.f.'s on $\mathbb{R}^p$. Then we say F_n *converges weakly* to F and write $F_n \Rightarrow F$, if $F_n(t)$ converges to $F(t)$ for every point $t \in \mathbb{R}^p$ at which F is continuous.

Recall, for any $x \in \mathbb{R}^p$, $\|x\|$ denotes the Euclidean norm of x.

Definition 2.10.2. Let $W, W_1, W_2, \cdots$ be p-dimensional random vectors and $F_W, F_{W_1}, F_{W_2}, \cdots$ denote their respective d.f.'s.

1. We say W_n converges *almost surely* to W and write $W_n \to_{a.s.} W$, if

$$P(\lim_{n\to\infty} W_n = W) = 1.$$

2. We say W_n converges *in probability* to W, and write $W_n \to_p W$ or $W_n = W + o_p(1)$, if

$$\lim_{n\to\infty} P\big(\|W_n - W\| > \epsilon\big) = 0, \quad \forall \epsilon > 0.$$

3. We say W_n converges *in distribution* to W, and write $W_n \to_D W$, if $F_{W_n} \Rightarrow F_W$.
4. We say W_n is *bounded in probability*, and write $W_n = O_p(1)$, if for every $\epsilon > 0$ there is a finite positive k such that

$$\limsup_{n\to\infty} P\big(\|W_n\| > k\big) \le \epsilon.$$

The first three lemmas presented below describe the hierarchy between these modes of convergence. The almost sure convergence implies convergence in probability, convergence in probability implies convergence in distribution, and convergence in distribution implies boundedness in probability. The proofs of many results stated below appear in Loève (1963) or Athreya and Lahiri (2006).

Lemma 2.10.1. *If $W_n \to_{a.s.} W$, then $W_n \to_p W$.*

Lemma 2.10.2. *If $W_n \to_p W$, then $W_n \to_D W$. The converse holds if $P(W = c) = 1$ for some constant $c \in \mathbb{R}$.*

Lemma 2.10.3. *If $W_n \to_D W$, then $W_n = O_p(1)$.*

The following result shows that all three modes of convergence are preserved under continuous transformations.

Theorem 2.10.1. [Continuous Mapping Theorem]. *Let h be a real-valued function defined on $\mathbb{R}^p$ that is continuous at each point in the Borel set B of $\mathbb{R}^p$. Suppose that $P(W \in B) = 1$. Then, $W_n \to_{a.s.} W$ implies $h(W_n) \to_{a.s.} h(W)$, $W_n \to_p W$ implies $h(W_n) \to_p h(W)$, and $W_n \to_D W$ implies $h(W_n) \to_D h(W)$.*

Example 2.10.1. Let $X, X_1, X_2, \cdots$ and $Y, Y_1, Y_2, \cdots$ be r.v.'s such that $X_n \to_p X$ and $Y_n \to_p Y$. Then $X_n + Y_n \to_p X + Y$, $X_n - Y_n \to X - Y$ and $X_n Y_n \to_p XY$. This follows as $(X_n, Y_n) \to_p (X, Y)$ and the continuity of the functions $h_1(x, y) = x + y$, $h_2(x, y) = x - y$ and $h_2(x, y) = xy$.

Next, we discuss the connection between the modes of convergence of a sequence of random vectors and those of their coordinates. For $n = 0, 1, 2, \cdots$, write $W_{n,j}$ for the j-coordinate of W_n so that $W_n = \big(W_{n,1}, \cdots, W_{n,p}\big)'$. The following results are easy to check.

Lemma 2.10.4. *The following are true.*

(1) *$W_n \to_{a.s.} W_0$, if and only if, $W_{n,j} \to_{a.s.} W_{0,j}$, for all $j = 1, \cdots, p$.*

(2) *$W_n \to_p W_0$, if and only if, $W_{n,j} \to_p W_{0,j}$, for all $j = 1, \cdots, p$.*

Unlike the results of Lemma 2.10.4, in general, the convergence of marginal distributions does not imply that of the joint distribution. An exception is provided by the following result.

Lemma 2.10.5. *Suppose the components $W_{n,1}, \cdots, W_{n,p}$ of W_n are independent for each nonnegative integer n. Then, $W_n \to_D W_0$ if and only if $W_{n,j} \to_D W_{0,j}$, for each $j = 1, \cdots, p$.*

For any random vector ζ, let ψ_ζ denote its characteristic function. The following theorem characterizes convergence in distribution.

Theorem 2.10.2. *The following are equivalent.*

(1) $W_n \to_D W_0$.

(2) *$E\big(h(W_n)\big) \to E\big(h(W_0)\big)$, for every bounded continuous function h from $\mathbb{R}^p$ to $\mathbb{R}$.*

(3) *$\psi_{W_n}(t) \to \psi_{W_0}(t)$, for all $t \in \mathbb{R}^p$.*

The next result characterizes convergence in distribution of random vectors in terms of convergence in distribution of random variables. It follows from the equivalence of (1) and (3) of Theorem 2.10.2.

Corollary 2.10.1. [Cramér-Wold Device.] The following are equivalent.

(1) $W_n \to_D W_0$.

(2) $u'W_n \to_D u'W_0$, for every $u \in \mathbb{R}^p$ with $\|u\| = 1$.

The following theorem is used to obtain the limiting distribution of a function of a sequence of random vectors.

Theorem 2.10.3. [Delta Method]. *Let q be a positive integer, $a_n, n \geq 1$ be a sequence of positive numbers satisfying $a_n \to \infty$ and h be a function from $\mathbb{R}^p$ to $\mathbb{R}^q$ that is differentiable at μ with derivative $\dot{h}(\mu)$, a $q \times p$ matrix. Then, $a_n(W_n - \mu) \to_D W_0$ implies $a_n\big(h(W_n) - h(\mu)\big) \to_D \dot{h}(\mu)W_0$.*

We shall now discuss some sufficient conditions for the convergence in probability. In view of part (2) of Lemma 2.10.4, it suffices to study this for r.v.'s only. Let now $X, X_1, X_2, \cdots$ be r.v.'s.

Lemma 2.10.6. *Suppose $E(|X_n - X|^r) \to 0$ for some positive r. Then $X_n \to_p X$.*

Proof. Fix $\epsilon > 0$. By (2.4.8), $P(|X_n - X| > \epsilon) \le \epsilon^{-r} E(|X_n - X|^r) \to 0$. □

The case of $r = 2$ is of practical importance.

Corollary 2.10.2. *If $E(X_n) \to E(X)$ and $\mathrm{Var}(X_n - X) \to 0$, then $X_n \to_p X$.*

Proof. The assumptions imply $E(|X_n - X|^2) = \mathrm{Var}(X_n - X) + \big(E(X_n) - E(X)\big)^2 \to 0$ and $X_n \to_p X$ by Corollary 2.10.2. □

If, for some $c \in \mathbb{R}$, $P(X = c) = 1$, then we have the following important corollary.

Corollary 2.10.3. *Let $c \in \mathbb{R}$. Suppose $E(X_n) \to c$ and $\mathrm{Var}(X_n) \to 0$, then $X_n \to_p c$.*

The next result relaxes the conditions in expectations to those in conditional expectations.

Lemma 2.10.7. *Let $c \in \mathbb{R}$ and V_n be a sequence of p-dimensional random vectors. Then $E(X_n|V_n) \to_p c$ and $\mathrm{Var}(X_n|V_n) \to_p 0$ imply $X_n \to_p c$.*

Proof. It suffices to show $Y_n := X_n - E(X_n|V_n) \to_p 0$. Fix $\epsilon > 0$. Then, $E\big(I(|Y_n| > \epsilon)|V_n\big) \le \epsilon^{-2} E(Y_n^2|V_n) \to_p 0$. Moreover, $E\big(I(|Y_n| > \epsilon)|V_n\big) \le 1$, for all $n \ge 1$. Hence, by Theorem 2.10.4 below, $P(|Y_n| > \epsilon) = E\Big(E\big(I(|Y_n| > \epsilon)|V_n\big)\Big) \to 0$. □

The following theorem gives some sufficient conditions for interchanging limit and expectation.

Theorem 2.10.4. [Dominated Convergence Theorem]. *Suppose $X_n \to_p X$ and there is a non-negative r.v. V with $E(V) < \infty$ such that $\|X_n\| \le V$, for all $n \ge 1$. Then, $E(X_n) \to E(X)$.*

The following theorem will be repeatedly used to derive the asymptotic distribution of random variables.

Theorem 2.10.5. [Slutsky's Theorem]. *Let $X, X_1, X_2, \cdots, Y_1, Y_2, \cdots$ be r.v.'s such that $X_n \to_D X$, $Y_n \to_p c$, for some constant c. Then, $X_n + Y_n \to_D X + c$ and $X_n Y_n \to_D cX$. If $c \ne 0$, then $X_n/Y_n \to_D X/c$.*

2.10.1 Some Limit Theorems

In this subsection we collect some additional important asymptotic results found useful in the book. Most of them will be presented without proof. Their proofs can be found in Loève (1963) or Athreya and Lahiri (2006). The next three theorems are the classical limit theorems for the sample mean. Let $X_1, X_2, \cdots$ be r.v.'s. and set $\bar{X}_n := n^{-1} \sum_{i=1}^n X_i$.

Theorem 2.10.6. [Weak Law of Large Numbers.] *If $X_1, X_2, \cdots$ are i.i.d. with finite mean μ, then $\bar{X}_n \to_p \mu$.*

Theorem 2.10.7. [Strong Law of Large Numbers.] *If $X_1, X_2, \cdots$ are i.i.d. with finite mean μ, then $\bar{X}_n \to_{a.s.} \mu$.*

Theorem 2.10.8. [Central Limit Theorem.] *Let $X_1, X_2, \cdots$ be i.i.d. r.v.'s with finite mean μ and finite positive variance σ^2. Then, $n^{-1/2} \sum_{i=1}^n (X_i - \mu) \to_D \mathcal{N}(0, \sigma^2)$.*

Next we give some extensions of this result.

Theorem 2.10.9. **[Lindeberg-Feller CLT].** *For each* $n \geq 1$, *let* $\xi_{ni}, 1 \leq i \leq n$ *be independent r.v.'s with* $E(\xi_{ni}) = 0$ *and* $\sigma_{ni}^2 := E(\xi_{ni}^2) < \infty$, *for all* $1 \leq i \leq n$. *Let* $S_n := \sum_{i=1}^n \xi_{ni}$ *and* $s_n^2 := \sum_{i=1}^n \sigma_{ni}^2$. *Then the following statements* (a) *and* (b) *are equivalent.*

(a) $s_n^{-1} S_n \to_D \mathcal{N}(0,1)$ *and* $\max_{1\leq i\leq n} P\big(|\xi_{ni}| > \epsilon s_n\big) \to 0, \forall \epsilon > 0.$

(b) *The Lindeberg condition holds: For every* $\epsilon > 0$,

$$g_n(\epsilon) := s_n^{-2} \sum_{i=1}^n E\Big(\xi_{ni}^2 I(|\xi_{ni}| > \epsilon s_n)\Big) \to 0. \tag{2.10.1}$$

The following corollaries treat some special cases that are used throughout the book.

Corollary 2.10.4. *For each positive integer* n, *let* $Y_{n1}, \cdots, Y_{nn}$ *be independent r.v.'s with zero means. Suppose* $\sum_{i=1}^n E(Y_{ni}^2) \to \sigma^2$ *for some* $\sigma > 0$ *and* $\sum_{i=1}^n E(Y_{ni}^2 I(|Y_{nj}| > \epsilon)) \to 0$ *for every* $\epsilon > 0$. *Then* $\sum_{i=1}^n Y_{ni} \to_D Y \sim \mathcal{N}(0, \sigma^2)$.

Corollary 2.10.5. *Let* F *be a d.f. with mean zero and finite positive variance* σ^2. *For each positive integer* n, *let* $\zeta_{n1}, \cdots, \zeta_{nn}$ *be i.i.d.* F *and* $c_{n1}, \cdots, c_{nn}$ *be real numbers satisfying the following conditions.*

$$\sum_{i=1}^n c_{ni}^2 = 1, \qquad c_n^* := \max_{1\leq i\leq n} |c_{ni}| \to 0. \tag{2.10.2}$$

Then, $\sum_{i=1}^n c_{ni}\zeta_i \to_D \mathcal{N}(0, \sigma^2)$.

Proof. This follows from Corollary 2.10.4 applied with $Y_{ni} = c_{ni}\zeta_{ni}$. Clearly, $Y_{n1}, \cdots, Y_{nn}$ are independent, have mean zero, and $\sum_{i=1}^n E(Y_{ni}^2) = \sum_{i=1}^n c_{ni}^2 \sigma^2 = \sigma^2$. Finally, for $\epsilon > 0$ and $\zeta \sim_D F$, we calculate

$$\sum_{i=1}^n E(Y_{ni}^2 I(|Y_{ni}| > \epsilon)) \leq \sum_{i=1}^n c_{ni}^2 E(\zeta_{ni}^2 I(c_n^* |\zeta_{ni}| > \epsilon)) = E\{\zeta^2 I(|\zeta| > \epsilon / c_n^*)\} \to 0,$$

by Theorem 2.10.4 as ζ has a finite second moment and $\epsilon/c_n^* \to_p \infty$. Hence, the assumptions of Corollary 2.10.4 are met and $\sum_{i=1}^n c_{ni}\zeta_{ni} = \sum_{i=1}^n Y_{ni} \to_D \mathcal{N}(0, \sigma^2)$. □

The choice $c_{ni} = n^{-1/2}$ and $\zeta_{ni} = W_i - \mu$ yields the classical CLT, Theorem 2.10.8.

Corollary 2.10.6. *Let* $X_{n1}, \cdots, X_{nn}$ *be i.i.d.* Ber(p_n) *r.v.'s. Suppose* $p_n \to p$, *for some* $0 < p < 1$. *Then* $Y_n := n^{-1/2} \sum_{i=1}^n (X_{ni} - p_n) \to_D \mathcal{N}(0, p(1-p))$.

Proof. The r.v.'s $\xi_{nj} = (X_{nj} - p_n)/n^{1/2}, 1 \leq j \leq n$, are independent with mean 0 and common variance $p_n(1-p_n)/n$ and are bounded by $n^{-1/2}$. Then we have $s_n^2 = p_n(1-p_n)$ and $g_n(\epsilon) \leq s_n^{-2} I(1 > \epsilon n^{1/2} s_n)$. Thus $g_n(\epsilon) = 0$ for large enough n, $Y_n/s_n := \sum_{i=1}^n \xi_{nj} \to_D \mathcal{N}(0,1)$ and $Y_n \to_D \mathcal{N}(0, p(1-p))$ by Slutsky's Theorem 2.10.5. □

Corollary 2.10.7. *Let* $\xi_{n1}, \cdots, \xi_{nn}$ *be independent r.v.'s satisfying the following condition for some* $\sigma^2 > 0$:

$$L_n(\epsilon) := \sum_{i=1}^n E(\xi_{ni}^2 I(|\xi_{ni}| > \epsilon)) \to 0, \qquad \forall \epsilon > 0, \tag{2.10.3}$$

$$\sum_{i=1}^n \big(E(\xi_{ni})\big)^2 \to 0, \qquad \sum_{i=1}^n E(\xi_{ni}^2) \to \sigma^2.$$

Then, the following three assertions hold.

$$\max_{1\le i\le n} |\xi_{ni}| \to_p 0, \qquad \sum_{i=1}^n (\xi_{ni} - E(\xi_{ni})) \to_D \mathcal{N}(0,\sigma^2), \qquad \sum_{i=1}^n \xi_{ni}^2 \to_p \sigma^2.$$

In particular, if $\sum_{i=1}^n E(\xi_{ni}) \to \mu$ *for some* $\mu \in \mathbb{R}$, *then* $\sum_{i=1}^n \xi_{ni} \to_D \mathcal{N}(\mu,\sigma^2)$.

Proof. By the Markov inequality (2.4.8) and (2.10.3), $\forall\, \epsilon > 0$,

$$P\Big(\max_{1\le i\le n} |\xi_{ni}| > \epsilon\Big) = P\Big(\bigcup_{i=1}^n \{|\xi_{ni}| > \epsilon\}\Big) \le \sum_{i=1}^n P(|\xi_{ni})| > \epsilon) \le \frac{1}{\epsilon^2} L_n(\epsilon) \to 0.$$

This proves the first assertion. The third assertion follows if we show $T_n := \sum_{i=1}^n (\xi_{ni}^2 - E(\xi_{ni}^2)) \to_p 0$. To show this, we proceed as follows. For a $\delta > 0$, set $Y_{ni}(\delta) = \xi_{ni}^2 I(|\xi_{ni}| \le \delta)$ and $Z_{ni}(\delta) = \xi_{ni}^2 I(|\xi_{ni}| > \delta)$. Note that $\sum_{i=1}^n E\big(Z_{ni}(\delta)\big) = L_n(\delta)$. Now rewrite

$$T_n = \sum_{i=1}^n (Y_{ni}(\delta) - E(Y_{ni}(\delta)) + \sum_{i=1}^n Z_{ni}(\delta) - L_n(\delta).$$

By the Markov inequality (2.4.8), we have

$$\begin{aligned} P(|T_n| > 3\eta) &\le P\Big(\Big|\sum_{i=1}^n (Y_{ni}(\delta) - E(Y_{ni}(\delta))\Big| > \eta\Big) + P\Big(\sum_{i=1}^n Z_{ni}(\delta) > \eta\Big) + I\big(L_n(\delta) > \eta\big) \\ &\le \frac{\mathrm{Var}\big(\sum_{i=1}^n Y_{nj}(\delta)\big)}{\eta^2} + \frac{L_n(\delta)}{\eta} + I\big(L_n(\delta) > \eta\big)\big), \qquad \forall\, \eta > 0. \end{aligned}$$

Since $\mathrm{Var}\big(\sum_{i=1}^n Y_{ni}(\delta)\big) \le \sum_{i=1}^n E\big(Y_{ni}^2(\delta)\big) = \sum_{i=1}^n E\big(\xi_{ni}^4 I(|\xi_{ni}| \le \delta)\big) \le \delta^2 \sum_{i=1}^n E\big(\xi_{ni}^2\big)$, in view of (2.10.3), we conclude that $\limsup_n P\big(|T_n| > 3\eta\big) \le \delta^2\sigma^2/\eta^2$. Since this is true for all $\eta > 0, \delta > 0$, we obtain the desired result of $T_n \to_p 0$ by letting $\delta \to 0$ in this bound.

With $\mu_{ni} := E\big(\xi_{ni}\big)$, we verify the following two assertions under (2.10.3):

$$\sum_{i=1}^n \mathrm{Var}\big(\xi_{ni}\big) = \sum_{i=1}^n E\big(\xi_{ni}^2\big) - \sum_{i=1}^n \mu_{ni}^2 \to \sigma^2$$

$$\sum_{i=1}^n E\big((\xi_{ni} - \mu_{ni})^2 I(|\xi_{ni} - \mu_{ni}| > \epsilon)\big) \le 4L_n(\epsilon/2) + 4\sum_{i=1}^n \mu_{ni}^2 \to 0.$$

In the last step we used the inequality that $\forall\, \epsilon > 0, a, b \in \mathbb{R}$,

$$\begin{aligned} (a-b)^2 I\big(|a-b| > \epsilon\big) &\le \big(2\max(|a|,|b|)\big)^2 I\big(2\max(|a|,|b|) > \epsilon\big) \\ &\le 4a^2 I\big(|a| > \epsilon/2\big) + 4b^2 I\big(|b| > \epsilon/2\big). \end{aligned} \tag{2.10.4}$$

Hence, Corollary 2.10.4 applied with $Y_{ni} = \xi_{ni} - \mu_{ni}$ implies $\sum_{i=1}^n \big(\xi_{ni} - \mu_{ni}\big) \to_D \mathcal{N}(0,\sigma^2)$ which is the second assertion. This assertion and $\sum_{i=1}^n \mu_{ni} \to \mu$ give the desired result $\sum_{i=1}^n \xi_{ni} \to_D \mathcal{N}\big(\mu,\sigma^2\big)$. □

Lemma 2.10.8. *Supose* $\xi_1, \xi_2, \cdots,$ *are i.i.d. r.v.'s with* $E\big(|\xi_1|^k\big) < \infty$, *for some* $k > 0$. *Then,* $\max_{1\le i\le n} |\xi_i| = o_p\big(n^{1/k}\big)$.

Proof. Fix $\epsilon > 0$. Then,

$$\begin{aligned} P\Big(\max_{1\le i\le n}|\xi_i| > \epsilon n^{1/k}\Big) &= P\Big(\bigcup_{i=1}^{n}\{|\xi_i| > \epsilon n^{1/k}\}\Big) \le \sum_{i=1}^{n} P\Big(|\xi_i| > \epsilon n^{1/k}\Big) \\ &= nP\Big(|\xi_1| > \epsilon n^{1/k}\Big) \le nE\Big(\frac{|\xi_1|^k I(|\xi_1| > \epsilon\, n^{1/k})}{\epsilon^k\, n}\Big) \\ &= \frac{1}{\epsilon^k} E\Big(|\xi_1|^k I(|\xi_1| > \epsilon\, n^{1/k})\Big) \to 0, \end{aligned}$$

by Theorem 2.10.4. □

Finally, we state a central limit theorem for martingale difference arrays. Let $(\Omega, \mathcal{F}, P)$ be a probability space; $\{\mathcal{F}_{n,i},\, 0 \le i \le n\}$ be an array of sub-σ-fields such that $\mathcal{F}_{n,i} \subset \mathcal{F}_{n,i+1}$, $0 \le i \le n-1$; Y_{ni} be $\mathcal{F}_{n,i}$ measurable r.v. with $E(Y_{ni}^2) < \infty$ and $E(Y_{ni}|\mathcal{F}_{n,i-1}) = 0$, $1 \le i \le n$. Then $\{Y_{ni}, \mathcal{F}_{n,i}, 1 \le i \le n\}$ is called a *square-integrable martingale difference array*. Let $s_n^2 := \sum_{i=1}^{n} E(Y_{ni}^2)$.

The central limit theorem for martingale differences which we find useful is obtained by combining results of Brown (1971) and Hall and Heyde (1980), and stated as the following lemma for an easy reference.

Lemma 2.10.9. *Let $\{Y_{ni}, \mathcal{F}_{n,i}; 1 \le i \le n\}$ be a martingale difference array for each $n \ge 1$ satisfying the following conditions.*

$$s_n^{-2}\sum_{i=1}^{n} E\big[Y_{ni}^2 I(|Y_{ni}| > \epsilon\, s_n)\big|\mathcal{F}_{n,i-1}\big] = o_p(1), \qquad \forall \epsilon > 0. \tag{2.10.5}$$

$$s_n^{-2}\sum_{i=1}^{n} E\big[Y_{ni}^2|\mathcal{F}_{n,i-1}\big] \to_p 1. \tag{2.10.6}$$

Then, $s_n^{-1}\sum_{i=1}^{n} Y_{ni} \to_D Z$, where Z is a $\mathcal{N}(0,1)$ random variable.

Remark 2.10.1. A sufficient condition for (2.10.5) is the unconditional Lindeberg condition:

$$L_n(\epsilon) := s_n^{-2}\sum_{i=1}^{n} E\big[Y_{ni}^2 I(|Y_{ni}| > \epsilon\, s_n)\big] \to 0, \quad \forall \epsilon > 0.$$

This follows from the Markov inequality (2.4.8) because $L_n(\epsilon)$ is the expected value of the sum in (2.10.5).

Because $I(|x| > y) \le (|x|/y)^\alpha I(|x| > y), \forall x \in \mathbb{R}, y > 0, \alpha > 0$, we obtain that

$$L_n(\epsilon) \le s_n^{-2}\sum_{i=1}^{n} E\Big(Y_{ni}^2 \frac{|Y_{ni}|^\alpha}{(\epsilon\, s_n)^\alpha} I(|Y_{ni}| > \epsilon s_n)\Big) \le \frac{1}{\epsilon^\alpha s_n^{2+\alpha}}\sum_{i=1}^{n} E\Big(|Y_{ni}|^{2+\alpha}\Big), \quad \forall \epsilon > 0, \alpha > 0.$$

Hence, another sufficient condition for (2.10.5) is

$$s_n^{-(2+\alpha)}\sum_{i=1}^{n} E\Big(|Y_{ni}|^{2+\alpha}\Big) \to 0, \qquad \text{for some } \alpha > 0.$$

2.10.2 Some Convergence Results for Integrals

In this subsection we shall mention some useful results about integrals $\int \gamma_n(x)\,dx$, where γ_n, $n \geq 1$ is a sequence of real-valued measurable functions defined on $\mathbb{R}^k$, $x \in \mathbb{R}^k$ is the k-tuple $(x_1, \cdots, x_k)'$, $dx = \prod_{i=1}^k dx_i$ and the range of integration is $\mathbb{R}^k$, k a positive integer. In other words, $\int \gamma_n(x)\,dx$ is the Lebesgue integral $\int \gamma_n\,d\lambda_k$. Again the proofs of these results can be found in Loève (1963) or Athreya and Lahiri (2006).

For the next three theorems we need the concept of almost everywhere convergence. Let $\gamma, \gamma_1, \gamma_2, \cdots$ be real-valued measurable functions defined on $\mathbb{R}^k$. Then we say γ_n converges to γ almost everywhere and write $\gamma_n \to_{a.e.} \gamma$, if $\lambda_k\big(\{x \in \mathbb{R}^k : \lim_{n\to\infty} \gamma_n(x) \neq \gamma(x)\}\big) = 0$. An extension of Theorem 2.10.4 is given in the following theorem.

Theorem 2.10.10. [Lebesgue Dominated Convergence Theorem]. *Let $g, \gamma_0, \gamma_1, \gamma_2, \cdots$ be measurable real-valued functions defined on $\mathbb{R}^k$ such that $\gamma_n \to_{a.e.} \gamma_0$, $|\gamma_n(x)| \leq g(x)$, $\forall x \in \mathbb{R}^k, n \geq 0$, and $\int g(x)\,dx < \infty$. Then $\int \gamma_n(x)\,dx \to \int \gamma_0(x)\,dx$ and $\int |\gamma_n(x) - \gamma_0(x)|\,dx \to 0$.*

The proof of the following theorem is similar to that of Theorem II.4.2, p6 of Hájek and Šidák (1967). It also provides an extension of Theorem 2.10.10.

Theorem 2.10.11. *Let k be a positive integer and $\gamma_0, \gamma_1, \gamma_2, \cdots$ be real-valued functions on $\mathbb{R}^k$ such that $\gamma_n \to_{a.e.} \gamma_0$ and $\limsup_{n\to\infty} \int \big|\gamma_n(x)\big|\,dx \leq \int \big|\gamma_0(x)\big|\,dx < \infty$. Then, for all measurable functions φ from $\mathbb{R}^k$ to $[0,1]$, $\int \varphi(x)\big|\gamma_n(x) - \gamma_0(x)\big|\,dx \to 0$.*

The following lemma is Theorem 9.5 of Rudin (1974) in the case $k = 1$. The proof of this theorem extends to $k > 1$. Thus we have the following result.

Lemma 2.10.10. *Let $1 \leq p < \infty$, g be a real-valued measurable function on $\mathbb{R}^k$ such that $\int |g(x)|^p dx$ is finite and let $W_g(s) := \int |g(x-s) - g(x)|^p\,dx,\ s \in \mathbb{R}^k$. Then, the function W_g is bounded and uniformly continuous on $\mathbb{R}^k$.*

This lemma and Theorem 2.10.10 together yield the following corollary.

Corollary 2.10.8. *Let $1 \leq p < \infty$ and g, v, and w be measurable real-valued functions on $\mathbb{R}^k$ satisfying $\int \big(|g(x)|^p + |v(x)| + |w(x)|\big)\,dx < \infty$. Then, as $h \to 0$,*

$$\iint \Big|g(x - hs) - g(x)\Big|^p\,dx|v(s)|\,ds \to 0, \tag{2.10.7}$$

$$\iiint \Big|g(x - hu(s,t)) - g(x)\Big|^p\,dx|v(s)w(t)|\,ds\,dt \to 0, \tag{2.10.8}$$

with u is a measurable function from $\mathbb{R}^{2k}$ into $\mathbb{R}^k$.

In applications, we shall take $u(s,t)$ to be $s + t$ or $s - t$ and, in the case $k = 1$, also to be st.

2.11 Some Properties of Densities

In this section we collect some smoothness properties of densities. For later use we first state that continuous distribution functions are uniformly continuous. A function h from $\mathbb{R}$ to $\mathbb{R}$ is said to be uniformly continuous if for every $\epsilon > 0$, there is a $\delta > 0$ such that $|h(y) - h(x)| < \epsilon$, for all $x, y \in \mathbb{R}$ with $|y - x| < \delta$.

Lemma 2.11.1. *Suppose F is a continuous d.f. on $\mathbb{R}$. Then F is uniformly continuous on $\mathbb{R}$.*

Proof. Fix $\epsilon > 0$. We need to show that $\forall\, \epsilon > 0$, there exist a $\delta > 0$ such that $F(x) - \epsilon < F(y) < F(x) + \epsilon$, uniformly for all $|y - x| < \delta$. Fix an $\varepsilon > 0$. Since F is nondecreasing, it suffices to find a $\delta > 0$ such that $F(y) - F(x) < \varepsilon$ for all x and y with $x < y < x + \delta$. Let M be a positive integer such that $2/(M+1) < \varepsilon$. Let $t_1, \cdots, t_M$ be real numbers such that $F(t_i) = i/(M+1)$, $i = 1, \cdots, M$. Set $\delta = \min_{2 \le i \le M} \big(t_i - t_{i-1}\big)$. Now let x and y be real numbers such that $x < y < x + \delta$. It suffices to verify the inequality $F(y) - F(x) \le 2/(M+1)$. If $x < t_1$, then $y < x + \delta \le t_1 + t_2 - t_1 = t_2$ and $F(y) - F(x) \le F(y) \le F(t_2) = 2/(M+1)$. If $x \in [t_{k-1}, t_k)$ for some $k = 2, \cdots, M-1$, then $y \le t_{k+1}$ and $F(y) - F(x) < F(t_{k+1}) - F(t_{k-1}) \le 2/(M+1)$. If $x \ge t_{M-1}$, then $F(y) - F(x) \le 1 - F(t_{M-1}) \le 2/(M+1)$. □

The next lemma gives properties of a uniformly continuous density.

Lemma 2.11.2. *Suppose f is a uniformly continuous function from $\mathbb{R}$ into $\mathbb{R}$ with $\int |f(x)|\, dx < \infty$. Then $f(x) \to 0$, as $|x| \to \infty$ and this implies that f is bounded.*

Proof. Let $\eta := \limsup_{x \to \infty} |f(x)|$. We need to show $\eta = 0$. Suppose the contrary holds, i.e., suppose $\eta > 0$. Then there are positive numbers $x_1, x_2, \cdots$, such that $x_{n+1} > x_n + 1$ and $\big|f(x_n)\big| > 2\eta/3$. By the uniform continuity of f, there is a $0 < \delta < 1/2$ such that for all $|x - y| < \delta$, $\big||f(x)| - |f(y)|\big| \le |f(x) - f(y)| < \eta/3$. Thus, for $|x - x_n| < \delta$, $|f(x)| > |f(x_n)| - \eta/3 > \eta/3$. From this we obtain the contradiction, viz,

$$\int |f(x)|\, dx \ge \sum_{n=1}^{\infty} \int_{x_n - \delta}^{x_n + \delta} |f(x)|\, dx \ge \sum_{n=1}^{\infty} 2\delta\eta/3 = \infty.$$

This proves $\eta = 0$. A similar argument yields $\limsup_{x \to -\infty} |f(x)| = 0$. Hence, $f(x) \to 0$ as $|x| \to \infty$ and $|f(x)| \le 1$ for all x with $|x| > M$ and some large M. Since $|f|$ is continuous on $[-M, M]$, $|f|$ has a maximal value on $[-M, M]$. Hence f is bounded. □

Next we recall definitions of Fisher information in location, scale and location-scale models. Let f be a density. The density f is *absolutely continuous* if there is a function h such that

$$f(b) - f(a) = \int_a^b h(x)\, dx, \quad \forall a, b \in \mathbb{R}, a < b.$$

This implies that f is differentiable almost everywhere and that $h(x)$ can be taken to be the derivative of f at x whenever the latter exists. In view of this the function h is called an a.e. derivative of f and often denoted as $\dot{f}$.

In various integrals appearing below and elsewhere in this book, we interpret $\dot{f}(x)/f(x)$ as 0 if $f(x) = 0$. Indeed, $f(x) = 0$ implies that f has a minimum at x and thus f is either differentiable at x with derivative 0 or f is not differentiable at x in which case we can and do take $\dot{f}(x) = 0$.

The location model generated by f is the parametric family of densities $f_\theta(x) = f(x - \theta)$, $x, \theta \in \mathbb{R}$. Fisher information in one observation about the location parameter in this model is described by the following definition.

Definition 2.11.1. The density f is said to have finite Fisher information for location if f is absolutely continuous and its a.e. density $\dot{f}$ satisfies

$$I_L(f) := \int \Big(\frac{\dot{f}(x)}{f(x)}\Big)^2 f(x)\, dx < \infty. \tag{2.11.1}$$

Lemma 2.11.3. *Let f have finite Fisher information for location. Then f is uniformly continuous and $\dot{f}$ is integrable and square integrable and integrates to zero, i.e.,*

$$\int \dot{f}(x)\,dx = \int \frac{\dot{f}(x)}{f(x)} f(x)\,dx = 0. \tag{2.11.2}$$

Moreover, f is Lipschitz$(1/2)$, i.e.,

$$|f(y) - f(x)| \le I_L^{3/4}(f)|y - x|^{1/2}, \quad \forall\, x, y \in \mathbb{R}. \tag{2.11.3}$$

Proof. We derive the above properties from the inequality

$$|f(y) - f(x)| \le \int_x^y |\dot{f}(x)|\,dx \le \Big(I_L(f)(F(y) - F(x))\Big)^{1/2}, \quad x < y, \tag{2.11.4}$$

where F is the d.f. with density f, and we used (2.4.6) with $p = q = 2$, $g_1 = (\dot{f}/f)\sqrt{f}$ and $g_2 = \sqrt{f}$ in the last step. Since F is continuous, it is uniformly continuous by Lemma 2.11.1. This and the above inequality imply that f is uniformly continuous, and hence bounded by Lemma 2.11.2. Taking $x = -n$ and $y = n$ in the above inequality and letting n go to infinity establishes the integrability of $\dot{f}$. Because f is bounded, the square integrability of $\dot{f}$ follows from

$$\int \big(\dot{f}(x)\big)^2\,dx = \int \Big(\frac{\dot{f}(x)}{f(x)}\Big)^2 f^2(x)\,dx \le \|f\|_\infty I_L(f).$$

Next, we prove (2.11.2). For $t \ne 0$, we have

$$\begin{aligned} 0 &= \frac{1}{t}\int (f(x+t) - f(x))\,dx = \int\int_0^1 \dot{f}(x+st)\,ds\,dx \\ &= \int \dot{f}(x)\,dx + \iint (\dot{f}(x+st) - \dot{f}(x))\,dx I(0 < s < 1)\,ds, \end{aligned}$$

where the last double integral tends to zero as t tends to zero by Lemma 2.10.10. This establishes (2.11.2).

The inequality (2.11.4) implies that $|f(y) - f(x)| \le I_L^{1/2}(f)$. Since, by Lemma 2.11.2, $f(x) \to 0$, as $x \to \infty$, upon letting $x \to \infty$ in the last bound we obtain $\|f\|_\infty \le I_L^{1/2}(f)$. Now use this fact, (2.11.4) and the fact that $|F(y) - F(x)| \le \|f\|_\infty |y - x|$ to conclude (2.11.3). □

The scale model generated by f is the parametric family of densities $f_\theta(x) = f(x/\theta)/\theta$, $x \in \mathbb{R}, \theta > 0$. The following definition describes Fisher information in one observation about the scale parameter in this model.

Definition 2.11.2. The density f is said to have finite Fisher information for scale if f is absolutely continuous and its a.e. density $\dot{f}$ satisfies

$$I_S(f) := \int \Big(1 + x\frac{\dot{f}(x)}{f(x)}\Big)^2 f(x)\,dx < \infty.$$

The location and scale model generated by f corresponds to the density

$$f_\theta(x) = \frac{1}{\theta_2} f\Big(\frac{x - \theta_1}{\theta_2}\Big), \quad x \in \mathbb{R}, \theta = (\theta_1, \theta_2)' \in \Theta = \mathbb{R} \times (0, \infty).$$

Fisher information in one observation about the location and scale parameters in this model is described by the following definition.

Definition 2.11.3. The density f is said to have finite Fisher information for location and scale if f is absolutely continuous and the integrals $I_L(f)$ and $I_S(f)$ are finite.

2.12 Stochastic Processes

In this section we shall briefly define stochastic processes. These constitute an important area of probability theory. We mention some of their definitions and facts as we need them later in the book for studying the large sample properties of some statistics. Expanded versions of the many notions defined below can be found in the books of Loève (1963), Billingsley (1968) and Athreya and Lahiri (2006). Our discussion below is necessarily brief.

Let $(\Omega, \mathscr{A}, P)$ be a probability space and T be a subinterval of $\mathbb{R}$. A real-valued function X defined on $T \times \Omega$ is said to be a stochastic process if for every $t \in T$, $X(t) := X(t, \cdot)$ is a r.v., i.e., $\{\omega \in \Omega; X(t,\omega) \in B\} \in \mathscr{A}$, for every $B \in \mathscr{B}$, where $\mathscr{B}$ is the Borel σ-field on $\mathbb{R}$. For each $\omega \in \Omega$, the function $X(\cdot, \omega)$, which maps $t \in T$ to $X(t, \omega)$, is called a sample path of the process. We say X is continuous if it has continuous sample path. Many processes of interest in statistics have sample paths that are right continuous and have left limits. We often write $\{X(t), t \in T\}$ for a stochastic process X.

Let $X = \{X(t), t \in T\}$ be a stochastic process. We say X is stationary if, for any positive integer k and all numbers $t_1, \cdots, t_k$ and $s_1, \cdots, s_k$ in T, the joint distribution of $X(s_1), \cdots, X(s_k)$ is the same as that of $X(t_1), \cdots, X(t_k)$ whenever the differences $s_1 - t_1, \cdots, s_k - t_k$ are identical. We say X has independent increments if for every positive k and all $t_0 \le t_1 \le \cdots \le t_k$ in T, the differences $X(t_1) - X(t_0), \cdots, X(t_k) - X(t_{k-1})$ are independent. We call X a second order process if $E\big(X^2(t)\big)$ is finite for each $t \in T$. In this case we can define a mean function ν by $\nu(t) := E\big(X(t)\big)$, $t \in T$, and a covariance function $\mathcal{C}$ by $\mathcal{C}(s,t) := \operatorname{Cov}(X(s), X(t))$, $s, t \in T$. If $\nu(t) = 0$ for all t, then we call the process centered. The process X is called a Gaussian stochastic process if for every positive integer k and every set of k numbers $t_1, \cdots, t_k$ in T, the joint distribution of $X(t_1), \cdots, X(t_k)$ is a k-variate normal distribution. A Gaussian process is characterized by its mean function and its covariance function.

A process $\{X(t), t \in [0, \infty)\}$ is said to be a standard Brownian motion if it has independent increments, $X(0) = 0$, wp1, and $X(t) \sim_D \mathcal{N}(0, t)$ for all $t > 0$. Such a process is a centered Gaussian process with covariance function $\mathcal{C}(s,t) := s \wedge t$, $s, t \ge 0$. Moreover, the process can be chosen to have continuous sample paths. We shall denote a standard Brownian motion on $[0, \infty)$ by $\mathbb{B}$. Given the standard Brownian motion $\{\mathbb{B}(t), t \ge 0\}$, the process $\{\mathbb{B}^0(t) := \mathbb{B}(t) - t\mathbb{B}(1), 0 \le t \le 1\}$ is called a Brownian bridge. Clearly, this is a centered Gaussian process with $\mathbb{B}^0(0) = 0 = \mathbb{B}^0(1)$ and $\operatorname{Cov}\big(\mathbb{B}^0(s), \mathbb{B}^0(t)\big) = s \wedge t - st$, $0 \le s, t \le 1$.

The empirical distribution function. Let $X_1, X_2, \cdots$ be i.i.d. r.v.'s with d.f. F. An important process is given by the empirical d.f. $F_n := \{F_n(x),\ x \in \mathbb{R}\}$ defined at (2.4.2), i.e.,

$$F_n(x) = n^{-1} \sum_{i=1}^{n} I(X_i \le x), \quad x \in \mathbb{R}.$$

It is easy to check that the process F_n has mean function F and covariance function $\mathcal{C}(x, y) = F(x \wedge y) - F(x)F(y)$, $x, y \in \mathbb{R}$. Also $nF_n(x) = \sum_{i=1}^{n} I(X_i \le x)$ has a $\operatorname{Bin}(n, F(x))$ distribution.

We shall now present some properties of the maximum error

$$\|F_n - F\|_\infty = \sup_{x \in \mathbb{R}} \big|F_n(x) - F(x)\big|$$

of estimating a d.f. F by F_n. We start with an exponential bound on the tail probability that is uniformly in F. Let $\mathcal{F}$ denote the set of all distribution functions on $\mathbb{R}$ and P_F denote the probability measure when $X_1, \cdots, X_n$ are i.i.d. F, for some $F \in \mathcal{F}$.

Lemma 2.12.1. *There exist a positive constant c such that, for all* $n \geq 1$ *and all* $z > 0$,

$$\sup_{F \in \mathcal{F}} P_F\Big(\big\|F_n - F\big\|_\infty > z\Big) \leq c\, e^{-2nz^2}. \tag{2.12.1}$$

This inequality is proved by Dvoretzky, Kiefer and Wolfowitz (1956). It was shown by Massart (1990) that the constant c can be chosen to be 2.

The bound (2.12.1) shows that the probability of the maximum error of estimating F by F_n decreases to zero exponentially fast. Lehmann (1959) shows that under some conditions, F_n is uniformly minimum variance unbiased estimator of F.

Theorem 2.10.7 implies $F_n(x) \to_{a.s.} F(x)$, for every $x \in \mathbb{R}$. Lemma 2.12.1 yields a much stronger result stated in the next lemma.

Lemma 2.12.2. **[Extended Glivenko-Cantelli Lemma]** *For all* $0 \leq \alpha < 1$, $n^{\alpha/2}\|F_n - F\|_\infty \to_{a.s.} 0$.

Another important implication of Lemma 2.12.1 is as follows.

Lemma 2.12.3. *For every* $0 < \epsilon < 1$ *there is a finite positive* t_ϵ *such that*

$$\sup_{F \in \mathcal{F}} P_F\Big(n^{1/2}\|F_n - F\|_\infty > t_\epsilon\Big) \leq \epsilon, \quad \forall n \geq 1.$$

Proof. Apply the bound (2.12.1) with $z = s/n^{1/2}$, $s > 0$, to obtain

$$\sup_{F \in \mathcal{F}} P_F\Big(n^{1/2}\big\|F_n - F\big\|_\infty > s\Big) \leq c\, e^{-2s^2}.$$

Now $c\, e^{-2s^2} = \epsilon$ holds for $s = t_\epsilon := \sqrt{-(1/2)\log(\epsilon/c)}$. Hence the claim. □

The previous result implies that $n^{1/2}\|F_n - F\|_\infty$ is bounded in probability under P_F for each F. If F is continuous, we have the following result which will be used frequently in the book.

$$n^{1/2}\big\|F_n - F\big\|_\infty \to_D \big\|\mathbb{B}^0\big\|_\infty. \tag{2.12.2}$$

A proof of this result is available in Billingsley (1968) or van der Vaart and Wellner (1996). We should mention that the limiting d.f. G can be expressed as an infinite series as follows.

$$G(t) := P\Big(\|\mathbb{B}^0\|_\infty \leq t\Big) = \sum_{k=-\infty}^{\infty} (-1)^k e^{-2k^2t^2} = 1 - 2\sum_{k=1}^{\infty}(-1)^{k-1}e^{-2k^2t^2}, \quad t > 0. \tag{2.12.3}$$

The weak convergence result (2.12.2) with limiting d.f. G was first obtained by Kolmogorov (1933) who established the uniform convergence

$$\lim_{n\to\infty} \sup_{t>0} \Big| P_F(n^{1/2}\big\|F_n - F\big\|_\infty < t) - G(t)\Big| = 0.$$

Remark 2.12.1. The above results have extensions to random vectors. Let the k-dimensional random vectors $X_1, \cdots, X_n$ be i.i.d. with d.f. F. Their empirical d.f. is defined as

$$F_n(x) = \frac{1}{n}\sum_{i=1}^{n} I(X_i \leq x), \quad x \in \mathbb{R}^k,$$

where $s \le t$ for s and t in $\mathbb{R}^k$ is understand to mean that the jth coordinate of s is less than or equal to the jth coordinate of t, for each $j = 1, \cdots, k$. Let $\mathcal{F}_k$ denote the set of all d.f.'s on $\mathbb{R}^k$. Kiefer and Wolfowitz (1958) obtained the following generalization of (2.12.1).

$$\sup_{F \in \mathcal{F}_k} P_F\Big(\sup_{x \in \mathbb{R}^k} |F_n(x) - F(x)| > z\Big) \le c_1(k) e^{-c_2(k)nz^2}, \quad z > 0. \tag{2.12.4}$$

for constants $c_1(k)$ and $c_2(k)$ that depend on the dimension k. Naaman (2021) established an upper bound of the form $k(n+1)e^{-2nz^2}$ and proved the following result. If the marginal distribution functions of F are continuous, then

$$n^{1/2} \sup_{x \in \mathbb{R}^k} \Big| F_n(x) - F(x) \Big| \to_D \max\Big(\|\mathbb{B}_1^0\|_\infty, \cdots, \|\mathbb{B}_k^0\|_\infty \Big) \tag{2.12.5}$$

where $\mathbb{B}_1^0, \cdots, \mathbb{B}_k^0$ are independent Brownian bridges.

Weighted empirical distribution function. Let $(V_1, W_1)', (V_2, W_2)', \cdots$ be i.i.d. bivariate random vectors and set

$$G_n(w) = \frac{1}{n} \sum_{i=1}^{n} V_j I(W_j \le w), \quad w \in \mathbb{R}.$$

Then G_n is called a weighted empirical distribution function. Lemma 2.12.2 with $\alpha = 0$ is known as the classical Glivenko-Cantelli Lemma. The next result is an extension of this classical result to weighted empirical processes.

Lemma 2.12.4. **[Weighted Empirical Glivenko-Cantelli Lemma.]** *If* $E\big(|V_1|\big)$ *is finite, then*

$$\sup_{w \in \mathbb{R}} \big| G_n(w) - E\big(G_n(w)\big) \big| \to_{a.s.} 0.$$

Proof. The proof mimics that of the Glivenko-Cantelli Lemma. Write a $v \in \mathbb{R}$ as $v = v^+ - v^-$, where $v^+ := vI(v > 0)$ and $v^- := -vI(-v > 0)$. Let $G_n^+(w) := n^{-1} \sum_{i=1}^n V_i^+ I(W_i \le w)$ and $G_n^-(w) := n^{-1} \sum_{i=1}^n V_i^- I(W_i \le w)$. Then, $G_n(w) = G_n^+(w) - G_n^-(w)$, $w \in \mathbb{R}$, and by the triangle inequality,

$$\sup_{w \in \mathbb{R}} \big| G_n(w) - E\big(G_n(w)\big) \big| \le \sup_{w \in \mathbb{R}} \big| G_n^+(w) - E\big(G_n^+(w)\big) \big| + \sup_{w \in \mathbb{R}} \big| G_n^-(w) - E\big(G_n^-(w)\big) \big|.$$

Since V_i^+ and V^- are nonnegative, it suffices to prove the theorem for nonnegative V_i's.

Assume from now on that $V_i \ge 0$, $i = 1, 2, \cdots$. Let (V, W) be a copy of (V_1, W_1). There is nothing to prove if $E\big(V\big)$ is zero, as then $V_i = 0$, $i = 1, 2, \cdots$ and hence $G_n(w) = 0$, for all $w \in \mathbb{R}$, wp1. Assume $E\big(V\big) > 0$ and, without loss of generality, equals 1; otherwise replace V by $V/E\big(V\big)$. Then, the function $G(w) := E\big(G_n(w)\big) = E\big(VI(W \le w)\big)$, $w \in \mathbb{R}$, is a d.f. on $\mathbb{R}$ and $G(-\infty) = 0$, $G(\infty) = E\big(V\big) = 1$. These properties are established using Theorem 2.10.4. Hence, for every $\delta > 0$, there is a positive integer k and a partition of $\mathbb{R}$, $-\infty = w_0 < w_1 < \cdots < w_k < w_{k+1} = \infty$ such that $G(w_{j+1\_}) - G(w_j) < \delta$, for every $j = 0, \cdots, k$. This allows us to write

$$\sup_{w \in \mathbb{R}} \big| G_n(w) - E\big(G_n(w)\big) \big| = \max_{j=0,\cdots,k} \sup_{w_j \le w < w_{j+1}} \big| G_n(w) - G(w) \big|.$$

For $w_j \le w < w_{j+1}$, we have the inequalities $G_n(w_j) \le G_n(w) \le G_n(w_{j+1\_})$ and

$$G(w_{j+1\_}) - \delta \le G(w_j) \le G(w) \le G(w_{j+1\_}) \le G(w_j) + \delta.$$

Therefore,

$$G_n(w) - G(w) \geq G_n(w_j) - G(w_j) - \delta \geq -|G_n(w_j) - G(w_j)| - \delta,$$
$$G_n(w) - G(w) \leq G_n(w_{j+1_-}) - G(w_{j+1_-}) + \delta \leq |G_n(w_{j+1_-}) - G(w_{j+1_-})| + \delta.$$

Hence,

$$\sup_{w\in\mathbb{R}} \left|G_n(w) - G(w)\right| \leq \max_{0\leq j\leq k} \left|G_n(w_j) - G(w_j)\right| + \max_{0\leq j\leq k} \left|G_n(w_{j+1_-}) - G(w_{j+1_-})\right| + \delta.$$

By the strong law of large numbers, see Theorem 2.10.7, for each $w \in \mathbb{R}$, $G_n(w) - G(w) \to_{a.s.} 0$ and $G_n(w_-) - G(w_-) \to_{a.s.} 0$. Hence, $\limsup_n \sup_{w\in\mathbb{R}} \left|G_n(w) - G(w)\right| \leq \delta$, wp1. Since this is true for all $\delta > 0$, we obtain $\sup_{w\in\mathbb{R}} \left|G_n(w) - G(w)\right| \to 0$, wp1. □

Autoregressive Process. An important stochastic process in applications is the so-called *linear autoregressive time series*, defined as follows. Let p be a known positive integer. The given stochastic process $X_i, i \in \mathbb{Z} := \{0, \pm 1, \cdots\}$ is said to form a linear autoregressive time series model of order p $\left(\text{AR}(p)\right)$, if it satisfies the following equations. For some real numbers $\rho_1, \cdots, \rho_p$,

$$X_i = \rho_1 X_{i-1} + \cdots + \rho_p X_{i-p} + \varepsilon_i, \qquad i \in \mathbb{Z}, \tag{2.12.6}$$

where $\varepsilon_i, i \in \mathbb{Z}$ are i.i.d. r.v.'s with $E(\varepsilon_0) = 0$ and $0 < \text{Var}(\varepsilon_0) < \infty$, and for each $i \in \mathbb{Z}$, ε_i is independent of $X_{i-1}, \cdots, X_{i-p}$. In this book we often mention various results available in the literature for stationary and ergodic linear autoregressive time series. The requirement of stationarity and ergodicity puts some restrictions on the values of the parameters $\rho_1, \cdots, \rho_p$. For example, an AR(1) process satisfies these two conditions if $|\rho_1| < 1$, c.f., Brockwell and Davis (1991).

2.13 Exercises

1. Let S and T be nonempty sets and $\mathscr{A}_T$ be a sigma field on T. Let f be a function from S into T. Let $\mathscr{A}_S$ be the collection of all subset A of S of the form $A = \{f \in B\}$ for some B in $\mathscr{A}_T$. Show that $\mathscr{A}_S$ is a sigma field on S. This sigma-field is the sigma field generated by f.
2. Let $\mathscr{A}_1$ and $\mathscr{A}_2$ be two sigma-fields on Ω. Show that $\mathscr{A} = \mathscr{A}_1 \cap \mathscr{A}_2$ is a sigma field on Ω. Note that $\mathscr{A}$ contains all subsets of Ω that belong to both $\mathscr{A}_1$ and $\mathscr{A}_2$.
3. Let P_1 and P_2 be two probability measures on the σ-field $\mathscr{A}$ on Ω. Show that the function P defined on $\mathscr{A}$ by $P(A) = (P_1(A) + P_2(A))/2$, $A \in \mathscr{A}$, is a probability measure on $\mathscr{A}$.
4. Let $(\Omega, \mathscr{A}, P)$ be a probability space and B be an event with $P(B) > 0$. Show that the function Q defined by $Q(A) = P(A \cap B)/P(B)$, $A \in \mathscr{A}$, is a probability measure on $\mathscr{A}$.
5. Let X and Y be random variables. Show that $\{X < Y\}$, $\{X \leq Y\}$ and $\{X = Y\}$ are events.
6. Let X and Y be independent. Verify the inequality

$$P(|X - Y| < t) \leq E(G(X + t) - G(X - t)), \quad t > 0,$$

with G the d.f. of Y.

7. Let X and Y be i.i.d. with continuous d.f. F. Verify the identities $P(X = Y) = 0$ and $P(X > Y) = 1/2$.
8. Let X have finite expectation and $-X =_D X$. Find $E(X)$.
9. Let X and Y be i.i.d. positive r.v.'s. Verify the identity $E\big(X/(X+Y)\big) = 1/2$.
10. Let $(\Omega, \mathscr{A}, P)$ be a probability space. Let $A_1, \ldots, A_m$ be disjoint events such that $\bigcup_{i=1}^{m} A_i = \Omega$ and $P(A_i) > 0$ for $i = 1, \ldots, m$.
(a) Show that $\mathscr{A}_0 = \{\bigcup_{j \in J} A_j, J \subset \{1, \ldots, m\}\}$ is a sub sigma-field of $\mathscr{A}$.
(b) Let X be a r.v. with finite expectation. Find $E\big(X|\mathscr{A}_0\big)$.
11. Let X have d.f. F. Express the d.f.'s of $|X|$, X^2 and X^3 in terms of F.
12. Let X and Y have joint d.f. F.
(a) Find the d.f. G of $\max(X,Y)$.
(b) Find the d.f. H of $\min(X,Y)$.
13. Let X and Y be r.v.'s with $\mathrm{Var}(X) = \mathrm{Var}(Y) = 10$ and $\mathrm{Var}(X+Y) = 30$. Find $\mathrm{Cov}(X,Y)$.
14. Let the bivariate random vector (X,Y) have joint f given by
$$f(x,y) = e^{-y} I(0 < x < y).$$
(a) Find a marginal density of X. (b) Find a marginal density of Y. (c) Find $E(X|Y)$ and $E(Y|X)$. (d) Find a joint density of X and $Y - X$.
15. Let X, Y, Z be r.v.'s with joint density
$$f(x,y,z) = ce^{-z} I(0 < x < y < z).$$
(a) Find the constant c.
(b) Find the marginal densities of X, Y and Z.
(c) Find the dispersion matrix of the random vector $(X,Y,Z)'$.
16. Let X_1, X_2, X_3 be independents r.v.'s with $X_i \sim_D \mathrm{Gamma}(\alpha_i, \beta)$, $i = 1, 2, 3$. Find a joint density of
$$Y_1 = X_1, \quad Y_2 = X_1 + X_2, \quad Y_3 = X_1 + X_2 + X_3.$$
17. Let X_1, X_2, X_3 be independents r.v.'s with $X_i \sim_D \mathrm{Gamma}(\alpha_i, \beta)$, $i = 1, 2, 3$. Find a joint density of
$$Y_1 = \frac{X_1}{X_1 + X_2}, \quad Y_2 = \frac{X_1 + X_2}{X_1 + X_2 + X_3}, \quad Y_3 = X_1 + X_2 + X_3.$$
18. Let X_1, X_2, X_3 be independents r.v.'s with $X_i \sim_D \mathrm{Gamma}(\alpha_i, \beta)$, $i = 1, 2, 3$. Find a joint density of
$$Y_1 = \frac{X_1}{X_1 + X_2 + X_3}, \quad Y_2 = \frac{X_2}{X_1 + X_2 + X_3}, \quad Y_3 = X_1 + X_2 + X_3.$$
19. Let $X \sim_D \mathrm{Unif}(a,b)$ with $-\infty < a < b < \infty$. Show that $U = (X-a)/(b-a) \sim_D \mathrm{Unif}(0,1)$, $E(X) = (a+b)/2$, $\mathrm{Var}(X) = (b-a)^2/12$.
20. Let $U \sim_D \mathrm{Unif}(0,1)$ and $-\infty < a \le b < \infty$. Show that $X := aU + b \sim_D \mathrm{Unif}(a,b)$.

21. Suppose X_1, X_2 are independent Unif$(0,1)$ r.v.'s. Determine the distribution of the sum $S = X_1 + X_2$ and the difference $D = X_1 - X_2$.

22. Suppose $X_i \sim_D \text{Gamma}(a_i, b), 1 \le i \le n$ are independent r.v.'s, for some $a_i > 0, b > 0, 1 \le i \le n$. Let $S_n := X_1 + \cdots + X_n$. Show that $S_n \sim_D \text{Gamma}\big(\sum_{i=1}^n a_i, b\big)$.

23. Suppose $X_i \sim_D \mathcal{N}(\mu_i, \sigma_i^2), i = 1, 2$ and that X_1, X_2 are independent. Show that $X_1 + X_2 \sim_D \mathcal{N}(\mu_1 + \mu_2, \sigma_1^2 + \sigma_2^2)$. Show, by induction, that if $X_1, \cdots, X_n$ are independent r.v.'s and $X_i \sim_D \mathcal{N}(\mu_i, \sigma_i^2)$, $1 \le i \le n$, then $S_n := \sum_{i=1}^n X_i \sim_D \mathcal{N}(\mu, \sigma^2)$, where $\mu = \sum_{i=1}^n \mu_i, \sigma^2 = \sum_{i=1}^n \sigma_i^2$.

24. Suppose Z is standard normal. Use the results of Example 2.9.2 to show that $Z^2 \sim_D \chi_1^2$ and $\Gamma(1/2) = \sqrt{\pi}$.

25. Suppose Z is standard normal. Find the moment generating function of Z^2. Use this to show that $Z^2 \sim_D \chi_1^2$.

26. Suppose X, X_1, X_2, X_3 are i.i.d. $\mathcal{N}(0,1)$ r.v.'s.

    (a) Compute $P(X_1 + X_2 \le 2X_3)$.

    (b) Show that $X_1 + X_2$ and $X_1 - X_2$ are independent.

    (c) Show that $E(X^k) = 0$, for all odd positive integers k.

    (d) Suppose $h(x), x \in \mathbb{R}$ is a real-valued skew-symmetric function, i.e., $h(-x) = -h(x), \forall x \in \mathbb{R}$ and $E(|h(X)|) < \infty$. Show that $E\big(h(X)\big) = 0$ without integration.

27. Let g be a density, and h be a function satisfying $|h(x)| < 1, x \in \mathbb{R}$, $\int h(x)g(x)\,dx = 0$ and $\int h^2(x)g(x)\,dx > 0$.
    (a) Show that the function f defined by

    $$f(x,y) = g(x)g(y)\big(1 + h(x)h(y)\big), \quad x, y \in \mathbb{R},$$

    is a bivariate density.
    (b) If X and Y have joint density f, then X and Y have marginal density g.

28. Let f be as in the previous problem with g the $\mathcal{N}(0,1)$ density and h also symmetric about 0. Let X and Y have joint density f. Show that X and Y are uncorrelated standard normal r.v.'s, but are not independent.

29. Suppose the r.v. Y is symmetrically distributed around zero. Let h be a skew-symmetric function as in exercise 8(d) above and $E(|h(Y)|) < \infty$. Show, without integration, that $E\big(h(Y)\big) = 0$.

30. Suppose X takes values $-1, 0, 1$ with respective probabilities $1/4$, $1/2, 1/4$. Let $Y = X^2$. Show that $\text{Cov}(X,Y) = 0$ but X and Y are not independent.

31. Suppose X is symmetrically distributed on $\mathbb{R}$ and $\ell(x), x \in \mathbb{R}$ is a real-valued function such that $\ell(-x) = \ell(x)$, for all $x \in \mathbb{R}$. Assume $E(|X| + |X\ell(X)| + |\ell(X)|) < \infty$. Show, without integration, that $\text{Cov}(X, \ell(X)) = 0$ but X and $\ell(X)$ are not independent.

32. Let F_{X_1,X_2} be a bivariate d.f. with joint density f_{X_1,X_2} of the pair of r.v.'s (X_1, X_2). Show that the joint d.f. and density of $Y_1 = aX_1 + b, Y_2 = cX_2 + d$, $a > 0, c > 0, b, d \in \mathbb{R}$ are as follows.

    $$F_{Y_1,Y_2}(y_1, y_2) = F_{X_1,X_2}\Big(\frac{y_1 - b}{a}, \frac{y_2 - c}{c}\Big),$$

    $$f_{Y_1,Y_2}(y_1, y_2) = f_{X_1,X_2}\Big(\frac{y_1 - b}{a}, \frac{y_2 - d}{c}\Big)\frac{1}{ac}, \quad y_1, y_2 \in \mathbb{R}.$$

33. Prove that $I\big(|x+y| > 2\eta\big) \le I\big(|x| > \eta\big) + I\big(|y| > \eta\big)$, $\forall\, x, y \in \mathbb{R}$ and $\eta > 0$.

34. Prove the inequality (2.10.4).

35. Let $X_1, X_2, \cdots, X_n$ be i.i.d. r.v.'s with finite mean μ and finite positive variance σ^2. Set $\bar{X}_n = \sum_{i=1}^n X_i/n$ and let Z be a $\mathcal{N}(0,1)$ r.v.. Use Theorems 2.10.3 and 2.10.8 to solve the following.
(a) Show that $n^{1/2}(\bar{X}_n^2 - \mu^2) \to_D 2\mu\sigma Z \sim_D \mathcal{N}\big(0, 4\mu^2\sigma^2\big)$ if $\mu \neq 0$.
(b) Show that $n\bar{X}_n^2 \to_D \sigma^2 Z^2$ if $\mu = 0$.

36. Let $X_1, X_2, \cdots, X_n$ be i.i.d. Ber(π) r.v.'s, for some $0 < \pi < 1$. Set $\bar{X}_n := \sum_{i=1}^n X_i/n$. Use Theorems 2.10.3 and 2.10.8 to prove the following statements.
(a) If $\pi \neq 1/2$, then $n^{1/2}\big[\bar{X}_n(1-\bar{X}_n) - \pi(1-\pi)\big] \to_D \mathcal{N}(0, \sigma^2)$, where $\sigma^2 = (1-2\pi)^2\pi(1-\pi)$.
(b) If $\pi = 1/2$, then $4n\big[\bar{X}_n(1-\bar{X}_n) - 1/4\big] = -4n(\bar{X} - 1/2)^2 \to_D -Z^2$, where Z is standard normal.

37. Theorem 2.10.11 is useful in solving this problem. Let $f_n, n \ge 0$ be densities on $\mathbb{R}$ such that $f_n \to_{a.e.} f_0$, as $n \to \infty$. Then show the following.

(A) For any measurable function φ from $\mathbb{R}$ to $[0,1]$, $\int \varphi(x)|f_n(x) - f_0(x)|\, dx \to 0$. This result with $\varphi = 1$ is known as Scheffé's Theorem. See Scheffé (1947).

(B) For any event A in the Borel sigma field $\mathscr{B}$, define $P_n(A) := \int I(x \in A)\, f_n(x)\, dx$. Then, $\lim_{n\to\infty} \sup_{A \in \mathscr{B}} \big|P_n(A) - P_0(A)\big| = 0$.

3

Order Statistics and Ranks

3.1 Introduction

Order statistics are obtained from a given sample by arranging the given observations in an ascending order. They are very useful in constructing estimates of the population percentiles. In this chapter, we shall discuss some properties of order statistics and derive finite sample confidence intervals for population percentiles based on order statistics that are valid for a large class of populations. We shall also discuss large sample confidence intervals for population percentiles.

Another set of simple statistics in the nonparametric setup are the relative ranks of the given observations. They are especially useful in constructing tests of various hypotheses. An advantage of using ranks of observations coming from a continuous population is that the distribution of ranks does not depend on the underlying population, which makes many tests of hypotheses based on them easily implementable. Some distributional properties of ranks are also discussed in this chapter.

3.2 Distribution of Order Statistics and Ranks

Hajék and Šidák (1967) discuss various properties of ranks and order statistics very elegantly. For the sake of completeness, we present some of their results about the distribution of order statistics and ranks in this section.

For any vector of real numbers $\boldsymbol{y} = (y_1, \cdots, y_n)'$, let $y_{(i)}$ denote the i-th smallest among $y_1, \cdots, y_n$, $1 \leq i \leq n$. Thus, we have

$$y_{(1)} \leq y_{(2)} \leq \cdots \leq y_{(n)}. \tag{3.2.1}$$

Let $\boldsymbol{y}_{(\cdot)} := (y_{(1)}, y_{(2)}, \cdots, y_{(n)})'$ denote the vector of ordered values of $\boldsymbol{y}$.

The rank of y_i is denoted by r_i and defined to be the number of given y's less than or equal to y_i, i.e., $r_i = \sum_{j=1}^{n} I(y_j \leq y_i)$, where $I(A)$ is the indicator function of the set A, see (2.3.1). Suppose $y_i \neq y_j$, for all $1 \leq i \neq j \leq n$. Then all the inequalities in (3.2.1) are strict, $\boldsymbol{r} := (r_1, r_2, \cdots, r_n)$ is a permutation of $1, \cdots, n$, and $y_i = y_{(r_i)}$, $1 \leq i \leq n$. Note that $r_i < r_j$ implies $y_{(r_i)} = y_i < y_j = y_{(r_j)}$. Let

$$\Pi_n := \text{set of all permutations of } 1, \cdots, n,$$
$$\mathcal{Y}_n := \{(y_1, \cdots, y_n)' \in \mathbb{R}^n, -\infty < y_1 < y_2 < \cdots < y_n < \infty\}.$$

Given a vector $(y_1, \cdots, y_n)'$, with $y_i \neq y_j$, $i \neq j$ and their ranks $\boldsymbol{r} = (r_1, \cdots, r_n)$, a permutation $\boldsymbol{d} = (d_1, \cdots, d_n)$ of the integers $1, \cdots, n$ is said to be a vector of *antiranks* or *inverse ranks*,

DOI: 10.1201/9781003585770-3

if d_i is the index of the y coordinate whose rank is i, i.e., $d_{r_i} = i$, $1 \le i \le n$. If we interpret $\boldsymbol{r}$ and $\boldsymbol{d}$ as one-to-one functions from $\{1, \cdots, n\}$ onto $\{1, \cdots, n\}$ by setting $\boldsymbol{r}(i) = r_i$ and $\boldsymbol{d}(i) = d_i$, $i = 1, \cdots, n$, then $\boldsymbol{d}$ is the inverse function of $\boldsymbol{r}$.

In the above definitions, if the components of $\boldsymbol{y}$ are r.v.'s $Y_1, \cdots, Y_n$, then $\boldsymbol{Y} := (Y_1, \cdots, Y_n)'$ is a random vector, rank r_i and antirank d_i are denoted by R_i and D_i, respectively, for each $1 \le i \le n$. The vectors $\boldsymbol{R} := (R_1, \cdots, R_n)$ and $\boldsymbol{D} := (D_1, \cdots, D_n)$ are called the vectors of ranks and antiranks, respectively. The components of the vector $\boldsymbol{Y}_{(\cdot)}$ are known as the order statistics of the given vector $\boldsymbol{Y}$.

Note that the empirical d.f. based on the observations $Y_1, \cdots, Y_n$ can be expressed as a function of their order statistics, i.e.,

$$G_n(y) := \frac{1}{n}\sum_{i=1}^{n} I(Y_i \le y) = \frac{1}{n}\sum_{i=1}^{n} I(Y_{(i)} \le y), \quad y \in \mathbb{R}.$$

Since G_n is piecewise constant with jumps only at the order statistics, we can recover the order statistics from the jump locations and jump sizes of the sample paths of the empirical process $\{G_n(y), y \in \mathbb{R}\}$.

If the random vector $\boldsymbol{Y}$ has a density, then the probability of any two coordinates being equal is 0, and hence $R_1, \cdots, R_n$ are well defined and $\boldsymbol{R}, \boldsymbol{D} \in \Pi_n$ and $\boldsymbol{Y}_{(\cdot)} \in \mathcal{Y}_n$ hold with probability 1 (wp1). Under the restriction that $\boldsymbol{Y}$ has a density, precise distributions of these entities are given in Theorem II.1.1 of Hajék and Šidák (1967). For the sake of completeness, we reproduce their result here with a few alterations. Let P_g (E_g) denote the probability distribution (expectation) when g is the density of the underlying random vector $\boldsymbol{Y}$.

Theorem 3.2.1. *Suppose $\boldsymbol{Y}$ has a density g. Then $\boldsymbol{Y}_{(\cdot)}$ has density*

$$\bar{g}(y_1, \cdots, y_n) := \sum_{\boldsymbol{r} \in \Pi_n} g(y_{r_1}, \cdots, y_{r_n}) I\big((y_1, \cdots, y_n)' \in \mathcal{Y}_n\big). \tag{3.2.2}$$

Moreover, for all $\boldsymbol{r} \in \Pi_n, \boldsymbol{y} = (y_1, \cdots, y_n)' \in \mathcal{Y}_n$,

$$P_g\big(\boldsymbol{R} = \boldsymbol{r} \big| \boldsymbol{Y}_{(\cdot)} = \boldsymbol{y}\big) = \frac{g(y_{r_1}, \cdots, y_{r_n})}{\bar{g}(y_1, \cdots, y_n)} \tag{3.2.3}$$

if $\bar{g}(y_1, \cdots, y_n)$ is positive.

Proof. For a permutation $\boldsymbol{r}$ in Π_n, we let $f_{\boldsymbol{r}}$ denote the function from $\mathbb{R}^n$ to $\mathbb{R}^n$ defined by $f_{\boldsymbol{r}}(y_1, \cdots, y_n) = (y_{r_1}, \cdots, y_{r_n})'$ and let $\boldsymbol{d}$ denote the vector of antiranks. This transformation is linear and invertible with inverse $f_{\boldsymbol{r}}^{-1} = f_{\boldsymbol{d}}$ and its Jacobian has absolute value 1. Since $\boldsymbol{Y}$ has density g, $f_{\boldsymbol{r}}^{-1}(\boldsymbol{Y})$ has density $g_{\boldsymbol{r}}(x) = g(f_{\boldsymbol{r}}(x))$. On the event $\{\boldsymbol{R} = \boldsymbol{r}\}$, we have the identities $\boldsymbol{Y} = f_{\boldsymbol{r}}(\boldsymbol{Y}_{(\cdot)})$ and $\boldsymbol{Y}_{(\cdot)} = f_{\boldsymbol{r}}^{-1}(\boldsymbol{Y})$. Fix a Borel subset B of $\mathcal{Y}_n$. Then the event $\{\boldsymbol{R} = \boldsymbol{r}, \boldsymbol{Y}_{(\cdot)} \in B\} = \{f_{\boldsymbol{r}}^{-1}(\boldsymbol{Y}) \in B\}$ and

$$P_g\big(\boldsymbol{R} = \boldsymbol{r}, \boldsymbol{Y}_{(\cdot)} \in B\big) = P_g(f_{\boldsymbol{r}}^{-1}(\boldsymbol{Y}) \in B) = \int_B g\big(f_{\boldsymbol{r}}(x)\big)\, dx.$$

Summing this expression over all permutations $\boldsymbol{r}$ yields (3.2.2). The desired (3.2.3) now follows from Example 2.7.7. □

We shall now discuss several important consequences of the above theorem. We shall do so under the assumption that the joint density g is symmetric under permutations, i.e., for all $\boldsymbol{y} \in \mathbb{R}^n$ and $\boldsymbol{r} \in \Pi_n$,

$$g(y_{r_1}, \cdots, y_{r_n}) = g(y_1, \cdots, y_n). \tag{3.2.4}$$

Then, by (3.2.2), $\boldsymbol{Y}_{(\cdot)}$ has density

$$\bar{g}(\boldsymbol{y}) = n!g(\boldsymbol{y})I(\boldsymbol{y} \in \mathcal{Y}_n).$$

Moreover,

$$\frac{g(y_{(r_1)},\cdots,y_{(r_n)})}{\bar{g}(y_{(1)},\cdots,y_{(n)})} = \frac{g(y_{(1)},\cdots,y_{(n)})}{n!g(y_{(1)},\cdots,y_{(n)})} = \frac{1}{n!}.$$

This together with (3.2.3) implies that ranks $\boldsymbol{R}$ and order statistics $\boldsymbol{Y}_{(\cdot)}$ are independent and $P_g(\boldsymbol{R} = \boldsymbol{r}) = 1/(n!)$.

Distribution free statistic. An important consequence of this fact is that for all densities g of $\boldsymbol{Y} = (Y_1,\cdots,Y_n)'$ that are symmetric under permutation, the distribution of any statistic $S(\boldsymbol{R})$ based on the ranks $\boldsymbol{R}$ of $Y_1,\cdots,Y_n$ does not depend on the underlying density g. In this case, we say that the statistic $S(\boldsymbol{R})$ is *distribution free.*

Next, let $T(\boldsymbol{Y})$ be a statistic such that $E_g\big(|T(\boldsymbol{Y})|\big) < \infty$. Then, the fact that on the event $\{\boldsymbol{R} = \boldsymbol{r}\}$, Y_i equals $Y_{(r_i)}$ for all $1 \le i \le n$ and the independence of $\boldsymbol{Y}_{(\cdot)}$ and $\boldsymbol{R}$ give

$$E_g\{T(\boldsymbol{Y})|\boldsymbol{R} = \boldsymbol{r}\} = E_g\{T\big(Y_{(r_1)},\cdots,Y_{(r_n)}\big)\}, \quad \boldsymbol{r} \in \Pi_n.$$

Finally, we give a formula for the distribution of ranks under another density that may not be symmetric under permutation. Let h be a density on $\mathbb{R}^n$ such that $g(\boldsymbol{y}) = 0$ implies $h(\boldsymbol{y}) = 0$, i.e., h vanishes, whenever g vanishes, or equivalently $h > 0$ implies $g > 0$. Then,

$$P_h\big(\boldsymbol{R} = \boldsymbol{r}\big) = \frac{1}{n!}E_g\Big(\frac{h(Y_{(r_1)},\cdots,Y_{(r_n)})}{g(Y_{(r_1)},\cdots,Y_{(r_n)})}\Big), \quad \boldsymbol{r} \in \Pi_n.$$

This formula is found useful in calculating the power of some rank tests. It follows from the calculations

$$P_h(\boldsymbol{R} = \boldsymbol{r}) = \int_{\mathcal{Y}_n} h(f\boldsymbol{r}(x))\,dx = \int_{\mathcal{Y}_n} \frac{h(f\boldsymbol{r}(x))}{n!g(f\boldsymbol{r}(x))}\bar{g}(x)\,dx,$$

where the first identity was established in the proof of Theorem 3.2.1 and the second follows from the permutation symmetry of g.

An important special case of (3.2.4) is obtained when $Y_1,\cdots,Y_n$ are i.i.d. with density f. Then $\boldsymbol{Y}$ has density

$$g(y_1,\cdots,y_n) := \prod_{i=1}^{n} f(y_i), \quad y_1,\cdots,y_n \in \mathbb{R},$$

which satisfies (3.2.4). Because of the importance of this case, we summarize the analogs of the above facts in this case in the following corollary, where now P_f and E_f denote the probability distribution and expectation, respectively, when $Y_1,\cdots,Y_n$ are i.i.d. f.

Corollary 3.2.1. *Suppose $Y_1,\cdots,Y_n$ are i.i.d. with density f. Then the following facts hold.*
(a) *The order statistic vector $\boldsymbol{Y}_{(\cdot)} = (Y_{(1)},\cdots,Y_{(n)})'$ has density*

$$n!\prod_{i=1}^{n} f(y_i)I\big(y_1 < \cdots < y_n\big). \tag{3.2.5}$$

(b) *The vector of ranks $\boldsymbol{R}$ of $\boldsymbol{Y}$ is independent of the order statistics $\boldsymbol{Y}_{(\cdot)}$.*
(c) *The random vector $\boldsymbol{R}$ is uniformly distributed over Π_n, i.e.,*

$$P_f\big(\boldsymbol{R} = \boldsymbol{r}\big) = \frac{1}{n!}, \quad \boldsymbol{r} \in \Pi_n. \tag{3.2.6}$$

(d) *Any rank statistic based on* $\boldsymbol{R}$ *is distribution free.*

(e) *Let* $\psi(\boldsymbol{Y})$ *be a real-valued statistic based on* $\boldsymbol{Y}$ *such that* $E_f\big(|\psi(\boldsymbol{Y})|\big) < \infty$. *Then*

$$E_f\Big(\psi(\boldsymbol{Y})\big|\boldsymbol{R}=\boldsymbol{r}\Big) = E_f\Big(\psi\big(Y_{(r_1)},\cdots,Y_{(r_n)}\big)\Big). \tag{3.2.7}$$

(f) *Let* q *be a density on* $\mathbb{R}$ *such that* $f(y)=0$ *implies* $q(y)=0$. *Then*

$$P_q(\boldsymbol{R}=\boldsymbol{r}) = \frac{1}{n!}E_f\Big[\prod_{i=1}^{n}\frac{q(Y_{(r_i)})}{f(Y_{(r_i)})}\Big]. \tag{3.2.8}$$

Parts of the previous corollary continue to hold when the common d.f. F is only continuous.

Lemma 3.2.1. *Suppose* $Y_1,\cdots,Y_n$ *are i.i.d. with continuous d.f.* F. *Then parts (b)–(e) of Corollary 3.2.1 hold.*

Proof. Since F is continuous, $P\big(Y_i=Y_j\big)=0$ for $1\le i<j\le n$. Thus $P\big(\boldsymbol{R}\in\Pi_n\big)=1$. Let $\boldsymbol{Y}=(Y_1,\cdots,Y_n)$ and $f_{\boldsymbol{r}}$ be as in the proof of Theorem 3.2.1. Then $f_{\boldsymbol{r}}(\boldsymbol{Y}) =_D \mathbf{Y}$ holds for every $\boldsymbol{r}\in\Pi_n$ as $f_{\boldsymbol{r}}(\mathbf{Y})$ has d.f.

$$G(s_1,\cdots,s_n) := F(s_1)\cdots F(s_n), \quad s_1,\cdots,s_n\in\mathbb{R}.$$

Let $B\subseteq\mathcal{Y}_n$ be a Borel subset of $\mathbb{R}^n$. Then $P\big(\boldsymbol{R}=\boldsymbol{r},\boldsymbol{Y}_{(\cdot)}\in B\big) = P\big(f_{\boldsymbol{r}}^{-1}(\boldsymbol{Y})\in B\big) = P\big(\boldsymbol{Y}\in B\big)$ for all $\boldsymbol{r}\in\Pi_n$. From this we conclude $P\big(\boldsymbol{Y}_{(\cdot)}\in B\big) = \sum_{\boldsymbol{r}\in\Pi_n}P\big(\boldsymbol{R}=\boldsymbol{r},\boldsymbol{Y}_{(\cdot)}\in B\big) = n!P\big(\boldsymbol{Y}\in B\big)$. Combining the above yields $P\big(\boldsymbol{R}=\boldsymbol{r},\boldsymbol{Y}_{(\cdot)}\in B\big) = (1/n!)P\big(\boldsymbol{Y}_{(\cdot)}\in B\big)$. Taking $B=\mathcal{Y}_n$, we obtain $P(\boldsymbol{R}=\boldsymbol{r}) = 1/(n!)$. Hence, we conclude that $\boldsymbol{R}$ and $\boldsymbol{Y}_{(\cdot)}$ are independent and $\boldsymbol{R}$ is uniformly distributed over Π_n. Thus (b) and (c) of Corollary 3.2.1 hold, and these imply (d) and (e). □

Lemma 3.2.2. *Suppose* $Y_1,\cdots,Y_n$ *are i.i.d. with d.f.* F. *Then, for each* $j=1,\cdots,n$,

$$P(Y_{(j)}\le y) = \sum_{k=j}^{n}\binom{n}{k}\big(F(y)\big)^k\big(1-F(y)\big)^{n-k}, \quad y\in\mathbb{R}.$$

Proof. Now fix $j=1,\cdots,n$ and $y\in\mathbb{R}$. The lemma follows because $\{Y_{(j)}\le y\}$ equals $\big\{\sum_{i=1}^n I(Y_i\le y)\ge j\big\}$ and the sum $\sum_{i=1}^n I(Y_i\le y)$ has a $\mathrm{Bin}(n,F(y))$ distribution. □

The next corollary discusses the mean and covariances of ranks, their marginal distributions, and mean and variance of a simple linear rank statistic. Let $\{c_{n1},c_{n2},\cdots,c_{nn}\}$ be a given array of real numbers and $\psi_n, n\ge1$ be a sequence of real-valued functions defined on the set of integers $1,\cdots,n$. The statistic

$$S_c := \sum_{i=1}^{n} c_{ni}\psi_n(R_i) \tag{3.2.9}$$

is called *simple linear rank statistic*. Let $\bar{c}_n := n^{-1}\sum_{i=1}^n c_{ni}$ and

$$\bar{\psi}_n := n^{-1}\sum_{i=1}^{n}\psi_n(i), \quad \sigma_c^2 := \sum_{i=1}^{n}\big(c_{ni}-\bar{c}_n\big)^2, \quad \sigma_\psi^2 := \sum_{i=1}^{n}\big(\psi_n(i)-\bar{\psi}_n\big)^2. \tag{3.2.10}$$

We are now ready to state and prove the following corollary, where, for any finite set A, $\mathrm{Unif}(A)$ denotes the uniform distribution over the set A, which means that each element of A has probability $1/m$ with m the cardinality of A.

Corollary 3.2.2. *Suppose* $\boldsymbol{R} = (R_1, \cdots, R_n) \sim_D \text{Unif}(\Pi_n)$. *Then the following results hold for all integers* i, j, r, s *in* $\{1, \cdots, n\}$.

$$P(R_j = r) = \frac{1}{n}, \qquad E(R_j) = \frac{n+1}{2}, \qquad \text{Var}(R_j) = \frac{n^2-1}{12}, \tag{3.2.11}$$
$$P(R_i = r, R_j = s) = \frac{1}{n(n-1)}, \quad i \neq j, r \neq s.$$

Moreover,

$$E(S_c) = n\bar{c}_n\bar{\psi}_n, \qquad \text{Var}(S_c) = \frac{1}{n-1}\sigma_c^2\sigma_\psi^2. \tag{3.2.12}$$

Proof. Let i, j, r, s belong to $A := \{1, \cdots, n\}$. The assumption $\boldsymbol{R} \sim_D \text{Unif}(\Pi_n)$ means

$$P(\boldsymbol{R} = \boldsymbol{r}) = \frac{1}{n!}, \qquad \boldsymbol{r} = (r_1, \cdots, r_n) \in \Pi_n.$$

Since there are $(n-1)!$ permutations $\boldsymbol{r}$ with $r_j = r$, we have $P(R_j = r) = (n-1)!/n! = 1/n$. Hence, $R_j \sim_D \text{Unif}(A)$ and

$$E(R_j) = \frac{n+1}{2}, \qquad \text{Var}(R_j) = \frac{n^2-1}{12}, \quad j = 1, \cdots, n.$$

These formulas are derived using the identities

$$\sum_{i=1}^n i = \frac{n(n+1)}{2}, \qquad \sum_{i=1}^n i^2 = \frac{n(n+1)(2n+1)}{6}.$$

Similarly, for $i \neq j$ and $r \neq s$, there are $(n-2)!$ permutations $\boldsymbol{r}$ with $r_i = r$ and $r_j = s$ and hence

$$P(R_i = r, R_j = s) = \frac{(n-2)!}{n!} = \frac{1}{n(n-1)}.$$

This completes the proof of (3.2.11).

Next, consider

$$E(\psi_n(R_j)) = n^{-1}\sum_{i=1}^n \psi_n(i) = \bar{\psi}_n, \qquad E(\psi_n^2(R_j)) = n^{-1}\sum_{i=1}^n \psi_n^2(i),$$
$$\text{Var}(\psi_n(R_j)) = n^{-1}\sum_{i=1}^n \psi_n^2(i) - (\bar{\psi}_n)^2 = n^{-1}\sigma_\psi^2, \quad 1 \le j \le n,$$
$$E(S_c) = \sum_{j=1}^n c_{nj}E(\psi_n(R_j)) = \sum_{j=1}^n c_{nj}\bar{\psi}_n = n\bar{c}_n\bar{\psi}_n.$$

Let $a(k) := \psi_n(k) - \bar{\psi}_n$. Then $\sum_{k=1}^n a(k) = 0$. Using this, we calculate, for $i \neq j$,

$$\text{Cov}(\psi_n(R_i), \psi_n(R_j)) = E(a(R_i), a(R_j)) \tag{3.2.13}$$
$$= \frac{1}{n(n-1)}\sum_{r=1}^n\sum_{s=1}^n I(s \neq r)\,a(r)a(s) = \frac{1}{n(n-1)}\sum_{r=1}^n a(r)\Big[\sum_{s=1}^n a(s) - a(r)\Big]$$
$$= -\frac{1}{n(n-1)}\sum_{r=1}^n a^2(r) = -\frac{1}{n(n-1)}\sigma_\psi^2.$$

Let $d_i = c_{ni} - \bar{c}_n$. Using $\sum_{i=1}^n d_i = 0$ and $\sum_{i=1}^n d_i^2 = \sigma_c^2$, we calculate

$$\begin{aligned}
\text{Var}(S_c) &= \text{Var}\Big(\sum_{j=1}^n d_j \psi_n(R_j)\Big) \\
&= \sum_{j=1}^n d_j^2 \text{Var}\big(\psi_n(R_j)\big) + \sum_{i=1}^n \sum_{j=1}^n I(i \neq j)\, d_i d_j \, \text{Cov}\big(\psi_n(R_i), \psi_n(R_j)\big) \\
&= n^{-1}\sigma_c^2 \sigma_\psi^2 - \frac{1}{n(n-1)}\sigma_\psi^2 \sum_{i=1}^n d_i \big[\sum_{j=1}^n d_j - d_i\big] \\
&= \frac{1}{n}\sigma_c^2\sigma_\psi^2 + \frac{1}{n(n-1)}\sigma_\psi^2\sigma_c^2 = \frac{1}{n-1}\sigma_c^2\sigma_\psi^2.
\end{aligned}$$

This completes the proof of (3.2.12) and also of the corollary. □

If in the above we take $\psi_n(r) = r$, then by (3.2.12) and (3.2.13),

$$\sigma_\psi^2 = \frac{n(n^2-1)}{12}, \qquad \text{Var}\Big(\sum_{i=1}^n c_{ni} R_i\Big) = \frac{n(n+1)}{12}\sigma_c^2. \tag{3.2.14}$$

Furthermore, if $c_{ni} = I(i \in A_n)$ for some nonempty subset A_n of $\{1, \cdots, n\}$ with $m_n < n$ elements, then

$$\bar{c}_n = \frac{m_n}{n}, \qquad \sigma_c^2 = \frac{m_n(n-m_n)}{n}. \tag{3.2.15}$$

The above formulas will be used later in the text.

Next, we describe marginal densities of individual order statistics and joint densities of pairs of order statistics. Let F denote the d.f. with density f. Upon performing repeated integration on the joint density of $\boldsymbol{Y}_{(\cdot)}$ given at (3.2.5), we obtain the following formulas.

Corollary 3.2.3. *Let $Y_1, \cdots, Y_n$ be i.i.d. with density f. Then, for $1 \le j \le n$, a density of $Y_{(j)}$ is given by*

$$f_j(x) = n!\,\frac{F(x)^{j-1}\big(1-F(x)\big)^{n-j} f(x)}{(j-1)!(n-j)!}, \qquad x \in \mathbb{R}.$$

For $1 \le j < k \le n$, a joint density of $Y_{(j)}$ and $Y_{(k)}$) is given by

$$f_{j,k}(x,y) = n!\,\frac{F(x)^{j-1}\big(F(y)-F(x)\big)^{k-j-1}\big(1-F(y)\big)^{n-k}}{(j-1)!(k-j-1)!(n-k)!} f(x)f(y)I(x<y).$$

Suppose $U_1, U_2, \cdots, U_n$ is a random sample from the Unif$(0,1)$ distribution, see (2.8.1). Let $U_{(1)}, \cdots, U_{(n)}$ denote the corresponding order statistics and set $\boldsymbol{U}_{(\cdot)} := \big(U_{(1)}, \cdots, U_{(n)}\big)$. Then, by Corollary 3.2.3, $U_{(i)}$ has density

$$f_i(x) = \frac{n!}{(i-1)!(n-i)!} x^{i-1}(1-x)^{n-i} I(0<x<1), \tag{3.2.16}$$

for $i = 1, \cdots, n$, while $U_{(j)}$ and $U_{(k)}$ for $1 \le j < k \le n$ have joint density

$$f_{j,k}(x,y) = n!\frac{x^{j-1}(y-x)^{k-j-1}(1-y)^{n-k}}{(j-1)!(k-j-1)!(n-k)!} I(0<x<y<1). \tag{3.2.17}$$

Thus $U_{(i)}$ has a Beta$(i, n+1-i)$ distribution, and we obtain

$$E(U_{(i)}) = \frac{i}{n+1}, \quad E(U_{(i)}^2) = \frac{i(i+1)}{(n+1)(n+2)}, \quad \mathrm{Var}(U_{(i)}) = \frac{i(n-i+1)}{(n+1)^2(n+2))}.$$

Let $U \sim_D$ Unif$(0,1)$. Recall the definition of F^{-1} from (2.9.1) and that $F^{-1}(U)$ has d.f. F, see Theorem 2.9.1. Thus, if $Y_1, \cdots, Y_n$ are i.i.d. F, then $\boldsymbol{Y} = (Y_1, \ldots, Y_n)' =_D \big(F^{-1}(U_1), \cdots, F^{-1}(U_n)\big)'$ and $\boldsymbol{Y}_{(\cdot)} =_D \big(F^{-1}(U_{(1)}), \cdots, F^{-1}(U_{(n)})\big)'$. Hence, for a statistic $h(Y_{(j)})$ with finite mean, we have

$$E\big(h(Y_{(j)})\big) = E\big(h(F^{-1}(U_{(j)})\big) = \frac{n!}{(j-1)!(n-j)!} \int_0^1 h(F^{-1}(u)) u^{j-1}(1-u)^{n-j}\, du.$$

If F has a density f, then we also have

$$E\big(h(Y_{(j)})\big) = \frac{n!}{(j-1)!(n-j)!} \int_{-\infty}^{\infty} h(y) F^{j-1}(y)(1-F(y))^{n-j} f(y)\, dy.$$

Next, we shall derive the density of the sample range $R := U_{(n)} - U_{(1)}$ of a random sample $U_1, \cdots, U_n$ from uniform distribution on $(0,1)$. By (3.2.17) applied with $j = 1, k = n$, a joint density of $U_{(1)}, U_{(n)}$ is

$$f_{1,n}(x,y) = n(n-1)\big(y-x\big)^{n-2}, \qquad 0 < x < y < 1.$$

Now put $r = y - x, v = x$. Then $x = v, y = v + r$, the Jacobian is the constant 1, and

$$f_{R,V}(r,v) = n(n-1)r^{n-2}, \qquad 0 < v < v + r < 1.$$

Integrating out v on the interval $0 < v < 1 - r$ gives

$$f_R(r) = n(n-1)r^{n-2}(1-r), \qquad 0 < r < 1. \tag{3.2.18}$$

Thus the sample range R has the same distribution as $U_{(n-1)}$, and we have

$$E(R) = \frac{n-1}{n+1}, \quad E(R^2) = \frac{n(n-1)}{(n+1)(n+2)}, \quad \mathrm{Var}(R) = \frac{2(n-1)}{(n+1)^2(n+2)}. \tag{3.2.19}$$

Let $-\infty < a < b < \infty$ and $X_1, \cdots, X_n$ be i.i.d. Unif(a,b) r.v.'s. Then $X_i = a + (b-a)U_i$, where $U_i = (X_i - a)/(b-a) \sim_D$ Unif$(0,1)$. Hence, the sample range $X_{(n)} - X_{(1)}$ of the random sample $X_1, \cdots, X_n$ has the same distribution as $\widetilde{R} := (b-a)[U_{(n)} - U_{(1)}] = (b-a)R$, and a density of R is given in (3.2.18). Hence, a density of $\widetilde{R}$ is

$$\begin{aligned} f_{\widetilde{R}}(x) &:= \frac{n(n-1)}{b-a}\Big(\frac{x}{b-a}\Big)^{n-2}\Big(\frac{b-a-x}{b-a}\Big) \\ &= \frac{n(n-1)}{(b-a)^n} x^{n-2}(b-a-x), \quad a \le x \le b. \end{aligned}$$

From (3.2.19), $E\big(\widetilde{R}\big) = (b-a)[(n-1)/(n+1)]$ and $T_n := (n+1)\widetilde{R}/(n-1)$ is an unbiased estimator of the population range $b - a$. Moreover, by (3.2.19),

$$\mathrm{Var}\big(T_n\big) = \frac{(n+1)^2}{(n-1)^2}(b-a)^2 \mathrm{Var}(R) = \frac{2(b-a)^2}{(n-1)(n+2)}.$$

3.3 Percentiles

In this section we shall use order statistics for developing point and interval estimates of population percentiles, without assuming any knowledge of the underlying population.

Let $0 < p < 1$. Let F be a d.f. on $\mathbb{R}$. Recall that $F(x_-) := P(X < x)$, for all $x \in \mathbb{R}$. For an $x \in \mathbb{R}$, let

$$\lfloor x \rfloor \text{ denote the integer part of } x \text{ and } \lceil x \rceil \text{ the smallest integer larger than or equal to } x. \tag{3.3.1}$$

Definition 3.3.1. Percentile. A real number ξ_p is called a p-th percentile (or p-quantile) of F if

$$F(\xi_{p-}) \leq p \leq F(\xi_p). \tag{3.3.2}$$

If F represents a population d.f., then ξ_p is called a p-th population percentile (or p-th population quantile).

Lemma 3.3.1. *The set of all p-th percentiles is the closed interval $[q_1, q_2]$, where $q_1 = \inf\{x : F(x) \geq p\}$ and $q_2 = \sup\{x : F(x_-) \leq p\}$.*

Proof. For $s < q_1$ we have $F(s) < p \leq F(q_1)$, by the definition of q_1 and the right-continuity of F, and thus also $F(q_{1-}) \leq p$. This shows that q_1 is a p-th percentile and no $s < q_1$ can be a p-th percentile. Similarly, for $t > q_2$ we have $F(q_{2-}) \leq p < F(t_-)$, by the definition of q_2 and the left-continuity of $x \mapsto F(x_-)$, and thus $F(q_2) \geq p$. This shows that q_2 is a p-th percentile, $q_2 \geq q_1$, and no $t > q_2$ can be a p-th percentile. For $q_1 \leq u \leq q_2$, we have $F(u_-) \leq F(q_{2-}) \leq p \leq F(q_1) \leq F(u)$, and this establishes that u is a p-th percentile. The above also shows that $F(u) = p$ for all $q_1 \leq u < q_2$. □

Definition 3.3.2. Sample percentile. Let $X_1, X_2, \cdots, X_n$ be a random sample from F and F_n be the empirical d.f. as in (2.4.2). A r.v. $\hat{\xi}_{n,p}$ based on the random sample $X_1, X_2, \cdots, X_n$ is said to be a p-th sample percentile if it satisfies

$$F_n(\hat{\xi}_{n,p-}) \leq p \leq F_n(\hat{\xi}_{n,p}). \tag{3.3.3}$$

Just as population percentiles, sample percentiles need not be unique. Let $X_{(1)} \leq X_{(2)} \leq \cdots \leq X_{(n)}$ denote the order statistics of the sample $X_1, \cdots, X_n$. Recall (3.3.1). Let $m = \lfloor np \rfloor$. If $m = 0$, then $\xi_{n,p} = X_{(1)}$, while for $m > 0$, $\xi_{n,p}$ belongs to the interval $[X_{(m)}, X_{(m+1)}]$ and we can take $\xi_{n,p}$ to be the midpoint of this interval. Thus a possible choice of a p-th sample percentile is as follows.

$$\hat{\xi}_{n,p} = \begin{cases} X_{(1)}, & np < 1, \\ (X_{(\lfloor np \rfloor)} + X_{(\lfloor np \rfloor + 1)})/2, & 1 \leq np. \end{cases}$$

Of particular interest is a population median $\theta = \xi_p$, i.e., θ and F satisfy

$$F(\theta_-) \leq 1/2 \leq F(\theta). \tag{3.3.4}$$

A sample median $\tilde{X}_n := \hat{\xi}_{n,0.5}$ is given by

$$F_n(\tilde{X}_{n-}) \leq 1/2 \leq F_n(\tilde{X}_n).$$

A unique choice of sample median is made as follows, which depends on whether the sample size is even or odd. Let m be a positive integer. Then the unique sample median $\tilde{X}_n$ is defined as follows.

$$\tilde{X}_n = \begin{cases} X_{(m)}, & \text{if } n = 2m-1 \\ (X_{(m)} + X_{(m+1)})/2, & \text{if } n = 2m. \end{cases} \tag{3.3.5}$$

Because $nF_n(x)$ denotes the number of observations that are less than or equal to x, we have

$$nF_n(\tilde{X}_{n-}) = \begin{cases} m-1, & \text{if } n = 2m-1 \\ m, & \text{if } n = 2m \end{cases}$$

$$nF_n(\tilde{X}_n) = \begin{cases} m, & \text{if } n = 2m-1 \\ m, & \text{if } n = 2m. \end{cases}$$

Next, we mention some basic facts that connect events pertaining to order statistics and the empirical distribution function. For any $1 \le r,\, s \le n$ and any $x \in \mathbb{R}$,

$$\begin{aligned} \{X_{(s)} \le x\} &= \text{at least } s\ X\text{'s are less than or equal to } x = \{nF_n(x) \ge s\}, \\ \{X_{(s)} < x\} &= \text{at least } s\ X\text{'s are less than } x = \{nF_n(x_-) \ge s\}, \\ \{X_{(r)} > x\} &= \text{at most } r-1\ X\text{'s are less than or equal to } x = \{nF_n(x) \le r-1\}. \end{aligned} \tag{3.3.6}$$

Let $1 \le k_n \le n$ be a sequence of integers satisfying

$$v_n := \frac{k_n - np}{\sqrt{n}} \to 0. \tag{3.3.7}$$

The next theorem describes the asymptotic distribution of $n^{1/2}(X_{(k_n)} - \xi_p)$.

Theorem 3.3.1. *If F has a positive derivative d_p at ξ_p and k_n satisfies* (3.3.7), *then*

$$n^{1/2}(X_{(k_n)} - \xi_p) \to_D \mathcal{N}\Big(0, \frac{p(1-p)}{d_p^2}\Big).$$

Proof. Fix a $t \in \mathbb{R}$. Let $Z \sim_D \mathcal{N}(0, p(1-p))$. We need to show that

$$p_{n,k_n}(t) := P\Big(n^{1/2}(X_{(k_n)} - \xi_p) \le t\Big) \to p(t) := P(Z/d_p \le t).$$

Set $q_n = \xi_p + tn^{-1/2}$, $Y_n := n^{1/2}(F_n(q_n) - F(q_n))$ and $u_n = n^{1/2}(F(q_n) - F(\xi_p))$. The assumption that F has positive derivative at ξ_p implies that F is continuous and strictly increasing at ξ_p and hence $F(\xi_p) = p$. Using (3.3.6) and $F(\xi_p) = p$, we have

$$p_{n,k_n}(t) = P(X_{(k_n)} \le q_n) = P(nF_n(q_n) \ge k_n) = P(Y_n + u_n - v_n \ge 0).$$

Corollary 2.10.6 applied with $X_{nj} = I(X_j \le q_n)$ and $p_n = F(q_n)$ yields $Y_n \to_D Z$. The differentiability of F at ξ_p yields $u_n \to td_p$. Of course, $v_n \to 0$ holds by assumption. Hence, $Y_n + u_n - v_n \to_D Z + td_p$. Since the limiting variable has a continuous distribution function, we derive

$$P\Big(n^{1/2}(X_{(k_n)} - \xi_p) \le t\Big) \to P\Big(Z + td_p \ge 0\Big) = P(-Z/d_p \le t).$$

This is the desired result as $-Z$ has the same distribution as Z. □

The next theorem gives the asymptotic normality of $\hat{\xi}_{n,p}$.

Theorem 3.3.2. *If F has a positive derivative d_p at ξ_p, then*

$$n^{1/2}\big(\hat{\xi}_{n,p} - \xi_p\big) \to_D \mathcal{N}\Big(0, \frac{p(1-p)}{d_p^2}\Big).$$

Proof. Fix $t \in \mathbb{R}$. Let $p_{n,r_n}(t)$ and $p(t)$ be as in the previous proof. We need to show

$$p_n(t) := P\big(n^{1/2}(\hat{\xi}_{n,p} - \xi_p) \le t\big) \to p(t).$$

Recall (3.3.1). Let $m_n = \lfloor np \rfloor$. The inequality $X_{(m_n)} \le \hat{\xi}_{n,p} \le X_{(m_n+1)}$ yields $p_{n,m_n+1}(t) \le p_n(t) \le p_{n,m_n}(t)$. Note that (3.3.7) holds with $k_n = m_n$ and $k_n = m_n + 1$. Thus $p_{n,m_n}(t)$ and $p_{n,m_n+1}(t)$ converge to $p(t)$, and this gives the desired $p_n(t) \to p(t)$. □

Because of its broad applications, we state the asymptotic normality of sample medians obtained upon taking $p = 1/2$ in the above theorem, in the following corollary.

Corollary 3.3.1. *Let $X_i = \theta + \varepsilon_i$, where $\varepsilon_i, 1 \le i \le n$ are i.i.d. F r.v's. Assume that F has median 0 and a positive derivative $f(0)$ at 0. Then, $n^{1/2}(\widetilde{X}_n - \theta) \to_D \mathcal{N}(0, 1/(4f^2(0)))$.*

3.3.1 Confidence Intervals for Population Medians

In this subsection we shall derive confidence intervals for population medians that are valid for all underlying population distributions. Before addressing this problem, we define the notion of a confidence interval for a general real parameter. We do so for a finite sample size and as the sample size tends to infinity.

Let $X_1, \cdots, X_n$ be a random sample from a d.f. F, θ be a real parameter pertaining to F and $0 < \alpha < 1$ be given. Let $C_{n,1} \le C_{n,2}$ be r.v.'s constructed from $X_1, \cdots, X_n$. Then the interval $[C_{n,1}, C_{n,2}]$ is called a finite sample confidence interval for θ of confidence level $1-\alpha$ if it captures θ with probability at least $1 - \alpha$, for all d.f.'s F and for all $n \ge 1$, i.e.,

$$P\big(C_{n,1} \le \theta \le C_{n,2}\big) \ge 1 - \alpha, \quad \text{for all d.f.'s } F, n \ge 1. \tag{3.3.8}$$

It is said to be a confidence interval of asymptotic confidence level $1 - \alpha$ or an asymptotic $1 - \alpha$ confidence interval for θ, if

$$\liminf_{n\to\infty} P\big(C_{n,1} \le \theta \le C_{n,2}\big) \ge 1 - \alpha$$

holds for all d.f.'s F. For example, if $\hat{\theta}_n$ and $\hat{s}_n > 0$ are statistics such that $Z_n := n^{1/2}(\hat{\theta}_n - \theta)/\hat{s}_n$ converges in distribution to a standard normal random r.v. Z, then $[\hat{\theta}_n - n^{-1/2}\hat{s}_n z_{\alpha/2}, \hat{\theta}_n + n^{-1/2}\hat{s}_n z_{\alpha/2}]$ is an asymptotic $1 - \alpha$ confidence interval for θ, where z_α is defined by

$$P\big(Z > z_\alpha\big) = \alpha. \tag{3.3.9}$$

This follows as

$$P\big(\hat{\theta}_n - n^{-1/2}\hat{s}_n z_{\alpha/2} \le \theta \le \hat{\theta}_n + n^{-1/2}\hat{s}_n z_{\alpha/2}\big) = P\big(-z_{\alpha/2} \le Z_n \le z_{\alpha/2}\big)$$

converges to $P\big(-z_{\alpha/2} \le Z \le z_{\alpha/2}\big) = 1-\alpha$, by the continuity of the standard normal distribution function.

Now we return to the problem of finding confidence intervals for median θ of a d.f. F, where θ and F satisfy (3.3.4). Let $X_1, \cdots, X_n$ be a random sample from F. Let $0 < \alpha < 1$ be given. We shall find fixed and large sample confidence intervals. The following lemma about binomial tail probabilities is useful in solving this problem. Let

$$b_{k,n}(x) := \sum_{j=k}^{n} \binom{n}{j} x^j (1-x)^{n-j}, \quad 0 < x < 1.$$

Lemma 3.3.2. *Let k and n be positive integers, $k \leq n$. Then $b_{k,n}(x)$ is increasing in x.*

Proof. Let $U_1, \cdots, U_n$ be i.i.d. Unif$(0,1)$ r.v.'s. Fix $0 < x < 1$. Then $N(x) = \sum_{i=1}^{n} I(U_i \leq x)$ has a Bin(n, x) distribution and $b_{k,n}(x) = P\big(N(x) \geq k\big)$. By (3.3.6) we have the identity $\{N(x) \geq k\} = \{U_{(k)} \leq x\}$ and hence, by (3.2.16),

$$b_{k,n}(x) = \int_0^x \frac{n!}{(k-1)!(n-k)!} y^{k-1}(1-y)^{n-k}\, dy.$$

The desired result is now immediate. □

Next, recall

$$nF_n(x_-) = \sum_{i=1}^{n} I(X_i < x), \qquad nF_n(x) = \sum_{i=1}^{n} I(X_i \leq x).$$

By Lemma 3.3.2, for all $k = 1, \cdots, n$, the inequalities $F(\theta_-) \leq 1/2 \leq F(\theta)$ imply

$$P\big(nF_n(\theta) \geq k\big) = b_{k,n}\big(F(\theta)\big) \geq b_{k,n}(1/2) = \sum_{j=k}^{n} \binom{n}{j} \left(\tfrac{1}{2}\right)^n,$$

$$P\big(nF_n(\theta_-) \geq k\big) = b_{k,n}\big(F(\theta_-)\big) \leq b_{k,n}(1/2) = \sum_{j=k}^{n} \binom{n}{j} \left(\tfrac{1}{2}\right)^n.$$

Therefore, using (3.3.6), for any $1 \leq r \leq s \leq n$ and for all θ and d.f.'s F satisfying (3.3.4),

$$\begin{aligned}
P\big(X_{(r)} \leq \theta \leq X_{(s)}\big) &= P\big(X_{(r)} \leq \theta\big) - P\big(X_{(r)} \leq \theta, X_{(s)} < \theta\big) \\
&= P\big(X_{(r)} \leq \theta\big) - P\big(X_{(s)} < \theta\big) \\
&= P\big(nF_n(\theta) \geq r\big) - P\big(nF_n(\theta_-) \geq s\big) \\
&\geq \sum_{j=r}^{n} \binom{n}{j} \left(\tfrac{1}{2}\right)^n - \sum_{j=s}^{n} \binom{n}{j} \left(\tfrac{1}{2}\right)^n = \sum_{j=r}^{s-1} \binom{n}{j} \left(\tfrac{1}{2}\right)^n.
\end{aligned}$$

Let $Y_n \sim_D$ Bin$(n, 1/2)$. Then the above inequality becomes

$$P\big(X_{(r)} \leq \theta \leq X_{(s)}\big) \geq P(r \leq Y_n \leq s-1).$$

Thus if we choose $r \leq s$ so that $P\big(r \leq Y_n \leq s-1\big) \geq 1-\alpha$, then the interval $\big[X_{(r)}, X_{(s)}\big]$ will be a confidence interval of confidence level $1-\alpha$ for θ. Note that this interval is valid even when θ is not unique. Here, $C_{n,1} = X_{(r)}, C_{n,2} = X_{(s)}$.

Recall that the Bin$(n, 1/2)$ distribution is symmetric around $n/2$. Using this property, it is relatively simple to choose r and s. First find the largest positive integer r such that $P(Y < r) \leq$

$\alpha/2$. Then take $s = n - r + 1$ and the required interval for the median θ is $[X_{(r)}, X_{(n-r+1)}]$. Thus, in this example, $C_{n,1} = X_{(r)}$, $C_{n,2} = X_{(n-r+1)}$.

Next, we shall describe large sample confidence intervals for the median θ. Recall (3.3.1). Let Z denote a $\mathcal{N}(0,1)$ r.v.. By Corollary 2.10.6 applied with $p = 1/2$, $Z_n = 2(Y_n - 0.5n)/n^{1/2} \to_D Z$. Using the continuity correction, we thus have

$$
\begin{aligned}
P(r \le Y_n \le s-1) &= P(r - 0.5 \le Y_n \le s - 0.5)\\
&= P\Big(\frac{2(r - 0.5 - 0.5n)}{n^{1/2}} \le Z_n \le \frac{2(s - 0.5 - 0.5n)}{n^{1/2}}\Big)\\
&\approx P\Big(\frac{2(r - 0.5 - 0.5n)}{n^{1/2}} \le Z \le \frac{2(s - 0.5 - 0.5n)}{n^{1/2}}\Big).
\end{aligned}
$$

We wish the last probability to be at least $1-\alpha$. Choose $r = r_n$ to be the largest integer and $s = s_n$ to be the smallest integer such that

$$
\begin{aligned}
&\frac{2(r_n - 0.5 - 0.5n)}{n^{1/2}} \le -z_{\alpha/2}, \qquad \frac{2(s_n - 0.5 - 0.5n)}{n^{1/2}} \ge z_{\alpha/2}, \qquad \text{or}\\
&r_n = \Big\lfloor -\frac{1}{2}n^{1/2}z_{\alpha/2} + \frac{1+n}{2}\Big\rfloor, \qquad s_n = \Big\lceil \frac{1}{2}n^{1/2}z_{\alpha/2} + \frac{1+n}{2}\Big\rceil.
\end{aligned}
$$

Then the confidence interval $[X_{(r_n)}, X_{(s_n)}]$ for θ will have asymptotic confidence level $1-\alpha$. Thus, here $C_{n,1} = X_{(r_n)}$, $C_{n,2} = X_{(s_n)}$. This approximation is reasonably good for $n \ge 18$.

3.3.2 Confidence Intervals for Population Percentiles

In this subsection we shall derive confidence intervals for a p-th population percentile ξ_p for a given $0 < p < 1$. These intervals are valid for all underlying population distributions.

Using Lemma 3.3.2 and $F(\xi_{p}-) \le p \le F(\xi_p)$, we obtain that

$$
\begin{aligned}
P\big(nF_n(\xi_p) \ge k\big) &= b_{k,n}(F(\xi_p)) \ge b_{k,n}(p) = \sum_{j=k}^{n}\binom{n}{j}p^j(1-p)^{n-j},\\
P\big(nF_n(\xi_{p}-) \ge k\big) &= b_{k,n}(F(\xi_{p}-)) \le b_{k,n}(p) = \sum_{j=k}^{n}\binom{n}{j}p^j(1-p)^{n-j}.
\end{aligned}
$$

Therefore, with $q := 1-p$,

$$
\begin{aligned}
P\big(X_{(r)} \le \xi_p \le X_{(s)}\big) &= P\big(X_{(r)} \le \xi_p\big) - P\big(X_{(s)} < \xi_p\big)\\
&= P\big(nF_n(\xi_p) \ge r\big) - P\big(nF_n(\xi_{p}-) \ge s\big)\\
&\ge \sum_{j=r}^{n}\binom{n}{j}p^jq^{n-j} - \sum_{j=s}^{n}\binom{n}{j}p^jq^{n-j} = \sum_{j=r}^{s-1}\binom{n}{j}p^jq^{n-j}.
\end{aligned}
$$

Now, choose $r \le s$ so that the above lower bound is at least $1-\alpha$. Then, the interval

$$
[X_{(r)}, X_{(s)}] \tag{3.3.10}
$$

is the finite sample confidence interval of confidence level $1-\alpha$ for population percentile ξ_p, for all d.f.'s F and for all sample sizes $n \ge 2$.

Large sample confidence intervals for population percentiles. Fix a $0 < p < 1$. Let Y_n denote a r.v. having a $\text{Bin}(n,p)$ distribution. From the above derivation we have,

$$P\big(X_{(r)} \le \xi_p \le X_{(s)}\big) \ge P\big(r \le Y_n \le s-1\big), \qquad \forall 1 \le r \le s \le n.$$

Proceeding as in the case of a median, by Corollary 2.10.6, $Z_n = \big(Y_n - np)/\sqrt{np(1-p)}\big)$ converges in distribution to Z. Using the continuity correction, we thus have

$$\begin{aligned} P\big(r \le Y_n \le s-1\big) &= P\big(r-0.5 \le Y_n \le s-0.5\big) \\ &= P\Big(\frac{r-0.5-np}{\sqrt{np(1-p)}} \le Z_n \le \frac{s-0.5-np}{\sqrt{np(1-p)}}\Big) \\ &\approx P\Big(\frac{r-0.5-np}{\sqrt{np(1-p)}} \le Z \le \frac{s-0.5-np}{\sqrt{np(1-p)}}\Big). \end{aligned}$$

Choose $r = r_n$ to be the largest integer and $s = s_n$ to be the smallest integer such that

$$\begin{aligned} &\frac{r-0.5-np}{\sqrt{np(1-p)}} \le -z_{\alpha/2}, \qquad \frac{s-0.5-np}{\sqrt{np(1-p)}} \ge z_{\alpha/2}, \quad \text{or} \qquad (3.3.11) \\ &r_n = \Big\lfloor -z_{\alpha/2}\sqrt{np(1-p)} + 0.5 + np\Big\rfloor, \qquad s_n = \Big\lceil z_{\alpha/2}\sqrt{np(1-p)} + 0.5 + np\Big\rceil. \end{aligned}$$

Then the interval $[X_{(r_n)}, X_{(s_n)}]$ will be a confidence interval for every population percentile ξ_p of the asymptotic confidence level $1-\alpha$. This approximation is reasonably good for sample sizes of 20 or larger when $0.3 \le p \le 0.7$.

It is important to note that the above confidence intervals are broadly applicable, without knowing anything about the population d.f. F. Here is an alternative approach under a stronger assumption. Suppose F has a positive derivative d_p at ξ_p. Then, by Theorem 3.3.2, the asymptotic distribution of $n^{1/2}(\xi_{n,p} - \xi_p)$ is $\mathcal{N}(0, p(1-p)/d_p^2)$. Let $\hat{d}_p$ be a consistent estimator of d_p, i.e., $\hat{d}_p \to_p d_p$. It then follows from Slutsky's Theorem 2.10.5 that $n^{1/2}\hat{d}_p(\xi_{n,p} - \xi_p)/\sqrt{p(1-p)}$ is asymptotically standard normal. This result is now used to describe alternative large sample confidence intervals for ξ_p. For $0 < \alpha < 1$, with z_α is as in (3.3.9), let

$$L_{n,\alpha} := \xi_{n,p} - n^{-1/2}\frac{\sqrt{p(1-p)}}{\hat{d}_p} z_{\alpha/2}, \quad U_{n,\alpha} := \xi_{n,p} + n^{-1/2}\frac{\sqrt{p(1-p)}}{\hat{d}_p} z_{\alpha/2}. \qquad (3.3.12)$$

Then, by the above asymptotic normality result, the interval $\big[L_{n,\alpha}, U_{n,\alpha}\big]$ is a confidence interval of asymptotic confidence level $1-\alpha$ for the unique population percentile ξ_p.

A possible choice for $\hat{d}_p$ is $\hat{d}_p := \big[F_n(\xi_{n,p} + h_n) - F_n(\xi_{n,p} - h_n)\big]/(2h_n)$, where F_n is the empirical d.f. and $h_n, n \ge 1$ is a sequence of positive numbers such that $h_n \to 0$ and $\liminf_{n\to\infty} n^\beta h_n > 0$, for some $0 < \beta < 1/2$. To prove $\hat{d}_p \to_p d_p$, proceed as follows. Let $\tilde{d}_p := \big[F(\xi_{n,p} + h_n) - F(\xi_{n,p} - h_n)\big]/2h_n$. By Lemma 2.12.2, $D_n := n^\beta \|F_n - F\|_\infty \to_p 0$. This shows that $|\hat{d}_p - \tilde{d}_p| \le D_n/(n^\beta h_n) \to_p 0$. Now, use the differentiability of F at ξ_p and $\xi_{np} - \xi_p = O_p(n^{1/2}) = o(h_n)$, to conclude $F(\xi_{np} + h_n) - F(\xi_p) = d_p(\xi_{np} + h_n - \xi_p) + o_p(\xi_{np} + h_n - \xi_p) = d_p h_n + o(h_n)$ and $F(\xi_{np} - h_n) - F(\xi_p) = -d_p h_n + o(h_n)$. From these two conclusions we derive $\tilde{d}_p \to_p d_p$. The above shows that $\hat{d}_p$ is a consistent estimator of d_p.

3.4 Exercises

1. Let $0 \le \pi_n \le 1, n \ge 1$, a sequence of numbers such that $n\pi_n \to \lambda$, for some $\lambda > 0$, as $n \to \infty$. Let Y_n be a Binomial(n, π_n) r.v. Show that

$$\lim_{n\to\infty} P(Y_n = k) \to \frac{\lambda^k e^{-\lambda}}{k!}, \quad k = 0, 1, 2, \cdots.$$

2. Let $U_{(1)} \le U_{(2)} \le \cdots \le U_{(n)}$ denote the order statistics of a random sample of size n from the Unif$(0, 1)$ distribution. Prove the following.

   (a) The joint density of $R = U_{(n)} - U_{(1)}$ and $V = (U_{(n)} + U_{(1)})/2$ is

$$f_{R,V}(r, v) = n(n-1)r^{n-2}, \quad 0 < r < 1, \ r/2 < v < 1 - r/2.$$

   (b) The r.v.'s $U_{(1)}/U_{(n)}$ and $U_{(n)}$ are independent.

   (c) Show that for $1 \le j < k \le n$, $U_{(k)} - U_{(j)} =_D U_{(k-j)}$.

   (d) Compute $E(R), \text{Var}(R), E(V), \text{Var}(V)$ and $\text{Cov}(R, V)$.

   (e) Let r $\in \{1, 2, \cdots, n\}$ be fixed (not a function of n). Show that

$$\lim_{n\to\infty} P(nU_{(r)} \le z) = \frac{1}{(r-1)!}\int_0^z e^{-t}t^{r-1}dt$$

   (f) Fix an $1 \le r < s \le n$. Determine the conditional density of $U_{(s)}$, given $U_{(r)}$.

3. Let $X_{(1)} \le X_{(2)} \le \cdots \le X_{(n)}$ denote the order statistics of a random sample of size n from the Unif$(0, \theta)$ distribution for some $\theta > 0$. Use part (a) of Problem 2 to prove the following.

   (a) The joint density of $R_1 = X_{(n)} - X_{(1)}$ and $V_1 = (X_{(n)} + X_{(1)})/2$ is

$$f_{R_1,V_1}(r, v) = \frac{n(n-1)r^{n-2}}{\theta^n}, \quad 0 < r < \theta, \ r/2 < v < \theta - r/2.$$

   (b) The r.v.'s $X_{(1)}/X_{(n)}$ and $X_{(n)}$ are independent.

4. Let F be a continuous d.f. and $X_{(1)} \le X_{(2)} \le \cdots \le X_{(n)}$ denote the order statistics of a random sample of size n from F and let $\tilde{R} := F(X_{(n)}) - F(X_{(1)})$. Show that

$$E(\tilde{R}) = \frac{n-1}{n+1}, \qquad \text{Var}(\tilde{R}) = \frac{2(n-1)}{(n+1)^2(n+2)}.$$

5. Let X_1, X_2 be i.i.d. $\mathcal{N}(0, 1)$ r.v.'s. Let $X_{(1)}, X_{(2)}$ denote their order statistics. Find $E(X_{(1)}), E(X_{(2)}), \text{Var}(X_{(1)}), \text{Var}(X_{(2)}), \text{Cov}(X_{(1)}, X_{(2)}),\ \text{Corr}(X_{(1)}, X_{(2)})$.
Answers: $-1/\sqrt{\pi},\ 1/\sqrt{\pi},\ 1 - 1/\pi,\ 1 - 1/\pi,\ 1/\pi,\ 1/(\pi - 1)$.

6. Let $U_1, U_2, \cdots$ be i.i.d. $U(0, 1)$ r.v.'s and let N be a positive integer-valued r.v. with the following distribution.

$$P(N = n) = \frac{1}{(e-1)n!}, \quad n = 1, 2, \cdots.$$

   Let $Z = \min\{U_1, U_2, \cdots, U_N\}$. Assume U_i's are independent of N. Show that

$$P(Z > z) = \frac{e^{1-z} - 1}{e - 1}, \quad 0 < z < 1.$$

7. Let $X_{(1)} \leq X_{(2)} \leq \cdots \leq X_{(n)}$ denote the order statistics of a random sample of size n from the exponential distribution with density $f(x) = \alpha e^{-\alpha x} I(x > 0)$ for some $\alpha > 0$. Prove the following.

   (a) For any $1 \leq r < s \leq n$, the r.v.'s $X_{(r)}$ and $X_{(s)} - X_{(r)}$ are independent.

   (b) $P\big(X_{(n)} - \log(n)/\alpha \leq x\big) \to e^{-e^{-\alpha x}}$ for all $x \in \mathbb{R}$.

8. Let F be a d.f. with density f. Fix $x \in \mathbb{R}$ such that $F(x) < 1$. Define $g_x(y) := f(y)/(1-F(x))I(y \geq x)$.

   (a) Show that g_x is a density.

   (b) Show that the d.f. G_x with density g_x is

$$G_x(y) := \frac{F(y) - F(x)}{1 - F(x)} I(y \geq x).$$

   (c) Let $X_{(1)} \leq X_{(2)} \leq \cdots \leq X_{(n)}$ denote the order statistics of a random sample of size n from the density f. Let $1 \leq r < s \leq n$. Show that the conditional density of $X_{(s)}$, given $X_{(r)} = x$, is

$$f_{s|r}(y|x) = \frac{(n-r)!}{(s-r-1)!((n-s)!} G_x^{s-r-1}(y)\big(1 - G_x(y)\big)^{n-s} g_x(y).$$

   (d) Let $n = 4$ in part (b) and assume F has finite mean. Compute the conditional expectation $E\big(X_{(4)} \big| X_{(1)} = x\big)$.

   Verify that when f is the $\text{Unif}(0,1)$ density, then $E\big(X_{(4)} \big| X_{(1)} = x\big) = (x+3)/4$, $0 < x < 1$.

9. Recall, if 0 is a median of the r.v. X, then $P(X \leq 0) \geq 1/2$. Let X_1, X_2 be i.i.d. r.v.'s. Show that $P(X_1 - X_2 \leq 0) \geq 1/2$.

10. Let F be a d.f. on $\mathbb{R}$. Show that $\int F(x)\, dF(x) \geq 1/2$.

11. Let the r.v.'s X_1, X_2, X_3 be i.i.d. with continuous d.f. F. Show that $P(X_1 < X_2 < X_3) = 1/6$.

12. Let $Y \sim_D \text{Bin}(n, 1/2)$ and $0 < \alpha < 1$. Suppose $0 \leq r \leq n$ is the largest positive integer such that $P(Y < r) \leq \alpha/2$. Show that $P(r \leq Y \leq n - r) \geq 1 - \alpha$.

13. Consider the following random sample from some population: 6.5, 5.8, 4.5, 6.2, 4.8, 7.3, 4.6, 3.9, 4.4, 5.5, 5.2, 6.7. Obtain the finite sample 95% confidence level confidence interval (3.3.10) for 50th and 75th population percentiles.

14. Consider the following data: 3.145, 0.292, 0.298, -0.071, -0.005, -0.637, 0.751, -1.401, -0.397, 0.675, 1.274, -0.893, -0.848, 1.191, -1.919, 1.237, 0.483, -2.712.

    Obtain 0.95 confidence level confidence intervals (3.3.10) and (3.3.11) for $\xi_{0.7}$. Which interval is shorter?

15. Let $\zeta_1, \cdots, \zeta_n$ be i.i.d. r.v.'s with mean ν and variance τ^2. Let $X_1, \cdots, X_n$ be i.i.d. r.v.'s with density f and independent of $\zeta_1, \cdots, \zeta_n$, and let $\psi(1), \cdots, \psi(n)$ be numbers such that $\sum_{j=1}^n \psi(j) = 0$. Let R_i denote the rank of X_i among $X_1, \cdots, X_n$ and define the statistic $S := \sum_{i=1}^n \zeta_i \psi(R_i)$. Show that $\text{Var}(S) = \tau^2 \sum_{j=1}^n \psi^2(j)$.

4

Testing of Hypotheses in Location Models

4.1 Introduction

In this chapter, we shall briefly review some concepts associated with the testing of hypotheses. This is followed by discussing some tests of hypotheses about the location parameters in the one, two and k sample location models without assuming any parametric form of the error distribution in these models. We shall discuss sign and the Wilcoxon signed rank tests and the corresponding confidence intervals for the one sample location parameter. For the two sample location parameter, the Wilcoxon rank sum test and the corresponding confidence intervals will be discussed in detail. We shall also discuss normal scores test in these two models. For the k-sample location model or the one way analysis of variance, we shall discuss the Kruskal-Wallis rank test, which is an extension of the two sample Wilcoxon rank sum test. The chapter also contains a simple linear rank test for testing hypotheses about regression parameters in the linear regression model.

4.2 Testing of Hypotheses

In this section we shall discuss briefly the notion of testing of hypotheses and some related concepts. Lehmann (1959) and Lehmann and Romano (2005) contain an elaborate discussion on this topic.

Let Θ_0, Θ_1 be two disjoint nonempty subsets of the parameter set Θ. Let $\{P_\theta, \theta \in \Theta\}$ be a family of probability distributions defined on the sample space $(\mathcal{X}, \mathscr{A})$ and ξ be an observation taking values in $\mathcal{X}$. The testing of hypothesis problem is concerned with deciding between two states of the parameter values, i.e., to decide between $H_0 : \theta \in \Theta_0$ and $H_1 : \theta \in \Theta_1$, based on the given observation ξ. The null hypothesis is called simple if Θ_0 contains only one element and is called composite otherwise. The same terminology applies to the alternative hypothesis.

A *test* of H_0 versus the alternative H_1 is any r.v. $\varphi = \varphi(\xi)$ taking values in $[0, 1]$ with the interpretation that $\varphi(x)$ is the conditional probability of rejecting H_0, given $\xi = x$. The test φ is called non-randomized if φ takes only the values 0 and 1, and is called randomized otherwise. A non-randomized test is of the form $I(\xi \in C)$ for some event C in $\mathscr{A}$. The set C is called the critical region or the rejection region of the test; it can be taken to be $C := \{x \in \mathcal{X} : \varphi(x) = 1\}$. A randomized test rejects the null hypothesis if $U \le \varphi(\xi)$, where $U \sim_D \text{Unif}(0, 1)$ is independent of ξ and generated by the researcher.

Often in applications, $\mathcal{X} = \mathbb{R}^n$, $\mathscr{A}$ is the Borel σ-field on $\mathbb{R}^n$, and $\xi = (X_1, \cdots, X_n)$, where $X_1, \cdots, X_n$ is a *random sample* from F_θ, i.e., the r.v.'s $X_1, \cdots, X_n$ are i.i.d. having d.f. F_θ. A random variable T is called a *statistic* if it is of the form $T = t(X_1, \cdots, X_n)$ for some measurable function t from $\mathbb{R}^n$ into $\mathbb{R}$ that does not depend on the parameter. In other words, T is constructed from the observations $X_1, \cdots, X_n$ without the knowledge of θ. Examples of statistics include the sample mean, the sample variance and order statistics. A random vector T is called a (vector-valued) statistic if each coordinate of T is a real-valued statistic. Examples of vector-valued statistic are a vector of sample moments and a vector of some order statistics.

During the process of deciding between the two hypotheses H_0 and H_1, it is possible to commit two types of errors. A type I error is committed when the null hypothesis is falsely rejected, while a type II error is committed when the false null hypothesis is not rejected, i.e.,

$$\text{Type I Error} = \text{Test rejects } H_0, \text{ when } H_0 \text{ is true},$$
$$\text{Type II Error} = \text{Test fails to reject } H_0, \text{ when } H_1 \text{ is true}.$$

The quality of any given test is assessed on the basis of the probabilities of committing these errors. Let E_θ represent the expectation when θ is the true parameter. The power or power function π_φ of a test φ is defined by

$$\pi_\varphi(\theta) := E_\theta\big(\varphi(\xi)\big), \quad \theta \in \Theta.$$

Note that the power function gives the probability of rejecting the null hypothesis for each parameter value. Thus, for $\theta \in \Theta_0$, $\pi_\varphi(\theta)$ is the probability of a type I error, while for $\theta \in \Theta_1$, $1 - \pi_\varphi(\theta)$ is the probability of a type II error. The size of a test φ is the largest type I error probability:

$$\text{size of } \varphi := \sup_{\theta\in\Theta_0} \pi_\varphi(\theta).$$

Unbiased tests. Let $0 \le \alpha \le 1$. A test φ is said to have level of significance α, or simply to be of level α, if $\pi_\varphi(\theta) \le \alpha$, for all $\theta \in \Theta_0$. It is said to be an *unbiased test of level* α if

$$\pi_\varphi(\theta) \le \alpha, \quad \forall\,\theta \in \Theta_0, \quad \text{and} \quad \pi_\varphi(\theta) \ge \alpha, \quad \forall\,\theta \in \Theta_1.$$

It is said to be an *unbiased test of size* α if

$$\sup_{\theta\in\Theta_0} \pi_\varphi(\theta) = \alpha \quad \text{and} \quad \pi_\varphi(\theta) \ge \alpha, \quad \forall\,\theta \in \Theta_1.$$

Most powerful level α test. Let $\mathcal{C}_\alpha$ denote the class of all level α tests. A test ψ is said to be most powerful at level α if it maximizes the power among all level α tests, i.e., if the following holds.

$$\psi \in \mathcal{C}_\alpha, \qquad \pi_\varphi(\theta) \le \pi_\psi(\theta), \quad \forall\,\theta \in \Theta_1,\ \varphi \in \mathcal{C}_\alpha. \tag{4.2.1}$$

The famous Neyman-Pearson Lemma gives some conditions under which such tests exist in the case of two simple hypotheses. Suppose $\{f_\theta, \theta \in \mathbb{R}\}$ is a family of densities or probability mass functions and $X_1, \cdots, X_n$ is a random sample from f_θ, for some $\theta \in \mathbb{R}$. Given two distinct values θ_0, θ_1 of θ, consider the problem of testing the simple hypothesis $H_0 : \theta = \theta_0$ versus the simple alternative $H_1 : \theta = \theta_1$. Let $\xi := (X_1, \cdots, X_n)$ and $\Lambda_\theta(\xi) := \prod_{i=1}^n f_\theta(X_i), \theta \in \mathbb{R}$ and define

$$\begin{aligned}
\psi(\xi) &:= 1, && \text{if } \Lambda_{\theta_1}(\xi) > k\Lambda_{\theta_0}(\xi),\\
&:= \gamma, && \text{if } \Lambda_{\theta_1}(\xi) = k\Lambda_{\theta_0}(\xi),\\
&:= 0, && \text{if } \Lambda_{\theta_1}(\xi) < k\Lambda_{\theta_0}(\xi),
\end{aligned}$$

where $0 \le \gamma \le 1$ and k are such that $E_{\theta_0}\big(\psi(\xi)\big) = \alpha$. According to the Neyman-Pearson Lemma, the above test ψ satisfies (4.2.1), i.e., ψ is the most powerful among all level α tests for testing H_0 versus H_1.

In practice this result is of limited use partly because one typically encounters composite hypotheses and partly because the likelihood is rarely known. It is thus important to find tests for a testing problem that are useful in practice. These tests may not be most powerful at a given level, but may have some other desirable properties, like being broadly applicable. This chapter discusses some such tests in some location models.

We shall now address some aspects of the behavior of tests when the sample size n increases to infinity. In the sequel, all limits are taken as $n \to \infty$, unless specified otherwise. Let Θ be an open interval in $\mathbb{R}$ and θ_0 be a fixed point in Θ. Let $\{P_\theta,\ \theta \in \Theta\}$ be a family of probability models and $X_1, \cdots, X_n$, $n \geq 1$, be a random sample from P_θ for some $\theta \in \Theta$. Consider the problem of testing

$$H_0 : \theta = \theta_0, \qquad \text{vs.} \qquad H_1 : \theta > \theta_0,$$

based on the given sample. For each n, let φ_n be a test, and T_n be an estimator of the parameter, both based on the sample $X_1, \cdots, X_n$. An *estimator* of the parameter $\theta \in \mathbb{R}$ is any real-valued statistic based on the given observations.

Definition 4.2.1. The sequence of tests φ_n based on $X_1, \cdots, X_n$, $n \geq 1$, is said to be *consistent for testing* H_0 vs. H_1 at the asymptotic size α if

$$E_{\theta_0}(\varphi_n) \to \alpha, \qquad E_\theta(\varphi_n) \to 1, \qquad \forall\, \theta > \theta_0.$$

The sequence of estimators T_n is said to be *consistent for the parameter* θ if,

$$P_\theta\big(|T_n - \theta| > \delta\big) \to 0, \quad \forall \theta \in \Theta,\ \delta > 0. \tag{4.2.2}$$

The following lemma shows that consistent estimators lead to consistent tests for testing H_0 versus H_1.

Lemma 4.2.1. *Suppose T_n is a sequence of consistent estimators and $d_{n,\alpha}$ is a sequence of constants such that*

$$P_{\theta_0}(T_n - \theta_0 \geq d_{n,\alpha}) \to \alpha \tag{4.2.3}$$

holds for some $0 < \alpha \leq 1$. Then the sequence of tests $\varphi_n := I\big(T_n - \theta_0 \geq d_{n,\alpha}\big)$ is consistent for the above testing problem at the asymptotic size α.

Proof. The assumption (4.2.3) implies that $d_{n,\alpha} \to 0$ as $n \to \infty$. Otherwise, there would exist an $\epsilon > 0$ and a subsequence m_n such that $d_{m_n,\alpha} > \epsilon$ for all n. This in turn implies

$$\begin{aligned} 0 < \alpha &= \lim_n P_{\theta_0}\big(T_{m_n} - \theta_0 \geq d_{m_n,\alpha}\big) \\ &\leq \lim_n P_{\theta_0}\big(T_{m_n} - \theta_0 > \epsilon\big) \leq \lim_n P_{\theta_0}\big(|T_{m_n} - \theta_0| > \epsilon\big) = 0, \end{aligned}$$

by (4.2.2), thereby contradicting (4.2.3).

Now fix a $\theta > \theta_0$ and set $\delta := (\theta - \theta_0)/2$. Because $d_{n,\alpha} \to 0$, there exists an N_0 such that $d_{n,\alpha} < (\theta - \theta_0)/2$, for all $n > N_0$. Moreover, for such n, we have $d_{n,\alpha} - \theta + \theta_0 < -\delta < 0$ and

$$\begin{aligned} E_\theta(\varphi_n) &= P_\theta\big(T_n - \theta_0 \geq d_{n,\alpha}\big) = P_\theta\big(T_n - \theta \geq d_{n,\alpha} - \theta + \theta_0\big) \\ &\geq P_\theta\big(T_n - \theta \geq -\delta\big) \geq 1 - P_\theta\big(|T_n - \theta| > \delta\big). \end{aligned}$$

Thus (4.2.2) implies $E_\theta(\varphi_n) \to 1$. □

Next, we consider the asymptotic power of a sequence of tests under local alternatives $\theta_n > \theta_0$ satisfying $n^{1/2}(\theta_n - \theta_0) \to \delta$, for some $\delta \geq 0$. We shall see later that many of our tests have the following property, where Φ denotes the standard normal d.f. and z_α defined in (3.3.9) satisfies $1 - \Phi(z_\alpha) = \alpha$.

Definition 4.2.2. A sequence of tests φ_n, $n \geq 1$, is said to be *locally regular at θ_0 with slope γ* if $\gamma > 0$ and $\pi_{\varphi_n}(\theta_0 + n^{-1/2}\delta_n) \to 1 - \Phi(z_\alpha - \gamma\delta)$ holds for every $\delta \geq 0$ and for every sequence $0 \leq \delta_n \to \delta$.

The following lemma gives some sufficient conditions for a class of tests to be locally regular at θ_0. Let $\mu(\theta)$ be a real-valued function of $\theta \in \Theta$, $\sigma_n(\theta), n \geq 1$, be a sequence of positive r.v.'s for each $\theta \in \Theta$, T_n, $n \geq 1$, be a sequence of real-valued statistics and define

$$Z_n(\theta) := \frac{T_n - \mu(\theta)}{\sigma_n(\theta)}, \quad \theta \in \Theta; \qquad \varphi_n := I\big(Z_n(\theta_0) \geq z_\alpha\big).$$

We are ready to state and prove the following lemma.

Lemma 4.2.2. *Suppose μ has a positive derivative $\dot{\mu}(\theta_0)$ at θ_0 and there is a positive constant σ such that the following hold for each $\delta \geq 0$ and every sequence δ_n of nonnegative numbers converging to δ. Under P_{θ_n} with $\theta_n := \theta_0 + n^{-1/2}\delta_n$,*

$$\text{(a)}\quad Z_n(\theta_n) \to_D \mathcal{N}(0,1) \qquad \text{(b)}\quad n^{1/2}\sigma_n(\theta_n) \to_p \sigma \qquad \text{(c)}\quad n^{1/2}\sigma_n(\theta_0) \to_p \sigma. \tag{4.2.4}$$

Then the sequence of tests $\varphi_n = I\big(Z_n(\theta_0) \geq z_\alpha\big)$ is locally regular at θ_0 with slope $\gamma = \dot{\mu}(\theta_0)/\sigma$.

Proof. Fix $\delta \geq 0$ and $0 \leq \delta_n \to \delta$. Then,

$$Z_n(\theta_0) = \frac{T_n - \mu(\theta_n)}{\sigma_n(\theta_0)} + \frac{\mu(\theta_n) - \mu(\theta_0)}{\sigma_n(\theta_0)} = Z_n(\theta_n)\frac{\sigma_n(\theta_n)}{\sigma_n(\theta_0)} + \delta_n \frac{\mu(\theta_n) - \mu(\theta_0)}{(\theta_n - \theta_0)n^{1/2}\sigma_n(\theta_0)},$$

where we interpret the last fraction as $\dot{\mu}(\theta_0)/(n^{1/2}\sigma_n(\theta_0))$ when $\delta_n = 0$. Using the assumptions (4.2.4), we conclude that

$$Z_n(\theta_0) = Z_n(\theta_n) + \delta\dot{\mu}(\theta_0)/\sigma + R_n,$$

where $R_n \to_p 0$ under P_{θ_n}, i.e. $P_{\theta_n}\big(|R_n| > \epsilon\big) \to 0$ for every $\epsilon > 0$. Thus, the asymptotic normality of $Z_n(\theta_n)$ under P_{θ_n} and Slutsky's Theorem 2.10.5 yield

$$P_{\theta_n}\big(Z_n(\theta_0) \geq z_\alpha\big) = P_{\theta_n}\big(Z_n(\theta_n) + R_n \geq z_\alpha - \delta\dot{\mu}(\theta_0)/\sigma\big) \to 1 - \Phi\big(z_\alpha - \delta\dot{\mu}(\theta_0)/\sigma\big).$$

This is the desired result. □

4.3 One Sample Location Model

Let $\mathcal{F}$ be a family of d.f.'s on the real line. By the location model generated by $\mathcal{F}$, we mean the family of d.f.'s $\mathcal{G} = \{G_{t,F} : t \in \mathbb{R}, F \in \mathcal{F}\}$, where $G_{t,F}$ is the d.f. defined by

$$G_{t,F}(x) = F(x - t), \quad x \in \mathbb{R}.$$

Sometimes we refer to the class $\mathcal{F}$ as an error model. An example of a location model is the family of all normal d.f.'s. It is generated by the class of normal d.f.'s with mean zero.

If X is a r.v. with d.f. F_X belonging to $\mathcal{G}$, then for some $\theta \in \mathbb{R}$ and a d.f. $F \in \mathcal{F}$, we have

$$F_X(x) = F(x - \theta), \quad \forall x \in \mathbb{R}. \tag{4.3.1}$$

Alternatively, there is a $\theta \in \mathbb{R}$ and a d.f. $F \in \mathcal{F}$ such that

$$X = \theta + \varepsilon, \qquad \varepsilon \sim_D F. \tag{4.3.2}$$

In order to identify the location parameter, one needs to impose restrictions on the class $\mathcal{F}$. If $\mathcal{F}$ is a class of d.f.'s which have mean zero and finite variance, then the location parameter is identified as the mean. If $\mathcal{F}$ is a class of d.f.'s F which satisfy the symmetry condition $F(x) = 1 - F(-x_-)$, $\forall x \in \mathbb{R}$, then the location parameter θ is the center of symmetry of the d.f. F_X, i.e., $F_X(x + \theta) = 1 - F_X\big((-x + \theta)_-\big)$, for all $x \in \mathbb{R}$. If $\mathcal{F}$ is a class of d.f.'s which have unique median equal to 0, then the location parameter is the median of F_X.

In this section, we are interested in testing hypotheses about the location parameter based on a random sample $X_1, \cdots, X_n$ from some member of a location model $\mathcal{G}$. A test is said to be distribution free if its null distribution does not depend on the underlying d.f. F and the parameter values under the null hypothesis. We shall discuss three popular distribution free tests for these hypotheses, viz, the sign test, the Wilcoxon signed rank test and the normal scores signed rank test.

4.3.1 Sign Test

In this subsection we consider the location model generated by $\mathcal{F}_0$, the class of all d.f.'s F that have unique median 0 and are continuous at 0. Thus, each F in $\mathcal{F}_0$ satisfies

$$F(x) < F(0) = 1/2 < F(y), \quad x < 0 < y. \tag{4.3.3}$$

Let $X_1, \cdots, X_n$ be a random sample from the location model generated by $\mathcal{F}_0$, i.e., for some real number θ and some $F \in \mathcal{F}_0$,

$$X_i = \theta + \varepsilon_i, \quad 1 \le i \le n, \tag{4.3.4}$$

where $\varepsilon_1, \cdots, \varepsilon_n$ are i.i.d. F. Set

$$S_n(t) := \sum_{i=1}^{n} I(X_i > t), \qquad t \in \mathbb{R}. \tag{4.3.5}$$

Let θ_0 be a specified real number and consider the problem of testing

$$H_0 : \theta = \theta_0 \qquad \text{versus} \qquad H_1 : \theta > \theta_0.$$

One of the simplest tests for this problem is the sign test based on the number of positive differences $X_i - \theta_0$, i.e., based on $S_n(\theta_0)$. We shall often write S_n for $S_n(\theta_0)$.

To stress the dependence on the location parameter θ, we write P_θ, E_θ for probability and expectation, respectively. For the sake of simplicity of the notation, we do not exhibit the dependence on F.

We have

$$p(\theta) := P_\theta(X > \theta_0) = 1 - F(\theta_0 - \theta).$$

Since every d.f. is nondecreasing, the probability $p(\theta)$ is nondecreasing in θ. The condition (4.3.3) implies $p(\theta_0) = 1/2$ and $p(\theta) > 1/2$ for $\theta > \theta_0$. Under P_θ, the r.v.'s $I(X_1 > \theta_0), \cdots, I(X_n > \theta_0)$ are i.i.d. Bernoulli$(p(\theta))$ implying $S_n \sim_D \text{Bin}(n, p(\theta))$ and $E_\theta(S_n) = np(\theta)$. Since S_n has mean $n/2$ under H_0 and mean greater than $n/2$ under H_1, large values of S_n provide evidence in support of H_1 and the sign test rejects H_0, in favor of H_1, at a given level of significance α, if $S_n \geq c_\alpha$, where $0 \leq c_\alpha \leq n$ is the smallest nonnegative integer such that the probability of a type I error is at most α, i.e.,

$$P_{\theta_0}(S_n \geq c_\alpha) = \sum_{k=c_\alpha}^{n} \binom{n}{k} \left(\frac{1}{2}\right)^n \leq \alpha. \tag{4.3.6}$$

Note that the critical value c_α equals $1 + q$, where q is the $1 - \alpha$ quantile of the Bin$(n, 1/2)$ distribution and can be obtained via the R command `qbinom(n,.5,1-alpha)`. Here $\xi = (X_1, \cdots, X_n)$, the test function is $\varphi_S(\xi) = I(S_n \geq c_\alpha)$ and the size of this test is $\pi_{\varphi_S}(\theta_0) = E_{\theta_0}(\varphi_S(\xi)) = P_{\theta_0}(S_n \geq c_\alpha)$. The sign test of size α is given by the randomized sign test $\psi_n := aI(S_n = q) + I(S_n > q)$ with $a = (\alpha - P_{\theta_0}(S_n > q))/P_{\theta_0}(S_n = q)$.

Since the determination of c_α does not depend on the error d.f. F, the above sign test is a distribution free test.

One could also use the p-value approach. Let s be the observed value of S_n based on the given data. The p-value is then defined as

$$\hat{p} := P_{\theta_0}(S_n \geq s) = \sum_{k=s}^{n} \binom{n}{k} \left(\frac{1}{2}\right)^n.$$

Then one rejects H_0 at level α if $\hat{p} \leq \alpha$.

For testing $H_0 : \theta = \theta_0$ against the two sided alternatives $H_1 : \theta \neq \theta_0$, the sign test rejects H_0, in favor of this H_1, if $S_n \leq n - c$ or $S_n \geq c$, where c is determined to have a type I error probability at most α, i.e., $P_{\theta_0}(S_n \leq n - c) + P_{\theta_0}(S_n \geq c) \leq \alpha$. Under H_0, $S_n \sim \text{Bin}(n, 1/2)$, which is symmetric about $n/2$. Thus c can be taken to be $c_{\alpha/2}$ and the two sided sign test rejects H_0 if $S_n \leq n - c_{\alpha/2}$ or $S_n \geq c_{\alpha/2}$. In this case, the p-value is

$$\hat{p} = 2 \min \left\{ P_{\theta_0}(S_n \leq s), P_{\theta_0}(S_n \geq s) \right\},$$

where s is the observed value of S_n.

Binomial probabilities can be obtained using R code. The R commands

`dbinom(x,n,p)` and `pbinom(x,n,p)`

yield $P(X = x)$ and $P(X \leq x)$, respectively, for a Bin(n, p) r.v. X. For example, if $n = 5, p = 0.6$, then `dbinom(2,5,0.6)`$= 0.2304 = P(X = 2)$, while `pbinom(2,5,0.6)`$= 0.3174 = P(X \leq 2)$. From this we have $P(X > 2) = 1 - P(X \leq 2) = 0.6826$. One can also use Excel to compute these probabilities.

Example 4.3.1. The number 1.618 is called golden ratio, and a rectangle with width equal to $1.618 \times$ length is called a golden rectangle. Let X denote the ratio of width and length of a beaded basket and θ denote the median of X. Test $H_0 : \theta = 1.618$ vs. $H_1 : \theta \neq 1.618$, using the sign test.

The sign test rejects H_0 if $S_n(1.618) \leq c$ or $S_n(1.618) \geq n - c$. A sample of 20 beaded baskets yielded the number of beaded baskets with ratio greater than 1.618 to be 9, i.e., the observed value of $S_n \equiv S_n(1.618) = 9$. Under H_0, $S_n \sim_D \text{Bin}(20, 0.5)$. Also, the p-value $= 2P_{\theta_0}(S_n \leq 9) = 0.8238$. Hence, there is no strong evidence to reject H_0.

Power function of sign test. We shall now discuss some important properties of the power function of the sign test.

Theorem 4.3.1. *The power function of the sign test for testing $H_0 : \theta = \theta_0$ vs. $H_1 : \theta > \theta_0$ is nondecreasing in θ.*

Proof. Let c_α be as in (4.3.6) so that $\{S_n(\theta_0) \geq c_\alpha\}$ is the critical region of the test. Then the power function of the level α sign test is

$$\pi_{S_n}(\theta) = P_\theta\big(S_n(\theta_0) \geq c_\alpha\big), \quad \theta \in \mathbb{R}.$$

When θ is the true parameter, then $S_n(\theta_0) \sim_D \text{Bin}(n, p(\theta))$ with $p(\theta) = 1 - F(\theta_0 - \theta)$ and

$$\pi_{S_n}(\theta) = \sum_{j=c_\alpha}^{n} \binom{n}{j} p(\theta)^j (1 - p(\theta))^{n-j}.$$

Now if $\theta_1 \leq \theta_2$, then $p(\theta_1) \leq p(\theta_2)$ and by Lemma 3.3.2, $\pi_{S_n}(\theta_1) \leq \pi_{S_n}(\theta_2)$, thereby completing the proof. □

An alternative proof of this claim is as follows. For any $x \leq y$, $I(X_i > y) \leq I(X_i > x)$, for all $1 \leq i \leq n$. Hence, for any $\theta_1 \leq \theta_2$,

$$S_n(\theta_0 - \theta_1) = \sum_{i=1}^{n} I(X_i > \theta_0 - \theta_1) \leq \sum_{i=1}^{n} I(X_i > \theta_0 - \theta_2) = S_n(\theta_0 - \theta_2).$$

Also,

$$\begin{aligned} \pi_{S_n}(\theta) &:= P_\theta(S_n(\theta_0) \geq c_\alpha) = P_\theta\Big(\sum_{i=1}^{n} I(X_i > \theta_0) \geq c_\alpha\Big) \qquad (4.3.7) \\ &= P_\theta\Big(\sum_{i=1}^{n} I(X_i - \theta > \theta_0 - \theta) \geq c_\alpha\Big) \\ &= P_0\Big(\sum_{i=1}^{n} I(X_i > \theta_0 - \theta) \geq c_\alpha\Big) = P_0(S_n(\theta_0 - \theta) \geq c_\alpha). \end{aligned}$$

Therefore, for $\theta_1 \leq \theta_2$,

$$\begin{aligned} \pi_{S_n}(\theta_1) &= P_{\theta_1}\big(S_n(\theta_0) \geq c_\alpha\big) = P_0(S_n(\theta_0 - \theta_1) \geq c_\alpha) \\ &\leq P_0\big(S_n(\theta_0 - \theta_2) \geq c_\alpha\big) = P_{\theta_2}\big(S_n(\theta_0) \geq c_\alpha\big) = \pi_{S_n}(\theta_2). \end{aligned}$$

Some of the implications of the above result are as follows. Suppose one is interested in testing the composite null hypothesis

$$H_0 : \theta \leq \theta_0 \quad \text{vs} \quad H_1 : \theta > \theta_0. \qquad (4.3.8)$$

The monotonicity of the power function implies

$$\sup_{\theta \leq \theta_0} \pi_{S_n}(\theta) = \pi_{S_n}(\theta_0) \leq \alpha.$$

Thus the level α sign test for testing the simple null hypothesis $H_0 : \theta = \theta_0$ versus the alternative $H_1 : \theta > \theta_0$ continues to be a level α test for testing the composite null hypothesis in (4.3.8). The monotonicity of the power function also implies that the randomized sign test of size α is unbiased.

Large sample sign test. For $\theta \in \mathbb{R}$ with $0 < p(\theta) < 1$, let

$$Z_n(\theta) := \big(S_n(\theta) - np(\theta)\big) / \big[np(\theta)(1 - p(\theta))\big]^{1/2}.$$

Under P_θ, $Z_n(\theta)$ is a standardized sum of bounded i.i.d. r.v.'s and by the CLT Theorem 2.10.8, $Z_n(\theta) \to_D \mathcal{N}(0,1)$. Since $p(\theta_0) = 1/2$, this result in particular yields that under H_0, $2n^{-1/2}(S_n(\theta_0) - n/2) \to_D \mathcal{N}(0,1)$. Therefore, the large sample level α sign test rejects H_0 if $S_n(\theta_0) \geq c_\alpha$, where c_α is such that $P_{\theta_0}(S_n(\theta_0) \geq c_\alpha) \to \alpha$ or equivalently

$$P_{\theta_0}\Big(2n^{-1/2}(S_n(\theta_0) - \frac{n}{2}) \geq 2n^{-1/2}(c_\alpha - \frac{n}{2})\Big) \to \alpha.$$

This implies $c_\alpha \approx n/2 + \sqrt{n} z_\alpha / 2$, where z_α is as in (3.3.9). Thus, the large sample level α sign test rejects H_0, in favor of the alternative $\theta > \theta_0$, whenever $S_n \geq c_\alpha$, with

$$c_\alpha := \frac{n}{2} + \frac{\sqrt{n}}{2} z_\alpha. \tag{4.3.9}$$

The next lemma establishes the local regularity of the sign test under an additional assumption.

Lemma 4.3.1. *If the error d.f. F has a positive derivative $f(0)$ at 0, then the sign test $\varphi_n := I\big(S_n(\theta_0) \geq c_\alpha\big)$ with c_α as in* (4.3.9) *for testing $H_0 : \theta = \theta_0$ versus $H_1 : \theta > \theta_0$ is locally regular at θ_0 with slope $2f(0)$.*

Proof. Fix nonnegative numbers $\delta, \delta_1, \delta_2, \cdots$ such that $\delta_n \to \delta$. Set $\theta_n := \theta_0 + n^{-1/2}\delta_n$. By (4.3.7), the power of the sign test at θ_n is

$$\pi_{S_n}(\theta_n) = P_0\Big(2n^{-1/2}\big(S_n(\theta_0 - \theta_n) - n/2\big) \geq z_\alpha\Big).$$

But $S_n(\theta_0 - \theta_n) = \sum_{i=1}^n I(\varepsilon_i > -n^{-1/2}\delta_n)$ is the sum of independent Ber$(1 - F(-n^{-1/2}\delta_n))$ r.v.'s. Accordingly, define

$$T_n := n^{-1/2} \sum_{i=1}^n \Big[I(\varepsilon_i > -n^{-1/2}\delta_n) - 1 + F(-n^{-1/2}\delta_n)\Big],$$
$$D_n := n^{1/2}\big(1 - F(-n^{-1/2}\delta_n) - 1/2\big) = n^{1/2}\big(F(0) - F(-n^{-1/2}\delta_n)\big).$$

Then, $n^{-1/2}(S_n(\theta_0 - \theta_n) - n/2) = T_n + D_n$. The differentiability of F at 0 and $\delta_n \to \delta$ yield $D_n \to \delta f(0)$. An application of Corollary 2.10.6 with $X_{nj} = I(\varepsilon_i > -n^{-1/2}\delta_n)$ and $p_n = 1 - F(-n^{-1/2}\delta_n)$ yields $T_n \to_D \mathcal{N}(0, 1/4)$. This and Slutsky's Theorem 2.10.5 then give $2(T_n + D_n) \to_D Z + 2\delta f(0)$ with $Z \sim_D \mathcal{N}(0,1)$. Hence,

$$\pi_{S_n}(\theta_n) = P_0\big(2(T_n + D_n) \geq z_\alpha\big) \to P(Z + 2\delta f(0)) \geq z_\alpha) = 1 - \Phi\big(z_\alpha - 2\delta f(0)\big). \tag{4.3.10}$$

This is the desired result. □

Sign test when there are zeros in the data. The continuity assumption on F implies that none of the X_i's are equal to θ_0, i.e., none of the differences $X_i - \theta_0$ equal zero, with a positive probability. In practice, we may encounter the case when some observations are equal to θ_0. To handle this case, we replace $\mathcal{F}_0$ by $\mathcal{F}_*$ which consists of all nondegenerate d.f.'s F that satisfy

$$F(0_-) + F(0) = 1, \qquad F(x) < F(0_-) \leq F(0) < F(y), \quad x < 0 < y.$$

In other words we now consider the location model (4.3.2) generated by $\mathcal{F}_*$. If $\varepsilon \sim_D F$, $F \in \mathcal{F}_*$, then the conditional d.f. of ε, given $\varepsilon \neq 0$, is

$$G(x) = \frac{F(x) - I(x \geq 0)(F(0) - F(0_-))}{1 - F(0) + F(0_-)}, \quad x \in \mathbb{R},$$

which belongs to $\mathcal{F}_0$.

Assuming the observations are coming from the location model generated by $\mathcal{F}_*$, we may see observations that equal θ_0 under the null hypothesis. In this case, first remove those X_i's that equal to θ_0 and then apply the sign test with the remaining observations. Let $N = \sum_{i=1}^n I(X_i \neq \theta_0)$ denote the number of remaining observations. Then the above facts about the level and power continue to hold conditionally, given N.

Symmetric errors. So far in this subsection we have assumed that the location model is generated by the error model $\mathcal{F}_0$, the class of d.f.'s on $\mathbb{R}$ with unique median 0. Let $\mathcal{F}_c$ denote a class of all continuous d.f.'s. The above results remain true if we consider instead the location model generated by the class of all continuous d.f.'s which are symmetric about 0 denoted by

$$\mathcal{F}_s := \{F \in \mathcal{F}_c : F(x) = 1 - F(-x), \ \forall x \in \mathbb{R}\}. \tag{4.3.11}$$

For this model we still have $F(0) = 1/2$, but 0 may no longer be the unique median. In view of this, the location parameter is now the center of symmetry of the d.f. F_X. Recall that a real number θ is said to be the center of symmetry of the distribution of a r.v. X if $X - \theta \sim_D -X + \theta$ or if $F_X(x + \theta) = 1 - F_X\big((-x + \theta)_-\big)$, for all $x \in \mathbb{R}$.

4.3.2 Wilcoxon Signed Rank Test

We shall now discuss one of the most popular distribution free tests for testing hypotheses about the center of symmetry. The sign test uses only the signs of the observations, which amounts to the loss of some information. The Wilcoxon signed rank test uses the signs and the ranks of the absolute values of the observations, which contains more information about the data than simply the signs.

We now consider the location model generated by $\mathcal{F}_s$ of (4.3.11). Let $X_1, \cdots, X_n$ be a random sample from the location model generated by $\mathcal{F}_s$ so that (4.3.4) holds for some $\theta \in \mathbb{R}$ and with $\varepsilon_1, \cdots, \varepsilon_n$ i.i.d. F, $F \in \mathcal{F}_s$. Consider the problem of testing

$$H_0 : \theta = 0, \quad \text{versus} \quad H_1 : \theta > 0.$$

To describe the Wilcoxon's signed rank test for this testing problem, we proceed as follows. Assume none of the X_i's is zero and there are no ties among $|X_i|$'s. Since F is continuous, this holds with probability 1. Rank the absolute values $|X_1|, |X_2|, \cdots, |X_n|$ from the smallest to the largest. Let R_i^+ denote the rank of $|X_i|$, i.e.,

$$R_i^+ := \sum_{j=1}^n I\big(|X_j| \leq |X_i|\big), \quad 1 \leq i \leq n.$$

The ranks $R_1^+, \cdots, R_n^+$ will be sometimes referred to as absolute ranks. Define the sign function

$$\begin{aligned} \operatorname{sgn}(x) &:= -1, \quad && x < 0, \\ &:= 0, && x = 0, \\ &:= 1, && x > 0. \end{aligned}$$

The Wilcoxon's signed rank test statistic is defined to be

$$W_s = \sum_{i=1}^{n} \text{sgn}(X_i) R_i^+. \tag{4.3.12}$$

Intuitively speaking, under H_0, about half of the observations are expected to be positive and the other half to be negative. Under H_1, more than half of the X_i's are expected to be positive, which would receive higher absolute ranks and result in larger values of W_s. Thus large values of W_s provide evidence in support of H_1. The Wilcoxon signed rank test rejects H_0 in favor of H_1, if

$$W_s \geq w_\alpha,$$

where w_α is such that the test has level α. To determine w_α we need to know the null distribution of W_s, i.e., the distribution of W_s under H_0. For this purpose we first prove the independence of $\text{sgn}(X)$ and $|X|$ under H_0 as is done in the following lemma.

Lemma 4.3.2. *If the d.f. F of the r.v. X belongs to $\mathcal{F}_s$, then* $\text{sgn}(X)$ *and $|X|$ are independent, and sgn(X) has mean zero and variance one.*

Proof. Fix $x > 0$. To establish independence it suffices to show

$$P\big(|X| \leq x, \text{sgn}(X) = s\big) = P\big(|X| \leq x\big) P\big(\text{sgn}(X) = s\big), \quad s = -1, 0, 1.$$

This holds for $s = 0$ as $P(\text{sgn}(X) = 0) = P(X = 0) = 0$, by the continuity of F at zero. The symmetry property of F implies $-X =_D X$, and this in turn implies $(|X|, \text{sgn}(X)) =_D (|-X|, \text{sgn}(-X)) = (|X|, -\text{sgn}(X))$. Thus $P(\text{sgn}(X) = -1) = P(\text{sgn}(X) = 1) = 1/2$ and

$$P\big(|X| \leq x, \text{sgn}(X) = -1\big) = P\big(|X| \leq x, \text{sgn}(X) = 1\big) = (1/2) P\big(|X| \leq x\big).$$

These identities give the desired result for $s = -1$ and $s = 1$, thereby completing the proof of the claimed independence of $\text{sgn}(X)$ and $|X|$. Since $\text{sgn}(X)$ takes values -1, 0 and 1 with probability $1/2, 0, 1/2$, respectively, it has mean zero and variance one. This completes the proof. □

Lemma 4.3.2 and X_i's being i.i.d. imply that $\text{sgn}(X_1), \cdots, \text{sgn}(X_n)$ are independent of R_1^+, $\cdots, R_n^+$. Since F is continuous, the distribution of $|X_i|$'s is also continuous, and $P(|X_i| = |X_j|$, for some $1 \leq i < j \leq n) = 0$. Hence, by Lemma 3.2.1, the vector of absolute ranks $(R_1^+, \cdots, R_n^+)$ is uniformly distributed on Π_n, the set of permutations of the integers $1, 2, \cdots, n$ and for each $1 \leq i \leq n$, R_i^+ is uniformly distributed on the set of integers $\{1, \cdots, n\}$.

We now use the above lemma to derive the mean and variance of W_s under the null hypothesis. Let P_0, E_0 and Var_0 denote probability, expectation and variance under H_0, respectively.

Lemma 4.3.3. *Under the null hypothesis, $E_0(W_s) = 0$ and*

$$\text{Var}_0(W_s) = \frac{n(n+1)(2n+1)}{6}. \tag{4.3.13}$$

Proof. By Lemma 4.3.2,

$$E_0(W_s) = \sum_{i=1}^{n} E_0\big(\text{sgn}(X_i)\big) E_0\big(R_i^+\big) = 0,$$

$$\begin{aligned}\text{Var}_0(W_s) = E_0(W_s^2) &= \sum_{i=1}^{n}\sum_{j=1}^{n} E_0\big(\text{sgn}(X_i)\text{sgn}(X_j)\big) E_0\big(R_i^+ R_j^+\big) \\ &= \sum_{i=1}^{n} E_0\big((R_i^+)^2\big) = \sum_{i=1}^{n} i^2 = n(n+1)(2n+1)/6.\end{aligned}$$

This completes the proof. □

The above lemma can also be derived using antiranks as shown below. To proceed further we need the following definition.

Definition 4.3.1 (Antiranks). Let $\boldsymbol{X} = (X_1, \cdots, X_n)$. The r.v. $D_i = D_i(\boldsymbol{X})$ taking values among the integers $1, \cdots, n$ is called the antirank of $|X_i|$ if $R^+_{D_i} = i$. In other words, D_i is the index of that observation whose absolute rank is i. The vector $(D_1, \cdots, D_n)$ is called the vector of antiranks of $(|X_1|, \cdots, |X_n|)$. By definition,

$$|X_{D_1}| < |X_{D_2}| < \cdots < |X_{D_n}|. \tag{4.3.14}$$

Example 4.3.2. The following table gives the quantities $|X_i|$, R_i^+ and D_i for a set of six observations.

		X_1	X_2	X_3	X_4	X_5	X_6
X_i	:	-4.8	3.2	5.4	2.1	-6.2	1.5
$\|X_i\|$	:	4.8	3.2	5.4	2.1	6.2	1.5
R_i^+	:	4	3	5	2	6	1
D_i	:	6	4	2	1	3	5

By definition, $(R^+_{D_1}, R^+_{D_2}, \ldots, R^+_{D_n}) = (1, 2, \cdots, n)$ and hence,

$$W_s = \sum_{i=1}^{n} \text{sgn}(X_i) R_i^+ = \sum_{j=1}^{n} j\, \text{sgn}(X_{D_j}).$$

We use this representation to derive some properties of the distribution of W_s under H_0 as follows. Clearly, the independence of $\big(\text{sgn}(X_1), \cdots, \text{sgn}(X_n)\big)$ and $\big(R_1^+, \cdots, R_n^+\big)$ implies that of $\big(\text{sgn}(X_1), \cdots, \text{sgn}(X_n)\big)$ and $\big(D_1, \cdots, D_n\big)$. Moreover, under H_0, $\big(-X_{(D_1)}, \cdots, -X_{(D_n)}\big) =_D \big(X_{(D_1)}, \cdots, X_{(D_n)}\big)$ and

$$-W_s = \sum_{j=1}^{n} j\, \text{sgn}(-X_{D_j}) =_D \sum_{j=1}^{n} j\, \text{sgn}(X_{D_j}) = W_s.$$

That is, the null distribution of W_s is symmetric about 0 and $E_0(W_s) = 0$. Independence of $\text{sgn}(X_i)$'s and D_j's yields $E\big(\text{sgn}(X_{D_i})\text{sgn}(X_{D_j})\big) = 0$, for all $1 \le i \ne j \le n$. Hence, one derives (4.3.13) from

$$\text{Var}_0(W_s) = \text{Var}_0\Big(\sum_{j=1}^{n} j\, \text{sgn}(X_{D_j})\Big) = \sum_{j=1}^{n} j^2 \text{Var}_0\big(\text{sgn}(X_{D_j})\big) = \sum_{j=1}^{n} j^2.$$

By symmetry, the higher-order odd moments of W_s are zero, while the even moments can be obtained from the moment generating function $E_0\left(e^{tW_s}\right)$ of W_s given as follows. Because under H_0, $\text{sgn}(X_{D_j}), 1 \le j \le n$ are i.i.d. r.v.'s, the moment generating function

$$E_0\left(e^{tW_s}\right) = E_0\left(e^{\sum_{j=1}^{n} tj\text{sgn}(X_{D_j})}\right) = \prod_{j=1}^{n} E_0\left(e^{tj\text{sgn}(X_{D_j})}\right) = \prod_{j=1}^{n} \frac{e^{-tj} + e^{tj}}{2}, \quad t \in \mathbb{R},$$

does not depend on F.

An equivalent formulation of the above signed-rank test is based on the statistic

$$W^+ = \sum_{i=1}^{n} I(X_i > 0) R_i^+. \tag{4.3.15}$$

Note that W^+ is the sum of the absolute ranks corresponding to the positive X_i's. Clearly, wp1,

$$\begin{aligned} W_s &= \sum_{i=1}^{n} \operatorname{sgn}(X_i) R_i^+ = \sum_{X_j>0} R_j^+ - \sum_{X_j<0} R_j^+ \\ &= \sum_{X_j>0} R_j^+ + \sum_{X_j>0} R_j^+ - \Big(\sum_{X_j<0} R_j^+ + \sum_{X_j>0} R_j^+\Big) \\ &= 2\sum_{X_j>0} R_j^+ - \sum_{j=1}^{n} j = 2W^+ - \frac{n(n+1)}{2}. \end{aligned}$$

Or, equivalently, wp1,

$$W^+ = \frac{W_s}{2} + \frac{n(n+1)}{4}.$$

Thus W^+ is linearly related to W_s, wp1, and the test based on W^+ is equivalent to the test based on W_s. Moreover,

$$E_0(W^+) = \frac{n(n+1)}{4}, \qquad \operatorname{Var}_0(W^+) = \frac{n(n+1)(2n+1)}{24}, \tag{4.3.16}$$

Under H_0, W^+ is symmetrically distributed about $n(n+1)/4$.

The test based on W^+ rejects H_0 if $W^+ \geq w_\alpha^+$, where, for small to moderate values of n, the critical value w_α^+ is calculated from the exact null distribution of W^+, which we shall now describe.

The smallest and the largest values of W^+ are 0 and $1 + \cdots + n = n(n+1)/2$, respectively. By the symmetry of the null distribution of W^+ around $n(n+1)/4$, it suffices to compute the probability distribution of W^+ from the largest value to the middle value $n(n+1)/4$ as the remaining probability distribution can be determined by symmetry. Under H_0, the 2^n allocations of plus and minus signs are all equally likely. Thus,

$$P_0(W^+ = w) = \text{\# of sign arrangements for which } \{W^+ = w\}/2^n.$$

Let $u(x) := I(x > 0)$ and $\boldsymbol{u} := (u(X_{D_1}), \cdots, u(X_{D_n}))$. Then, $W^+ = \sum_{j=1}^{n} j u(X_{D_j})$. We illustrate the computation of the null distribution of W^+ by an example for $n = 4$. The middle and the largest values here are 5 and 10, respectively. Tables 4.1 and 4.2 contain half and complete null distribution of W^+, respectively, for $n = 4$.

The computation of the null distribution of W^+ in general is available in R under the command `psirank(c,n)`. This computation gets complicated quickly with the increase in n. Using Theorem 2.10.9, we obtain the following theorem that gives the asymptotic normality of W^+ under H_0.

Theorem 4.3.2. *Suppose $X_1, \cdots, X_n$ is a random sample from the location model (4.3.2) with the error d.f. $F \in \mathcal{F}_s$. Then, under H_0, W_s is distribution free and*

$$\frac{W_s}{\sqrt{\frac{n(n+1)(2n+1)}{6}}} \to_D \mathcal{N}(0,1), \qquad \frac{W^+ - n(n+1)/4}{\sqrt{\frac{n(n+1)(2n+1)}{24}}} \to_D \mathcal{N}(0,1).$$

Table 4.1 Half of the null distribution of W^+ for $n = 4$

j	1	2	3	4	W^+	P_0
u	1	1	1	1	10	1/16
	0	1	1	1	9	1/16
	1	0	1	1	8	1/16
	1	1	0	1	7	1/16
	1	1	1	0	6	1/16
	0	0	1	1	7	1/16
	0	1	0	1	6	1/16
	0	1	1	0	5	1/16
	1	0	0	1	5	1/16

Table 4.2 Null distribution of W^+ for $n = 4$

W^+	0	1	2	3	4	5	6	7	8	9	10
P_0	$\frac{1}{16}$	$\frac{1}{16}$	$\frac{1}{16}$	$\frac{2}{16}$	$\frac{2}{16}$	$\frac{2}{16}$	$\frac{2}{16}$	$\frac{2}{16}$	$\frac{1}{16}$	$\frac{1}{16}$	$\frac{1}{16}$

For $n \geq 15$, one may use the normal approximation given by Theorem 4.3.2 with or without the continuity correction (cc), i.e.,

$$\begin{aligned} w_\alpha^+ &\approx z_\alpha \sqrt{\frac{n(n+1)(2n+1)}{24}} + \frac{n(n+1)}{4}, \quad \text{without cc,} \qquad (4.3.17) \\ &\approx z_\alpha \sqrt{\frac{n(n+1)(2n+1)}{24}} + \frac{n(n+1)}{4} + 0.5, \quad \text{with cc.} \end{aligned}$$

Example 4.3.3. We now give an application of the above test. Let X, Y denote the height of a cross-fertilized and self-fertilized plant. Suppose the effect of self-fertilization over the cross fertilization is additive, i.e., suppose $D := Y - X = \theta + \varepsilon$, for some $\theta \in \mathbb{R}$, where the random error ε is assumed to be symmetrically distributed around zero and continuous. A problem of interest is to test the hypothesis that there is no difference in the two fertilization methods against the alternatives that self-fertilization increases the heights of the plants over all, i.e., we wish to test the hypothesis $H_0 : \theta = 0$ versus the alternative $H_1 : \theta > 0$.

To test this hypothesis, suppose an experiment is conducted that yields the 15 heights $X_i, 1 \leq i \leq 15$ of 15 cross-fertilized plants and 15 heights $Y_i, 1 \leq i \leq 15$ of self-fertilized plants. Further, suppose $(X_i, Y_i), 1 \leq i \leq n$ are guaranteed to be i.i.d.. Suppose it is observed that the number of positive differences $D_i := Y_i - X_i$ is 13, and $W^+ = \sum_{i=1}^n I(D_i > 0)R_i^+ = 96$. Using exact distribution (R-command : `1-psignrank (95,15)`) the p-value is

$$\hat{p} = P_0\big(W^+ \geq 96\big) = 0.021.$$

Thus this test would reject H_0, in favor of H_1, at all levels $\alpha \geq 0.021$.

With $Z \sim_D \mathcal{N}(0,1)$, the asymptotic distribution of W^+ gives the p-value

$$P_0\big(W^+ \geq 96\big) \approx P\Big(Z \geq \frac{96 - 60}{\sqrt{\frac{15\times16\times31}{24}}}\Big) = P(Z \geq 2.016) = 0.022,$$

which is close to the exact value 0.021.

This example shows that the approximation of the exact null distribution of W^+ by the $\mathcal{N}\big(E_0(W^+), \mathrm{Var}_0(W^+)\big)$ distribution is effective even for sample sizes as small as 15.

4.3.3 Wilcoxon Signed Rank Confidence Interval for θ

Here we shall discuss confidence intervals for the location parameter θ based on the Wilcoxon signed rank statistic. But before that we mention a general method for obtaining confidence sets or intervals from test statistics.

Let $\theta \in \mathbb{R}$ be a parameter of interest and, for each $t \in \mathbb{R}$, let $T_n(t) := T_n(X_1, \cdots, X_n; t)$ be a real-valued statistic. Let $0 < \alpha < 1$. Suppose, for each t in $\mathbb{R}$, the level α test based on $T_n(t)$ rejects the null hypothesis $\theta = t$, whenever either $T_n(t) < a_{n,\alpha}(t)$ or $T_n(t) > b_{n,\alpha}(t)$, where $a_{n,\alpha}(t) < b_{n,\alpha}(t)$ are such that $P_t\big(T_n(t) < a_{n,\alpha}(t)$ or $T_n(t) > b_{n,\alpha}(t)\big) \le \alpha$. In other words, this test accepts the hypothesis $\theta = t$, whenever $a_{n,\alpha}(t) \le T_n(t) \le b_{n,\alpha}(t)$. The set of all acceptable values of θ at level α, $\big\{t \in \mathbb{R}; a_{n,\alpha}(t) \le T_n(t) \le b_{n,\alpha}(t)\big\}$, is called the *confidence set* of confidence level $1 - \alpha$. If this set is an interval, then it is called *confidence interval* for θ of the confidence level $1 - \alpha$. We shall use this principle in the rest of this chapter to derive confidence sets or intervals for the one and two sample location parameters.

The above principle of confidence sets applies even when θ is higher-dimensional and T_n and $X_1, \cdots, X_n$ are vector-valued. See Lehmann (1959) and Lehmann and Romano (2005) for more on this concept.

To proceed further, let $X_1, \cdots, X_n$ be a random sample from the location model generated by $\mathcal{F}_s$ so that (4.3.4) holds for some $\theta \in \mathbb{R}$ and with $\varepsilon_1, \cdots, \varepsilon_n$ i.i.d. F, $F \in \mathcal{F}_s$. Let $X_{(1)} < X_{(2)} < \cdots < X_{(n)}$ denote the ordered X_i's and define

$$V(t) := \sum_{1\le i\le j\le n} I\big(X_i + X_j > 2t\big) = \sum_{1\le i\le j\le n} I\big(X_{(i)} + X_{(j)} > 2t\big), \quad t \in \mathbb{R}. \tag{4.3.18}$$

For $t \in \mathbb{R}$, let $R_j^+(t)$ denote the absolute rank of $X_{(j)} - t$ and

$$W^+(t) := \sum_{j=1}^{n} I(X_{(j)} > t) R_j^+(t).$$

Note that $W^+(t)$ is well defined only for those $t \in \mathbb{R}$ for which there are no ties among $|X_{(1)} - t|, \cdots, |X_{(n)} - t|$. One can verify that for all $t \notin A := \big\{\big((X_{(i)} + X_{(j)}\big)/2; 1 \le i < j \le n\big\}$, $W^+(t) = V(t)$. See Exercise 4 at the end of this chapter. Let P_t denote the distribution of $X_1, \cdots, X_n$, when t is the true location parameter. By the continuity of F and the location invariance, $P_t(t \notin A) = P_0(0 \notin A) = 1$, because $P_0\big((X_{(i)} + X_{(j)})/2 = 0\big) = 0, \forall 1 \le i \le j \le n$. Hence, $P_t\big(W^+(t) = V(t)\big) = 1$, for all t. Therefore, the null distribution of $V = V(0)$ is the same as that of $W^+ \equiv W^+(0)$. Let $M := n(n+1)/2$. The range of V is the set of integers $\{0, \cdots, M\}$ and the null distribution of V is symmetric about $M/2$. From the null distribution of V, obtain the largest integer r such that $P_0(V < r) \le \alpha/2$. By the symmetry property, $P_0(V > M - r) \le \alpha/2$.

We shall apply the above general principle with $T_n(\theta) = V(\theta)$. The Wilcoxon signed rank test of level α accepts θ to be the true parameter value whenever $r \le V(\theta) \le M - r$. Here $a_{n,\alpha}(\theta) = r, b_{n,\alpha}(\theta) = M - r$ do not depend on θ. Therefore, the set $\mathcal{C}_{1,\alpha} := \big\{\theta; r \le V(\theta) \le M - r\big\}$ of all acceptable values of the parameter θ at level α is the Wilcoxon signed rank confidence set of confidence level $1 - \alpha$ for θ. This confidence set is indeed an interval as $V(\theta)$ is a nonincreasing step function of θ. Let $a_{(1)} \le a_{(2)} < \cdots \le a_{(M)}$ denote the M ordered pairwise averages $[X_{(i)} +$

$X_{(j)}]/2, 1 \le i \le j \le n$. Then, analogous to (3.3.6), we have

$$\{a_{(j)} \le x\} = \text{at least } j \text{ } a\text{'s are less than or equal to } x = \{V(x) \le M - j\}, \qquad (4.3.19)$$
$$\{a_{(s)} > x\} = \text{at least } (M - s + 1) \text{ } a\text{'s are} > x = \{V(x) \ge M - s + 1\}.$$

Apply (4.3.19) with $x = \theta$, $j = r$, $s = M - r + 1$ to yield

$$C_{1,\alpha} := \{\theta; a_{(r)} \le \theta < a_{(M-r+1)}\}. \qquad (4.3.20)$$

Example 4.3.4. Here we shall illustrate the computation of the above confidence interval when data consists of the five observations $2, -3, 10, 8, -5$. Let $\alpha = .125$, $\alpha/2 = .0625$. Here $M = (5 \times 6)/2 = 15$ and $M/2 = 7.5$. To begin with we need to compute r such that $P_0(V < r) \le .0625$. For this purpose, it suffices to obtain a few values of the null d.f. of V. Argue as for Table 4.2 to obtain Table 4.3 that gives some values of the null d.f. of V. From this table we see that

Table 4.3 $P_0(V \le v)$ for $n = 5$

v	0	1	2	3	4	5	6
$P_0(V \le v)$	.0313	.0625	.0938	.1563	.2188	.3125	.4063

$r = 2, M - r + 1 = 14$ and the 87.5% confidence level confidence interval for the location parameter is the interval $[a_{(2)}, a_{(14)})$. The $a_{(j)}$'s for the given data appear in Table 4.4. From this table we see the required interval is $[-4, 9)$.

Table 4.4 Values of $(X_{(i)} + X_{(j)})/2, i \le j$

$X_{(\cdot)} \setminus X_{(\cdot)}$:	-5	-3	2	8	10
-5	-5	-4	-3/2	3/2	5/2
-3		-1	-1/2	5/2	7/2
2			2	5	6
8				8	9
10					10

Similarly, the 93.74% confidence level confidence interval for θ is $[-5, 10)$. In fact this is the confidence interval of level $1 - \alpha$ for all $\alpha < .0626$. In general, for a given n, the interval $[a_{(1)}, a_{(M)})$ is the confidence interval of exact confidence level $1/2^{n-1}$ and of the confidence level $1 - \alpha$ for all $\alpha < 1/2^{n-1}$.

To describe the large sample confidence intervals, recall (3.3.1) and let

$$r_n := \frac{n(n+1)}{4} - \sqrt{\frac{n(n+1)(2n+1)}{24}} z_{\alpha/2} + 0.5,$$
$$s_n := \frac{n(n+1)}{4} + \sqrt{\frac{n(n+1)(2n+1)}{24}} z_{\alpha/2} - 0.5.$$

Then, by Theorem 4.3.2, the confidence interval $[a_{(\lfloor r_n \rfloor + 1)}, a_{(\lfloor s_n \rfloor)})$ is the Wilcoxon signed rank confidence interval of the asymptotic level $1 - \alpha$.

4.3.4 Wilcoxon Signed Rank Test in the Presence of Ties

So far we have assumed there are no zeros among $X_1, \cdots, X_n$ and no ties in $|X_1|, \cdots, |X_n|$. Although under the assumption of continuity of the underlying error distribution, these events occur with zero probability, in practice one often comes across data with zeros and ties in the absolute values. Here we shall describe how to modify the W_s test in this situation. The source for the material in this subsection is Chapter VII of Hájek (1969) and the paper of Vorličková (1972).

Let ν be the number of nonzero observations denoted by $\tilde{X}_i, 1 \le i \le \nu$. For the ease of writing, set $Y_i = |\tilde{X}_i|, 1 \le i \le \nu$. Suppose there are g groups of tied observations among the Y's. Let t_j denote the length of the jth group, $1 \le j \le g$ and $t = (t_1, \cdots, t_g)$. Note that $1 \le t_j \le \nu$ and $\sum_{j=1}^{g} t_j = \nu$. Thus

$$\begin{aligned} Y_{(1)} = \cdots = Y_{(t_1)} < Y_{(t_1+1)} &= \cdots = Y_{(t_1+t_2)} \\ &< \cdots < Y_{(t_1+\cdots+t_{g-1}+1)} = \cdots = Y_{(\nu)}. \end{aligned}$$

Define the average absolute ranks as follows. In the jth group the ranks of all observations are equal to the average of the ranks which they would have gotten had there been no ties. More precisely, let $\gamma_0 = 0, \gamma_k := t_1 + \cdots + t_k$ and

$$\bar{R}_i^+ := \frac{1}{t_k} \sum_{j=\gamma_{k-1}+1}^{\gamma_k} j = \gamma_{k-1} + \frac{t_k+1}{2}, \quad \gamma_{k-1}+1 \le i \le \gamma_k,$$

for $1 \le k \le g$. Then the analog of the Wilcoxon signed rank test is

$$\bar{W}^+ := \sum_{j=1}^{\nu} I(\tilde{X}_j > 0) \bar{R}_j^+. \tag{4.3.21}$$

To illustrate, consider the data consisting of the 10 observations X_i : $-1, -2, -10, 0, 1, 5, 5, 2, 6, 5$. Omitting the observation 0 leaves 9 observations $\tilde{X}_i$: $-1, -2, -10, 1, 5, 5, 2, 6, 5$; $Y_i = |\tilde{X}_i|$: $1, 2, 10, 1, 5, 5, 2, 6, 5$, $Y_{(i)}$: $1, 1, 2, 2, 5, 5, 5, 6, 10$. Thus there are $g = 5$ tied groups among the absolute values Y_i's. The first group consists of the two 1's, which gives $t_1 = 2$; the second group consists of the two 2's with $t_2 = 2$. Like this $t_3 = 3, t_4 = 1 = t_5$ and the average absolute ranks are

$$\begin{aligned} \bar{R}_j^+ &= \frac{t_1+1}{2} = 1.5, \quad 1 \le j \le 2, \\ &= \gamma_1 + \frac{t_2+1}{2} = 2 + 1.5 = 3.5, \;\; 3 \le j \le 4, \\ &= \gamma_2 + \frac{t_3+1}{2} = 4 + 2 = 6, \;\; 5 \le j \le 7, \\ &= \gamma_3 + \frac{t_4+1}{2} = 7 + 1 = 8, \;\; j = 8, \\ &= \gamma_4 + \frac{t_5+1}{2} = 8 + 1 = 9, \;\; j = 9. \end{aligned}$$

In other words, the vector of average absolute ranks is $1.5, 1.5, 3.5, 3.5, 6, 6, 6, 8, 9$, which yields

$$\begin{aligned} \bar{W}^+ &= \text{the sum of the average absolute ranks of the positive } \tilde{X}_i\text{'s} \\ &= 1.5 + 3.5 + 6 + 6 + 6 + 8 = 31. \end{aligned}$$

In general the exact null distribution of this statistic is not known. The test based on $\bar{W}^+$ is no longer distribution free. The exact conditional distribution of $\bar{W}^+$, given ν and $t = (t_1, \cdots, t_g)$, can be computed by direct counting. From it we obtain the conditional mean and variance of W^+ as $E_0(\bar{W}^+|\nu, t) = \nu(\nu+1)/4$ and

$$\sigma^2_{\bar{W}^+}(\nu, t) = \frac{1}{24}\Big[\nu(\nu+1)(2\nu+1) - \frac{1}{2}\sum_{j=1}^{g} t_j(t_j^2 - 1)\Big].$$

Note that the mean is the same as in the case of no ties except in terms of the number of non zero observations. Also if there are no ties then each t_j equals 1 and the above variance agrees with that of the no ties case among the non zero absolute observations.

We have the following large sample result available. As $\nu \to_{a.s.} \infty$, under $H_0 : \theta = 0$, $\sigma^{-1}_{\bar{W}^+}(\nu, t)(\bar{W}^+ - \nu(\nu+1)/4) \to_D \mathcal{N}(0, 1)$. This result may be used to determine the approximate α level critical values w^+_α of the $\bar{W}^+$-test as follows.

$$w^+_\alpha \approx z_\alpha\Big(\frac{1}{24}\Big[\nu(\nu+1)(2\nu+1) - \frac{1}{2}\sum_{j=1}^{g} t_j(t_j^2 - 1)\Big]\Big)^{1/2} + \frac{\nu(\nu+1)}{4}.$$

The proof of this result appears in Vorličková (1972).

4.3.5 Normal Scores Signed Rank Test

Another important distribution free test for testing about the one sample location parameter is the so called normal scores signed rank test. To describe this test, let Φ denote the standard normal d.f. and $\psi := \Phi^{-1}$ its inverse, the so-called normal quantile function, i.e., $\psi(t) = \inf\{x; \Phi(x) \geq t\}$, $0 < t < 1$. The test is based on

$$S_\psi := \sum_{i=1}^{n} u(X_i)\psi\Big(\frac{R_i^+}{n+1}\Big) = \sum_{j=1}^{n} u(X_{D_j})\psi\Big(\frac{j}{n+1}\Big), \tag{4.3.22}$$

where $u(x) := I(x > 0)$. The continuity and strict monotonicity of Φ yields that $\Phi(\Phi^{-1}(t)) = t$, $\Phi^{-1}(\Phi(x)) = x, \forall 0 < t < 1, x \in \mathbb{R}$. Because $\Phi(x) = 1 - \Phi(-x)$, we have $\psi(1-t) = -\psi(t)$, for all $0 < t < 1$ and $\psi(1/2) = 0$. These facts readily imply the following identities.

$$\psi\Big(\frac{i}{n+1}\Big) = -\psi\Big(\frac{n-i+1}{n+1}\Big), \quad \forall 1 \leq i \leq n, \tag{4.3.23}$$

$$\sum_{i=1}^{n} \psi\Big(\frac{i}{n+1}\Big) = 0. \tag{4.3.24}$$

For the sake of brevity we set $\psi_n(i) := \psi(i/(n+1))$. To compute the statistic S_ψ, one needs to know the normal scores $\psi_n(i)$'s. This is an extra step needed to implement this test in comparison to the W^+ test. The following R code

```
p=array() n
for (i in 1:n) p[i]=qnorm(i/(n+1),0,1)
round(p,4)
```

was used to obtain the normal scores $\psi_n(i)$'s given in Table 4.5 below.

As an example for the computation of S_ψ, suppose a random sample of size 10 yielded $X_i : -6, -4, -10, -9, 1, 5, 12, 18, 7, 11$. Then the absolute ranks of the positive observations are $1, 3, 9, 10, 5, 8$. From the normal scores Table 4.5, the corresponding test statistic is

$$\begin{aligned} S_\psi &= \psi_{10}(1) + \psi_{10}(3) + \psi_{10}(5) + \psi_{10}(8) + \psi_{10}(9) + \psi_{10}(10) \\ &= -1.3352 + -0.6046 - 0.1145 + 0.6046 + 0.9085 + 1.3352 = 0.0940. \end{aligned}$$

Table 4.5 Normal scores

$n = 8$	$\psi_8(i)$	1.2206	-0.7647	-0.4307	-0.1397	0.1397	0.4307
		0.7647	1.2206				
$n = 9$	$\psi_9(i)$	-1.2816	-0.8416	-0.5244	-0.2533	0.0000	0.2533
		0.5244	0.8416	1.2816			
$n = 10$	$\psi_{10}(i)$	-1.3352	-0.9085	-0.6046	-0.3488	-0.1142	0.1142
		0.3488	0.6046	0.9085	1.3352		
$n = 11$	$\psi_{11}(i)$	-1.3830	-0.9674	-0.6745	-0.4307	-0.2104	0.0000
		0.2104	0.4307	0.6745	0.9674	1.3830	
$n = 12$	$\psi_{12}(i)$	-1.4261	-1.0201	-0.7363	-0.5024	-0.2934	-0.0966
		0.0966	0.2934	0.5024	0.7363	1.0201	1.4261
$n = 13$	$\psi_{13}(i)$	-1.4652	-1.0676	-0.7916	-0.5659	-0.3661	-0.1800
		0.0000	0.1800	0.3661	0.5659	0.7916	1.0676
		1.4652					
$n = 14$	$\psi_{14}(i)$	-1.5011	-1.1108	-0.8416	-0.6229	-0.4307	-0.2533
		-0.0837	0.0837	0.2533	0.4307	0.6229	0.8416
		1.1108	1.5011				
$n = 15$	$\psi_{15}(i)$	-1.5341	-1.1503	-0.8871	-0.6745	-0.4888	-0.3186
		-0.1573	0.0000	0.1573	0.3186	0.4888	0.6745
		0.8871	1.1503	1.5341			
$n = 16$	$\psi_{16}(i)$	-1.5647	-1.1868	-0.9289	-0.7215	-0.5414	-0.3774
		-0.2230	-0.0738	0.0738	0.2230	0.3774	0.5414
		0.7215	0.9289	1.1868	1.5647		
$n = 17$	$\psi_{17}(i)$	-1.5932	-1.2206	-0.9674	-0.7647	-0.5895	-0.4307
		-0.2822	-0.1397	0.0000	0.1397	0.2822	0.4307
		0.5895	0.7647	0.9674	1.2206	1.5932	
$n = 18$	$\psi_{18}(i)$	-1.6199	-1.2521	-1.0031	-0.8046	-0.6336	-0.4795
		-0.3360	-0.1992	-0.0660	0.0660	0.1992	0.3360
		0.4795	0.6336	0.8046	1.0031	1.2521	1.6199
$n = 19$	$\psi_{19}(i)$	-1.6449	-1.2816	-1.0364	-0.8416	-0.6745	-0.5244
		-0.3853	-0.2533	-0.1257	0.0000	0.1257	0.2533
		0.3853	0.5244	0.6745	0.8416	1.0364	1.2816
		1.6449					
$n = 20$	$\psi_{20}(i)$	-1.6684	-1.3092	-1.0676	-0.8761	-0.7124	-0.5659
		-0.4307	-0.3030	-0.1800	-0.0597	0.0597	0.1800
		0.3030	0.4307	0.5659	0.7124	0.8761	1.0676
		1.3092	1.6684				

The exact null distribution of S_ψ is obtained in a similar way as that of W^+. Recall from Lemma 4.3.2 that under H_0, $\{u(X_i), 1 \le i \le n\}$ and $\{R_j^+, 1 \le j \le n\}$ are independent and each

R_i^+ is uniformly distributed over the integers $1, \cdots, n$. Hence, by (4.3.24),

$$E_0(\psi_n(R_i^+)) = \frac{1}{n}\sum_{j=1}^{n} \psi_n(j) = 0, \qquad E_0(\psi_n^2(R_i^+)) = \frac{1}{n}\sum_{i=1}^{n} \psi_n^2(i),$$

$$E_0(S_\psi) = \sum_{i=1}^{n} E_0(u(X_i))E\psi_n(R_i^+) = 0,$$

$$\mathrm{Var}_0(S_\psi) = \sum_{i=1}^{n} E_0(u(X_i))E(\psi_n^2(R_i^+)) = \frac{1}{2}\sum_{i=1}^{n} \psi_n^2(i).$$

Moreover, by (4.3.24), the distribution of S_ψ is symmetric about its expected value 0. Thus it suffices to compute the distribution for half the values of S_ψ. To illustrate, suppose $n = 4$. Then half of the null distribution consisting of half of the $2^4 = 16$ possible values is given in Table 4.6, where $\boldsymbol{u} = \big(u(X_{(D_1)}), \cdots, u(X_{(D_4)})\big)$, while the full null distribution is given in Table 4.7.

Table 4.6 Half of the null distrn. of S_ψ for $n = 4$

i	1	2	3	4	S_ψ	P_0
$\psi_4(i)$	-0.8416	-0.2533	0.2533	0.8416		
$\boldsymbol{u}$	1	1	1	1	0	1/16
	0	1	1	1	0.84	1/16
	1	0	1	1	0.25	1/16
	1	1	0	1	-0.25	1/16
	1	1	1	0	-0.84	1/16
	0	0	1	1	1.09	1/16
	0	1	0	1	0.59	1/16
	0	1	1	0	0	1/16

Table 4.7 Null distrn. of S_ψ for $n = 4$

S_c	-1.09	-0.84	-0.59	-0.25	0	0.25	0.59	0.84	1.09
P_0	$\frac{1}{16}$	$\frac{2}{16}$	$\frac{1}{16}$	$\frac{2}{16}$	$\frac{4}{16}$	$\frac{2}{16}$	$\frac{1}{16}$	$\frac{2}{16}$	$\frac{1}{16}$

The computation of the null distribution gets complicated rapidly with increasing n. Fortunately, the asymptotic null distribution of S_ψ is available in Hájek and Šidák (1967). Under H_0, $S_\psi/\sqrt{\mathrm{Var}(S_\psi)} \to_D \mathcal{N}(0,1)$. In view of the identity $n^{-1}\mathrm{Var}(S_\psi) = \frac{1}{2n}\sum_{i=1}^{n} \psi^2(i/(n+1))$ and the fact

$$\frac{1}{n}\sum_{i=1}^{n} \psi^2\Big(\frac{i}{n+1}\Big) \to \int_0^1 \psi^2(u)\, du = \int_0^1 \left(\Phi^{-1}(u)\right)^2 du = 1,$$

we obtain that $n^{-1}\mathrm{Var}_0(S_\psi) \to 1/2$. Hence, under H_0, the asymptotic distribution of $n^{-1/2}S_\psi$ is $\mathcal{N}(0, 1/2)$. This approximation is good even for sample sizes as small as 15. Thus the large sample normal scores test rejects $H_0 : \theta = 0$ at level α, in favor of the alternative $H_1 : \theta > 0$, whenever $S_\psi > z_\alpha n^{1/2}/2^{1/2}$, where z_α is as in (3.3.9).

Paired Comparison. When comparing the effect of a treatment on a disease, the practitioner takes an observation (Y_1) from a patient before the treatment and another observation (Y_2) from the same patient after the treatment and uses them to assess the effect of the treatment. An implicit assumption here is that the treatment effect is additive, i.e., for some real number θ, $Y_2 = Y_1 + \theta + \varepsilon$, where ε is the experimental error. For example, if the treatment is for reducing blood sugar levels, then $\theta < 0$ would mean that the treatment is being effective. In an agricultural study, treatment could be an application of a fertilizer to increase the yield, in which case $\theta > 0$ would support the effectiveness of the treatment. In general, let $(Y_{1i}, Y_{2i}), 1 \le i \le n$ denote the observations on n homogenous experimental units so that the differences $D_i := Y_{2i} - Y_{1i}, 1 \le i \le n$are i.i.d. $F(\cdot - \theta)$ r.v.'s, for some d.f. F and some real number θ. The problem of interest is to test $H_0 : \theta = 0$, without knowing F. The alternative could be $\theta < 0$ or $\theta > 0$ or $\theta \neq 0$. One can use any one of the above tests to test these hypotheses.

If F belongs to $\mathcal{F}_0$, then one may use the sign test based on the statistic $\tilde{S} := \sum_{i=1}^{n} I(D_i > 0)$. This test rejects H_0 in favor of the alternative $\theta > 0\,(\theta < 0)$ at level α if $\tilde{S} \ge c_\alpha\ (\tilde{S} \le n - c_\alpha)$, where c_α is as in (4.3.6). It rejects H_0 in favor of the two sided alternatives $\theta \neq 0$ at the level α, if $\tilde{S} \le n - c_{\alpha/2}$ or $\tilde{S} \ge c_{\alpha/2}$. For $n \ge 18$, one may use $c_\alpha = n/2 + \sqrt{n}/2z_\alpha$ as in (4.3.9).

If F belongs to $\mathcal{F}_s$, then one may use the Wilcoxon signed rank test based on $\tilde{W} := \sum_{1 \le i \le j \le n} I\big(D_{(i)} + D_{(j)} > 0\big)$. The null distribution of $\tilde{W}$ is the same as that of W^+. This test rejects H_0 in favor of the alternative $\theta > 0\,(\theta < 0)$ at level α if $\tilde{W} \ge w_\alpha\ (\tilde{W} \le n(n+1)/2 - w_\alpha)$, where w_α is such that under H_0, $P_0(\tilde{W} \ge w_\alpha) \le \alpha$. The $\tilde{W}$ test rejects the two-sided alternatives $\theta \neq 0$ at the level α if $\tilde{W} \le n(n+1)/2 - w_{\alpha/2}$ or $\tilde{W} \ge w_{\alpha/2}$. For $n \ge 18$, w_α is as in (4.3.17). One can also use the analog of the confidence intervals (4.3.20) for θ based on the pairwise averages $\big[D_{(i)} + D_{(j)}\big]/2, 1 \le i \le j \le n$. Similarly, a suitably modified normal scores signed rank test may be used where X_i's are replaced by D_i's.

4.4 Two Sample Location Model: Rank Sum Tests

In this section we consider testing about the location parameter in the two sample location model. More precisely, let F be a continuous d.f. on $\mathbb{R}$, not necessarily symmetric. In the two sample location model of interest here, one has two independent random samples $X_1, \cdots, X_{n_1}$ and $Y_1, \cdots, Y_{n_2}$ of sizes n_1, n_2, respectively. The first random sample is from F while the second is from $G(\cdot) := F(\cdot - \theta)$, for some $\theta \in \mathbb{R}$. The problem is to test

$$H_0 : \theta = 0 \quad \text{versus} \quad H_1 : \theta > 0.$$

A popular test for this problem is the two sample Wilcoxon rank sum test. Before describing this test, recall the classical setup where one is testing this hypothesis when $F = \Phi$, the standard normal d.f.. In this case the classical test is based on the difference of the two sample means $\bar{Y} - \bar{X}$. Since in general we do not assume the knowledge of F, it is hard to determine the null distribution of this statistic for finite sample sizes. Wilcoxon (1945, 1947) had the idea of replacing these averages by the average ranks as follows.

Rank X_i's and Y_j's together in an ascending order. Set $N = n_1 + n_2$ and let $R_1, \cdots, R_N$ denote these combined ranks. Without loss of generality let $R_1, \cdots, R_{n_1}$ denote the ranks of $X_1, \cdots, X_{n_1}$ and $R_{n_1+1}, \cdots, R_N$ denote the ranks of $Y_1, \cdots, Y_{n_2}$ in this combined ranking. Mimicking the two sample t-statistic, Wilcoxon suggested to replace the difference $\bar{Y} - \bar{X}$ in the two sample t-statistic by the difference between the averages of the ranks of the two samples. Set

$$\mathcal{D} := \frac{1}{n_2} \sum_{j=n_1+1}^{N} R_j - \frac{1}{n_1} \sum_{i=1}^{n_1} R_i, \qquad W := \sum_{j=n_1+1}^{N} R_j. \tag{4.4.1}$$

Because $\sum_{i=1}^{N} R_i = N(N+1)/2$,

$$\mathcal{D} = \frac{1}{n_2} W - \frac{1}{n_1} \Big[\frac{N(N+1)}{2} - W \Big] = \frac{N}{n_1 n_2} W - \frac{N(N+1)}{2n_1}.$$

The statistic W is called the two sample Wilcoxon rank sum statistic.

Under H_0, all arrangements of X's and Y's are equally likely while under H_1, the Y's will tend to be stochastically larger than X's, i.e.,

$$P_\theta(X \le Y) = \int F(y+\theta)\, dF(y) = 1/2, \quad \theta = 0,$$
$$> 1/2, \quad \theta > 0.$$

Thus under H_1, Y's will tend to get larger ranks than X's and $\mathcal{D}$ will tend to be positive and large. Since $\mathcal{D}$ is an increasing function of W, equivalently, the two sample Wilcoxon rank sum test rejects H_0, in favor of H_1, at the given level α, if $W \ge w_\alpha$, where w_α is found from the null distribution of W so that the $P_0(W \ge w_\alpha) \le \alpha$.

Let $\mathcal{W}$ denote the range of values of W. Note that the smallest value of W is $n_2(n_2+1)/2$, which corresponds to the arrangement of X and Y's where all Y's are to the left of all X's. The largest value $n_1 n_2 + n_2(n_2+1)/2$ of W is obtained when all Y's are to the right of all X's. The null distribution of W is given in the following lemma.

Lemma 4.4.1. *Let N_w denote the number of subsets of $\{1, \cdots, N\}$ of size n_2 whose members sum to w. Then, under $H_0 : \theta = 0$,*

$$P_0(W = w) = N_w \Big/ \binom{N}{n_2}, \qquad w \in \mathcal{W}, \tag{4.4.2}$$

$$E_0(W) = \frac{n_2(N+1)}{2}, \qquad \mathrm{Var}_0(W) = \frac{n_1 n_2 (N+1)}{12}. \tag{4.4.3}$$

Proof. Under H_0, $X_1, \cdots, X_{n_1}, Y_1, \cdots, Y_{n_2}$ is a random sample of size N from a continuous d.f. F. Hence by (3.2.6), the vector of ranks $(R_1, \cdots, R_N)$ is uniformly distributed on Π_N, the set of all permutations of the integers $1, \cdots, N$. Since W is the sum of the ranks of the Y's, it only depends on the set of n_2 elements formed by these ranks. There are $\binom{N}{n_2}$ such sets which are equally likely, and exactly N_w such sets that produce a given value w of W. This yields (4.4.2).

The formulas in (4.4.3) follow from (3.2.12), (3.2.14) and (3.2.15) applied with $n = N$, $A_n = \{n_1+1, \cdots, N\}$, $m_n = n_2$ and $\psi_n(r) = r$. □

Lemma 4.4.2. *Under $H_0 : \theta = 0$, the distribution of W is symmetric about $n_2(N+1)/2$.*

Proof. Suppose H_0 holds. The proof is facilitated by emphasizing the dependence of W on X's and Y's. Write $W(X,Y)$ for W. If $\boldsymbol{R} = (R_1, \cdots, R_N)$ denotes the combined ranks of $X_1, \cdots, X_{n_1}$, $Y_1, \cdots, Y_{n_2}$ then $\boldsymbol{S} := (N+1-R_1, \cdots, N+1-R_N)$ denotes the combined ranks of the vector $(-X_1, \cdots, -X_{n_1}; -Y_1, \cdots, -Y_{n_2})$. Both $\boldsymbol{R}$ and $\boldsymbol{S}$ are uniformly distributed on Π_N and $W(-X,-Y) =_D W(X,Y)$. Moreover,

$$\begin{aligned} W(-X,-Y) &= \sum_{i=n_1+1}^{N} (N+1-R_i) = n_2(N+1) - \sum_{i=n_1+1}^{N} R_i \\ &= n_2(N+1) - W(X,Y), \end{aligned}$$

which in turn yields

$$W(X,Y) - \frac{n_2(N+1)}{2} =_D W(-X,-Y) - \frac{n_2(N+1)}{2} =_D \frac{n_2(N+1)}{2} - W(X,Y),$$

thereby proving the claim. □

Example 4.4.1. In this example we show how the null distribution of W is obtained. Suppose $n_1 = 3, n_2 = 2$ so that $N = 5$ and $\binom{5}{2} = 10$. Table 4.8 lists the sets $\{R_4, R_5\}$ and the corresponding value of W. From this table we obtain the null distribution of W given in Table 4.9. Note that this

Table 4.8 Function table for W for $n_1 = 3, n_2 = 2$

$\{R_4, R_5\}$	{1,2}	{1,3}	{1,4}	{1,5}	{2,3}	{2,4}
W	3	4	5	6	5	6
$\{R_4, R_5\}$	{2,5}	{3,4}	{3,5}	{4,5}		
W	7	7	8	9		

Table 4.9 Null distribution of W for $n_1 = 3, n_2 = 2$

w	3	4	5	6	7	8	9
$P_0(W = w)$	1/10	1/10	2/10	2/10	2/10	1/10	1/10

distribution is symmetric about $n_2(N+1)/2 = 2(6)/2 = 6$.

Suppose for a given data, we observe $W = 8$. Then the p-value is

$$P_0(W \geq 8) = \frac{1}{10} + \frac{1}{10} = \frac{2}{10} = 0.2 > 0.05.$$

Hence, H_0 would not be rejected at the 0.05 level of significance.

Determination of the null distribution becomes rapidly involved as $\min(n_1, n_2)$ increases. We shall prove later, see (5.7.7) below, that under H_0, as $\min(n_1, n_2) \to \infty$,

$$\frac{W - n_2(N+1)/2}{\sqrt{n_1 n_2 (N+1)/12}} \to_D \mathcal{N}(0,1). \tag{4.4.4}$$

This result together with or without the continuity correction (cc) is used to determine that

$$\begin{aligned} w_\alpha &\approx z_\alpha \sqrt{\frac{n_1 n_2 (N+1)}{12}} + 0.5 + \frac{n_2(N+1)}{2}, && \text{with cc,} \\ &\approx z_\alpha \sqrt{\frac{n_1 n_2 (N+1)}{12}} + \frac{n_2(N+1)}{2}, && \text{without cc.} \end{aligned}$$

This approximation is reasonably good for $\min(n_1, n_2) \geq 12$.

Another test for the above problem proposed by Mann and Whitney (1947) is based on

$$U := \sum_{i=1}^{n_1} \sum_{j=1}^{n_2} I(X_i \leq Y_j).$$

But U is related to W as follows. By the definition of ranks,

$$R_{n_1+j} = \sum_{i=1}^{n_1} I(X_i \leq Y_j) + \sum_{k=1}^{n_2} I(Y_k \leq Y_j), \quad j = 1, \cdots, n_2, \tag{4.4.5}$$

$$W = \sum_{j=1}^{n_2} R_{n_1+j} = \sum_{j=1}^{n_2} \Big[\sum_{i=1}^{n_1} I(X_i \leq Y_j) + \sum_{k=1}^{n_2} I(Y_k \leq Y_j) \Big] = U + n_2(n_2+1)/2.$$

For this reason, the W-test is also called Mann-Whitney-Wilcoxon test. This linear relation and (4.4.4) trivially imply

$$\frac{U - (n_1 n_2)/2}{\sqrt{n_1 n_2 (N+1)/12}} \to_D \mathcal{N}(0,1). \tag{4.4.6}$$

4.4.1 A Class of Two Sample Rank Sum Tests

In this short section we shall describe a class of rank sum tests for the above two sample testing problem. Let φ be a nondecreasing real-valued function defined on $(0,1)$ and define

$$T_\varphi := \sum_{j=n_1+1}^{N} \varphi\Big(\frac{R_j}{N+1}\Big).$$

Because φ is nondecreasing, larger ranks will make T_φ larger. Recall, under the alternative $\theta > 0$, the Y's tend to get larger ranks. Thus, the T_φ rank test rejects H_0, in favor of the right-sided alternative $\theta > 0$, whenever T_φ is large. There is one such test for each φ. The choice of $\varphi(u) := u - 1/2, 0 < u < 1$, makes the T_φ test equal to the two sample Wilcoxon rank sum test W. Another important example is that of the two sample normal scores rank sum test which corresponds to the choice $\varphi(u) := \psi(u) = \Phi^{-1}(u), 0 < u < 1$, where Φ is the standard normal d.f.

The null distribution of T_φ is determined in the same way as that of W, where each value of T_φ is taken with the probability $1/\binom{N}{n_2}$. Let

$$\bar{\varphi}_N := \frac{1}{N} \sum_{i=1}^{N} \varphi\Big(\frac{i}{N+1}\Big), \qquad \sigma^2_{N,\varphi} := \frac{1}{N} \sum_{i=1}^{N} \varphi^2\Big(\frac{i}{N+1}\Big) - \bar{\varphi}^2_N.$$

Then, by (3.2.12) applied with $n = N$, $c_{ni} = I(i > n_1)$ and $\psi_n(i) := \varphi(i/(N+1))$, under H_0,

$$\mu_{N,T_\varphi} := E_0(T_\varphi) = n_2 \bar{\varphi}_N, \qquad \sigma^2_{N,T_\varphi} := \mathrm{Var}_0(T_\varphi) = \frac{n_1 n_2}{N-1} \sigma^2_{N,\varphi}.$$

Under H_0 and as $\min(n_1, n_2) \to \infty$, $(T_\varphi - \mu_{N,T_\varphi})/\sigma_{N,T_\varphi} \to_D \mathcal{N}(0,1)$.

Suppose $\varphi(u) = -\varphi(1-u)$, for all $0 < u < 1$ and $\int_0^1 \varphi^2(u)\,du < \infty$. Then, $\bar{\varphi}_N = 0$, $\int_0^1 \varphi(u)\,du = 0$, the null distribution of T_φ is symmetric about its mean zero, and the Riemann sum $\sigma^2_{N,\varphi}$ converges to $\sigma^2_\varphi := \int_0^1 \varphi^2(u)\,du$. If, in addition, n_1/N tends to some limit $0 < \lambda < 1$, then under H_0, $n_2^{-1/2}T_\varphi \to_D \mathcal{N}(0, \lambda\sigma_\varphi^2)$, as $\min(n_1, n_2) \to \infty$. For a proof of these statements, see Chapter V in Hájek and Šidák (1967).

In particular, if $\varphi = \Phi^{-1}$, then by (4.3.24), $\bar{\varphi}_N = 0$, $\int_0^1 \varphi^2(u)\,du = 1$ and the normal scores test statistic satisfies $n_2^{-1/2}T_{\Phi^{-1}} \to_D \mathcal{N}(0, \lambda)$, under H_0.

4.4.2 Two Sample Rank Sum Confidence Interval for θ

In this subsection we shall describe confidence intervals for the two sample location parameter based on the Mann-Whitney-Wilcoxon rank sum test.

For a given value t of θ, let $R_{i,t}, 1 \le i \le N$ denote the combined ranks of $X_1, \cdots, X_{n_1}$ and $Y_1 - t, \cdots, Y_{n_2} - t$. Write $W(t)$ for the W based on these combined ranks and let

$$U(t) := \sum_{j=1}^{n_2}\sum_{i=1}^{n_1} I(Y_j - X_i \ge t).$$

Then, as in (4.4.5),

$$W(t) := U(t) + \frac{n_2(n_2+1)}{2}, \qquad \forall t \in \mathbb{R}.$$

Here we shall apply the general method of finding confidence intervals described in Subsection 4.3.3 with $T_n(\theta) = U(\theta)$, where again $a_{n,\alpha}(\theta)$ and $b_{n,\alpha}(\theta)$ do not depend on θ. Note that the distributions of $W(\theta)$ and $U(\theta)$, when θ is the true parameter, are the same as those of $W(0)$ and $U(0)$, when $\theta = 0$ is the true parameter, respectively. Hence, by Lemma 4.4.2, $W(\theta)$ and $U(\theta)$ are symmetrically distributed around $n_2(N+1)/2$ and $m/2$, respectively, where $m = n_1 n_2$.

Let $u_{\alpha/2}$ be such that $P_0\big(U(0) > m/2 + u_{\alpha/2}\big) \le \alpha/2$. Then, the Mann-Whitney-Wilcoxon (MWW) test accepts θ at level α if $(m/2) - u_{\alpha/2} \le U(\theta) \le (m/2) + u_{\alpha/2}$. Thus here $a_{n,\alpha}(\theta) = (m/2) - u_{\alpha/2}$ and $b_{n,\alpha}(\theta) = (m/2) + u_{\alpha/2}$ do not depend on θ. Hence, the set

$$\mathcal{C}_{2,\alpha} := \big\{\theta \in \mathbb{R};\ m/2 - u_{\alpha/2} \le U(\theta) \le u_{\alpha/2} + m/2\big\}$$

is the MWW confidence set of confidence level $1-\alpha$ for the two sample location parameter θ. This set is indeed an interval as we shall show now.

Let $X_{(1)}, \cdots, X_{(n_1)}$ denote the order statistics of $X_1, \cdots, X_{n_1}$ and $Y_{(1)}, \cdots, Y_{(n_2)}$ denote the order statistics of $Y_1, \cdots, Y_{n_2}$. Let $D_1, D_2, \cdots, D_m$ denote the $m = n_1 n_2$ differences $Y_{(j)} - X_{(i)}, 1 \le i \le n_1, 1 \le j \le n_2$ and $D_{(1)} \le D_{(2)} \le \cdots \le D_{(m)}$ denote their order statistics. Then,

$$U(t) = \sum_{k=1}^{m} I\big(D_{(k)} \ge t\big), \quad t \in \mathbb{R}.$$

Now, argue as for (4.3.20) to conclude that

$$\begin{aligned} \mathcal{C}_{2,\alpha} &= \Big\{\theta;\ D_{(m/2-u_{\alpha/2})} \le \theta < D_{(m/2+u_{\alpha/2}+1)}\Big\}, && m \text{ even}, \\ &= \Big\{\theta;\ D_{(\lfloor m/2-u_{\alpha/2}\rfloor)} \le \theta < D_{(\lfloor m/2+u_{\alpha/2}+1\rfloor)}\Big\}, && m \text{ odd}. \end{aligned} \tag{4.4.7}$$

where, for any $x \in \mathbb{R}$, $\lfloor x \rfloor$ is as in (3.3.1).

As an example consider the case when $n_1 = 4, n_2 = 6$ and $\alpha = 0.04$. Then $\alpha/2 = .02$ and from the null distribution table of the $U(0)$ statistic, $P_0(U(0) > 21) = 0.019 \leq .02$. Because $m/2 = 12$, we obtain $12 + u_{0.02} = 21$ or $u_{0.02} = 21 - 12 = 9$. Hence $m/2 - u_{\alpha/2} = 12 - 9 = 3$, and $m/2 + u_{\alpha/2} + 1 = 12 + 9 + 1 = 22$. Thus the MWW confidence interval of confidence level 0.04 for the two sample location parameter is the interval $\big[D_{(3)}, D_{(22)}\big)$.

Suppose the data consists of the four X_i : 3, 5, 10, -5 and six Y_j : $4, 6, 2, -10, 20, 12$. Order the X's and Ys and compute the ordered differences D's as in Table 4.10.

Table 4.10 Values of $Y(j) - X_{(i)}$

$X_{(\cdot)} \setminus Y_{(\cdot)}$:	-10	2	4	6	12	20
-5	-5	7	9	11	17	25
3	-13	-1	1	3	9	17
5	-15	-3	-1	1	7	15
10	-20	-8	-6	-4	2	10

Observe that the largest difference $D_{(m)} = Y_{(n_2)} - X_{(1)} = 25$ and the smallest difference $D_{(1)} = Y_{(1)} - X_{(n_1)} = -20$. In Table 4.10, these are the numbers in the north east and south west corners, respectively. Counting from the north east corner, we see that the largest four differences are $D_{(24)} = 25, D_{(23)} = 17 = D_{(22)}, D_{(21)} = 15$. Similarly counting from the south west corner, we see the four smallest differences are $D_{(1)} = -20, D_{(2)} = -15, D_{(3)} = -13, D_{(4)} = -8$. Hence, the required CI is the interval $[D_{(3)}, D_{(22)}) = [-13, 17)$.

By (4.4.6), the approximate MWW confidence interval for the two sample location model is given by $\big[D_{(\lfloor r \rfloor)}, D_{(\lceil s \rceil)}\big)$, where

$$r := -z_{\alpha/2}\sqrt{\frac{n_1 n_2 (N+1)}{12}} + \frac{n_1 n_2}{2}, \quad s := z_{\alpha/2}\sqrt{\frac{n_1 n_2 (N+1)}{12}} + \frac{n_1 n_2}{2}.$$

4.4.3 Rank Sum Test and Ties

In the above discussion we have assumed that the error distribution F is continuous so that the probability of any two observations being the same is zero, i.e., almost surely there are no ties in the given data. However, in practice, one sometimes encounters ties in the data. Here we shall discuss a modification of the Wilcoxon rank sum test when there are ties in the data. The discussion below is based on Chapter VII of the book by Hájek (1969) and the work of Vorličková (1969, 1972).

Suppose there are g groups of tied data. Let t_j be the length of the jth group, $j = 1, \cdots, g$. Note that $1 \leq t_j \leq N$ and $\sum_{j=1}^{g} t_j = N$, where $N = n_1 + n_2$. Then the order statistics of $Z_1 = X_1, \cdots, Z_{n_1} = X_{n_1}, Z_{n_1+1} = Y_1, \cdots, Z_N = Y_{n_2}$ are

$$Z_{(1)} = \cdots = Z_{(t_1)} < Z_{(t_1+1)} = \cdots = Z_{(t_1+t_2)} < \cdots\cdots < Z_{(t_1+\cdots+t_{g-1}+1)} = \cdots = Z_{(N)}.$$

As far as ranking the data is concerned, we shall use the average ranking method, i.e., in the jth group the ranks of all observations are equal to the average of the ranks which they would have gotten had there been no ties. More precisely, with $s_0 = 0$, $s_k := t_1 + \cdots + t_k$, $1 \leq k \leq g$, the average rank of the kth groups is

$$a_k := \frac{1}{t_k}\sum_{j=1}^{t_k}\big(s_{k-1} + j\big) = s_{k-1} + \frac{t_k + 1}{2}.$$

Now let $\bar{R}_{n_1+j}, 1 \le j \le n_2$ denote the average ranks corresponding to the data from the second sample. Then the two sample Wilcoxon rank sum test statistic in this case is

$$\bar{W} := \sum_{j=1}^{n_2} \bar{R}_{n_1+j}. \tag{4.4.8}$$

For example, suppose the combined two samples data consists of the numbers $1.5, 6, 1.5, 6, 1.5, 3, 10, 3$, where the first three observations $1.5, 6, 1.5$ came from the first sample and the remaining ones from the second sample. Then the combined ordered data is $1.5, 1.5, 1.5, 3, 3, 6, 6, 10$. Thus there are $g = 4$ tied groups with the group lengths $t_1 = 3$, $t_2 = 2$, $t_3 = 2$, $t_4 = 1$. Moreover, $s_1 = t_1 = 3$, $s_2 = t_1 + t_2 = 5$, $s_3 = 5 + t_3 = 7$, $s_4 = 8$ resulting in $a_1 = (1+2+3)/3 = 2$, $a_2 = (4+5)/2 = 4.5$, $a_3 = (6+7)/2 = 6.5$ and $a_4 = 8$. Thus the original ordered observations would be replaced by the average ranks $2, 2, 2, 4.5, 4.5, 6.5, 6.5, 8$. Now recall the second sample consisted of $6, 1.5, 3, 10, 3$ (belonging to the groups 3, 1, 2, 4, 2) with the respective average ranks $\bar{R}_{n_1+1} = 6.5, \bar{R}_{n_1+2} = 2, \bar{R}_{n_1+3} = 4.5, \bar{R}_{n_1+34} = 8, \bar{R}_{n_1+5} = 4.5$, resulting in

$$\bar{W} = \sum_{j=1}^{n_2} \bar{R}_{n_1+j} = 6.5 + 2 + 4.5 + 8 + 4.5 = 25.5.$$

The null distribution of $\bar{W}$ is not known in general. But the conditional null distribution, given g and $t := (t_1, \cdots, t_g)$, is obtained by direct counting, which gets complicated rapidly with the increase in n_1, n_2, g. In particular from Theorem 29B of Hájek (1969), we obtain $E(\bar{W}|g,t)) = n_2(N+1)/2$ and

$$\sigma^2_{\bar{W}}(g,t) := \mathrm{Var}(\bar{W}|g,t)) = \frac{n_1 n_2}{12}\Big[N + 1 - \frac{\sum_{k=1}^{g} t_k(t_k^2 - 1)}{N(N-1)}\Big].$$

Note that the mean is the same as in the case of no ties, while the conditional variance is affected by the length of ties in each group. In the case of no ties, each t_k is 1, and this variance is the same as that in the case of no ties.

The asymptotic null distribution of $\bar{W}$ is normal. See Theorem 29C in Hájek (1969). More precisely, the following fact holds. As $n_1 \wedge n_2 \to \infty$,

$$\frac{\bar{W} - n_2(N+1)/2}{\sigma_{\bar{W}}(g,t)} \to_D \mathcal{N}(0,1).$$

Using this fact, the asymptotic level α critical value $\bar{w}_\alpha$ of the right-sided $\bar{W}$ test is

$$\begin{aligned}\bar{w}_\alpha &\approx z_\alpha \sigma_{\bar{W}} + (n_2(N+1)/2)\\ &= z_\alpha\Big(\frac{n_1 n_2}{12}\Big[N + 1 - \frac{\sum_{k=1}^{g} t_k(t_k^2 - 1)}{N(N-1)}\Big]\Big)^{1/2} + \frac{n_2(N+1)}{2}.\end{aligned}$$

4.5 Testing in the k-sample Location Model

In this section we consider the problem of testing for the equality of the location parameters in the k-sample problem. More precisely, let F be a continuous d.f., $n_1, \cdots, n_k$ be given sample sizes and $N = \sum_{j=1}^{k} n_j$. In the k sample location model data consists of $X_{ij}, 1 \le i \le n_j, 1 \le j \le k$

such that for some θ_j, $\varepsilon_{ij} := X_{ij} - \theta_j, 1 \le i \le n_j$ are i.i.d. F r.v.'s and all observations $\{X_{ij}\}$ are mutually independent. This model is also known as the one-way classification design.

The problem of interest is to test the hypothesis

$$H_0 : \theta_1 = \theta_2 = \cdots = \theta_k, \quad \text{versus} \quad H_1 : \ H_0 \text{ is not true.} \tag{4.5.1}$$

In the context of the one-way classification, this amounts to testing for treatment effects. The null hypothesis claims that there are no treatment effects, while the alternative states that there are treatment effects. Let

$$\bar{X}_{.j} := n_j^{-1} \sum_{i=1}^{n_j} X_{ij}, \qquad \bar{X}_{..} := N^{-1} \sum_{j=1}^{k} n_j \bar{X}_{.j} = N^{-1} \sum_{j=1}^{k} \sum_{i=1}^{n_j} X_{ij},$$

$$V := \frac{\sum_{j=1}^{k} n_j (\bar{X}_{.j} - \bar{X}_{..})^2}{\sum_{j=1}^{k} \sum_{i=1}^{n_j} (X_{ij} - \bar{X}_{.j})^2}, \qquad \mathbb{F} := \frac{N-k}{k-1} V.$$

In the case of $N(0, \sigma^2)$ errors, the classical F-test is based on the statistic $\mathbb{F}$, with its large values being significant. In the case of non-Gaussian errors, the null distribution of this statistic is not known, and hence implementing test based on this is infeasible. Like Wilcoxon, Kruskal and Wallis (1952) proposed to use an analog of the $\mathbb{F}$ statistic based on ranks as follows.

Rank all $N := \sum_{j=1}^{k} n_j$ observations from the smallest to the largest. Let R_{ij} denote the rank of X_{ij} among these combined ranks and $\boldsymbol{R} = (R_{11}, \cdots, R_{n_1 1}; \cdots; R_{1k}, \cdots, R_{n_k k})$ be the combined rank vector. Let

$$\bar{R}_{.j} := n_j^{-1} \sum_{i=1}^{n_j} R_{ij} = n_j^{-1} R_{.j}, \quad 1 \le j \le k,$$

$$\bar{R}_{..} := N^{-1} \sum_{j=1}^{k} n_j \bar{R}_{.j} = N^{-1} \sum_{j=1}^{k} \sum_{i=1}^{n_j} R_{ij} = N^{-1} R_{..} = \frac{N+1}{2}.$$

An analog of the statistic V in terms of ranks obtained after replacing original observations by their ranks is

$$T := \frac{\sum_{j=1}^{k} n_j (\bar{R}_{.j} - \bar{R}_{..})^2}{\sum_{j=1}^{k} \sum_{i=1}^{n_j} (R_{ij} - \bar{R}_{.j})^2} = \frac{T_1}{T_2}, \quad \text{say.}$$

Recall the analysis of variance decomposition, say from Rao (1973): Total sum of squares = treatment sum of squares + the sum of squares due to error, i.e.,

$$\sum_{j=1}^{k} \sum_{i=1}^{n_j} (R_{ij} - \bar{R}_{..})^2 = \sum_{j=1}^{k} n_j (\bar{R}_{.j} - \bar{R}_{..})^2 + \sum_{j=1}^{k} \sum_{i=1}^{n_j} (R_{ij} - \bar{R}_{.j})^2 = T_1 + T_2.$$

On the other hand,

$$\sum_{j=1}^{k} \sum_{i=1}^{n_j} (R_{ij} - \bar{R}_{..})^2 = \sum_{j=1}^{k} \sum_{i=1}^{n_j} R_{ij}^2 - \frac{R_{..}^2}{N} = \frac{N(N+1)(2N+1)}{6} - \frac{N(N+1)^2}{4}$$
$$= N(N^2 - 1)/12.$$

Let

$$H := \frac{12}{N(N^2-1)} T_1 = \frac{12}{N(N^2-1)} \sum_{j=1}^{k} n_j \Big(\bar{R}_{\cdot j} - \frac{N+1}{2}\Big)^2.$$

The above decomposition gives $T_2 = N(N^2-1)/12 - T_1$ and

$$T = \frac{T_1}{\frac{N(N^2-1)}{12} - T_1} = \frac{H}{1-H} = \frac{1}{1-H} - 1.$$

Thus T is an increasing function of H, and the test that rejects H_0 for large values of T is equivalent to the test that rejects H_0 for large values of H. Under H_0 the distribution of this statistic is determined from that of the combined ranks, which in turn does not depend on the error d.f. F. Thus tests based on T are distribution free.

Just as in the two sample location model, under H_0,

$$P(\boldsymbol{R} = \boldsymbol{r}) = \frac{1}{\binom{N}{n_1, \cdots, n_k}} = \frac{n_1! \cdots n_k!}{N!}, \quad \boldsymbol{r} \in \Pi_N.$$

Indeed, under H_0, all orderings of the observations are equally likely, so that all partitions of the set $\{1, \cdots, N\}$ into k subsets of respective sizes $n_1, \cdots, n_k$ are equally likely. The above formula is used to compute the null distribution of H. This gets quickly computationally involved. For large sample sizes, $H_* := (N-1)H$ is approximately a chi-squared r.v. with $k-1$ degrees of freedom. Indeed, Kruskal and Wallis have shown that under H_0, H_* converges in distribution to a chi-squared r.v. with $k-1$ degrees of freedom as $\min\{n_1, \cdots, n_k\} \to \infty$. The statistic H_* is called the Kruskal–Wallis rank test.

Note that

$$H_* := (N-1)H = \frac{12}{N(N+1)} \sum_{j=1}^{k} n_j \Big(\bar{R}_{\cdot j} - \frac{N+1}{2}\Big)^2.$$

Because of the identity

$$\sum_{j=1}^{k} n_j \Big(\bar{R}_{\cdot j} - \frac{N+1}{2}\Big)^2 = \sum_{j=1}^{k} \frac{R_{\cdot j}^2}{n_j} - \frac{N(N+1)^2}{4},$$

we obtain the formula

$$H_* = \frac{12}{N(N+1)} \sum_{j=1}^{k} \frac{R_{\cdot j}^2}{n_j} - 3(N+1), \tag{4.5.2}$$

which provides an efficient way for computing H_*.

Example 4.5.1. As an example suppose three brands A, B and C of drugs were tested at random: A was assigned to two units, B to three units, and C to two units, and the data produced the following combined ranking (higher ranking means greater improvement):

A	B	C
2,4	3,5,7	1,6
$R_{11} < R_{21}$	$R_{12} < R_{22} < R_{32}$	$R_{13} < R_{23}$

Here, $n_1 = 2, n_2 = 3, n_3 = 2, N = 7$.

The aim is to test: H_0: there is no difference between the brands. Note

$$P(\boldsymbol{R} = \boldsymbol{r}) = \frac{1}{\binom{N}{n_1, n_2, n_3}} = \frac{1}{\binom{7}{2,3,2}} = \frac{2 \cdot 6 \cdot 2}{7!} = \frac{1}{210}.$$

Also, for the observed data $R_{.1} = 6$, $R_{.2} = 15$ and $R_{.3} = 7$,

$$H_* = \frac{12}{N(N+1)} \sum_{j=1}^{k} \frac{R_{.j}^2}{n_j} - 3(N+1) = \frac{12}{7 \cdot 8} \left[\frac{36}{2} + \frac{225}{3} + \frac{49}{2} \right] - 24 = 1.18.$$

Thus, we can compute the values of H_* for all the 210 cases which occur with equal probability $1/210$, and compute the null distribution of H_*. For moderately large N, it becomes very laborious. For $k = 3$, and $n_j \le 5$, the critical values of H_* test are given in tables by Kruskal and Wallis (1952), see Table I in Lehmann (1975). From the table, $P_0(H_* \ge 1.18) > 0.105$. Thus H_0 would not be rejected at any $\alpha < .105$.

For a positive integer ν and $0 \le \alpha \le 1$, let

$$\chi^2_{\alpha,\nu} \text{ denote the } 100(1-\alpha)\text{th percentile of the } \chi^2_\nu \text{ distribution.} \tag{4.5.3}$$

For $k \ge 3$ and/or $n_j \ge 5$, one uses the large sample chi-squared approximation of the null distribution of H_*, which says to reject H_0 at level the approximate level α whenever $H_* \ge \chi^2_{\alpha,k-1}$.

4.6 Rank Tests in Linear Regression

In this section we shall describe a class of rank tests for testing a hypothesis about the underlying regression parameters in the multiple linear regression model. To begin with we focus on a simple linear regression model. Let $x_{n1}, \cdots, x_{nn}$ be known real numbers denoting the values of a design variable and are allowed to depend on the sample size. Set

$$\bar{x}_n = \frac{1}{n} \sum_{i=1}^{n} x_{ni}, \qquad \sigma_x^2 = \sum_{i=1}^{n} (x_{ni} - \bar{x}_n)^2.$$

We require that σ_x^2 is positive which rules out that the x_{ni}'s are identical and allows us to introduce their standardized versions $c_{ni} = \sigma_x^{-1}(x_{ni} - \bar{x}_n)$, $i = 1, \cdots, n$, which satisfy $\sum_{i=1}^n c_{ni} = 0$ and $\sum_{i=1}^n c_{ni}^2 = 1$.

In simple linear regression, the response variable Y_{ni} is related to the design variable x_{ni} via the formula

$$Y_{ni} = \beta_0 + \beta_1 x_{ni} + \varepsilon_{ni}, \quad 1 \le i \le n, \tag{4.6.1}$$

for some unknown real numbers β_0, β_1, where the error variables $\varepsilon_{n1}, \cdots, \varepsilon_{nn}$ are i.i.d. with continuous d.f. F.

Let b be a known real number and consider the problem of testing $H_0 : \beta_1 = b$ against the alternative $H_1 : \beta_1 > b$, when nothing is specified about β_0. A class of rank tests for this problem can be based on simple linear rank statistics of the type given in (3.2.9). More precisely, let R_i

denote the rank of $Y_{ni} - bx_{ni}$ among $Y_{n1} - bx_{n1}, \cdots, Y_{nn} - bx_{nn}$ and let J be a nondecreasing real-valued function defined on $(0, 1)$. Define

$$T_x := \sum_{i=1}^{n} c_{ni} J\Big(\frac{R_i}{n+1}\Big).$$

Because R_i is also the rank of $Y_{ni} - \beta_0 - bx_{ni}$ among $Y_{n1} - \beta_0 - bx_{n1}, \cdots, Y_{nn} - \beta_0 - bx_{nn}$, not knowing β_0 does not affect the statistic T_x. An α level test based on T_x rejects H_0 whenever $T_x > \tau_\alpha$, where τ_α is determined from the null distribution of T_x such that $P_0(T_x > \tau_\alpha) \le \alpha$.

Under H_0, for each n, $\varepsilon_{ni} = Y_{ni} - \beta_0 - bx_{ni}, 1 \le i \le n$ are i.i.d. F r.v.'s. Since F is assumed to be continuous, $(R_1, R_2, \cdots, R_n)$ is uniformly distributed on the set of permutations Π_n of $1, 2, \cdots, n$ and hence

$$P_0(T_x = t) = \frac{\#\text{ of rank permutations for which } T_x = t}{n!},$$

for t in the range of T_x. In other words this is a distribution free statistic. Because its values depend on the design variables $\{x_{ni}\}$, it is not possible to have one table giving the distribution of T_x for each n. However, for large sample sizes, we can use the following limiting distribution result.

Let

$$\bar{J} := n^{-1}\sum_{i=1}^{n} J\Big(\frac{i}{n+1}\Big), \qquad \sigma_{nJ}^2 := \sum_{i=1}^{n}\Big(J\Big(\frac{i}{n+1}\Big) - \bar{J}\Big)^2.$$

Assume the following about $\{x_{ni}\}$ and the function J.

J is a nondecreasing real-valued function on $(0, 1)$ satisfying (4.6.2)

$$\int_0^1 J^2(u)\,du < \infty, \quad \sigma_J^2 := \int_0^1 J^2(u)\,du - \Big(\int_0^1 J(u)\,du\Big)^2 > 0.$$

$$\max_{1\le i\le n} \sigma_x^{-1}\big|x_{ni} - \bar{x}_n\big| \to 0. \tag{4.6.3}$$

By (3.2.12) applied with $c_{ni} = (x_{ni} - \bar{x}_n)/\sigma_x$, $\psi_n(i) = J(i/(n+1))$, $E_0(T_x) = 0$, $\mathrm{Var}_0(T_x) = \sigma_{nJ}^2/(n-1) \to \sigma_J^2$, by (4.6.2). Moreover, under H_0, (4.6.2) and (4.6.3),

$$\big(\mathrm{Var}_0(T_x)\big)^{-1/2} T_x \to_D \mathcal{N}(0,1), \quad T_x \to_D \mathcal{N}(0, \sigma_J^2). \tag{4.6.4}$$

For a proof of this result, see Hájek and Šidák (1967), Chapter V. For the special case of $J(u) = u$, see Theorem 5.4.3 of Chapter 5 of this book. In view of this asymptotic normality result, the test that rejects H_0 whenever $T_x > z_\alpha \sigma_J$ has asymptotic size α, where z_α is as in (3.3.9).

Example 4.6.1. Here we consider the case of $x_{ni} = i$, $i = 1, \cdots, n$. Then, the regression model, (4.6.1) is known as a trend model, and the above testing problem with $\beta_0 = 0$ amounts to testing for no trend versus a positive trend. In this case,

$$\bar{x}_n = \frac{n+1}{2}, \quad \sigma_x^2 = \sum_{i=1}^{n} i^2 - \frac{(n+1)^2}{4} = \frac{n(n^2-1)}{12},$$

$$\max_{1\le i\le n} c_{ni}^2 = \frac{12}{n(n^2-1)}\frac{(n-1)^2}{4} = \frac{3(n-1)}{n(n+1)} \to 0.$$

Hence the above result is applicable in this case.

Example 4.6.2. Suppose $J(u) = u$, $0 < u < 1$. Then,

$$T_x := \frac{1}{n+1} \sum_{i=1}^{n} c_{ni} R_i. \tag{4.6.5}$$

Let n_1 and n_2 be the two positive integers such that $n_1 + n_2 = n$ and suppose $x_{ni} = 0$ for $i = 1, \cdots, n_1$ and $x_{ni} = 1$ for $i = n_1 + 1, \cdots, n_1 + n_2$. Then, $\bar{x}_n = n_2/n$, $\sigma_x^2 = n_1 n_2 / n$, $\max_{1 \le i \le n} c_{ni}^2 \le 1/\min\{n_1, n_2\}$, so that (4.6.3) holds if $\min(n_1, n_2) \to \infty$, and the model (4.6.1) becomes

$$\begin{aligned} Y_{ni} &= \beta_0 + \varepsilon_{ni}, & 1 \le i \le n_1, \\ &= \beta_0 + \beta_1 + \varepsilon_{ni}, & n_1 + 1 \le i \le n_1 + n_2. \end{aligned} \tag{4.6.6}$$

This is then the two sample location model where $Y_{ni}, 1 \le i \le n_1$ is a random sample from the first population and $Y_{ni}, n_1 + 1 \le i \le n$ is a random sample form the second population. These two populations differ only by their location β_1. Direct calculations show that the test based on T_x is equivalent to the two sample Wilcoxon rank sum test.

Now consider the more general multiple linear regression model where $\boldsymbol{x}_{ni} \in \mathbb{R}^p$ and for some $\boldsymbol{\beta} \in \mathbb{R}^p$,

$$Y_{ni} = \boldsymbol{\beta}' \boldsymbol{x}_{ni} + \varepsilon_{ni}, \quad 1 \le i \le n. \tag{4.6.7}$$

Let $\bar{\boldsymbol{x}}_n = (1/n) \sum_{i=1}^n \boldsymbol{x}_{ni}$. In order to identify the parameter $\boldsymbol{\beta}$, we require that the symmetric $p \times p$ matrix

$$\boldsymbol{\Sigma}_x = \frac{1}{n} \sum_{j=1}^{n} (\boldsymbol{x}_{nj} - \bar{\boldsymbol{x}}_n)(\boldsymbol{x}_{nj} - \bar{\boldsymbol{x}}_n)'$$

be positive definite. Then, there is a symmetric matrix $\boldsymbol{\Sigma}_x^{-1/2}$ such that $\boldsymbol{\Sigma}_x^{-1/2} \boldsymbol{\Sigma}_x \boldsymbol{\Sigma}_x^{-1/2} = \mathcal{I}_p$, the $p \times p$ identity matrix. Furthermore, we assume

$$\max_{1 \le i \le n} \left\| \boldsymbol{\Sigma}_x^{-1/2} (\boldsymbol{x}_{ni} - \bar{\boldsymbol{x}}_n) \right\| \to 0. \tag{4.6.8}$$

This is the analog of (4.6.3) in the present case.

Let $\boldsymbol{b} \in \mathbb{R}^p$ be given parameter vector and consider the problem of testing $\boldsymbol{\beta} = \boldsymbol{b}$ versus that $\boldsymbol{\beta} \ne \boldsymbol{b}$. Let R_i denote the rank of $Y_{ni} - \boldsymbol{b}' \boldsymbol{x}_{ni}$ among $Y_{n1} - \boldsymbol{b}' \boldsymbol{x}_{n1}, \cdots, Y_{nn} - \boldsymbol{b}' \boldsymbol{x}_{nn}$ and define

$$\boldsymbol{T} := \sum_{i=1}^{n} (\boldsymbol{x}_{ni} - \bar{\boldsymbol{x}}_n) J\left(\frac{R_i}{n+1}\right).$$

Note that $\boldsymbol{T}$ is a p-dimensional vector of simple linear rank statistics. Then, under H_0, $\boldsymbol{\Sigma}_x^{-1/2} \boldsymbol{T} \to_D \mathcal{N}_p(0, \sigma_J^2 \mathcal{I}_p)$. To see this, by the Cramér-Wold device (see Corollary 2.10.1), it suffices to show that for every unit vector $u \in \mathbb{R}^p$, $u' \boldsymbol{\Sigma}_x^{-1/2} \boldsymbol{T} \to_D \mathcal{N}(0, \sigma_J^2)$, under H_0. For such u, $u' \boldsymbol{\Sigma}_x^{-1/2} \boldsymbol{T}$ equals T_x with $c_{ni} = u' \boldsymbol{\Sigma}_x^{-1/2} (\boldsymbol{x}_{ni} - \bar{\boldsymbol{x}}_n)$, $i = 1, \cdots, n$. By assumption (4.6.8), these coefficients satisfy $\sum_{i=1}^n c_{ni} = 0$, $\sum_{i=1}^n c_{ni}^2 = 1$ and (4.6.3). Hence, by (4.6.4), under H_0, $T_x \to_D \mathcal{N}(0, \sigma_J^2)$.

Consequently, under H_0, $Q := \boldsymbol{T}' \boldsymbol{\Sigma}_x^{-1} \boldsymbol{T} \to_D \chi_p^2$ and the test that rejects H_0 whenever $Q > \chi_{\alpha,p}^2$ has asymptotic size α, where $\chi_{\alpha,p}^2$ as in (4.5.3).

Hettmansperger (1984) and the reference therein discuss numerous rank-based analysis of variance tests in a variety of models like the two-way classification and randomized block design.

The review paper by McKean and Hettmansperger (2016) nicely summarizes the development of rank-based analysis in the multiple linear regression model and various extensions like multivariate models and models with dependent error structure such as mixed models, time series models, and longitudinal data models. Kloke and McKean (2012, 2014) developed R libraries to implement this methodology. The monograph of Kloke and McKean (2014) illustrates the various rank based procedures using R language.

4.7 Multivariate Ranks

The main purpose of this short section is to provide information about some of the recent developments in the area of multivariate ranks. Let p be a positive integer. Often in practice one observes multivariate data where a random vector $\boldsymbol{X}$ has p components with $p \geq 2$. Examples include the investigations that study height, weight, age and blood pressure of a human being jointly or the effect of a fertilizer, soil conditions and crop of a certain type jointly. In these examples $p = 4$ and $p = 3$, respectively. The area of multivariate statistical inference is vast. Given the broad applicability of the inference procedures based on ranks and signed ranks in the univariate setup, it is desirable to develop their analogs in the multivariate setup.

As noted in Section 3.2 of this book, when data is a random sample from a continuous univariate distribution, then ranks are distribution free and independent of the order statistics. In the case of a symmetric distribution, signs and absolute ranks are independent. It is desirable to have a definition of ranks and signed ranks in the multivariate setup which will have similar properties. The lack of a natural order in $\mathbb{R}^p$ creates difficulty in defining ranks and signed ranks properly.

Puri and Sen (1971) discussed numerous inference procedures based on componentwise ranks, where components of n observations $\boldsymbol{X}_1, \cdots, \boldsymbol{X}_n$, each of dimension p, are ranked. In particular they develop analogs of the Wilcoxon rank sum and Kruskal-Wallis statistics in the multivariate setup. However, these ranks are not distribution free even asymptotically, unless the components are exchangeable. Much later, Chaudhuri (1996), Oja (1999, 2010), Marden (1999), and Oja and Randles (2004) developed spatial ranks and signs and some related concepts. However, none of them have the property of being distribution free. Multivariate signed ranks defined in Oja and Paindaveine (2005) are asymptotically distribution free. Liu (1992), Liu and Singh (1993), Zuo and He (2006), Zuo and Serfling (2000) and Serfling (2002a,b) used various measures of statistical depths to define ranks. A depth measure is a real-valued function, and the ranks of the depths of i.i.d. observations are well defined and distribution free. More recently, using Monge–Kantorovich measure transportation ideas, Chernozhukov et al. (2017), Ghosal and Sen (2022), Hallin (2022), Hallin, Hlubinka and Hudecová (2023), among others, developed concepts of distribution and quantile functions, ranks and signs in the multivariate setup that do satisfy the properties analogous to those in the univariate case.

4.8 Exercises

1. Let $X_1, \cdots, X_n$ be a random sample from a population with unknown mean μ and finite positive variance σ^2. Let μ_0 be a known real number, and consider the problem of testing

$H_0 : \mu = \mu_0$ versus the alternative $H_1 : \mu > \mu_0$. Let $\bar{X}_n$ and s_n denote the sample mean and s.d., respectively. Show that the test $\varphi_n := I(\bar{X}_n - \mu_0 > z_\alpha s_n/n^{1/2})$ is locally regular at μ_0 with slope $\gamma := 1/\sigma$ for this testing problem.

2. Let $-\infty < s < t < \infty$. Consider the d.f.

$$F_{s,t}(x) = (x - s)I(s < x < s + 1/2) + (1/2)I(s + 1/2 \le x) \\ + (x - t)I(t < x < t + 1/2) + (1/2)I(t + 1/2 \le x).$$

(a) Show that $F_{s,t}(x - \theta) = F_{s+\theta,t+\theta}(x), \forall\, x, \theta \in \mathbb{R}$.

(b) Show that $F_{s,t}$ belongs to the location model generated by $\mathcal{F}_s$ and find θ, the center of symmetry θ.

(c) Show that $F_{s,t}$ does not belong to the location model generated by $\mathcal{F}_0$.

3. Prove Theorem 4.3.2 for W_s by verifying the condition (2.10.1) with $\xi_{ni} = i\,\mathrm{sgn}(X_{D_i})$, $1 \le i \le n$.

4. Suppose $(x_1, \cdots, x_n) \in \mathbb{R}^n$ is such that $x_1 < x_2 < \cdots < x_n$. Let

$$w(t) := \sum_{i=1}^{n}\sum_{j=1}^{n} I(x_j > t)I(|x_i - t| \le |x_j - t|), \qquad t \in \mathbb{R}.$$

Show that for all $t \in \mathbb{R}$,

$$w(t) = \sum_{j=1}^{n} I(x_j > t) + \sum_{1\le i<j\le n} I(x_j > t)I(x_i + x_j \ge 2t).$$

Hence, for $t \notin \{(x_i + x_j)/2; 1 \le i < j \le n\}$,

$$w(t) = \sum_{1\le i\le j\le n} I(x_i + x_j > 2t).$$

5. Suppose the following data is a random sample from the one sample location model with the location parameter θ: 30, −5, 12, −6, −7, 15, 40, 25, −19, 10. Obtain the 95% finite sample and approximate Wilcoxon signed rank confidence intervals for θ. Few values of the null d.f. of V for $n = 10$ are given in Table 4.11.

Table 4.11 $P_0(V \le v)$ for $n = 10$

v	4	5	6	7	8	9
$P_0(V \le v)$	.0068	.0098	.0137	.0186	.0244	.0322

6. Read Subsection 4.4.2 for this problem. From the null distribution of $U(0)$, we obtain the following.

$$\text{I. } n_1 = 6, n_2 = 5, \quad P(U(0) < 4) = 0.025, \; P(U(0) < 2) = 0.005.$$
$$\text{II. } n_1 = 7, n_2 = 8, \quad P(U(0) < 9) = 0.025, \; P(U(0) < 7) = 0.005.$$

(a) Describe finite sample MWW confidence intervals of the confidence levels 95% and 99% for the two sample location parameter in terms of the ordered differences $D_{(j)}$'s in each of the above cases.

[Answers: 95% : I. $[D_{(3)}, D_{(27)})$, II. $[D_{(9)}, D_{(38)})$. 99% : I. $[D_{(2)}, D_{(29)})$, II. $[D_{(7)}, D_{(50)})$.]

(b) Describe the large sample analogs of these confidence intervals in the case $n_1 = 7, n_2 = 8$.

(c) What are the finite sample and approximate confidence intervals of confidence level 0.95 of the two sample location parameter for the following data.

$$X_i : 5, 10, 13, 15, 8, 7, 20; \quad Y_j : 30, 6, 9, 25, 2, 14, 11, 21.$$

7. Show that for the two sample location model (4.6.6), the statistic T_x of (4.6.5) is equal to the two sample Wilcoxon rank sum statistic W of (4.4.1).
8. Prove the identity (4.3.23).
9. Let $\nu, g, Y_1, \cdots, Y_n$ and $\bar{R}_i^+$ be as in Section 4.3.4. Let $N(y) = \sum_{j=1}^{\nu}[1 + I(Y_j < y) + I(Y_j \leq y)]/2$. Show that $\bar{R}_i^+ = N(Y_i)$, for all $1 \leq i \leq n$.

5

U-Statistics

5.1 Introduction

As is seen in previous chapters the computation of the null distributions of several test statistics gets complicated rapidly with increasing sample sizes. Moreover, the assessment of the finite sample behaviour of the power of distribution free tests is often not easily possible. There is thus a need to derive their asymptotic distributions under the null and alternate hypotheses in as much unified fashion as possible. Fortunately, many test statistics belong to a class of statistics, called U-statistics. For example, the sample mean, the sample variance and the Wilcoxon signed rank statistics belong to the class of one sample U-statistics while the two sample Wilcoxon rank sum test statistic belongs to the class of two sample U-statistics.

U-statistics were introduced by Hoeffding (1948). He provided a unified way of proving the consistency of estimators based on U-statistics and their asymptotic normality. Hájek (1968) crystallized the method of Hoeffding's proof of the asymptotic normality of U-statistics. In this chapter we shall discuss some statistical properties of U-statistics, Hájek's projection lemma and exhibit its usefulness in proving the asymptotic normality of U-statistics, the Wilcoxon signed rank and rank sum statistics in the one and two sample location models and the simple linear rank statistics used in testing in linear regression models. We also include a proof of the asymptotic normality of an estimator of the integral of a squared density.

The paper of Hall (1984) contains a fundamental result about the asymptotic normality of degenerate U-statistics, which he used to prove the asymptotic normality of integrated squared difference between a kernel density estimator and its expectation. Hall's result is also found useful in proving the asymptotic normality of some minimum distance regression parameter estimators. Section 5.9 discusses an extension of Hall's result to some weighted degenerate weighted U-statistics of independent r.v.'s.

5.2 Definition of U-Statistics

In this section we shall define the class of U-statistics and discuss some examples.

For any positive integer m, let Π_m denote the set of all permutations of $1, \cdots, m$. Let n and r be positive integers, $r \leq n$ and set

$$B_{r,n} = \{b = (b_1, \cdots, b_r) : b_1, \cdots, b_r \in \{1, \cdots, n\}, b_1 < \cdots < b_r\}.$$

The set $B_{r,n}$ has $\binom{n}{r}$ elements as is easily checked.

DOI: 10.1201/9781003585770-5

Let $X_1, \cdots, X_n$ be i.i.d. p-dimensional random vectors for some positive integer p. A U-statistic is a statistic of the form

$$U = \frac{1}{\binom{n}{r}} \sum_{b \in B_{r,n}} \varphi(X_{b_1}, \cdots, X_{b_r})$$

for some measurable function φ from $(\mathbb{R}^p)^r$ into $\mathbb{R}$ that is symmetric under permutations, i.e., the identity

$$\varphi(x_{i_1}, \cdots, x_{i_r}) = \varphi(x_1, \cdots, x_r) \tag{5.2.1}$$

holds for all permutations $(i_1, \cdots, i_r) \in \Pi_r$ and all $x_1, \cdots, x_r \in \mathbb{R}^p$. If $r = 2$, this simply means $\varphi(x, y) = \varphi(y, x)$ for all $x, y \in \mathbb{R}^p$. The function φ is called the kernel and r the order of the kernel and the U-statistic.

Suppose that $\varphi(X_1, \cdots, X_r)$ has finite mean $E\big(\varphi(X_1, \cdots, X_r)\big) = \nu$. Then U has mean

$$E(U) = E\big(\varphi(X_1, \cdots, X_r)\big) = \nu.$$

This follows from the fact that U is an average of the $\binom{n}{r}$ identically distributed random variables $\varphi(X_{b_1}, \cdots, X_{b_r})$, $b \in B_{r,n}$. Thus a U-statistic is an unbiased estimator of its mean. It is easy to see that U has a finite variance if $\varphi(X_1, \cdots, X_r)$ has a finite second moment. A formula for the variance of U will be derived in the next section. This variance will converge to zero as n increases. From this and (2.4.9) we conclude that U is a consistent estimator of its mean ν. Lehmann (1983) provides some conditions under which U is the minimum variance unbiased estimator of ν.

We end this section by providing some examples of U-statistics of order $r = 2$ which will be used in the sequel to illustrate some of the presented results. Note that a U-statistic of order $r = 2$ takes the form

$$U = \frac{2}{n(n-1)} \sum_{1 \le i < j \le n} \varphi(X_i, X_j)$$

where $\varphi(x, y) = \varphi(y, x)$ for all x and y in $\mathbb{R}^p$. In the following examples we take $p = 1$.

Example 5.2.1. Proportion of the positive pairwise sum. Consider the problem of estimating $\nu = P(X_1 + X_2 > 0)$, where X_1, X_2 are i.i.d. F r.v.'s. Clearly, $\varphi(X_1, X_2) := I(X_1 + X_2 > 0)$ is an unbiased estimator of this ν and satisfies (5.2.1). Therefore, φ is a kernel of order 2. The corresponding U-statistic is

$$U^{(1)} := \frac{2}{n(n-1)} \sum_{1 \le i < j \le n} I(X_i + X_j > 0). \tag{5.2.2}$$

Example 5.2.2. Sample variance as U-statistic. Given a random sample $X_1, \cdots, X_n$, the sample mean and sample variance are

$$\bar{X} = \frac{1}{n} \sum_{i=1}^{n} X_i \qquad s_n^2 = \frac{1}{n-1} \Big[\sum_{i=1}^{n} X_i^2 - n\bar{X}^2 \Big].$$

The sample mean is a U-statistic of order $r = 1$. Direct calculations show that the sample variance s_n^2 can be expressed as U-statistic

$$U^{(2)} := \frac{2}{n(n-1)} \sum_{1 \le i < j \le n} \frac{\left(X_i - X_j\right)^2}{2}$$

of order 2 with kernel $\varphi(x,y) = (x-y)^2/2$.

If X_1 has a finite variance σ^2, then $E(X_1 - X_2) = 0$ and $E\big((X_1 - X_2)^2\big) = \mathrm{Var}(X_1 - X_2) = \mathrm{Var}(X_1) + \mathrm{Var}(X_2) = 2\sigma^2$. Thus $U^{(2)}$ is an unbiased estimator of the population variance σ^2.

Sometimes one encounters U-statistics whose kernel depends on the sample size n. An example of this is given next.

Example 5.2.3. Let $X_1, \cdots, X_n$ be a random sample from a density f. Let h_n be a sequence of positive numbers and K be a density on $\mathbb{R}$ that is symmetric about zero. Let

$$U_n^{(3)} := \frac{2}{n(n-1)h_n} \sum_{1 \le i < j \le n} K\Big(\frac{X_i - X_j}{h_n}\Big). \tag{5.2.3}$$

Clearly, $U_n^{(3)}$ is a U-statistic of order $r = 2$ and its kernel $\varphi_n(x,y) := h_n^{-1} K\big((x-y)/h_n\big)$ depends on n. The property $\varphi_n(x,y) = \varphi_n(y,x)$, for all $x, y \in \mathbb{R}$, follows from the symmetry of K.

This type of U-statistic arises in connection with bandwidth selection in kernel density estimation as we shall see later in Chapter 7. It is also used in the estimation of $I_2(f) := \int f^2(x)\,dx$, whenever $I_2(f) < \infty$. To motivate this, note that $X_1 - X_2$ has density

$$g(y) := \int f(y+x) f(x) dx, \quad y \in \mathbb{R}.$$

Using the substitution $s/h_n = u$ one calculates

$$E\big(U_n^{(3)}\big) = h_n^{-1} \int K\big(\frac{s}{h_n}\big) g(s) ds = \int K(u) g(u h_n) du$$

and expects this expectation to converge to $g(0) = I_2(f)$ as $h_n \to 0$. We shall use the theory developed in this chapter to show that, under mild assumptions on the density f and proper choice of K and h_n, $U_n^{(3)}$ is a consistent and asymptotically normal estimator of $I_2(f)$.

5.3 The Variance of a U-Statistic

We shall now derive an expression for the variance of a general U-statistic. Since the case $r = 1$ is well understood, we focus on the case $r \ge 2$.

Let U be a U-statistic of order $r \ge 2$ with kernel φ and assume that $E(\varphi^2(X_1, \cdots, X_r))$ is finite. To proceed further, we introduce the following conditional expectation, where the last equality follows from the independence of $X_1, \cdots, X_n$:

$$\begin{aligned} \varphi_c(x_1, \cdots, x_c) &:= E\Big(\varphi(X_1, \cdots, X_c, X_{c+1}, \cdots, X_r)\Big| X_1 = x_1, \cdots, X_c = x_c\Big) \\ &= E\Big(\varphi(x_1, \cdots, x_c, X_{c+1}, \cdots, X_r)\Big), \qquad c = 1, \cdots, r. \end{aligned} \tag{5.3.1}$$

Note the identity $\varphi_r = \varphi$. Also define

$$\delta_c^2 := \mathrm{Var}\big(\varphi_c(X_1, \cdots, X_c)\big), \qquad c = 1, \cdots, r. \tag{5.3.2}$$

Assume $n \geq 2r-1$. Recall the formula $\mathrm{Var}(aX) = a^2\mathrm{Var}(X)$ and the fact that the variance of a sum is the sum of all covariances. Since U is the average of the $\binom{n}{r}$ identically distributed r.v.'s $Z_b := \varphi(X_{b_1}, \cdots, X_{b_r})$, $b \in B_{r,n}$, we have

$$\mathrm{Var}(U) = \frac{1}{\binom{n}{r}^2} \sum_{a,b \in B_{r,n}} \mathrm{Cov}(Z_a, Z_b).$$

For a and b in $B_{r,n}$, let $N(a,b)$ denote the number of elements in the intersection $\{a_1, \cdots, a_r\} \cap \{b_1, \cdots, b_r\}$ and let $M_c := \{(a,b); a, b \in B_{r,n}, N(a,b) = c\}$ for $c = 0, \cdots, r$. For $(a,b) \in M_0$, Z_a and Z_b are independent and $\sum_{(a,b)\in M_0} \mathrm{Cov}(Z_a, Z_b) = 0$. Hence the variance of U can be rewritten as

$$\mathrm{Var}(U) = \frac{1}{\binom{n}{r}^2} \sum_{c=1}^{r} \sum_{(a,b)\in M_c} \mathrm{Cov}(Z_a, Z_b).$$

Since there are $\binom{n}{r}$ choices for a, $\binom{r}{c}$ choices for the common indices, and then $\binom{n-r}{r-c}$ choices for the other indices of b, M_c has $N_c := \binom{n}{r}\binom{r}{c}\binom{n-r}{r-c}$ elements, provided $n-r \geq r-c$ or $n \geq 2r-c$ for all $1 \leq c \leq r$, which is guaranteed by the assumption $n \geq 2r-1$.

Suppose $N(a,b) = c$, for some $c = 1, \cdots, r$. Then the sets $A = \{a_1, \cdots, a_r\}$ and $B = \{b_1, \cdots, b_r\}$ have c elements in common. Denote the common elements by $d_1, \cdots, d_c$. Let $\alpha_{c+1}, \cdots, \alpha_r$ denote the other elements in A and $\beta_{c+1}, \cdots, \beta_r$ denote the other elements in B. Then $\boldsymbol{X}_d := (X_{d_1}, \cdots, X_{d_c})$, $\boldsymbol{X}_\alpha := (X_{\alpha_{d+1}}, \cdots, X_{\alpha_r})$ and $\boldsymbol{X}_\beta := (X_{\beta_{d+1}}, \cdots, X_{\beta_r})$ are independent. Moreover, $(\boldsymbol{X}_d, \boldsymbol{X}_\alpha)$ and $(\boldsymbol{X}_d, \boldsymbol{X}_\beta)$ are conditionally independent, given $\boldsymbol{X}_d$, and have the same distribution as $\boldsymbol{X} = (X_1, \cdots, X_r)$. Use these facts, the permutation symmetry of φ and the i.i.d. nature of $X_1, \cdots, X_n$ to compute

$$\begin{aligned} E(Z_a Z_b) &= E\big[\varphi(\boldsymbol{X}_d, \boldsymbol{X}_\alpha)\varphi(\boldsymbol{X}_d, \boldsymbol{X}_\beta)\big] \\ &= E\big[E(\varphi(\boldsymbol{X}_d, \boldsymbol{X}_\alpha)\big|\boldsymbol{X}_d)E(\varphi(\boldsymbol{X}_d, \boldsymbol{X}_\beta)\big|\boldsymbol{X}_d)\big] \\ &= E\big[\varphi_c^2(\boldsymbol{X}_d)\big] = E\big[\varphi_c^2(X_1, \cdots, X_c)\big]. \end{aligned}$$

Since $E\big(\varphi_c(X_1, \cdots, X_c)\big) = E\big(\varphi(X_1, \cdots, X_r)\big) = \nu$, we obtain

$$\mathrm{Cov}(Z_a, Z_b) = E(Z_a Z_b) - \nu^2 = \mathrm{Var}\big(\varphi_c(X_1, \cdots, X_c)\big) = \delta_c^2,$$

for all $(a,b) \in M_c$. This shows

$$\sum_{(a,b)\in M_c} \mathrm{Cov}(Z_a, Z_b) = N_c \delta_c^2.$$

Combining the above facts we arrive at the formula

$$\mathrm{Var}(U) = \binom{n}{r}^{-1} \sum_{c=1}^{r} \binom{r}{c}\binom{n-r}{r-c}\delta_c^2, \quad \forall n \geq 2r-1. \tag{5.3.3}$$

If we define $\binom{a}{b}$ to be zero for $b > a$, then formula (5.3.3) is still valid for $n = r, \cdots, 2r-2$.

For $r = 1$, this formula reduces to $\mathrm{Var}(\varphi(X_1))/n$. For $r = 2$, we have the identity $\varphi_1(x) = E\big(\varphi(x, X_2)\big)$ and the above formula becomes

$$\mathrm{Var}(U) = \frac{4(n-2)\delta_1^2 + 2\delta_2^2}{n(n-1)} = \frac{4\delta_1^2}{n} + \frac{2(\delta_2^2 - 2\delta_1^2)}{n(n-1)}, \quad n \geq 3. \tag{5.3.4}$$

We now use this formula to derive the variances of $U^{(1)}$ and $U^{(2)}$.

Example 5.2.1 continued. The U-statistic $U^{(1)}$ of Example 5.2.1 has kernel $\varphi(x,y) = I(x+y > 0)$. Direct calculations yield the identities

$$\varphi_1(x) = 1 - F(-x), \qquad \nu = \int (1 - F(-x))\, dF(x),$$
$$\delta_1^2 = \int (1 - F(-x))^2\, dF(x) - \nu^2, \qquad \delta_2^2 = \nu(1-\nu).$$

In particular, if F is continuous and $1 - F(-x) = F(x)$, for all $x \in \mathbb{R}$, then the above simplifies to $\nu = 1/2$, $\delta_1^2 = 1/12$, $\delta_2^2 = 1/4$ and

$$\mathrm{Var}(U^{(1)}) = \frac{1}{3n} + \frac{1}{6n(n-1)}, \qquad n \ge 3.$$

In the sequel, all limits are taken as $n \to \infty$, unless specified otherwise. Since the above variance tends to zero, by (2.4.9), it follows that $U^{(1)}$ is a consistent estimator of ν.

Example 5.2.2 continued. The U-statistic $U^{(2)}$ from Example 5.2.2 has kernel $\varphi(x,y) = (x-y)^2/2$. Assume that X_1 has mean μ, variance σ^2 and finite fourth central moment $\mu_4 = E\big[(X_1 - \mu)^4\big]$. Then one verifies $\nu = \sigma^2$, $\varphi_1(x) = (x-\mu)^2/2 + \sigma^2/2$, $\delta_1^2 = (\mu_4 - \sigma^4)/4$ and $\delta_2^2 = E[(X_1 - X_2)^4]/4 - \sigma^4 = (\mu_4 + 6\sigma^4 + \mu_4)/4 - \sigma^4 = (\mu_4 + \sigma^4)/2$. Thus the variance of $U^{(2)}$ computes to be

$$\mathrm{Var}(U^{(2)}) = \frac{1}{n}\Big[\mu_4 - \frac{n-3}{n-1}\sigma^4\Big], \quad n \ge 3.$$

This variance tends to zero. Hence, by (2.4.9), $U^{(2)}$ is a consistent estimator of σ^2.

If δ_1^2 is positive, then by (5.3.4), $n\mathrm{Var}(U)/(4\delta_1^2) \to 1$. An analog of this result holds for $r > 2$, i.e., $n\mathrm{Var}(U)/(r^2\delta_1^2) \to 1$ for $r > 2$, as will be shown next. To proceed with this, assume $n \ge 2r-1$ and introduce, for $1 \le c \le r$,

$$K_c := \frac{r!\binom{r}{c}}{(r-c)!}, \qquad a_{nc} := n^c \frac{(n-r)!}{n!}\,\frac{(n-r)!}{(n-2r+c)!}.$$

Verify $K_1 = r^2$ and for $c = 1, \cdots, r$,

$$\frac{(n-2r+c+1)^{r-c}}{n^{r-c}} \le a_{nc} \le \frac{n^{c-1}}{(n-r+1)^{c-1}} \le 2^{c-1}. \tag{5.3.5}$$

Theorem 5.3.1. *Suppose $r \ge 2$, $E(\varphi^2(X_1, \cdots, X_r)) < \infty$ and $\delta_1^2 > 0$. Then,*

$$\frac{n\mathrm{Var}(U)}{r^2\delta_1^2} = a_{n1} + \frac{1}{r^2\delta_1^2}\sum_{c=2}^{r} K_c \frac{a_{nc}}{n^{c-1}}\,\delta_c^2, \quad \forall n \ge 2r-1. \tag{5.3.6}$$

Consequently, the following hold.

$$\mathrm{Var}(U) \le \frac{r^2\delta_1^2}{n} + \sum_{c=2}^{r} \frac{K_c a_{n,c}\delta_c^2}{n^c}, \quad \forall n \ge 2r-1. \tag{5.3.7}$$

$$\frac{n\mathrm{Var}(U)}{r^2\delta_1^2} \to 1. \tag{5.3.8}$$

Proof. Recall that $K_1 = r^2$. Rewrite

$$\mathrm{Var}_F(U) = \sum_{c=1}^{r} \frac{r!\binom{r}{c}}{(r-c)!}\,\frac{(n-r)!}{n!}\,\frac{(n-r)!}{(n-2r+c)!}\,\delta_c^2 = \frac{r^2}{n}\delta_1^2 a_{n1} + \sum_{c=2}^{r} K_c \frac{a_{nc}}{n^c}\,\delta_c^2.$$

Now divide by $r^2\delta_1^2/n$ on both sides of this identity to obtain the identity (5.3.6). The claim (5.3.7) follows from the above identity and the fact that $a_{n1} \le 1$, for all $n \ge 2r - 1$. Use the fact that $a_{n1} \to 1$, as $n \to \infty$ to conclude that the right-hand side of (5.3.6) converges to 1. This proves (5.3.8) and completes the proof of the theorem. □

As seen in Example 5.2.3, sometimes the kernel varies with the sample size. It is thus of interest to derive an analog of the above theorem for a sequence of kernels φ_n. Proceeding a bit more generally, let $X_{n,1}, \cdots, X_{n,n}$ be i.i.d. p-dimensional random vectors. For each $n \ge 2r - 1$, define the U-statistic

$$U_n = \frac{1}{\binom{n}{r}} \sum_{1 \le i_1 < \cdots < i_r \le n} \varphi_n(X_{n,i_1}, \cdots, X_{n,i_r}), \qquad (5.3.9)$$

where φ_n satisfies (5.2.1) and $E(\varphi_n^2(X_{n,1}, \cdots, X_{n,r}))$ is finite for all $n \ge r$. To stress the dependence on n, write $\delta_{n,c}^2$ for δ_c^2. From the formula (5.3.6), which is valid even when φ depends on n, and the inequality $a_{nc} \le 2^c$, see (5.3.5), we derive the following corollary.

Corollary 5.3.1. *In addition to the above assumptions, suppose*

$$\delta_{n,1}^2 > 0, \ \forall n \ge r; \quad R_{nc} := \frac{\delta_{nc}^2}{n^{c-1}\delta_{n1}^2} \to 0, \ \forall c = 2, \cdots, r. \qquad (5.3.10)$$

Then, $n\mathrm{Var}(U_n)/(r^2\delta_{n1}) \to 1$.

5.4 A Projection Lemma

In this section we shall first state and prove a projection lemma, stemming from the work of Hoeffding (1948) and Hájek (1968). This lemma is of general interest and useful in deriving the limiting distribution of general statistics based on independent observations. In particular, we shall apply this lemma to derive the asymptotic normality of U-statistics and simple linear rank statistics.

Let m be a positive integer and $Y_1, \cdots, Y_m$ be m independent k-dimensional random vectors. For $j = 1, \cdots, m$, let $\mathcal{H}_j$ denote the set of all measurable functions ℓ from $\mathbb{R}^k$ into $\mathbb{R}$ such that $E(\ell^2(Y_j)) < \infty$ and set

$$\mathcal{H}(Y_1, \cdots, Y_m) = \Big\{ \sum_{j=1}^m \ell_j(Y_j); \ell_1 \in \mathcal{H}_1, \cdots, \ell_m \in \mathcal{H}_m \Big\}.$$

Let S be a r.v. with $E(S^2) < \infty$. Define

$$\widehat{S} := \sum_{j=1}^m E(S|Y_j) - (m-1)E(S). \qquad (5.4.1)$$

Then $\widehat{S}$ belongs to $\mathcal{H}(Y_1, \cdots, Y_m)$. The following lemma shows that $\widehat{S}$ is a projection of S into $\mathcal{H}(Y_1, \cdots, Y_m)$.

Lemma 5.4.1. Projection Lemma. *Under the above setup, the following facts hold.*

$$E(\widehat{S}) = E(S). \tag{5.4.2}$$

$$E\big[(S-L)^2\big] = E\big[(S-\widehat{S})^2\big] + E\big[(\widehat{S}-L)^2\big], \quad \forall L \in \mathcal{H}(Y_1,\cdots,Y_m). \tag{5.4.3}$$

$$E\big[(S-\widehat{S})^2\big] = \mathrm{Var}(S) - \mathrm{Var}(\widehat{S}). \tag{5.4.4}$$

Proof. Claim (5.4.2) follows, because by the first part of (2.7.6), $E\big(E(S|Y_i)\big) = E(S)$ for all i. Next, for a statistic L with finite second moment, we have the identity

$$\begin{aligned} E\big[(S-L)^2\big] &= E\big[(S-\widehat{S}+\widehat{S}-L)^2\big] \\ &= E\big[(S-\widehat{S})^2\big] + 2E\big[(S-\widehat{S})(\widehat{S}-L)\big] + E\big[(\widehat{S}-L)^2\big]. \end{aligned}$$

From this we see that the claim (5.4.3) will follow if we show that for $L = \sum_{j=1}^m \ell_j(Y_j) \in \mathcal{H}(Y_1,\cdots,Y_m)$ the cross product term is zero. Let $g_i(Y_i) := E(S|Y_i) - \ell_i(Y_i)$. Since $S - \widehat{S}$ has mean zero, we obtain

$$\begin{aligned} E\big[(S-\widehat{S})(\widehat{S}-L)\big] &= E\big[(S-\widehat{S})\big(\widehat{S}-L+(m-1)E(S)\big)\big] \\ &= \sum_{j=1}^m E\big[(S-\widehat{S})g_j(Y_j)\big] \\ &= \sum_{j=1}^m E\big[g_j(Y_j)E(S-\widehat{S}|Y_j)\big]. \end{aligned}$$

In the last step we used the second part of (2.7.6), which yields that for any two r.v.'s V_1 and $g(V_2)$ with $E(V_1^2) + E(g^2(V_2)) < \infty$, $E\big(V_1 g(V_2)\big) = E\big(g(V_2)E(V_1|V_2)\big)$. By the independence of $Y_1,\cdots,Y_m$, we obtain, for all $i,j = 1,\cdots,m$,

$$\begin{aligned} E\big(E(S|Y_i)|Y_j\big) &= E(S|Y_j), \qquad j = i, \\ &= E(S), \qquad j \neq i. \end{aligned}$$

Therefore, for all $j = 1,\cdots,m$, by the linearity of conditional expectation,

$$\begin{aligned} E\big(\widehat{S}|Y_j\big) &= \sum_{i=1}^m E\big(E(S|Y_i)|Y_j\big) - (m-1)E(S) \\ &= E\big(S|Y_j\big) + (m-1)E(S) - (m-1)E(S) = E\big(S|Y_j\big). \end{aligned}$$

Thus the cross product term vanishes, which proves (5.4.3). The proof of the lemma is completed upon taking $L = E(S)$ in (5.4.3) to obtain (5.4.4). □

We have the identity

$$\widehat{S} = E(S) + \sum_{j=1}^m \big[E(S|Y_j) - E(S)\big] = E(S) + \sum_{j=1}^m E\big(S - E(S)|Y_j\big).$$

With $S_0 = S - E(S)$, this becomes

$$\widehat{S} = E(S) + \widehat{S}_0. \tag{5.4.5}$$

This is sometimes helpful in computing $\widehat{S}$.

Example 5.4.1. Let X be a random vector. By the Projection Lemma applied with $m = 1$ and $Y_1 = X$, we derive $\widehat{S} = \mu(X) := E(S|X)$. In this case (5.4.3) yields the inequality

$$\begin{aligned} E\big[(S-\ell(X))^2\big] &= E\big[(S-\mu(X))^2\big] + E\big[(\mu(X)-\ell(X))^2\big] \\ &\geq E\big[(S-\mu(X))^2\big] \end{aligned}$$

for all r.v.'s $\ell(X)$ with finite second moment. This shows that $\mu(X)$ minimizes $E\big[(S-\ell(X))^2\big]$. Taking $\ell(X) = E(S) = E(\mu(X))$, we derive the identity

$$\begin{aligned} \mathrm{Var}(S) &= E\big[(S-E(S))^2\big] \\ &= E\Big[E\{(S-\mu(X))^2|X\}\Big] + E\Big[(\mu(X)-E(\mu(X)))^2\Big] \\ &= E\big(\mathrm{Var}(S|X)\big) + \mathrm{Var}\big(E(S|X)\big), \end{aligned}$$

where $\mathrm{Var}(S|X)$ is defined as in (2.7.5). This is known as the law of total variance.

Example 5.4.2. Let h be a measurable function from $\mathbb{R}^k \times \mathbb{R}^k$ to $\mathbb{R}$ satisfying $E\big(h^2(Y_1, Y_2)\big) < \infty$. Suppose $S = h(Y_1, Y_2)$. Set

$$h_1(Y_1) = E\big(h(Y_1, Y_2)\big|Y_1\big), \quad h_2(Y_2) = E\big(h(Y_1, Y_2)\big|Y_2\big).$$

Then the projection of S into $\mathcal{H}(Y_1, Y_2)$ is given by $\widehat{S} = h_1(Y_1) + h_2(Y_2) - E(S)$ and (5.4.4) yields

$$\begin{aligned} &E\Big(\big\{h(Y_1, Y_2) - h_1(Y_1) - h_2(Y_2) + E(S)\big\}^2\Big) \\ &\quad = \mathrm{Var}\big(h(Y_1, Y_2)\big) - \mathrm{Var}\big(h_1(Y_1)\big) - \mathrm{Var}\big(h_2(Y_2)\big). \end{aligned}$$

This shows that the term $\delta_2^2 - 2\delta_1^2$ appearing in the formula (5.3.4) equals $E\big((\varphi(X_1, X_2) - \varphi_1(X_1) - \varphi_1(X_2) + \nu)^2\big)$.

For any r.v. Y with $E\big(Y^2\big) < \infty$, let σ_Y denote the standard deviation of Y. An important consequence of (5.4.4) is the following corollary.

Corollary 5.4.1. *The following two statements* (a) *and* (b) *are equivalent.*

$$\text{(a)} \quad \frac{E\big[(S-\widehat{S})^2\big]}{\sigma^2_{\widehat{S}}} \to 0. \qquad \text{(b)} \quad \frac{\sigma_S^2}{\sigma^2_{\widehat{S}}} \to 1. \tag{5.4.6}$$

Either (5.4.6)(a) *or* (5.4.6)(b) *imply that*

$$\frac{|S-\widehat{S}|}{\sigma_S} + \frac{|S-\widehat{S}|}{\sigma_{\widehat{S}}} \to_p 0. \tag{5.4.7}$$

By Theorem 2.10.5, (5.4.7) implies that the asymptotic distribution of $\big(S - E(S)\big)/\sigma_S$ or $\big(S - E(S)\big)/\sigma_{\widehat{S}}$ is the same as that of $\big(\widehat{S} - E(\widehat{S})\big)/\sigma_{\widehat{S}}$. This is a rather important consequence. It says that under (a) or (b) of (5.4.6), the standardized version of any square-integrable statistic S can be approximated, in probability, by the standardized version of the sum $\widehat{S}$ of independent r.v.'s, to which typically a central limit theorem is applicable.

5.4.1 Asymptotic Normality of U-Statistics

We shall now illustrate the above projection method by proving the asymptotic normality of a sequence of U-statistics. Here we treat the case of a kernel which does not vary with the sample size. The more complicated case of a kernel which changes with the sample size will be treated in a later section. Since U-statistics of order one are averages and their behavior is well understood, we assume that r is greater than one.

Let the p-dimensional random vectors $X_1, X_2, \cdots$ be i.i.d. and let

$$U_n = \frac{1}{\binom{n}{r}} \sum_{b \in B_{r,n}} \varphi(X_{b_1}, \cdots, X_{b_r})$$

be a sequence of U-statistics of order r. Assume $E\big(\varphi^2(X_1, \cdots, X_r)\big)$ is finite and set $\nu = E\big(\varphi(X_1, \cdots, X_r)\big)$. Recall the definition of φ_1 from (5.3.1). We claim that the projection of U_n onto $\mathcal{H}(X_1, \cdots, X_n)$ is

$$\widehat{U}_n = \nu + \frac{r}{n} \sum_{i=1}^{n} \big(\varphi_1(X_i) - \nu\big). \tag{5.4.8}$$

Indeed, for $b \in B_{r,n}$, the conditional expectation of $\varphi(X_{b_1}, \cdots, X_{b_r}) - \nu$, given X_i, is 0 if i does not belong to the set $\{b_1, \cdots, b_r\}$ and equals $\varphi_1(X_i) - \nu$, otherwise. Note also that there are $\binom{n-1}{r-1}$ subsets of $\{1, \cdots, n\}$ with r elements that contain i. Use these properties and the linearity of conditional expectations to obtain that, for all $1 \le i \le n$,

$$E\big(U_n - \nu \big| X_i\big) = \frac{\binom{n-1}{r-1}}{\binom{n}{r}} \big(\varphi_1(X_i) - \nu\big) = \frac{r}{n}\big(\varphi_1(X_i) - \nu\big).$$

This establishes the claim (5.4.8) in view of the formula (5.4.5).

Theorem 5.4.1. *Suppose $E\big(\varphi^2(X_1, \cdots, X_r)\big) < \infty$ and $\delta_1^2 > 0$. Then the following hold:*

$$nE\big[(U_n - \widehat{U}_n)^2\big] \to 0, \qquad U_n = \widehat{U}_n + o_p\big(n^{-1/2}\big), \tag{5.4.9}$$

$$n^{1/2}\big(U_n - \nu\big)/\big(r\delta_1\big) \to_D \mathcal{N}(0,1). \tag{5.4.10}$$

Proof. The identity $\mathrm{Var}(\widehat{U}_n) = r^2\delta_1^2/n$ yields $\mathrm{Var}(U_n)/\mathrm{Var}(\widehat{U}_n) \to 1$, by (5.3.8). Hence, (5.4.9) follows from Corollary 5.4.1. By the classical CLT, $(\widehat{U} - \nu)/\big(\frac{r^2}{n}\delta_1\big)^{1/2} \to_D \mathcal{N}(0,1)$. The assertion (5.4.10) follows from this, the second conclusion in (5.4.9) and Theorem 2.10.5. □

Example 5.2.1 continued. Write $U_n^{(1)}$ for the $U^{(1)}$ of Example 5.2.1. We shall now derive the limiting distribution of $U_n^{(1)}$ in the symmetric location model. This means that $X_i = \theta + \varepsilon_i$, $i = 1, \cdots, n$ for some $\theta \in \mathbb{R}$ and the errors $\varepsilon_1, \varepsilon_2, \cdots$ are i.i.d. F, for some $F \in \mathcal{F}_s$. Thus F is continuous and $1 - F(-x) = F(x)$, $\forall x \in \mathbb{R}$. The statistic $U_n^{(1)}$ can then be expressed in terms of the errors as

$$U_n^{(1)} = \frac{2}{n(n-1)} \sum_{1 \le i < j \le n} I\big(\varepsilon_i + \varepsilon_j > -2\theta\big).$$

Here, the kernel is $\varphi(x,y) = I(x + y > -2\theta) = I\big(y > -(x + 2\theta)\big)$. Clearly, $\varphi(x,y) = \varphi(y,x)$, $\forall\, x, y \in \mathbb{R}$. By the continuity and symmetry of F, $P\big(\varepsilon_2 > -(x + 2\theta)\big) = F(x + 2\theta)$ and

$$\varphi_1(x) = F(x + 2\theta), \quad \nu = \int F(x + 2\theta)\, dF(x), \quad \delta_1^2 = \int F^2(x + 2\theta) dF(x) - \nu^2. \tag{5.4.11}$$

Hence the projection of $U_n^{(1)}$ is given by

$$\widehat{U}_n^{(1)} = \nu + \frac{2}{n}\sum_{i=1}^{n}\left[F(\varepsilon_i + 2\theta) - \nu\right].$$

Since F is continuous, we have $\delta_1^2 > 0$. Thus Theorem 5.4.1 yields $n^{1/2}\big(U_n^{(1)} - \nu\big) \to_D \mathcal{N}\big(0, 4\delta_1^2\big)$.

If $\theta = 0$, then $\nu = 1/2$, $\delta_1^2 = \int F^2(x)dF(x) - 1/4 = 1/12$, $4\delta_1^2 = 1/3$ and we obtain $(3n)^{1/2}(U^{(1)} - 1/2) \to_D \mathcal{N}(0,1)$. Hence, a large sample-level α test based on $U_n^{(1)}$ rejects H_0 : $\theta = 0$ in favor of $H_1 : \theta > 0$ if

$$U_n^{(1)} \geq \frac{1}{2} + \frac{z_\alpha}{(3n)^{1/2}}, \tag{5.4.12}$$

where z_α is as in (3.3.9). It will be shown later that this test is locally regular at 0 if F has a square-integrable density.

Example 5.2.2 continued. Recall that the sample variance s_n^2 can be expressed as the U-statistic $U^{(2)}$ which has kernel $\varphi(x,y) = (x-y)^2/2$. Assume that X_1 has a finite fourth moment. Let μ denote the mean, σ the standard deviation, and μ_4 the fourth central moment of X_1. We have already seen that $2\varphi_1(x) = (x-\mu)^2 - \sigma^2$. Assume that $4\delta_1^2 = \mathrm{Var}((X_1-\mu)^2) = \mu_4 - \sigma^4$ is positive. Then Theorem 5.4.1 implies the following asymptotic normality result for the sample variance,

$$n^{1/2}\big(s_n^2 - \sigma^2\big) \to_D \mathcal{N}\big(0, \mu_4 - \sigma^4\big).$$

5.4.2 Asymptotic Normality of W_s

We shall now use the Projection Lemma 5.4.1 to derive the limiting null distribution of the Wilcoxon signed rank statistic W_s of (4.3.12). To be precise, we shall prove the following theorem.

Theorem 5.4.2. *Suppose F is continuous and symmetric about zero and $X_1, \cdots, X_n$ are i.i.d. F. Then $W_s/\sqrt{\frac{n(n+1)(2n+1)}{6}} \to_D \mathcal{N}(0,1)$.*

Proof. Let F^+ denote the d.f. of $|X|$. Because F is continuous and symmetric about zero, $F^+(x) = \big(2F(x) - 1\big)I(x \geq 0)$. By Lemma 4.3.3, $E(W_s) = 0$ and

$$\sigma_{W_s}^2 := \mathrm{Var}(W_s) = \frac{n(n+1)(2n+1)}{6}.$$

Set $S_i := \mathrm{sgn}(X_i)$, $i = 1, \cdots, n$. We shall compute the projection of W_s onto $\mathcal{H}\big((S_1, |X_1|), \cdots, (S_n, |X_n|)\big)$. For this, rewrite

$$W_s = \sum_{k=1}^{n} S_k R_k^+ = \sum_{k=1}^{n} S_k \sum_{j=1}^{n} I(|X_j| \leq |X_k|) = \sum_{k=1}^{n}\sum_{j=1}^{n} S_k I(|X_j| \leq |X_k|).$$

By Lemma 4.3.2, $S_1, \cdots, S_n, |X_1|, \cdots, |X_n|$ are independent. Therefore, for all $1 \leq i,j,k \leq n$,

$$E\big[S_k I(|X_j| \leq |X_k|)\big|S_i, |X_i|\big] = E(S_k|S_i)E\big(I(|X_j| \leq |X_k|)\big||X_i|\big),$$
$$E(S_k|S_i) = S_i I(k=i), \quad E\big(I(|X_j| \leq |X_i|)\big||X_i|\big) = I(j=i) + F^+\big(|X_i|\big)I(j \neq i).$$

Here we also used $E(S_k) = 0$. The above facts give the identity

$$E(W_s|X_i) = S_i\big[1 + (n-1)F^+\big(|X_i|\big)\big], \quad 1 \leq i \leq n.$$

Therefore, in view of $E(W_s) = 0$, the projection of W_s into $\mathcal{H}\big((S_1, |X_1|), \cdots, (S_n, |X_n|)\big)$ is

$$\widehat{W}_s = \sum_{i=1}^{n} E(W_s|X_i) = \sum_{i=1}^{n} S_i\big[1 + (n-1)F^+(|X_i|)\big].$$

Since F^+ is continuous, $F^+(|X_1|), \cdots, F^+(|X_n|)$ are i.i.d. $\mathrm{Unif}(0,1)$ r.v.'s having mean $1/2$ and second moment $1/3$. Using these facts we calculate

$$\sigma^2_{\widehat{W}_s} := \mathrm{Var}(\widehat{W}_s) = nE\big[1 + (n-1)F^+(|X_1|)\big]^2 = \frac{n(n^2+n+1)}{3}.$$

Consequently, the ratio

$$\frac{\sigma^2_{W_s}}{\sigma^2_{\widehat{W}_s}} = \frac{3n(n+1)(2n+1)}{6n(n^2+n+1)} \to 1.$$

Now, let

$$Y_{n,i} = \frac{\sqrt{3}S_i\big[1 + (n-1)F^+(|X_i|)\big]}{\sqrt{n(n^2+n+1)}}, \quad i = 1, \cdots, n.$$

The r.v.'s $Y_{n,1}, \cdots, Y_{n,n}$ are independent, $E\big(Y_{ni}\big) = 0, i = 1, \cdots, n$, $\sum_{i=1}^{n} \mathrm{Var}\big(Y_{ni}\big) = 1$, and $\sum_{i=1}^{n} E\big(Y_{ni}^2 I(|Y_{ni}| > \epsilon\big) \to 0$, for every $\epsilon > 0$, because $Y_{n,i}^2 \le 3/n$ for $i = 1, \cdots, n$. Thus Corollary 2.10.4 implies $\widehat{W}_s/\sigma_{\widehat{W}_s} = \sum_{i=1}^{n} Y_{n,i} \to_D \mathcal{N}(0,1)$. Therefore, by Corollary 5.4.1 and Theorem 2.10.5, $W_s/\sigma_{W_s} \to_D \mathcal{N}(0,1)$. □

5.4.3 Asymptotic Normality of a Simple Linear Rank Statistic

In this subsection we shall derive the asymptotic normality of a sequence of simple linear rank statistics. Such a result is useful in tests about regression parameters in linear regression models.

Let the r.v.'s $Y_1, \cdots, Y_n$ be i.i.d. F. Let R_i be the rank of Y_i among $Y_1, \cdots, Y_n$. Let $\{c_{n,i}; 1 \le i \le n\}$ be a triangular array of real numbers and define

$$S := \frac{1}{n+1}\sum_{k=1}^{n} c_{n,k}R_k.$$

Theorem 5.4.3. *Suppose F is a continuous d.f. and $c_{n,1}, \cdots, c_{n,n}$ are real numbers satisfying*

$$\sum_{i=1}^{n} c_{n,i} = 0, \quad \sum_{i=1}^{n} c_{n,i}^2 = 1, \quad \max_{1\le i\le n} |c_{n,i}| \to 0. \tag{5.4.13}$$

Then, $S \to_D \mathcal{N}\big(0, 1/12\big)$.

Proof. . We shall use Corollary 5.4.1 to prove this lemma. To compute the projection $\widehat{S}$ onto $\mathcal{H}(Y_1, \cdots, Y_n)$, we begin by noting that $E\big(I(Y_j \le Y_i)\big|Y_i\big)$ equals 1 for $j = i$ and $F(Y_i)$ for $j \ne i$, and that for $k \ne i$, $E\big(I(Y_j \le Y_k)\big|Y_i\big)$ equals 1 for $j = k$, $1 - F(Y_i)$ for $j = i$, and $1/2$ otherwise. Use these identities and $R_k = \sum_{j=1}^{n} I(Y_j \le Y_k)$ to calculate

$$\begin{aligned} E\big(R_k\big|Y_i\big) &= 1 + (n-1)F(Y_i), & k = i, \\ &= 1 + (1 - F(Y_i)) + \frac{(n-2)}{2} = 1 + \frac{n - 2F(Y_i)}{2}, & k \ne i. \end{aligned}$$

Note that $\sum_{i=1}^n c_{ni} = 0$ implies that $\sum_{k=1}^n I(k \neq i)c_{nk} = -c_{ni}$, and in view of (3.2.12), also $E(S) = 0$. Using these properties we derive

$$\begin{aligned}
\widehat{S} &= \sum_{i=1}^n \sum_{k=1}^n c_{n,k} \frac{E(R_k|Y_i)}{n+1} = \sum_{i=1}^n \sum_{k=1}^n c_{n,k} \frac{E(R_k|Y_i) - 1}{n+1} \\
&= \sum_{i=1}^n c_{n,i} \frac{n-1}{n+1} F(Y_i) + \sum_{i=1}^n \sum_{k=1}^n I(k \neq i) c_{n,k} \frac{n - 2F(Y_i)}{2(n+1)} \\
&= \frac{n-1}{n+1} \sum_{i=1}^n c_{n,i} F(Y_i) - \sum_{i=1}^n c_{n,i} \frac{n - 2F(Y_i)}{2(n+1)} = \frac{n}{n+1} \sum_{i=1}^n c_{ni} \big(F(Y_i) - 1/2\big).
\end{aligned}$$

Since F is continuous, $F(Y_1), \cdots, F(Y_n)$ are i.i.d. Unif$(0,1)$ with mean $1/2$ and variance $1/12$. Hence, $\sigma^2_{\widehat{S}} = \big(n/(n+1)\big)^2 \sum_{i=1}^n c^2_{ni}/12 = \big(n/(n+1)\big)^2/12$. From (3.2.12) applied with $\psi_n(i) = i/(n+1)$ and the given c_{ni}'s,

$$\sigma^2_S = \frac{1}{n-1} \sum_{i=1}^n \Big(\frac{i}{n+1} - \frac{1}{2}\Big)^2 = \frac{n}{12(n+1)}.$$

Hence, $\sigma^2_{\widehat{S}}/\sigma^2_S = (n+1)/n \to 1$.

Next, we need to derive the limiting distribution of $\widehat{S}$. Apply Corollary 2.10.5 with the present c_{ni} and $\zeta_i = F(Y_i) - 1/2$ to conclude $\sum_{i=1}^n c_{ni}\big(F(Y_i) - 1/2\big) \to_D \mathcal{N}\big(0, 1/12\big)$. In view of $n/(n+1) \to 1$, Theorem 2.10.5 implies that $\widehat{S} \to_D \mathcal{N}(0, 1/12)$. By Corollary 5.4.1, the same conclusion holds for S. □

5.5 U-Statistics with Varying Kernels

In this section we study the asymptotic behavior of U-statistics when the kernel is allowed to vary with the sample size and the observations are coming from a triangular array with i.i.d. rows. We only consider the case $r > 1$ as the behavior of averages is well understood.

Let $r \geq 2$ be a known integer. For each $n \geq 2r - 1$, let φ_n be a kernel of order r satisfying (5.2.1), $X_{n,1}, \cdots, X_{n,n}$ be i.i.d. p-dimensional random vectors with joint distribution function F, and U_n be the U-statistic

$$U_n = \frac{1}{\binom{n}{r}} \sum_{b \in B_{r,n}} \varphi_n\big(X_{n,b_1}, \cdots, X_{n,b_r}\big).$$

Assume the kernel φ_n satisfies

$$E\big(\varphi_n^2(X_{n,1}, \cdots, X_{n,r})\big) < \infty, \quad \forall n \geq r. \tag{5.5.1}$$

Set $\nu_n := E(U_n) = E\big(\varphi_n(X_{n,1}, \cdots, X_{n,r_n})\big)$ and

$$\varphi_{n,1}(x) = E\big(\varphi_n(x, X_{n,2}, \cdots, X_{n,r})\big), \quad x \in \mathbb{R}; \qquad \delta^2_{n,1} = \mathrm{Var}\big(\varphi_{n,1}(X_{n,1})\big).$$

The projection $\widehat{U}_n$ of U_n onto $\mathcal{H}\big(X_{n,1}, \cdots, X_{n,n}\big)$ is given by

$$\widehat{U}_n = \nu_n + \frac{r}{n} \sum_{i=1}^n \Big(\varphi_{n,1}(X_{n,i}) - \nu_n\Big).$$

Under (5.3.10), $\mathrm{Var}(U_n)/\mathrm{Var}(\widehat{U}_n) \to 1$. Hence, Corollary 5.4.1 implies

$$\frac{nE\big[(U_n - \widehat{U}_n)^2\big]}{r^2\delta_{n,1}^2} \to 0, \qquad \frac{U_n - \nu_n}{r\delta_{n,1}} = \frac{\widehat{U}_n - \nu_n}{r\delta_{n,1}} + o_p\big(n^{-1/2}\big).$$

One can then use the Lindeberg-Feller CLT Theorem 2.10.9 to give sufficient conditions that imply $n^{1/2}\big(\widehat{U}_n - \nu_n\big)/\big(r\delta_{n,1}\big) \to_D \mathcal{N}\big(0,1\big)$. We shall not pursue this any further. Instead, we shall present a different approach as follows.

Let X have d.f. F. In practice there are situations where the sequence $\varphi_{n,1}(X)$ has a limit $\phi(X)$ in the mean squared error sense, i.e., for some measurable function ϕ with $E\big(\phi^2(X)\big) < \infty$,

$$\Delta_{n,2} := E\big[\big(\varphi_{n,1}(X) - \phi(X)\big)^2\big] \to 0.$$

Let $\mu := E(\phi(X))$ and $\tau_0^2 := \mathrm{Var}(\phi(X))$. Then it is possible to approximate the projection $\widehat{U}_n$ by its analog with $\varphi_{n,1}$ replaced by ϕ, namely

$$S_n := \mu + \frac{r}{n}\sum_{i=1}^{n}\Big(\phi(X_{n,i}) - \mu\Big).$$

The equations (5.4.3) and (5.4.4) give

$$\mathrm{Var}(U_n - S_n) = \mathrm{Var}(U_n) - \mathrm{Var}(\widehat{U}_n) + \mathrm{Var}(\widehat{U}_n - S_n). \tag{5.5.2}$$

As $\mathrm{Var}(\widehat{U}_n) = r^2\delta_{n,1}^2/n$ and $\mathrm{Var}(\widehat{U}_n - S_n) = r^2\mathrm{Var}\big(\varphi_{n,1}(X) - \phi(X)\big)/n$, the inequalities (5.3.5) and (5.3.7) then give the bound

$$\mathrm{Var}(U_n - S_n) \le \sum_{c=2}^{r}\frac{K_c 2^{c-1}\delta_{n,c}^2}{n^c} + \frac{r^2}{n}\Delta_{n,2}. \tag{5.5.3}$$

This can be used to obtain bounds on

$$E\big[\big(U_n - S_n - t_n\big)^2\big] = (\nu_n - \mu - t_n)^2 + \mathrm{Var}\big(U_n - S_n\big) \tag{5.5.4}$$

for any sequence $t_n \in \mathbb{R}$. If τ_0^2 is positive, then the CLT yields

$$n^{1/2}(S_n - \mu) \to \mathcal{N}(0, r^2\tau_0^2). \tag{5.5.5}$$

The next theorem makes the above derivations precise.

Theorem 5.5.1. *Suppose* (5.5.1) *holds and there are a* $\gamma \in \mathbb{R}$ *and a r.v.* $\phi(X)$ *with finite mean* μ *and positive and finite variance* τ_0^2 *such that the following conditions are met.*

$$\Delta_{n,1} := \sqrt{n}\big(\nu_n - \mu\big) \to \gamma. \tag{5.5.6}$$

$$\Delta_{n,2} := E\big[\big(\varphi_{n,1}(X) - \phi(X)\big)^2\big] \to 0. \tag{5.5.7}$$

$$R_{n,c} := n^{1-c}\delta_{n,c}^2 \to 0, \qquad c = 2, \cdots, r. \tag{5.5.8}$$

Then the following hold.

$$\mathcal{D}_n := nE\big[\big(U_n - S_n - n^{-1/2}\gamma\big)^2\big] \to 0. \tag{5.5.9}$$

$$n^{1/2}\big(U_n - \mu\big) = n^{1/2}\big(S_n - \mu\big) + \gamma + o_p(1), \tag{5.5.10}$$

$$n^{1/2}\big(U_n - \mu\big) \to_D \mathcal{N}\big(\gamma, r^2\tau_0^2\big). \tag{5.5.11}$$

Proof. The identity (5.5.4) with $t_n = n^{-1/2}\gamma$ and the inequality (5.5.3) give the bound

$$\mathcal{D}_n \le (\Delta_{n,1} - \gamma)^2 + \sum_{c=2}^{r} 2^{c-1} K_c R_{n,c} + r^2 \Delta_{n,2}.$$

Thus (5.5.9) follows from (5.5.6)–(5.5.8). It implies (5.5.10). Finally, (5.5.11) follows from (5.5.10) and (5.5.5). □

Remark 5.5.1. For $r = 2$, the condition (5.5.8) simplifies to $\delta^2_{n,2}/n \to 0$ and is implied by $E\big(\varphi_n^2(X_1, X_2)\big) = o(n)$. For $r = 2$, by (5.3.4), (5.5.2) and (5.5.4), the following inequality holds:

$$E\big[(U_n - S_n - t_n)^2\big] \le (\nu_n - \mu - t_n)^2 + \frac{2E(\varphi_n^2(X_1, X_2))}{n(n-1)} + \frac{4\Delta_{n,2}}{n}, \quad \forall t_n \in \mathbb{R}.$$

This inequality will be used in the next section to give rates of convergence for the U-statistic $U_n^{(3)}$.

Example 5.2.1 continued. We shall illustrate the above theorem by the following example. Suppose the observations $X_1, X_2, \cdots$ are from the symmetric location model (4.3.2) where the error d.f. $F \in \mathcal{F}_s$ is now assumed to have a square integrable density f. We shall now use the above theorem to show that the test $\psi_n := I\big((3n)^{1/2}(U_n^{(1)} - 1/2) \ge z_\alpha\big)$ given at (5.4.12) for testing $H_0 : \theta = 0$ versus $H_1 : \theta > 0$ is locally regular at 0 with slope $\gamma_0 = \sqrt{12} I_2(f) = \sqrt{12} \int f^2(x)\, dx$.

Let $\theta_n \ge 0$ be a sequence such that $\delta_n := n^{1/2}(\theta_n - 0) \to \delta$, for some $\delta \ge 0$. We need to show that

$$\lim_n P_{\theta_n}\Big((3n)^{1/2}(U_n^{(1)} - 1/2) \ge z_\alpha\Big) = 1 - \Phi(z_\alpha - \delta\gamma_0). \tag{5.5.12}$$

This will follow if we show that, under P_{θ_n},

$$n^{1/2}(U_n^{(1)} - 1/2) \to_D \mathcal{N}\big(2\delta I_2(f), 1/3\big). \tag{5.5.13}$$

Under P_{θ_n}, $\varepsilon_{n,1} := X_1 - n^{-1/2}\delta_n, \cdots, \varepsilon_{n,n} := X_n - n^{-1/2}\delta_n$ are i.i.d. with density f and

$$U_n^{(1)} = U_n := \frac{2}{n(n-1)} \sum_{1 \le i < j \le n} I\big(\varepsilon_{n,i} + \varepsilon_{n,j} > -2n^{-1/2}\delta_n\big).$$

The kernel $\varphi_n(x, y) = I\big(x + y > -2n^{-1/2}\delta_n\big)$ depends on n. Analogous to (5.4.11),

$$\begin{aligned}
\varphi_{n,1}(x) &= F(x + 2n^{-1/2}\delta_n), \quad x \in \mathbb{R},\\
\nu_n &= E_{\theta_n}\big(\varphi_n(\varepsilon_{n,1}, \varepsilon_{n,2})\big) = \int F(x + 2n^{-1/2}\delta_n) f(x)\, dx,\\
\delta^2_{n,1} &= \int F^2(x + 2n^{-1/2}\delta_n) f(x)\, dx - \nu_n^2.
\end{aligned}$$

Let ε be a r.v. with density f. To prove (5.5.13) we apply Theorem 5.5.1 with $X_{n,i} = \varepsilon_{n,i}$ and $\phi = F$ so that $\mu = E(F(\varepsilon)) = 1/2$ and $\tau_0^2 = \mathrm{Var}(F(\varepsilon)) = 1/12$. To apply the theorem we need to establish the following three assertions.

$$\sqrt{n}(\nu_n - \mu) \to 2\delta I_2(f), \quad \delta^2_{n,2}/n \to 0, \quad E\big[(F(\varepsilon + 2n^{-1/2}\delta_n) - F(\varepsilon))^2\big] \to 0.$$

The second assertion holds because $\delta^2_{n,2}$ is bounded by 1, the third assertion follows from Theorem 2.10.4 as F is bounded and continuous. To verify the first assertion, let

$$H(s) := \int \big|f(x + s) - f(x)\big| f(x) dx, \; s \in \mathbb{R}.$$

By (2.4.4), $H^2(s) \le \int [f(x+s) - f(x)]^2 dx \int f^2(x) dx$, $s \in \mathbb{R}$. Hence, by Lemma 2.10.10 and the square integrability of f, $H(s)$ is bounded in s and has limit 0 at 0.

Using this, we obtain the convergence

$$\begin{aligned}
&\sqrt{n}(\nu_n - \mu) - 2\delta_n \int f^2(x)\, dx \\
&= \int n^{1/2} \Big[F(x + 2n^{-1/2}\delta_n) - F(x) - 2n^{-1/2}\delta_n f(x) \Big] f(x)\, dx \\
&= 2\delta_n \int \int_0^1 \big[f(x + 2n^{-1/2}\delta_n s) - f(x) \big]\, ds f(x)\, dx \to 0.
\end{aligned}$$

The first assertion follows from this and $\delta_n \to \delta$. Thus Theorem 5.5.1 applies and yields (5.5.13). We summarize this result for an easy reference later as the following lemma.

Lemma 5.5.1. *Suppose $X_1, \cdots, X_n$ is random sample from the location model (4.3.2) with a symmetric about zero and square integrable error density f. Then, the sequence of tests $\psi_n := I\big((3n)^{1/2}(U_n^{(1)} - 1/2) \ge z_\alpha\big)$ for testing $H_0 : \theta = 0$ versus $\theta > 0$ is locally regular at 0 with slope $\gamma_0 := \sqrt{12} \int f^2(x) dx$.*

Note that we do not assume f to be continuous in the above derivation, i.e., f is only an almost everywhere derivative of F, not its derivative.

5.6 Asymptotic Behavior of $U_n^{(3)}$

In this section we derive various results about the large sample behavior of the U-statistic $U_n^{(3)}$ of (5.2.3), i.e.,

$$U_n^{(3)} = \frac{2}{n(n-1)h_n} \sum_{1 \le i < j \le n} K\Big(\frac{X_i - X_j}{h_n}\Big),$$

where K is a square integrable density symmetric about zero, h_n is a positive number, and the r.v.'s $X_1, \cdots, X_n$ are i.i.d. with density f. Then $X_1 - X_2$ has density g given by

$$g(y) = \int f(x+y) f(x)\, dx, \quad y \in \mathbb{R}.$$

We have already seen that $U_n^{(3)}$ has mean $\nu_n = \int g(h_n t) K(t)\, dt$. We calculate

$$\begin{aligned}
\phi_{n,1}(x) &= \int \frac{1}{h_n} K\Big(\frac{x-y}{h_n}\Big) f(y)\, dy = \int f(x - h_n t) K(t)\, dt \\
h_n E(\varphi_n^2(X_1, X_2)) &= \int \frac{1}{h_n} K^2\big(\frac{y}{h_n}\big) g(y)\, dy = \int g(h_n t) K^2(t)\, dt,
\end{aligned}$$

Set

$$S_n^{(3)} = g(0) + \frac{2}{n} \sum_{i=1}^n \big(f(X_i) - g(0)\big). \tag{5.6.1}$$

Now introduce the following functions:

$$\begin{aligned}
H_0(y) &:= \int \big(f(x+y) - f(x)\big)^2 f(x)\, dx, \quad y \in \mathbb{R}. \\
H_j(y) &:= \int |f(x+y) - f(x)|^j\, dx, \quad y \in \mathbb{R}, j = 1, 2, 3,
\end{aligned}$$

$$b_k(h) := \int H_k(ht)K(t)\,dt, \qquad h \geq 0,\ k = 0,1,2,3.$$

Moreover, for any square integrable function ψ, set $I_2(\psi) := \int \psi^2(x)\,dx$. The following two lemmas are key to our analysis.

Lemma 5.6.1. *If f^2 is integrable, then*

$$g(y) = g(0) - \frac{1}{2}H_2(y) = g(0) - \frac{1}{2}\int \big(f(x+y) - f(x)\big)^2\,dx$$

holds for all $y \in \mathbb{R}$. Thus g is bounded by $g(0)$ and is uniformly continuous.

Proof. Fix $y \in \mathbb{R}$. The assumption $I_2(f) < \infty$ guarantees the finiteness of the following integrals. The substitution $u = x + y$ yields

$$\begin{aligned} g(-y) - g(0) &= \int \big(f(u-y) - f(u)\big)f(u)\,du \\ &= \int \big(f(x) - f(x+y)\big)f(x+y)\,dx = g(y) - g(0). \end{aligned}$$

Therefore, $2(g(y) - g(0)) = g(y) - g(0) + g(-y) - g(0)$ equals

$$\begin{aligned} \int \big(f(x+y) - f(x)\big)f(x)\,dx &- \int \big(f(x+y) - f(x)\big)f(x+y)\,dx \\ &= -\int \big(f(x+y) - f(x)\big)^2\,dx. \end{aligned}$$

This proves the identity. Thus g is bounded by $g(0)$ and g inherits uniform continuity from H_2, see Lemma 2.10.10. □

Lemma 5.6.2. *If f^3 is integrable, then H_0 is bounded and uniformly continuous. Moreover, $b_k(h_n) \to 0$ as $h_n \to 0$ for each $k = 0,1,2,3$.*

Proof. Assume f^3 is integrable. An application of the Hölder inequality (2.4.5) with $p = 3/2$ and $q = 3$ yields the inequality

$$H_0(y) \leq \Big(\int |f(x+y) - f(x)|^3\,dx\Big)^{2/3}\Big(\int f^3(x)\,dx\Big)^{1/3}.$$

This gives the first claim as H_3 is bounded and uniformly continuous by Lemma 2.10.10. It follows from (2.4.4) that $f^2 = f^{3/2}f^{1/2}$ is integrable. Indeed, $(I_2(f))^2 \leq \int f^3(x)\,dx \int f(x)\,dx$. Similarly, from Lemma 2.10.10, it follows that the function H_k is bounded and continuous at 0, for each $k = 1,2$. Hence, by Theorem 2.10.10, $b_k(h_n) \to 0$ as $h_n \to 0$, for each $k = 0,1,2,3$. □

Let X have density f. Assume from now on that f^3 is integrable. Then $f(X)$ has finite second moment $\int f^3(x)\,dx$ and thus finite mean $g(0) = I_2(f)$ and finite variance

$$\tau^2 := \mathrm{Var}(f(X)) = \int f^3(x)dx - \Big(\int f^2(x)dx\Big)^2.$$

The formula for ν_n and Lemma 5.6.1 give the identity

$$\nu_n - g(0) = \int \big(g(h_n t) - g(0)\big)K(t)\,dt = -\frac{1}{2}\int H_2(h_n t)K(t)\,dt = -\frac{b_2(h_n)}{2}.$$

Since g is maximized by $g(0)$ and $I_2(K)$ is finite, we derive the bound

$$h_n E(\varphi_n^2(X_1,X_2)) = \int g(h_n t)K^2(t)\,dt \le g(0)I_2(K).$$

Next, we show

$$\Delta_{n,2} = E\Big[\big(\varphi_{n,1}(X) - f(X)\big)^2\Big] \le b_0(h_n).$$

Using the moment inequality, Fubini's theorem and the symmetry about zero of K, we calculate

$$\begin{aligned}\Delta_{n,2} &= \int \Big(\int (f(x-h_n t) - f(x))K(t)\,dt\Big)^2 f(x)\,dx \\ &\le \iint (f(x-h_n t) - f(x))^2 K(t)\,dt\, f(x)\,dx \\ &= \int H_0(-h_n t)K(t)\,dt = \int H_0(h_n t)K(t)\,dt = b_0(h_n).\end{aligned}$$

The inequality in Remark 5.5.1 applied with $t_n = 0$ gives

$$E\Big[\big(U_n^{(3)} - S_n^{(3)}\big)^2\Big] \le \Big(\frac{b_2(h_n)}{2}\Big)^2 + \frac{2g(0)I_2(K)}{n(n-1)h_n} + \frac{4b_0(h_n)}{n}. \tag{5.6.2}$$

The right-hand side converges to zero if $h_n \to 0$ and $n^2 h_n \to \infty$. For such h_n, the U-statistic $U_n^{(3)}$ is a consistent estimator of $I_2(f)$, i.e., $U_n^{(3)} \to_p I_2(f)$.

Theorem 5.6.1. *Suppose f^3 is integrable and the inequality*

$$H_2(y) = \int \big(f(x+y) - f(x)\big)^2\,dx \le C_2|y|^{2\beta}, \quad \forall\, y \in \mathbb{R}, \tag{5.6.3}$$

holds for some $0 < \beta \le 1$ and some constant C_2. Suppose furthermore that $\int |t|^{2\beta}K(t)\,dt$ is finite. Then the following hold.

(a) *If $\beta > 1/4$ and h_n satisfies $nh_n^{4\beta} \to 0$ and $nh_n \to \infty$, then*

$$\begin{aligned} n^{1/2}\big(U_n^{(3)} - I_2(f)\big) &\to_D \mathcal{N}(0, 4\tau^2), && \text{if } \tau^2 > 0, \\ &\to_p 0, && \text{if } \tau^2 = 0.\end{aligned}$$

(b) *If $\beta \le 1/4$ and h_n is proportional to $n^{-2/(4\beta+1)}$, then*

$$U_n^{(3)} = I_2(f) + O_p\big(n^{-4\beta/(4\beta+1)}\big).$$

Proof. From the inequality (5.6.3) we derive

$$b_2(h_n) \le C_2 h_n^{2\beta} \int |t|^{2\beta}K(t)\,dt.$$

Using this and $h_n \to 0$ we derive from (5.6.2) the rate

$$U_n^{(3)} = S_n^{(3)} + O_p(h_n^{2\beta}) + O_p(n^{-1}h_n^{-1/2}) + o_p(n^{-1/2}).$$

Note also that $T_n := n^{1/2}(S_n^{(3)} - I_2(f)) \to_D \mathcal{N}(0,\tau^2)$ holds by the classical CLT if τ^2 is positive, while $T_n = 0$ with probability one if $\tau^2 = 0$. Thus, under the assumptions of part (a), we have $U_n^{(3)} = S_n^{(3)} + o_p(n^{-1/2})$ and the conclusion of part (a) is immediate.

Under the assumptions of part (b), $h_n^{2\beta}$ and $n^{-1}h_n^{-1/2}$ are of order $n^{-4\beta/(4\beta+1)}$, and we have the bound $4\beta/(4\beta+1) \le 1/2$. This, in turn, yields $U_n^{(3)} = S_n^{(3)} + O_p\big(n^{-4\beta/(4\beta+1)}\big) = I_2(f) + O_p\big(n^{-4\beta/(4\beta+1)}\big)$. □

The next lemma provides a sufficient condition for (5.6.3) to hold with $\beta = 1$.

Lemma 5.6.3. *Suppose there is a square integrable function $\dot{f}$ such that*

$$f(x_2) - f(x_1) = \int_{x_1}^{x_2} \dot{f}(t)\,dt, \quad \forall x_1, x_2 \in \mathbb{R}, x_1 < x_2,$$

holds. Then f is bounded and $H_2(y) \le I_2(\dot{f})y^2$ holds for all $y \in \mathbb{R}$.

Proof. The Cauchy-Schwarz inequality (2.4.4) yields

$$\big(f(x_2) - f(x_1)\big)^2 \le (x_2 - x_1) I_2(\dot{f}), \quad x_1 < x_2.$$

Thus the density f is uniformly continuous and therefore bounded by Lemma 2.11.2. For $y \in \mathbb{R}$, we use the identity

$$f(x+y) - f(x) = y \int_0^1 \dot{f}(x+sy)\,ds$$

to compute

$$H_2(y) = y^2 \int \Big(\int_0^1 \dot{f}(x+sy)\,ds \Big)^2 dx \le y^2 \int \int_0^1 \big(\dot{f}(x+sy)\big)^2\,ds\,dx.$$

Now, use the Fubini Theorem 2.3.2, change the order of integration and use the substitution $u = x + sy$ to conclude $H_2(y) \le y^2 I_2(\dot{f})$. □

Under the assumptions of Lemma 5.6.3, f is bounded, and this implies the inequality $H_0(y) \le \|f\|_\infty H_2(y)$, $\forall y \in \mathbb{R}$. Thus, if K has a finite second moment κ_2, we derive $b_2(h_n) \le \kappa_2 I_2(\dot{f}) h_n^2$ and $b_0(h_n) \le \|f\|_\infty b_2(h_n)$ and obtain from (5.6.2) the inequality

$$E\Big[\big(U_n^{(3)} - S_n^{(3)}\big)^2\Big] \le \Big(\frac{\kappa_2 I_2(\dot{f})}{2}\Big)^2 h_n^4 + \frac{2g(0)I_2(K)}{n(n-1)h_n} + \frac{4\|f\|_\infty \kappa_2 I_2(\dot{f}) h_n^2}{n}.$$

If we take h_n proportional to $n^{-2/5}$, then we obtain

$$E\Big[\big(U_n^{(3)} - S_n^{(3)}\big)^2\Big] = O(n^{-8/5}).$$

To proceed further, we need the following definition. Any real-valued integrable function ψ defined on $\mathbb{R}$ is said to be Lipschitz in L_1, or in short L_1-Lipschitz, if

$$\int \big|\psi(x+y) - \psi(x)\big|\,dx \le C\,|y|, \quad \forall\, y \in \mathbb{R} \text{ and some constant } 0 < C < \infty. \tag{5.6.4}$$

If f is bounded by say B, and L_1-Lipschitz with constant C, then we have the inequality $H_0(y) \le BH_2(y) \le 2B^2 H_1(y) \le 2B^2 C|y|$, for all real y. If also $\kappa := \int |t| K(t)\,dt$ is finite, then these inequalities and (5.6.2) yield

$$E\Big[\big(U_n^{(3)} - S_n^{(3)}\big)^2\Big] \le (BC\,\kappa)^2 h_n^2 + \frac{2g(0)I_2(K)}{n(n-1)h_n} + \frac{8B^2 C \kappa h_n}{n}.$$

The right-hand side is of order $O(n^{-4/3})$ if we choose h_n proportional to $n^{-2/3}$. Thus we have the following result.

Theorem 5.6.2. *Assume f is bounded and L_1-Lipschitz and K has a finite mean. Then the following are true.*
(a) *If τ^2 is positive and h_n satisfies $nh_n \to \infty$ and $nh_n^2 \to 0$, then*

$$n^{1/2}\big(U_n^{(3)} - I_2(f)\big) \to_D \mathcal{N}(0, \tau^2).$$

(b) *If h_n is proportional to $n^{-2/3}$, then $U_n^{(3)} = S_n^{(3)} + O_p(n^{-2/3})$.*

Note that uniform densities are bounded and L_1-Lipschitz. But the asymptotic normality result is not valid for these densities as they are almost surely constant and thus $\mathrm{Var}(f(X)) = 0$. For such densities we have $(U_n^{(3)} - I_2(f)) = O_p(n^{-2/3})$ if h_n is proportional to $n^{-2/3}$.

Densities that are L_1-Lipschitz need not be continuous. Indeed, every density of bounded variation is bounded and L_1-Lipschitz. The verification of this claim is left as an exercise. Thus Theorem 5.6.2 is valid for such densities.

When discussing bandwidth selection for kernel density estimators in Chapter 7, we encounter a linear combination of the present U-statistic with different densities, namely

$$\tilde{U}_n = \frac{2}{n(n-1)h_n} \sum_{1 \le i < j \le n} \tilde{K}\Big(\frac{X_i - X_j}{h_n}\Big),$$

where $\tilde{K} = 2K - K * K$, with $K * K$ the convolution of K with itself, i.e., $(K * K)(t) := \int K(t-x)K(x)\,dx$, $t \in \mathbb{R}$. Note that $\tilde{K}$ is no longer a density, but is still symmetric about zero and square integrable, the latter since $K * K$ is integrable and bounded by $I_2(K)$. The kernel of the U-statistic $\tilde{U}_n$ is $\tilde{\varphi}_n(x, y) = \tilde{K}((x-y)/h_n)/h_n$. We then have

$$\tilde{\nu}_n = E(\tilde{U}_n) = \int g(h_n t)\tilde{K}(t)\,dt,$$

$$E(\tilde{\varphi}_n^2(X_1, X_2)) = \int g(h_n t)\tilde{K}^2(t)\,dt \le g(0)I_2(\tilde{K}),$$

$$\tilde{\varphi}_{n,1}(x) = \int \tilde{\varphi}_n(x, y)f(y)\,dy = f(x - h_n t)\tilde{K}(t)\,dt.$$

From now on assume that f has continuous and square integrable first and second derivatives which we denote by $\dot{f}$ and $\ddot{f}$, and that K has a finite fourth moment. The former implies that f is bounded in view of Lemma 5.6.3, while the latter implies that the integrals

$$\tilde{\kappa}_j = \int t^j \tilde{K}(t)\,dt, \qquad \tilde{\kappa}_j^* = \int |t^j \tilde{K}(t)|\,dt, \quad j = 0, 1, 2, 3, 4,$$

are finite. With σ_K the standard deviation of K, one obtains the identities

$$\tilde{\kappa}_0 = 1, \quad \tilde{\kappa}_1 = \tilde{\kappa}_2 = \tilde{\kappa}_3 = 0, \quad \tilde{\kappa}_4 = -6\sigma_K^4. \tag{5.6.5}$$

We leave their verification as an exercise.

A Taylor expansion yields

$$f(x+v) - f(x) = v\dot{f}(x) + \frac{v^2}{2}\ddot{f}(x) + \frac{v^2}{2}\int \big(\ddot{f}(x+sv) - \ddot{f}(x)\big)q(s)\,ds \tag{5.6.6}$$

with $q(s) = 2(1-s)I(0 < s < 1)$. Note that q is a Beta(1,2)-density. Taking $v = -h_n t$ and integrating with respect to $\tilde{K}$ yields

$$\tilde{\varphi}_{n,1}(x) - f(x) = h_n^2 \iint t^2\big(\ddot{f}(x - sh_n t) - \ddot{f}(x)\big)q(s)\,ds\,\tilde{K}(t)\,dt.$$

This identity, the Cauchy-Schwarz inequality (2.4.4) and Theorem 2.3.2 imply the inequality

$$\big(\tilde{\varphi}_{n,1}(x) - f(x)\big)^2 \le h_n^4 \tilde{\kappa}_2^* \iint \big(\ddot{f}(x - sh_n t) - \ddot{f}(x)\big)^2 t^2 |\tilde{K}(t)|\,dt\,q(s)\,ds.$$

Since f is bounded by, say B, and $\ddot{f}$ is square integrable, Corollary 2.10.8 yields

$$E((\tilde{\varphi}_{n,1}(X) - f(X))^2) \le B \int (\varphi_{n,1}(x) - f(x))^2 \, dx = o(h_n^4).$$

Next we shall show that there is a bounded function R with limit 0 at 0 such that

$$H_2(y) = y^2 I_2(\dot{f}) + y^3 J_0 - \frac{y^4}{12} I_2(\ddot{f}) + y^4 R(y), \quad y \in \mathbb{R}, \tag{5.6.7}$$

holds with $J_0 := \int \dot{f}(x)\ddot{f}(x)\,dx$. Using the Taylor expansion (5.6.6) with $v = y$ we calculate

$$H_2(y) = y^2 I_2(\dot{f}) + y^3 J_0 + \frac{y^4}{4} I_2(\ddot{f}) + y^3 J_1(y) + \frac{y^4}{2} J_2(y) + \frac{y^4}{4} J_3(y)$$

with

$$J_1(y) := \iint \dot{f}(x)(\ddot{f}(x+sy) - \ddot{f}(x))\,dx\,q(s)\,ds,$$

$$J_2(y) := \iint \ddot{f}(x)(\ddot{f}(x+sy) - \ddot{f}(x))\,dx\,q(s)\,ds,$$

$$J_3(y) := \iiint (\ddot{f}(x+sy) - \ddot{f}(x))(\ddot{f}(x+ty) - \ddot{f}(x))\,dx\,q(s)q(t)\,ds\,dt.$$

Write $J_1(y)$ as the difference of two intergrals, make the substitution $u = x + sy$ in the first integral, switch u to x and then combine the integrals to obtain

$$J_1(y) = \iint (\dot{f}(x - sy) - \dot{f}(x))\ddot{f}(x)\,dx\,q(s)\,ds.$$

Now use the Taylor expansion

$$\dot{f}(x - sy) - \dot{f}(x) = -sy\ddot{f}(x) - sy \int_0^1 (\ddot{f}(x - vsy) - \ddot{f}(x))\,dv,$$

a change of the order of integration, and the fact that a Beta(1,2)-density has mean $1/3$ to arrive at $J_1(y) = -(yI_2(\ddot{f}))/3 - yJ_4(y)$ with

$$J_4(y) = \int_0^1 \iint \ddot{f}(x)(\ddot{f}(x - vsy) - \ddot{f}(x))\,dx\,sq(s)\,ds\,dv.$$

This shows that (5.6.7) holds with

$$R(y) = \frac{J_2(y)}{2} + \frac{J_3(y)}{4} - J_4(y).$$

Applications of the Cauchy-Schwarz inequality yield

$$(J_2(y))^2 \le I_2(\ddot{f}) \int W(-sy)q(s)\,ds, \qquad |J_3(y)| \le \int W(-sy)q(s)\,ds.$$

$$(J_4(y))^2 \le I_2(\ddot{f}) \int_0^1 \int W(vsy)\,s^2 q(s)\,ds\,dv,$$

with $W(u) = \int (\ddot{f}(x-u) - \ddot{f}(x))^2\,dx$, $u \in \mathbb{R}$. By Lemma 2.10.10, the square integrability of $\ddot{f}$ implies that W is bounded and has limit 0 at 0. The above bounds and Theorem 2.10.10 imply that R is bounded and has limit 0 at 0. This completes the proof of (5.6.7).

Using (5.6.5) and (5.6.7) we obtain the expansion

$$\tilde{\nu}_n - g(0) = -\frac{1}{2} \int H_2(h_n t)\tilde{K}(t)\,dt = -\frac{h_n^4 \sigma_K^4 I_2(\ddot{f})}{4} + o(h_n^4).$$

In view of the above we have the following results.

Theorem 5.6.3. *Suppose f has continuous and square integrable first and second derivatives and K has a finite fourth moment. Then the following hold.*
(a) If $nh_n \to \infty$ and $nh_n^8 \to 0$, then $\tilde{U}_n = S_n^{(3)} + o_p(n^{-1/2})$.
(b) If h_n is proportional to $n^{-1/5}$, then

$$\tilde{U}_n = S_n^{(3)} - h_n^4 \sigma_K^4 I_2(\ddot{f})/4 + o_p(h_n^4).$$

5.7 Two sample U-Statistics

To keep exposition transparent we shall present the two sample U-statistics for the case of r.v.'s only. Accordingly, let F and G be two d.f.'s. Let n_1, n_2 be positive integers, $X_1, \cdots, X_{n_1}$ be a random sample of size n_1 from F and $Y_1, \cdots, Y_{n_2}$ be an independent random sample of size n_2 from G. Let (X, Y) denote a copy of (X_1, Y_1).

Let r, s be two positive integers and $\varphi(x_1, \cdots, x_r; y_1, \cdots, y_s)$ be a measurable function on $\mathbb{R}^r \times \mathbb{R}^s$ symmetric under permutations in the x-components and in the y-components, i.e.,

$$\varphi(x_{i_1}, \cdots, x_{i_r}; y_{j_1}, \cdots, y_{j_s}) = \varphi(x_1, \cdots, x_r; y_1, \cdots, y_s),$$

for all permutations $x_{i_1}, \cdots, x_{i_r}$ of $x_1, \cdots, x_r$ and for all permutations $y_{j_1}, \cdots, y_{j_s}$ of $y_1, \cdots, y_s$. Recall the definition of $B_{r,n}$. Set

$$\boldsymbol{X} := (X_1, \cdots, X_r), \qquad \boldsymbol{X}_\alpha := (X_{\alpha_1}, \cdots, X_{\alpha_r}), \quad \alpha \in B_{r,n_1},$$
$$\boldsymbol{Y} := (Y_1, \cdots, Y_s), \qquad \boldsymbol{Y}_\beta := (Y_{\beta_1}, \cdots, Y_{\beta_s}), \quad \beta \in B_{s,n_2}.$$

The two sample U-statistic based on the two samples of sizes (n_1, n_2) is

$$\mathbb{U} = \frac{1}{\binom{n_1}{r}\binom{n_2}{s}} \sum_{\alpha \in \mathscr{B}_{r,n_1}} \sum_{\beta \in \mathscr{B}_{s,n_2}} \varphi(\boldsymbol{X}_\alpha; \boldsymbol{Y}_\beta).$$

Assume $E(|\varphi(\boldsymbol{X}; \boldsymbol{Y})|) < \infty$ and let $\nu := E(\varphi(\boldsymbol{X}; \boldsymbol{Y}))$. Clearly, $\mathbb{U}$ is an unbiased estimator of ν based on the two samples.

Example 5.7.1. Consider the problem of estimating of $\nu = P(Y \geq X)$. A kernel for estimating this probability is $\varphi(x, y) = I(y \geq x)$ with $r = 1, s = 1$. The corresponding U-statistic is

$$\mathbb{U}_w = \frac{1}{n_1 n_2} \sum_{i=1}^{n_1} \sum_{j=1}^{n_2} I(Y_j \geq X_i).$$

From (4.4.5), it follows that the two sample Wilcoxon rank sum statistic of (4.4.1) is related to this $\mathbb{U}_w$ by the relation given at (4.4.5), viz,

$$\mathbb{U}_w = (n_1 n_2)^{-1}(W - n_2(n_2 + 1)/2). \tag{5.7.1}$$

We shall now give an expression for the variance of a general two sample U-statistic. For this we assume

$$E(\varphi^2(\boldsymbol{X}; \boldsymbol{Y})) < \infty. \tag{5.7.2}$$

Let $\delta_{0,0}^2 = 0$, and for $c = 1, \cdots, r$ and $d = 1, \cdots, s$, introduce the random variables

$$\varphi_{0,d}(Y_1, \cdots, Y_d) = E\big(\varphi(\boldsymbol{X}, \boldsymbol{Y})\big|Y_1, \cdots, Y_d\big),$$
$$\varphi_{c,0}(X_1, \cdots, X_c) = E\big(\varphi(\boldsymbol{X}, \boldsymbol{Y})\big|X_1, \cdots, X_c\big),$$
$$\varphi_{c,d}(X_1, \cdots, X_c, Y_1, \cdots, Y_d) = E\big(\varphi(\boldsymbol{X}, \boldsymbol{Y})\big|X_1, \cdots, X_c, Y_1, \cdots, Y_d\big),$$

and write $\delta_{c,d}^2$ for their variances. These are the analogues of φ_c and δ_c^2 from the one sample case. Analogous to the one sample case, under (5.7.2),

$$\mathrm{Var}(\mathbb{U}) = \frac{1}{\binom{n_1}{r}\binom{n_2}{s}} \sum_{c=0}^{r}\sum_{d=0}^{s} \binom{r}{c}\binom{n_1 - r}{r - c}\binom{s}{d}\binom{n_2 - s}{s - d}\delta_{c,d}^2.$$

Example 5.7.2. Suppose that $r = s = 1$. Then

$$\mathbb{U} = \frac{1}{n_1 n_2}\sum_{i=1}^{n_1}\sum_{j=1}^{n_2}\varphi(X_i, Y_j),$$

$$\varphi_{1,0}(x) = E\big(\varphi(x, Y_1)\big), \qquad \varphi_{0,1}(y) = E\big(\varphi(X_1, y)\big), \qquad \varphi_{1,1}(x, y) = \varphi(x, y),$$
$$\delta_{1,0}^2 = \mathrm{Var}\big(\varphi_{1,0}(X_1)\big), \qquad \delta_{0,1} = \mathrm{Var}\big(\varphi_{0,1}(Y_1)\big), \qquad \delta_{1,1} = \mathrm{Var}\big(\varphi(X_1, Y_1)\big),$$

and

$$\mathrm{Var}(\mathbb{U}) = \frac{1}{n_1 n_2}\Big[(n_2 - 1)\delta_{1,0}^2 + (n_1 - 1)\delta_{0,1}^2 + \delta_{1,1}^2\Big].$$

The following lemma describes the first-order asymptotic behavior of this variance. We shall do so under the following assumptions on n_1 and n_2, where $N := n_1 + n_2$.

$$n_1 \to \infty, \quad n_2 \to \infty, \quad \frac{n_1}{N} \to \lambda, \text{ for some } 0 < \lambda < 1. \tag{5.7.3}$$

Lemma 5.7.1. *Suppose* (5.7.2) *and* (5.7.3) *hold. Then*

$$N\mathrm{Var}(\mathbb{U}) \to \tau_* := \frac{r^2}{\lambda}\delta_{1,0} + \frac{s^2}{(1-\lambda)}\delta_{0,1}. \tag{5.7.4}$$

The following corollary is an immediate consequence of this lemma and (2.4.9).

Corollary 5.7.1. *Under* (5.7.2) *and* (5.7.3), $E\big[(\mathbb{U} - \nu)^2\big] \to 0$ *and* $\mathbb{U} \to_p \nu$. *That is,* $\mathbb{U}$ *is consistent and mean square consistent estimator of* ν.

To find the asymptotic distribution of the two sample U-statistic, we first obtain the projection of $\mathbb{U}$ into $\mathcal{H}(X_1, \cdots, X_{n_1}, Y_1, \cdots, Y_{n_2})$. Since

$$E(\mathbb{U} - \nu|X_i) = (r/n_1)\big(\varphi_{1,0}(X_i) - \nu\big), \qquad 1 \le i \le n_1$$
$$E(\mathbb{U} - \nu|Y_j) = (s/n_2)\big(\varphi_{0,1}(Y_j) - \nu\big), \qquad 1 \le j \le n_2,$$

the projection of $\mathbb{U}$ is given by

$$\widehat{\mathbb{U}} = \nu + \frac{r}{n_1}\sum_{i=1}^{n_1}\big(\varphi_{1,0}(X_i) - \nu\big) + \frac{s}{n_2}\sum_{j=1}^{n_2}\big(\varphi_{0,1}(Y_j) - \nu\big).$$

Note that $\widehat{\mathbb{U}} - \nu$ is the sum of centered independent r.v.'s. Direct calculations show that $N\mathrm{Var}(\widehat{\mathbb{U}}) \to \tau_*$. Moreover, by Theorem 2.10.9, $\sqrt{N}(\widehat{\mathbb{U}} - \nu) \to_D \mathcal{N}(0, \tau_*)$, if $\tau^* > 0$. Thus we have the following result.

Theorem 5.7.1. *Suppose (5.7.2) and (5.7.3) hold and τ_* of (5.7.4) is positive. Then* $\sqrt{N}(\mathbb{U}-\nu) \to_D \mathcal{N}(0, \tau_*)$.

Example 5.7.3. We shall now derive the limiting null distribution of the rank sum statistic W of (4.4.1). In view of (5.7.1), it suffices to derive the limiting null distribution of the statistic

$$\mathbb{U}_w = \frac{1}{n_1 n_2} \sum_{i=1}^{n_1} \sum_{j=1}^{n_2} I(Y_j \geq X_i).$$

Here we have $r = s = 1$, $\varphi(x, y) = I(y \geq x)$, and (5.7.2) is *a priori* satisfied. Assume F, G to be continuous. Then, $\varphi_{1,0}(x) = 1 - G(x)$, $\varphi_{0,1}(y) = F(y)$, $\delta_{1,0}^2 = \mathrm{Var}\big(G(X)\big)$, $\delta_{0,1}^2 = \mathrm{Var}\big(F(Y)\big)$. Thus by Theorem 5.7.1, under (5.7.3),

$$\sqrt{N}\big(\mathbb{U}_w - \nu\big) \to_D N\big(0, \frac{\delta_{1,0}^2}{\lambda} + \frac{\delta_{0,1}^2}{1-\lambda}\big). \tag{5.7.5}$$

Suppose $H_0 : F = G$ holds. Let E_0, Var_0 denote the expectation and variance under H_0. Then $F(Y)$ and $1-F(X)$ are uniformly distributed on $(0, 1)$, and we obtain $\nu = 1/2$, $\delta_{1,0}^2 = \delta_{0,1}^2 = 1/12$ and $\delta_{1,1}^2 = 1/4$ and thus

$$\sqrt{N}\big(\mathbb{U}_w - E_0(\mathbb{U}_w)\big) \to_D \mathcal{N}\big(0, \frac{1}{12\lambda(1-\lambda)}\big). \tag{5.7.6}$$

Now the relation (5.7.1) gives $W = n_1 n_2 \mathbb{U}_w + n_2(n_2+1)/2$. Recall also that $E_0(\mathbb{U}_w) = 1/2$, $E_0(W) = n_2(N+1)/2$ and $\mathrm{Var}_0(W) = n_1 n_2(N+1)/12$. Then,

$$\frac{1}{n_1 n_2}\big(W - E_0(W)\big) = \mathbb{U}_w - E_0(\mathbb{U}_w).$$

Therefore, by (5.7.6),

$$\frac{\sqrt{N}}{n_1 n_2}\big(W - E_0(W)\big) \to_D \mathcal{N}\big(0, \frac{1}{12\lambda(1-\lambda)}\big). \tag{5.7.7}$$

Rewrite

$$\frac{W - E_0(W)}{\sqrt{\mathrm{Var}_0(W)}} = \frac{W - E_0(W)}{\sqrt{\frac{n_1 n_2(N+1)}{12}}} = \frac{\sqrt{N}}{n_1 n_2}\big(W - E_0(W)\big) d_N,$$

where

$$d_N := \frac{n_1 n_2}{\sqrt{N}} \frac{1}{\sqrt{\frac{n_1 n_2(N+1)}{12}}} = \sqrt{12}\sqrt{\frac{n_1 n_2}{N(N+1)}} \to \sqrt{12\lambda(1-\lambda)}.$$

Hence, by Slutsky's Theorem 2.10.5 and (5.7.7), $\big(W - E_0(W)\big)/\sqrt{\mathrm{Var}_0(W)} \to_D \mathcal{N}(0, 1)$.

Remark 5.7.1. Analogous to two sample U-statistics, one can define k-sample U-statistics for an integer $k \geq 2$. Here we give an example of a k-sample U-statistic useful in testing for H_0 of (4.5.1). The Kruskal-Wallis test based on H_* of (4.5.2) was obtained by extending the two sample Wilcoxon rank sum test to the k-sample setting. Similarly, Bhapkar (1961) proposed another test of H_0 of (4.5.1) by extending the two sample Mann-Whitney U-statistic $\mathbb{U}_w$ to the analogous k-sample U-statistic. As in the description of the Kruskal-Wallis rank test, let F be a continuous d.f.,

$n_1, \cdots, n_k$ be given sample sizes and $N = \sum_{j=1}^k n_j$. In the k-sample location model data consists of independent r.v.'s $X_{i,j}, i = 1, \cdots, n_j, j = 1, \cdots, k$, such that for some $\theta_j \in \mathbb{R}$, $X_{i,j} - \theta_j$, $i = 1, \cdots, n_j$, $j = 1, \cdots, k$ are i.i.d. F. For $j = 1, \cdots, k$, let

$$\begin{aligned}\phi_j(x_1, \ldots, x_k) &:= 1, \quad \text{if } x_j < x_i, \ \forall i = 1, \cdots, k, i \neq j,\\ &:= 0, \quad \text{otherwise},\end{aligned}$$

$$V_j := \sum_{s_1=1}^{n_1} \cdots \sum_{s_k=1}^{n_k} \phi_j\big(X_{s_1,1}, X_{s_2,2}, \cdots, X_{s_k,k}\big).$$

The variable V_j is the number of k-tuples that are formed by choosing one observation from each sample such that the observation from the jth sample is the least. Let $\mathbb{U}_j := V_j/(n_1 \cdots n_k)$, $\lambda_j := n_j/N$, $j = 1, \cdots, k$ and set

$$T := N(2k-1)\Big[\sum_{j=1}^k \lambda_j(\mathbb{U}_j - 1/k)^2 - \Big\{\sum_{j=1}^k \lambda_j(\mathbb{U}_j - 1/k)\Big\}^2\Big].$$

Direct calculations show that under H_0 of (4.5.1), $E(\mathbb{U}_j) = 1/k$, for all $j = 1, \cdots, k$. This test thus rejects H_0 for large values of T. Using results from Hoeffding (1948), Bhapkar proved that as $\min(n_1, \cdots, n_k) \to \infty$ and $n_1/N, \cdots, n_k/N$ stay bounded away from 0 and 1, the asymptotic null distribution of T is chi-squared distribution with $k-1$ degrees of freedom. Hence, the test that rejects H_0 of (4.5.1) if $T > \chi^2_{\alpha,k-1}$ has asymptotic size α.

If $k = 2$, then $\phi_1(x, y) = I(x < y)$, $\phi_2(x, y) = I(y < x)$, $V_1 = \sum_{i=1}^{n_1} \sum_{j=1}^{n_2} I(X_{i,1} < X_{j,2})$ and $V_2 = \sum_{i=1}^{n_1} \sum_{j=1}^{n_2} I(X_{j,2} < X_{i,1})$. Note, under H_0, we have the identity

$$V_1 + V_2 = \sum_{i=1}^{n_1} \sum_{j=1}^{n_2} \big[I(X_{i,1} < X_{j,2}) + I(X_{j,2} < X_{i,1})\big] = n_1 n_2, \quad \text{wp1},$$

by the continuity of F. Hence $\mathbb{U}_1 = 1 - \mathbb{U}_2$. Also note that $\mathbb{U}_1$ is equivalent to the Mann-Whitney statistic $\mathbb{U}_w$ with $X_i = X_{i,1}, Y_j = X_{j,2}$, $i = 1, \cdots, n_1; j = 1, \cdots, n_2$. Hence, it follows that the test based on T is equivalent to the two sample Mann-Whitney test based on $\mathbb{U}_w$, and in general, the test based on T is an extension of this test to the k-sample setting. Bhapkar and Deshpande (1968) provide some other similar tests for testing H_0.

5.8 Hoeffding Decomposition

In this section we shall discuss Hoeffding's decomposition of a U-statistic. Roughly speaking, this decomposition says that a U-statistic of order r can be written as its mean plus a weighted sum of degenerate U-statistics of orders 1 to r. A U-statistic of order k is degenerate if its kernel ψ satisfies

$$E(\psi(X_1, \cdots, X_k)|X_1, \cdots, X_{k-1}) = 0,$$

almost surely. For the sake of clarity of exposition, we shall describe this decomposition only for U-statistics of order 2.

Consider such a U-statistic

$$U := \frac{2}{n(n-1)} \sum_{1 \le i < j \le n} \varphi(X_i, X_j),$$

where $X_1, \cdots, X_n$ are again i.i.d. p-dimensional random vectors and φ is a kernel of order 2, and assume $E(\varphi^2(X_1, X_2)) < \infty$. Set $\nu = E(\varphi(X_1, X_2))$ and $\varphi_1(x) = E\big(\varphi(X_1, X_2)|X_1 = x\big)$. We let δ_1^2 denote the variance of $\varphi_1(X_1)$ and δ_2^2 the variance of $\varphi(X_1, X_2)$.

Then φ_* defined by

$$\varphi_*(x_1, x_2) := \varphi(x_1, x_2) - \varphi_1(x_1) - \varphi_1(x_2) + \nu$$

is the degenerate kernel of order two corresponding to φ and $\varphi_1 - \nu$ is a degenerate kernel of order 1. The latter follows from $E(\varphi_1(X_1) - \nu) = 0$ and the former from

$$E\big(\varphi^*(X_1, X_2)|X_1 = x_1\big) = \varphi_1(x_1) - \varphi_1(x_1) - \nu + \nu = 0. \tag{5.8.1}$$

For all $x_1, x_2 \in \mathbb{R}^p$, we have the relation

$$\varphi(x_1, x_2) = \nu + \varphi_1(x_1) - \nu + \varphi_1(x_2) - \nu + \varphi_*(x_1, x_2). \tag{5.8.2}$$

Note that for any function ψ from $\mathbb{R}^p$ to $\mathbb{R}$

$$\sum_{1 \le i < j \le n} [\psi(X_i) + \psi(X_j)] = \sum_{i=1}^{n} (n - i)\psi(X_i) + \sum_{j=1}^{n} (j - 1)\psi(X_j) = (n - 1)\sum_{i=1}^{n} \psi(X_i).$$

Let

$$U_1 := \frac{1}{n}\sum_{i=1}^{n} (\varphi_1(X_i) - \nu), \qquad U_* := \frac{2}{n(n-1)} \sum_{1 \le i < j \le n} \varphi_*(X_i, X_j).$$

Combining the above gives the *Hoeffding decomposition* of U:

$$U = \nu + \frac{2}{n}\sum_{i=1}^{n} (\varphi_1(X_i) - \nu) + U_* = \nu + 2U_1 + U_*.$$

Since $\nu + 2U_1$ is the projection of U onto $\mathcal{H}(X_1, \cdots, X_n)$, we may think of U_* as the residual of this projection. In view of (5.8.1) and the definitions of φ_1, the r.v.'s $\varphi_*(X_i, X_j), 1 \le i < j \le n$ are uncorrelated. Thus

$$\mathrm{Var}(U_*) = \frac{2}{n(n-1)} E\big(\varphi_*^2(X_1, X_2)\big) = \frac{2(\delta_2^2 - 2\delta_1^2)}{n(n-1)},$$

where the second identity follows from Example 5.4.2. Also, the random variables $\varphi_*(X_i, X_j)$ and $\varphi_1(X_k)$ are uncorrelated for all $1 \le i < j \le n$ and $1 \le k \le n$. Hence U_* and U_1 are uncorrelated. From this we derive

$$\mathrm{Var}(U) = 4\mathrm{Var}(U_1) + \mathrm{Var}(U_*) = \frac{4\delta_1^2}{n} + \frac{2(\delta_2^2 - 2\delta_1^2)}{n(n-1)}$$

which is formula (5.3.4). Hence, $n\mathrm{Var}(U) \to 4\delta_1^2$, $\mathrm{Var}(U)/\mathrm{Var}(\widehat{U}) \to 1$ and we have again proved Theorem 5.4.1.

5.9 Asymptotic Normality of Degenerate Weighted U Statistics

In this section we shall prove a slight extension of a very useful lemma of Hall (1984), which is used to prove the asymptotic normality of certain classes of degenerate U-statistics. In particular it is useful in proving the asymptotic normality of integrated squared difference between a density estimators and its expectation and certain integrated squared differences between a nonparametric regression function estimators and the given regression function as illustrated in Chapter 9.

Let $\zeta_1, \zeta_2, \cdots$ be i.i.d. p-variate random vectors, $H_n, n \geq 2$ be a sequence of kernels of order 2 defined on $\mathbb{R}^p \times \mathbb{R}^p$ satisfying

$$E\big(H_n(\zeta_1, \zeta_2)|\zeta_1\big) = 0, \text{ a.s.}, \quad 0 < E\big(H_n^2(\zeta_1, \zeta_2)\big) < \infty, \quad \forall n \geq 2, \tag{5.9.1}$$

and let U_n denote the sequence of the corresponding degenerate U-statistics, i.e.,

$$U_n := \sum_{1 \leq i < j \leq n} H_n(\zeta_i, \zeta_j).$$

Let $\mathcal{G}_j := \sigma\text{-field}\{\zeta_i, 1 \leq i \leq j\}$. Hall (1984) used Lemma 2.10.9 with $\mathcal{F}_{n,j} \equiv \mathcal{G}_j$ to prove Theorem 1 of his paper, which we state here as the following lemma. It gives the asymptotic normality of a degenerate U-statistic with kernel depending on the sample size. Set

$$G_n(x, y) := E\big(H_n(\zeta_1, x)H_n(\zeta_1, y)\big), \quad x, y \in \mathbb{R}^p.$$

Lemma 5.9.1. Hall Lemma. *Suppose the following condition holds.*

$$\frac{E\big(G_n^2(\zeta_1, \zeta_2)\big) + n^{-1}E\big(H_n^4(\zeta_1, \zeta_2)\big)}{\Big[E\big(H_n^2(\zeta_1, \zeta_2)\big)\Big]^2} \to 0. \tag{5.9.2}$$

Then,

$$\frac{U_n}{\Big(\frac{1}{2}n^2 E\big(H_n^2(\zeta_1, \zeta_2)\big)\Big)^{1/2}} \to_D \mathcal{N}(0, 1).$$

In later sections we shall often use the following corollary.

Corollary 5.9.1. *Suppose the following conditions hold for some finite positive constant Γ.*

$$E\big(H_n^2(\zeta_1, \zeta_2)\big) \to \Gamma, \quad E\big(G_n^2(\zeta_1, \zeta_2)\big) \to 0, \quad E\big(H_n^4(\zeta_1, \zeta_2)\big) = o(n). \tag{5.9.3}$$

Then $2n^{-1}U_n \to_D \mathcal{N}(0, 2\Gamma)$.

Proof. Let $\tilde{H}_n := (2/n)H_n$. Then, (5.9.3) implies (5.9.2) with H_n replaced by $\tilde{H}_n$, since $(n^2/2)E\big(\tilde{H}_n^2(\zeta_1, \zeta_2)\big) = 2E\big(H_n^2(\zeta_1, \zeta_2)\big) \to 2\Gamma$. □

Hall (1984) used this lemma to prove the asymptotic normality of a class of integrated squared errors of nonparametric density estimators. Koul and Ni (2004) used it to prove the asymptotic normality of a class minimum integrated squared distances between a nonparametric estimator of a regression function and its conditional expectation, given the covariates, under the null parametric model. These statistics are used to fit a parametric regression model when covariates are random. To obtain similar results in the case of non-random predictors, one needs to extend the above Hall Lemma to degenerate weighted U-statistics

$$U_{nc} := \sum_{1 \leq i < j \leq n} c_{n,ij} H_n(\zeta_i, \zeta_j), \qquad n \geq 2,$$

where $c_{n,ij}$ is an array of real numbers such that $c_{n,ij} = c_{n,ji}$, for all $1 \leq i, j \leq n$, and $\sum_{1 \leq i < j \leq n} c_{n,ij}^2 > 0$, for all $n \geq 2$. Let $\tau_n^2 := \mathrm{Var}\big(U_{nc}\big)$. The following lemma gives the needed extension.

Lemma 5.9.2. *Suppose $\tau_n^2 \to \infty$ and the following three conditions hold.*

$$\frac{\sum_{j=2}^n \left(\sum_{i=1}^{j-1} c_{n,ij}^2\right)^2 E\left(H_n^4(\zeta_1,\zeta_2)\right)}{\left(\sum_{j=2}^n \sum_{i=1}^{j-1} c_{n,ij}^2\right)^2 \left[E\left(H_n^2(\zeta_1,\zeta_2)\right)\right]^2} \to 0. \tag{5.9.4}$$

$$\frac{\sum_{1\le i<k<n} \left(\sum_{j=k+1}^n c_{n,ij}c_{n,kj}\right)^2 E\left(G_n^2(\zeta_1,\zeta_2)\right)}{\left(\sum_{j=2}^n \sum_{i=1}^{j-1} c_{n,ij}^2\right)^2 \left[E\left(H_n^2(\zeta_1,\zeta_2)\right)\right]^2} \to 0, \tag{5.9.5}$$

$$\frac{\sum_{i=1}^{n-1} \left(\sum_{j=i+1}^n c_{n,ij}^2\right)^2 E\left(H_n^4(\zeta_1,\zeta_2)\right)}{\left(\sum_{j=2}^n \sum_{i=1}^{j-1} c_{n,ij}^2\right)^2 \left[E\left(H_n^2(\zeta_1,\zeta_2)\right)\right]^2} \to 0. \tag{5.9.6}$$

Then $\tau_n^{-1} U_{nc} \to_D \mathcal{N}(0,1)$.

Upon taking $c_{n,ij} = 1$ for all i, j, n, one readily sees that the assumptions (5.9.4) to (5.9.6) are equivalent to (5.9.2). Thus Lemma 5.9.2 includes Lemma 5.9.1. The following proof of Lemma 5.9.2 is similar to that of Lemma 5.9.1.

Proof. Let $Y_{n1} := 0$ and

$$Y_{nj} := \sum_{i=1}^{j-1} c_{n,ij} H_n(\zeta_i,\zeta_j), \quad 2 \le j \le n; \qquad S_{nk} := \sum_{j=1}^{k} Y_{nj}, \quad 1 \le k \le n.$$

Recall $\mathcal{G}_j := \sigma$-field$\{\zeta_i, 1 \le i \le j\}$ and note that $\{Y_{nj}, \mathcal{G}_j; j = 1, \cdots, n\}$ is a martingale difference array and

$$\tau_n^{-1} U_{nc} = \tau_n^{-1} \sum_{j=2}^{n} \sum_{i=1}^{j-1} c_{n,ij} H_n(\zeta_i,\zeta_j) = \tau_n^{-1} \sum_{j=1}^{n} Y_{nj} = \tau_n^{-1} S_{nn}.$$

Let

$$V_n^2 := \sum_{j=1}^{n} E\left(Y_{nj}^2 | \mathcal{G}_{j-1}\right).$$

Note that $\tau_n^2 = E\left(V_n^2\right) = \sum_{j=1}^n E\left(Y_{nj}^2\right)$.

In order to apply Lemma 2.10.9, we need to verify (2.10.5) and (2.10.6) in this case, i.e., we need to verify the following two conditions.

$$\tau_n^{-2} \sum_{j=1}^{n} E\left\{Y_{nj}^2 I\left(|Y_{nj}| > \epsilon\tau_n | \mathcal{G}_{j-1}\right)\right\} \to 0, \quad \forall \epsilon > 0. \tag{5.9.7}$$

$$\tau_n^{-2} V_n^2 \to_p 1. \tag{5.9.8}$$

Because $\zeta_1, \cdots, \zeta_n$ are i.i.d. , the properties (5.9.1) of H_n imply that $c_{n,ij} H_n(\zeta_i,\zeta_j)$, $1 \le i < j$, are uncorrelated for each $j = 2, \cdots, n$. Thus we obtain

$$E\left(Y_{nj}^2\right) = \sum_{i=1}^{j-1} c_{n,ij}^2 E\left(H_n^2(\zeta_j,\zeta_i)\right) = \sum_{i=1}^{j-1} c_{n,ij}^2 \, E\left(H_n^2(\zeta_1,\zeta_2)\right), \quad \forall\, 2 \le j \le n, \tag{5.9.9}$$

$$\tau_n^2 = \sum_{j=2}^{n} \sum_{i=1}^{j-1} c_{n,ij}^2 \, E\left(H_n^2(\zeta_1,\zeta_2)\right).$$

Note that the common denominator in the three expressions of (5.9.4)–(5.9.6) is equal to τ_n^4.

For the same reasons as above and by the Cauchy-Schwarz inequality (2.4.4),

$$E(Y_{nj}^4) = \sum_{i=1}^{j-1} c_{n,ij}^4 E\big(H_n^4(\zeta_j,\zeta_i)\big) + 3\sum_{i=1}^{j-1}\sum_{k\neq i=1}^{j-1} c_{n,ij}^2 c_{n,kj}^2 E\big(H_n^2(\zeta_j,\zeta_i)H_n^2(\zeta_j,\zeta_k)\big)$$

$$\le 3\Big(\sum_{i=1}^{j-1} c_{n,ij}^2\Big)^2 E\big(H_n^4(\zeta_1,\zeta_2)\big).$$

Hence, by (5.9.4),

$$\tau_n^{-4}\sum_{j=2}^{n} E(Y_{nj}^4) \le 3\frac{\sum_{j=2}^{n}\big(\sum_{i=1}^{j-1} c_{n,ij}^2\big)^2 E\big(H_n^4(\zeta_1,\zeta_2)\big)}{\Big(\sum_{j=2}^{n}\sum_{i=1}^{j-1} c_{n,ij}^2\Big)^2\big[E\big(H_n^2(\zeta_1,\zeta_2)\big)\big]^2} \to 0.$$

In view of Remark 2.10.1 applied with $\alpha = 2$, this verifies (5.9.7).

It remains to prove (5.9.8) or equivalently $V_n^2 - \tau_n^2 = o_p(\tau_n^2)$. By the fact

$$E\big(H_n(\zeta_i,\zeta_j)H_n(\zeta_k,\zeta_j)\big|\mathcal{G}_{j-1}\big) = G_n(\zeta_i,\zeta_k), \quad 1 \le i,k < j \le n,$$

we derive the identity

$$V_n^2 - \tau_n^2 = \sum_{j=2}^{n}\big[E(Y_{nj}^2|\mathcal{G}_{j-1}) - E(Y_{nj}^2)\big]$$

$$= \sum_{j=2}^{n}\sum_{i=1}^{j-1}\sum_{k=1}^{j-1} c_{n,ij}c_{n,kj}\big[G_n(\zeta_i,\zeta_k) - E\big(G_n(\zeta_i,\zeta_k)\big)\big].$$

In view of $E\big(G_n(\zeta_i,\zeta_k)\big) = 0$ for $i \neq k$, we obtain $V_n^2 - \tau_n^2 = \Gamma_{n,1} + 2\Gamma_{n,2}$ with

$$\Gamma_{n,1} = \sum_{i=1}^{n-1} a_{n,i}\big[G_n(\zeta_i,\zeta_i) - E\big(G_n(\zeta_i,\zeta_i)\big)\big], \qquad a_{n,i} = \sum_{j=i+1}^{n} c_{n,ij}^2,$$

$$\Gamma_{n,2} = \sum_{1\le i<k<n} b_{n,ik}G_n(\zeta_i,\zeta_k), \qquad b_{n,ik} = \sum_{j=k+1}^{n} c_{n,ij}c_{n,kj}.$$

Thus it suffices to show that $E(\Gamma_{n,\ell}^2) = o(\tau_n^4)$ for $\ell = 1,2$. Since $G_n(\zeta_1,\zeta_1),\cdots,G_n(\zeta_n,\zeta_n)$ are i.i.d. with finite second moment, by (2.4.4) applied with the conditional expectation, we obtain

$$E\big(G_n^2(\zeta_1,\zeta_1)\big) = E\big(\big[E\big(H_n^2(\zeta_1,\zeta_2)\big|\zeta_{n,1}\big)\big]^2\big) \le E\big(E(H_n^4(\zeta_1,\zeta_2)|\zeta_1)\big) = E\big(H_n^4(\zeta_1,\zeta_2)\big).$$

Hence,

$$E(\Gamma_{n,1}^2) = \sum_{i=1}^{n-1} a_{n,i}^2 \operatorname{Var}\big(G_n(\zeta_1,\zeta_1)\big) \le \sum_{i=1}^{n-1} a_{n,i}^2 E\big(H_n^4(\zeta_1,\zeta_2)\big),$$

and $E\big(\Gamma_{n,1}^2\big) = o(\tau_n^4)$ follows from (5.9.6). Since $G_n(\zeta_i,\zeta_k)$, $1 \le i < k < n$ are uncorrelated and identically distributed with zero mean, we derive

$$E(\Gamma_{n,2}^2) = \sum_{1\le i<k<n} b_{n,ik}^2 E\big(G_n^2(\zeta_1,\zeta_2)\big) = \sum_{1\le i<k<n}\Big(\sum_{j=k+1}^{n} c_{n,ij}c_{n,kj}\Big)^2 E\big(G_n^2(\zeta_1,\zeta_2)\big),$$

and $E\big(\Gamma_{n,2}^2\big) = o(\tau_n^4)$ follows from (5.9.5). □

5.10 Asymptotic Relative Efficiency of Tests

In this section we shall discuss a large sample measure for comparing any two tests for the same testing problem.

If there are two competing tests for a given testing problem, then a natural question that arises is which test should one use. The answer to this question depends on the requirements imposed on the tests. One typically requires a test to have given probability of type 1 error and power at the given alternative. For example, consider the one sample location model, where θ denotes the population mean and median, and the problem of testing $\theta = 0$ versus $\theta > 0$. The two tests of interest here are the sign test S and the classical test based on the sample mean $\bar{X}$. Suppose we wish to test these hypotheses at the level α and power β at a given alternative. Let N_1, N_2 be the sample size required by the S-test and $\bar{X}$-test, respectively, to achieve these goals. The ratio $e := N_2/N_1$ is known as the relative efficiency of the S-test, relative to the $\bar{X}$-test. Note that if $e > 1$, then the sign test is preferable to the $\bar{X}$-test. Unfortunately, in the case of finite sample sizes, the exact distribution of $\bar{X}$ is hard to determine in general. This in turn results in our inability to determine the ratio e when sample sizes are finite. However, Pitman argued that it is possible to determine this ratio when sample sizes tend to infinity as follows.

Proceeding a bit more precisely and generally, let Θ be an open interval in $\mathbb{R}$ and θ_0 be a fixed point in Θ. Let $\{P_\theta,\ \theta \in \Theta\}$ be a family of probability models and let $X_1, \cdots, X_n$ be a random sample from P_θ for some $\theta \in \Theta$. Consider the problem of testing

$$H_0 : \theta = \theta_0, \qquad \text{versus} \qquad H_1 : \theta > \theta_0,$$

based on the given sample. Let $0 < \alpha < 1 - \beta < 1$ and Φ denote the d.f. of the standard normal distribution and z_α be as in (3.3.9).

As in Chapter 4, a test is a statistic φ_n taking values in the interval $[0, 1]$ and $\pi_n(\theta) := \pi_{\varphi_n}(\theta) = E_\theta(\varphi_n)$, $\theta \in \Theta$ is its power function. Recall $\pi_n(\theta_0)$ is the probability of a type I error, i.e., the probability of rejecting the null hypothesis when the null hypothesis is true, while for $\theta > \theta_0$, $1 - \pi_n(\theta)$ is the probability of a type II error, i.e., the probability of failing to reject the null hypothesis when the null hypothesis is false. We say the test has asymptotic size α if $\pi_n(\theta_0) \to \alpha$.

Assume that the test φ_n has asymptotic size α. We are now looking for a sequence of positive integers N_n, $N_n \to \infty$, such that $\pi_{N_n}(\theta_0 + n^{-1/2}\delta_0) \to 1 - \beta$ for some fixed $\delta_0 > 0$. Then the test sequence φ_{N_n} has asymptotic size α as $\pi_{N_n}(\theta_0) \to \alpha$ and asymptotic power $1 - \beta$ under the local alternative $\theta_0 + n^{-1/2}\delta_0$. Thus N_n is a sample size required for the test φ_n to have asymptotic power $1 - \beta$ under the above local alternative.

Definition 5.10.1. Let $\varphi_{1,n}$ and $\varphi_{2,n}$ be two test sequences with asymptotic size α. For $i = 1, 2$, let $N_{i,n}$ be a sample size required for $\varphi_{i,n}$ to have the asymptotic power $1 - \beta$ under the sequence of local alternatives $\theta_0 + n^{-1/2}\delta_0$. If the limit

$$\text{ARE}(\varphi_{1,n}, \varphi_{2,n}) = \lim_{n\to\infty} \frac{N_{2,n}}{N_{1,n}}$$

exists, we call it Pitman's asymptotic relative efficiency (ARE) of the test sequence $\varphi_{1,n}$ relative to the test sequence $\varphi_{2,n}$.

Suppose φ_n is locally regular at θ_0 with slope γ (see Definition 4.2.2) and N_n is a sequence of positive integers satisfying $N_n/n \to \rho$ for some $\rho > 0$. Then $\delta_{N_n} = (N_n/n)^{1/2}\delta_0 \to \rho^{1/2}\delta_0$, and this implies

$$\pi_{N_n}\Big(\theta_0 + \frac{\delta_0}{n^{1/2}}\Big) = \pi_{N_n}\Big(\theta_0 + \frac{\delta_{N_n}}{N_n^{1/2}}\Big) \to 1 - \Phi(z_\alpha - \rho^{1/2}\delta_0\gamma).$$

Recall $\alpha < 1-\beta$ so that $z_\alpha > z_{1-\beta}$. If $\rho^{1/2} = (z_\alpha - z_{1-\beta})/(\delta_0\gamma)$, then the above limit equals $1-\Phi(z_{1-\beta}) = 1-\beta$ and N_n is a required sample size for the test φ_n.

Now assume that the tests $\varphi_{1,n}$ and $\varphi_{2,n}$ are locally regular at θ_0 with respective slopes γ_1 and γ_2. Then the respective required sample sizes $N_{1,n}$ and $N_{2,n}$ must satisfy $N_{i,n}/n \to (z_\alpha - z_{1-\beta})^2/(\delta_0^2\gamma_i^2)$, $i = 1,2$ so that

$$\lim_{n\to\infty} \frac{N_{2,n}}{N_{1,n}} = \frac{\gamma_1^2}{\gamma_2^2}.$$

This shows that $\mathrm{ARE}(\varphi_{1,n}, \varphi_{2,n}) = (\gamma_1/\gamma_2)^2$. Thus we have proved the following result.

Lemma 5.10.1. *If the tests $\varphi_{1,n}$ and $\varphi_{2,n}$ are locally regular at θ_0 with respective slopes γ_1 and γ_2, then $\mathrm{ARE}(\varphi_{1,n}, \varphi_{2,n}) = (\gamma_1/\gamma_2)^2$.*

Now we shall derive the ARE of three tests for testing H_0: $\theta = 0$ versus the alternatives H_1: $\theta > \theta_0$ in the one sample symmetric location model. They are the sign, t and Wilcoxon signed rank W^+ tests, denoted by φ_S, φ_t and φ_{W^+}, respectively. All three tests are locally regular at θ_0 under some minimal additional assumptions on the error distribution. For the t-test we require that F has a finite positive variance σ^2. By Exercise 1 in Chapter 4, the t-test is locally regular at θ_0 with slope $\gamma_t = 1/\sigma$. Next, assume also that F satisfies the conditions of Lemmas 4.3.1 and 5.5.1. Then, see (4.3.10) and (5.5.12), φ_S and φ_{W^+} are locally regular at θ_0 with respective slopes

$$\gamma_S = 2f(0) \qquad \gamma_{W^+} = \sqrt{12}I_2(f), \quad \text{with} \quad I_2(f) = \int f^2(x)\,dx. \tag{5.10.1}$$

Hence

$$\begin{aligned} \mathrm{ARE}(\varphi_S, \varphi_t) &= \Big(\frac{2f(0)}{1/\sigma}\Big)^2 = 4\sigma^2 f^2(0), \\ \mathrm{ARE}(\varphi_{W^+}, \varphi_t) &= \Big(\frac{\sqrt{12}I_2(f)}{1/\sigma}\Big)^2 = 12\sigma^2\big(I_2(f)\big)^2, \\ \mathrm{ARE}(\varphi_S, \varphi_{W^+}) &= \Big(\frac{2f(0)}{\sqrt{12}I_2(f)}\Big)^2 = \frac{f^2(0)}{3(I_2(f))^2}. \end{aligned} \tag{5.10.2}$$

We shall now evaluate these efficiencies at some error densities f. Suppose $f(x) = e^{-|x|}/2$. Then $2f(0) = 1$, $\sigma^2 = 2$, $I_2(f) = 1/4$ and

$$\begin{aligned} \mathrm{ARE}(\varphi_S, \varphi_t) &= 4\sigma^2 f^2(0) = 2, \\ \mathrm{ARE}(\varphi_{W^+}, \varphi_t) &= 12\sigma^2(I_2(f))^2 = 24/16 = 1.5, \\ \mathrm{ARE}(\varphi_S, \varphi_{W^+}) &= \frac{f^2(0)}{3(I_2(f))^2} = \frac{1/4}{3/16} = 4/3 \approx 1.33. \end{aligned}$$

In other words, if the underlying error distribution is double exponential, then the t-test would need twice as many observations than the sign test to achieve the same accuracy in terms of the asymptotic level and power. Similarly, the t test would need 150 observations for every 100 observations needed by the W^+ test to achieve this accuracy, and for every 100 observations used by sign test the W^+ test would need 133 observations to achieve the same asymptotic level and power. One sees that the sign test is the most efficient among these three tests. In fact, when the error distribution is double exponential, the sign test is asymptotically the most powerful among all tests of the asymptotic level α against the local alternatives $n^{-1/2}\delta, \delta > 0$. See Chapter VII, Hájek and Šidák (1967).

Next, suppose $F = \Phi$. Then $\sigma^2 = 1$, $f(0) = 1/\sqrt{2\pi}$ and $I_2(f) = 1/(2\sqrt{\pi})$ and hence, $\text{ARE}(\varphi_S, \varphi_t) = 4/2\pi = 2/\pi \approx 0.64$,

$$\text{ARE}(\varphi_{W^+}, \varphi_t) = \frac{3}{\pi} \approx 0.955, \quad \text{ARE}(\varphi_S, \varphi_{W^+}) = \frac{1/(2\pi)}{3(1/4\pi)} = \frac{2}{3}.$$

Based on these numbers, for the normal error distribution, the sign test based on 100 observations will achieve the same large sample accuracy as the t test based on 67 observations. Similarly, the W^+ test based on 100 observations will achieve the same large sample accuracy as the t test based on 96 observations. This fact says that the W^+ test needs only 4 more observations to compete against the optimal test at the Gaussian error distribution, in large samples. Moreover, using the method of undetermined multipliers, Hodges and Lehmann (1956) proved that

$$\text{ARE}(\varphi_{W^+}, \varphi_t) \geq 0.864,$$

for each square integrable density f with mean zero and variance 1, and this bound is achieved by the density $f(x) = 3\sqrt{5}\,(5 - x^2)I(x^2 \leq 5)/20$.

These facts fueled the research activity on signed rank statistics in the mid to late 1950s and led to the discovery of a signed rank test that is asymptotically as efficient as the t test at $F = \Phi$, and asymptotically more efficient than the t test at non-normal error distributions for testing $\theta = 0$ versus the alternatives $\theta = n^{-1/2}\delta, \delta > 0$, in the one sample location model. The test is based on the normal scores statistic S_ψ of (4.3.22).

Let ϕ denote a density of Φ. The ARE of the S_ψ-test φ_{S_ψ}, relative to the t-test, at an arbitrary error d.f. F, is

$$\text{ARE}(\varphi_{S_\psi}, \varphi_t) = \sigma^2\Big(\int_0^1 \frac{f(F^{-1}(u))}{\phi(\Phi^{-1}(u))}du\Big)^2. \tag{5.10.3}$$

Hodges and Lehmann (1956) conjectured that for each error density f with zero mean and finite variance σ^2,

$$\text{ARE}(\varphi_{S_\psi}, \varphi_t) \geq 1, \tag{5.10.4}$$

with the equality holding if, and only if, $f = \phi$. Chernoff and Savage (1958) were the first to prove (5.10.4). Gastwirth and Wolff (1968) used the Jensen inequality (2.4.10) to give an alternate proof of (5.10.4). It is a remarkable property of the S_ψ-test. Besides being applicable much more broadly than the t-test, the S_ψ-test is asymptotically as efficient as the optimal t-test at the normal error density and asymptotically more efficient than the t-test at non-normal error density. The only thing that has held back its use in applications relative to the W^+ test is that one needs to use the normal scores to implement the test.

5.11 Exercises

1. Verify the inequality (5.3.5) for $c = 1, \cdots, r$ and $n \geq 2r - 1$.

2. Let $X_1, \cdots, X_n$ be i.i.d. with density f. Assume that $\int x^2 f(x)\,dx$ is finite. Find mean and variance of the U-statistics with the following kernels (a) $\varphi(x_1, x_2) = |x_1 - x_2|$, (b) $\varphi(x_1, x_2) = \max(x_1, x_2)$ and (c) $\varphi(x_1, x_2) = \min(|x_1|, |x_2|)$.
3. Consider the U-statistic of order $r = 3$ with kernel $\varphi(x_1, x_2, x_3) = I(x_1 + x_2 + x_3 > 0)$. Find the functions φ_1 and φ_2 and compute the variance of this U-statistic.
4. Let $X_1, \cdots, X_n$ be a random sample from a d.f. F. Find mean, variance and Hoeffding decomposition of the U-statistic

$$U := \frac{2}{n(n-1)} \sum_{1 \le i < j \le n} I(X_i \le t) I(X_j \le t),$$

for some fixed $t \in \mathbb{R}$. For which t are the assumptions of Theorem 5.4.1 met?
5. Let $X_1, \cdots, X_n$ be a random sample from a d.f. F. Find mean and variance of the U-statistic

$$U := \frac{1}{\binom{n}{3}} \sum_{1 \le i < j < k} I(X_i \le t) I(X_j \le t) I(X_k \le t),$$

for some fixed $t \in \mathbb{R}$. For which t are the assumptions of Theorem 5.4.1 met?
6. Let $X_1, \cdots, X_n$ be a random sample from a d.f. F with finite mean μ and finite positive variance σ^2. Find mean, variance and Hoeffding decomposition of the U-statistic

$$U = \frac{2}{n(n-1)} \sum_{1 \le i < j \le n} X_i X_j.$$

7. Let $X_1, \cdots, X_n$ be a random sample from a d.f. F and let ψ be a measurable function from $\mathbb{R}$ to $\mathbb{R}$ such that $E(\psi^2(X)) < \infty$, where X is a copy of X_1. Find mean, variance and Hoeffding decomposition of the U-statistic

$$U = \frac{2}{n(n-1)} \sum_{1 \le i < j \le n} \psi(X_i) \psi(X_j).$$

8. Let U be a U-statistic of order $r = 2$ with $E(\varphi^2(X_1, X_2))$ finite. Show that $\text{Var}(U) \le (4/n) E(\varphi^2(X_1, X_2))$.
9. Let U_n be a U-statistic of order $r = 2$ with kernel φ_n and let ν be a real number. Show that $E(\varphi_n(X_1, X_2)) \to \nu$ and $E(\varphi_n^2(X_1, X_2)) = o(n)$ imply $U_n \to_p \nu$.
10. Show that $U_n^{(3)}$ is a consistent estimator of $g(0) = I_2(f)$ if f and K are square integrable, K is symmetric about zero, $h_n \to 0$ and $nh_n \to \infty$. Hint: Use the previous problem and Lemma 5.6.1.
11. Let U be a U-statistic of order 3 with kernel φ. Suppose

$$E(\varphi^2(X_1, X_2, X_3)) < \infty$$

and set $\nu = E(\varphi(X_1, X_2, X_3))$.

(a) Show that

$$\begin{aligned} \varphi_*(x_1, x_2, x_3) = {} & \varphi(x_1, x_2, x_3) - \varphi_2(x_1, x_2) - \varphi_2(x_1, x_3) \\ & - \varphi_2(x_2, x_3) + \varphi_1(x_1) + \varphi_1(x_2) + \varphi_1(x_3) - \nu, \end{aligned}$$

is a degenerate kernel of order three.

(b) Verify the Hoeffding decomposition

$$U = \nu + \frac{3}{n}\sum_{i=1}^{n}(\varphi_1(X_i) - \nu) + 3U_2 + U_3$$

where U_2 is the U-statistic of order 2 with degenerate kernel associated with φ_2, and U_3 is the U-statistic of order 3 with degenerate kernel φ_* from part (a).

12. Recall the definition (5.3.2). Let $\hat{\delta}_1$ be a consistent estimator of δ_1, i.e., $\hat{\delta}_1 \to_p \delta_1$. Under what additional conditions do we have $\sqrt{n}(U - \nu)/(r\hat{\delta}_1) \to_D \mathcal{N}(0,1)$?

13. Verify (5.6.5).

14. Let F be a d.f.. Verify the inequality

$$\int \big|F(x+t) - F(x)\big|\, dx \le |t|, \qquad \forall\, t \in \mathbb{R}.$$

15. Let ψ be a bounded nondecreasing function on the real line. Verify the inequality

$$\int \big|\psi(x+t) - \psi(x)\big|\, dx \le |t|\big(\psi(\infty) - \psi(-\infty)\big), \ \forall\, t \in \mathbb{R}.$$

16. A function φ from $\mathbb{R}$ to $\mathbb{R}$ is said to be L_1-Lipschitz if the inequality

$$\int |\varphi(x+t) - \varphi(x)|\, dx \le C|t|, \qquad \forall\, t \in \mathbb{R}$$

holds for some positive finite constant C.

Show that real-valued functions of bounded variation defined on $\mathbb{R}$ are L_1-Lipschitz. (Hint: A function of bounded variation can be written as a difference of two bounded nondecreasing functions.)

17. Let $g(y) = \int f(x+y)f(x)dx$, where f is a density on $\mathbb{R}$. Suppose f has continuous and square-integrable first and second order derivatives. Show that g is four times differentiable with $g^{(4)}(y) = \int \ddot{f}(x+y)\ddot{f}(x)\, dx$ and $g^{(4)}$ is bounded and uniformly continuous.

18. Let F, G be arbitrary d.f.'s on $\mathbb{R}$ having finite and positive variances. Let $X, X_1, \cdots, X_m$ be i.i.d. F r.v.'s, $Y, Y_1, \cdots, Y_n$ be i.i.d. G r.v.'s, and X's be independent of Y's.

(a) Let $\nu = \mathrm{Var}(Y) - \mathrm{Var}(X) = \sigma_Y^2 - \sigma_X^2$. Find a two sample U-statistic for estimating ν.

(b) Let $\nu := \mathrm{Var}(X) + \mathrm{Var}(Y)$. Find a two sample U-statistic for estimating ν and derive its variance.

(c) Let $\nu := P(X_1 + X_2 < Y_1 + Y_2)$. Find a two sample U-statistic $\mathbb{U}$ for estimating ν. Under $H_0 : F = G$, compute ν and, using Theorem 5.7.1, find an expression for the limiting variance of $\sqrt{N}(\mathbb{U} - \nu)$.

19. Let $F_1, \cdots, F_n$ be continuous d.f.'s, each symmetric about zero. Let $X_1, X_2, \cdots, X_n$ be independent r.v.'s, $X_i \sim_D F_i$, $1 \le i \le n$. Let $R_i^+ := \sum_{j=1}^n I(|X_j| \le |X_i|)$, $H_i(x) := P(|X_i| \le x), x \ge 0$, $1 \le i \le n$ and $W_s := \sum_{i=1}^n \mathrm{sgn}(X_i)R_i^+$. Let $Y_i := \big(\mathrm{sgn}(X_i), |X_i|\big)$.

(a) Show that the projection of W_s into $\mathcal{H}(Y_1, \cdots, Y_n)$ is

$$\widehat{W}_s := \sum_{i=1}^{n} E(W_s|Y_i) = \sum_{i=1}^{n} \operatorname{sgn}(X_i)\big[1 + (n-1)H_i(|X_i|)\big].$$

(b) Compute $\tau^2 := \operatorname{Var}\big(\widehat{W}_s\big)$. Is it the same as in the case of all F_i's being the same? Show that $\widehat{W}_s/\tau \to_D \mathcal{N}(0,1)$.

(c) Let $c_{ni}, 1 \le i \le n$, be arrays of real numbers such that $\sum_{i=1}^{n} c_{ni}^2 = 1$ and $\max_{1\le i\le n} |c_{ni}| \to 0$. Let $T := \sum_{i=1}^{n} c_{ni}\operatorname{sgn}(X_i)R_i^+$. Show that T has mean zero and its projection into $\mathcal{H}(Y_1, \cdots, Y_n)$ is

$$\widehat{T} := \sum_{i=1}^{n} E(T|Y_i) = \sum_{i=1}^{n} c_{ni} \operatorname{sgn}(X_i)\big[1 + (n-1)H_i(|X_i|)\big].$$

Compute $\sigma^2 := \operatorname{Var}(\widehat{T})$ and show that $\big(\widehat{T}/\sigma\big) \to_D \mathcal{N}(0,1)$.

20. Compute $\text{ARE}_{\varphi_{W^+}, \varphi_S}(F)$ for the following two f's:

$$f(x) = e^{-x}/\big(1+e^{-x}\big)^2, \qquad f(x) = 1/\pi(1+x^2), \;\; x \in \mathbb{R}.$$

Notes: Giné and Nickl (2008) obtained the results of Theorem 5.6.1 under the additional assumption that f is bounded. They also provide some uniformity for their results. It was shown by Bickel and Ritov (1988) that $\beta > 1/4$ is needed for the asymptotic normality result of this theorem. More about U-statistics can be learned from the books of Serfling (1980), Lee (1990), Koroljuk and Borovskich (1994) and the review paper of Sen (1992). Sen (1963b) discusses some properties of U-statistics based on some dependent r.v.'s.

A variant of Lemma 5.9.2 dealing with a class of weighted degenerate U-statistics first appeared in Koul (2011), where it was used it to derive the asymptotic null distribution of a class of minimum integrated squared distances between a nonparametric estimator of a regression function and its expectation under the parametric model being fitted, when design variables are nonrandom. O'Neil and Redner (1993), Major (1994) and Rifi and Utzet (2000) have also investigated asymptotic distributions of weighted degenerate U-statistics. But, the weights and kernels in these papers are not allowed to depend on n, and the weak limits are generally non-normal. Hsing and Wu (2004) prove the asymptotic normality of non-degenerate U-statistics of some dependent r.v.'s, where again the weights and kernels are not allowed to depend on n.

6

Estimation in Location Models

6.1 Introduction

Besides requiring the knowledge of likelihood, the classical estimation procedures based on the given likelihood are often sensitive to departures from the assumptions under which they are derived. These estimators of the underlying parameters in location and regression models are typically affected by gross errors. Data is said to have come from a population with d.f. F with δ-proportion of gross errors if it is a random sample from the population with d.f. $(1-\delta)F + \delta H$, for some $0 < \delta < 1$, where H is another d.f.. For example, the sample mean $\bar{X}$ is known to be a minimum variance unbiased estimator of the population mean μ when sampling from $\mathcal{N}(\mu, \sigma^2)$ population, $\sigma^2 > 0$. But it looses this optimality property even when just one percent of the observations is from double exponential distribution, i.e., if data comes from the population with the d.f. $0.99\Phi(\cdot/\sigma) + 0.01H$, where Φ (H) is the d.f. of a $\mathcal{N}(0,1)$ (double exponential) r.v.. Estimators that are not sensitive to gross errors are typically called robust estimators.

In this chapter we shall study several estimation methods of the location parameter in the one and two sample location models and of regression parameters in the linear regression model that are robust against gross errors. We shall discuss R estimators, which are based on signed rank and rank statistics, as developed by Hodges and Lehmann (1963), Sen (1963a) and Jurčková (1969), M estimators of Huber (1964) and some classes of minimum distance estimators based on weighted residual empirical processes, as developed in Fine (1966) and Koul (1992, 2002).

In general, let $\boldsymbol{X}_n := (X_1, \cdots, X_n)$ be given data and $\theta \in \mathbb{R}$ be the parameter of interest. Let $S(\boldsymbol{X}_n, \theta)$ be a given function such that either $\mu := E_\theta\big(S(\boldsymbol{X}_n, \theta)\big)$ exists and is completely known or the distribution of $S(\boldsymbol{X}_n, \theta)$, when θ is the true parameter value, is symmetric about a known point ν. Then an estimator of θ is defined to be that value of t for which $S(\boldsymbol{X}_n, t) = \mu$ or $S(\boldsymbol{X}_n, t) = \nu$. Such a function $S(\boldsymbol{X}_n, t)$ is called the estimating score, cf., Godambe and Thompson (1984), Godambe (1997) and the references therein.

A classical example of the above methodology is provided by the maximum likelihood estimation method, where $S(\boldsymbol{X}_n, t)$ is the likelihood score, and one typically solves the likelihood equation $S(\boldsymbol{X}_n, t) = 0$, provided the likelihood function is available. Other classical examples are those of the least squares and least absolute deviation estimators, where $S(\boldsymbol{X}_n, t)$ equals $\sum_{i=1}^n (X_i - t)$ and $\sum_{i=1}^n \text{sign}(X_i - t)$, respectively. In these cases, the knowledge of the underlying likelihood is not needed but one needs to know if θ is mean or median of the data generating population. We shall now describe this methodology in the one and two sample location models.

DOI: 10.1201/9781003585770-6

6.2 One and Two Sample Location Models

In this section we shall first discuss some general principles for deriving estimators and some of their properties based on the work of Hodges and Lehmann (1963) and then R, M and minimum distance estimators of the one and two sample location parameters.

6.2.1 Some General Principles

Let $\boldsymbol{X}_n = (X_1, \cdots, X_n)'$ be a random sample from the one sample location model $X = \theta + \varepsilon$, for some $\theta \in \mathbb{R}$, where ε is a r.v. with some d.f. F that is continuous and symmetric about zero. Let $\mathbf{1}_n := (1, \cdots, 1)'$ be the vector of 1's of dimension n. For any set of real numbers $x_1, \cdots, x_n$, let $\boldsymbol{x}_n := (x_1, \cdots, x_n)'$. Let T be a function from $\mathbb{R}^n$ to $\mathbb{R}$ satisfying the following two conditions, where μ is a known real number.

$$\text{Under } \theta = 0, \;\; T(\boldsymbol{X}_n) - \mu \sim_D \mu - T(\boldsymbol{X}_n). \tag{6.2.1}$$

$$T(\boldsymbol{x}_n - t\mathbf{1}_n) \text{ is nonincreasing in } t, \text{ for all } \boldsymbol{x}_n \in \mathbb{R}^n. \tag{6.2.2}$$

Define

$$\begin{aligned} \hat{\theta}_1 &:= \sup\{t : T(\boldsymbol{X}_n - t\mathbf{1}_n) > \mu\}, \qquad \hat{\theta}_2 := \inf\{t : T(\boldsymbol{X}_n - t\mathbf{1}_n) < \mu\}, \\ \hat{\theta} &:= (\hat{\theta}_1 + \hat{\theta}_2)/2. \end{aligned} \tag{6.2.3}$$

To emphasize the dependence of these entities on $\boldsymbol{X}_n$, we shall sometimes write $\hat{\theta}(\boldsymbol{X}_n)$, $\hat{\theta}_1(\boldsymbol{X}_n)$ and $\hat{\theta}_2(\boldsymbol{X}_n)$ for $\hat{\theta}$, $\hat{\theta}_1$ and $\hat{\theta}_2$, respectively. Note the inequality $\hat{\theta}_1 \le \hat{\theta}_2$. This inequality is strict if $\{t : T(\boldsymbol{X}_n - t\mathbf{1}_n) = \mu\}$ is an interval of positive length. By the change of variable $s = t - b$, for any $b \in \mathbb{R}$,

$$\begin{aligned} \hat{\theta}_1(\boldsymbol{X}_n + b\mathbf{1}_n) &= \sup\{t : T(\boldsymbol{X}_n - (t-b)\mathbf{1}_n) > \mu\} \\ &= b + \sup\{s : T(\boldsymbol{X}_n - s\mathbf{1}_n) > \mu\} = b + \hat{\theta}_1(\boldsymbol{X}_n), \\ \hat{\theta}_2(\boldsymbol{X}_n + b\mathbf{1}_n) &= \inf\{t : T(\boldsymbol{X}_n - (t-b)\mathbf{1}_n) > \mu\} = b + \hat{\theta}_2(\boldsymbol{X}_n). \end{aligned}$$

This implies that the estimator $\hat{\theta}$ is location invariant, i.e.,

$$\hat{\theta}(\boldsymbol{X}_n + b\mathbf{1}_n) = b + \hat{\theta}(\boldsymbol{X}_n), \quad \forall b \in \mathbb{R}. \tag{6.2.4}$$

Therefore, the distribution of $\hat{\theta} - \theta$ does not depend on θ, i.e., for all $x \in \mathbb{R}$,

$$P_\theta\Big(\hat{\theta}(\boldsymbol{X}_n) - \theta \le x\Big) = P_\theta\Big(\hat{\theta}(\boldsymbol{X}_n - \theta\mathbf{1}_n) \le x\Big) = P_0\big(\hat{\theta} \le x\big). \tag{6.2.5}$$

The following lemma is useful in determining the distribution of $\hat{\theta}$ from that of $T(\boldsymbol{X}_n)$.

Lemma 6.2.1. *Assume* (6.2.1) *and* (6.2.2) *hold and that $\hat{\theta}_1$ and $\hat{\theta}_2$ are continuous r.v.'s. Then the following relations hold for every $b \in \mathbb{R}$.*

$$P_\theta\big(T(\boldsymbol{X}_n - b\mathbf{1}_n) < \mu\big) \le P_\theta\big(\hat{\theta} \le b\big) \le P_\theta\big(T(\boldsymbol{X}_n - b\mathbf{1}_n) \le \mu\big). \tag{6.2.6}$$

Proof. Fix a $b \in \mathbb{R}$. Then

$$\{\hat{\theta}_2 < b\} \subseteq \{T(\boldsymbol{X}_n - b\mathbf{1}_n) < \mu\} \subseteq \{\hat{\theta}_2 \le b\},$$
$$\{\hat{\theta}_1 > b\} \subseteq \{T(\boldsymbol{X}_n - b\mathbf{1}_n) > \mu\} \subseteq \{\hat{\theta}_1 \ge b\}.$$

These bounds and the continuity of the d.f.'s of $\hat{\theta}_1, \hat{\theta}_2$ yield

$$P_\theta(\hat{\theta}_2 \le b) = P_\theta\big(T(\boldsymbol{X}_n - b\mathbf{1}_n) < \mu\big), \quad P_\theta(\hat{\theta}_1 \le b) = P_\theta\big(T(\boldsymbol{X}_n - b\mathbf{1}_n) \le \mu\big).$$

These facts together with the fact $\hat{\theta}_1 \le \hat{\theta} \le \hat{\theta}_2$ readily yield (6.2.6). □

Corollary 6.2.1. *Let* $\gamma := P_0\big(T(\boldsymbol{X}_n) = \mu\big)$. *Under the conditions of Lemma 6.2.1, the following inequalities hold.*

$$\frac{1}{2} - \frac{\gamma}{2} \le P_\theta(\hat{\theta} \le \theta) \le \frac{1}{2} + \frac{\gamma}{2}, \quad \forall\, \theta \in \mathbb{R}.$$

Proof. Write T for $T(\boldsymbol{X}_n)$. By (6.2.1), when $\theta = 0$, T is symmetrically distributed around μ. Hence $P_0(T < \mu) = P_0(T > \mu)$. We also have $P_0(T < \mu) + P_0(T > \mu) = 1 - P_0(T = \mu) = 1 - \gamma$. Hence

$$P_0(T < \mu) = \frac{1-\gamma}{2},$$
$$P_0(T \le \mu) = P_0(T < \mu) + P_0(T = \mu) = \frac{1-\gamma}{2} + \gamma = \frac{1+\gamma}{2}.$$

By (6.2.6) applied with $b = \theta$ and the location invariance of T,

$$\frac{1-\gamma}{2} = P_0(T < \mu) \le P_0(\hat{\theta} \le 0) \le P_0(T \le \mu) = \frac{1+\gamma}{2},$$

thereby completing the proof. □

Statisticians often demand an estimator to be unbiased, i.e., whenever the expected value of an estimator exists, its expectation must equal the given parameter value, for every value of the parameter. This restricts the class of estimators. There is another notion that is less restrictive. It requires that an estimator overestimates the parameter as often as it underestimates it. This is the so-called notion of *median unbiasedness*.

Definition 6.2.1. An estimator $\tilde{\theta}$ of a parameter θ is said to be median unbiased if

$$P_\theta(\tilde{\theta} \le \theta) = P_\theta(\tilde{\theta} \ge \theta), \quad \forall\, \theta. \tag{6.2.7}$$

By Corollary 6.2.1, if the distributions of T and $\hat{\theta}$ are continuous under P_θ, then $\gamma = 0$ and $\hat{\theta}$ is median unbiased for θ, i.e.,

$$P_\theta(\hat{\theta} \le \theta) = 1/2 = P_\theta(\hat{\theta} \ge \theta), \quad \forall\, \theta \in \mathbb{R}.$$

In some examples below we will see that the distribution of the underlying T is not continuous for finite n, but in the limit, so that γ tends to zero. In such cases, as $n \to \infty$, $P_\theta(\hat{\theta} \le \theta) - P_\theta(\hat{\theta} \ge \theta) \to 0$, for all $\theta \in \mathbb{R}$, and we say that $\hat{\theta}$ is *asymptotically median unbiased* for θ.

Next, we give some conditions under which the distribution of $\hat{\theta}$ is symmetric about θ.

Lemma 6.2.2. *Suppose F is continuous and $F(x) = 1 - F(-x), \forall\, x \in \mathbb{R}$, and T satisfies*

$$T(\boldsymbol{x}) + T(-\boldsymbol{x}) = 2\mu, \quad \text{for some known } \mu \in \mathbb{R} \text{ and all } \boldsymbol{x} \in \mathbb{R}^n. \tag{6.2.8}$$

Then the distribution of $\hat{\theta}$ is symmetric about θ. Consequently, if $E_\theta\big(|\hat{\theta}|\big)$ is finite for every $\theta \in \mathbb{R}$, then $\hat{\theta}$ is unbiased for θ.

Proof. By (6.2.5), without loss of generality we shall assume $\theta = 0$. Thus we have to prove that $\hat{\theta}(\boldsymbol{X}_n) \sim_D -\hat{\theta}(\boldsymbol{X}_n)$. The symmetry of F about zero implies that

$$\hat{\theta}(\boldsymbol{X}_n) \sim_D \hat{\theta}(-\boldsymbol{X}_n). \tag{6.2.9}$$

We claim that under (6.2.8),

$$\hat{\theta}(-\boldsymbol{X}_n) = -\hat{\theta}(\boldsymbol{X}_n). \tag{6.2.10}$$

Combining this fact with (6.2.9) will yield the claim of the lemma.

By (6.2.8), we have $\{T(-\boldsymbol{X}_n+t\mathbf{1}_n > \mu\} = \{2\mu - T(bfX_n - t\mathbf{1}_n > \mu\} = \{T(\boldsymbol{X}_n - t\mathbf{1}_n < \mu\}$ for all real t. Change $-t$ to t to obtain,

$$\begin{aligned}
\hat{\theta}_1(-\boldsymbol{X}_n) &= \sup\{t : T(-\boldsymbol{X}_n - t\mathbf{1}_n) > \mu\} \\
&= -\inf\{t : T(-\boldsymbol{X}_n + t\mathbf{1}_n) > \mu\} \\
&= -\inf\{t : T(\boldsymbol{X}_n - t\mathbf{1}_n) < \mu\} \\
&= -\hat{\theta}_2(\boldsymbol{X}_n).
\end{aligned}$$

Similarly, $\hat{\theta}_2(-\boldsymbol{X}_n) \equiv -\hat{\theta}_1(\boldsymbol{X}_n)$ and hence the claim (6.2.10). □

We shall now see how the asymptotic distribution of $\hat{\theta}$ is obtained from that of $T(\boldsymbol{X}_n)$. Let a_n be a sequence of positive real numbers, $a_n \to \infty$, and let $x_n := a_n^{-1}x$. By (6.2.5) and (6.2.6),

$$\begin{aligned}
P_0\big(T(\boldsymbol{X}_n - x_n\mathbf{1}_n) < \mu\big) &\le P_\theta\big(a_n(\hat{\theta} - \theta) \le x\big) \\
&= P_0\big(\hat{\theta} \le x_n\big) \le P_0\big(T(\boldsymbol{X}_n - x_n\mathbf{1}_n) \le \mu\big).
\end{aligned} \tag{6.2.11}$$

The distribution of $T(\boldsymbol{X}_n - x_n\mathbf{1}_n)$ under P_0 is equivalent to the distribution of $T(\boldsymbol{X}_n)$ under P_{-x_n}. Therefore,

$$P_0\big(T(\boldsymbol{X}_n - x_n\mathbf{1}_n) \le \mu\big) = P_{-x_n}\big(a_n(T(\boldsymbol{X}_n) - \mu) \le 0\big).$$

Suppose there is a continuous d.f. H having zero mean and unit variance, and such that for some $b > 0$ and $\tau > 0$,

$$P_{-x_n}\Big(a_n\big(T(\boldsymbol{X}_n) - \mu\big) \le t\Big) \to H\Big(\frac{t + xb}{\tau}\Big), \quad \forall\, t, x \in \mathbb{R}. \tag{6.2.12}$$

Then from (6.2.11) and the continuity of H we readily obtain

$$P_\theta\Big(a_n(\hat{\theta} - \theta) \le x\Big) \to H\Big(\frac{xb}{\tau}\Big). \tag{6.2.13}$$

We shall now discuss some R and M estimators of the location parameter in the one and two sample location models, where the above methodology is found useful.

6.2.2 R Estimators

First, consider the one sample location model. Let $X_1, \cdots, X_n$ be a random sample from the one sample location model (4.3.2) for some $\theta \in \mathbb{R}$ and some d.f. F that is continuous and symmetric about zero. In (6.2.3), if T is based on $\boldsymbol{X}_n := (X_1, \cdots, X_n)'$ only through their signed ranks, then the estimator $\hat{\theta}$ is said to be an R estimator. A class of R estimators is developed as follows.

For $t \in \mathbb{R}$, let R_{it}^+ be the rank of $|X_i - t|$ among $|X_1 - t|, \cdots, |X_n - t|$. Let φ be a nondecreasing nonnegative function on $(0, 1)$, $\mathbf{1}_n$ be the n-dimensional vector of 1's, and define

$$S(\boldsymbol{X}_n - t\mathbf{1}_n) := \sum_{i=1}^{n} \operatorname{sgn}(X_i - t)\varphi(R_{it}^+/(n+1)). \tag{6.2.14}$$

Assume $\theta = 0$. Then, by Lemma 4.3.2, $E_0(\operatorname{sgn}(X)) = 0$, $\operatorname{sgn}(X_1), \cdots, \operatorname{sgn}(X_n)$ are independent of $R_{10}^+, \cdots, R_{n0}^+$ and hence $E_0(S(\boldsymbol{X}_n)) = 0$. Also, $-S(\boldsymbol{X}_n) = S(-\boldsymbol{X}_n) \sim_D S(\boldsymbol{X}_n)$. van Eeden (1972) has shown that $S(\boldsymbol{x}_n - t\mathbf{1}_n)$ is a nonincreasing function of t for every $\boldsymbol{x}_n \in \mathbb{R}^n$. Therefore (6.2.1) and (6.2.2) are satisfied by the estimating score function $S(\boldsymbol{X}_n - t\mathbf{1}_n)$. The analog of the estimator defined at (6.2.3), where T is replaced by this S, is an R estimator. For each φ, there is one such estimator. An R estimator that is easy to compute is the one with $\varphi(u) \equiv u$, which corresponds to the Wilcoxon signed rank statistic and which we shall now discuss in some detail. Let

$$V(\boldsymbol{X}_n) = \sum_{1 \le i \le j \le n} I(X_i + X_j > 0).$$

We shall apply the above principle with $S_n(\boldsymbol{X}_n, t) = V(\boldsymbol{X}_n - t\mathbf{1}_n)$.

Because of the location model structure, the distribution of $V(\boldsymbol{X}_n - \theta\mathbf{1}_n)$ under θ is the same as that of $V(\boldsymbol{X}_n)$ when $\theta = 0$, and $V(\boldsymbol{X}_n)$ is symmetrically distributed about $n(n+1)/4$. Also,

$$V(\boldsymbol{x}_n - t\mathbf{1}_n) = \sum_{1 \le i \le j \le n} I(x_i + x_j > 2t)$$

is clearly nonincreasing in t. Thus $V(\boldsymbol{x}_n)$ satisfies (6.2.1) and (6.2.2) with $\mu = n(n+1)/4$.

Let $a_{(1)} \le a_{(2)} \le \cdots \le a_{(M)}$ denote the order statistics of the $M = n(n+1)/2$ pairwise averages $(X_i + X_j)/2$, $1 \le i \le j \le n$. Then

$$V(\boldsymbol{X}_n - t\mathbf{1}_n) = \sum_{1 \le i \le j \le n} I(X_i + X_j > 2t) = \sum_{i=1}^{M} I(a_{(i)} > t).$$

The estimator $\hat{\theta}$ is that value of t for which $V(t) = M/2$. More precisely, suppose M is odd, i.e., $M = 2k + 1$, k a positive integer. Then $\mu = M/2 = k + 1/2$ and by (4.3.19)

$$\begin{aligned} \hat{\theta}_1 &:= \sup\{t : V(\boldsymbol{X}_n - t\mathbf{1}_n) > k + 1/2\} \qquad (6.2.15)\\ &= \sup\{t; \text{More than } k + 1/2 \text{ of the pairwise averages exceed } t\}\\ &= \sup\{t; a_{(k+1)} > t\} = a_{(k+1)},\\ \hat{\theta}_2 &= \inf\{t; V(\boldsymbol{X}_n - t\mathbf{1}_n) < k + 1/2\}\\ &= \inf\{t; \text{Fewer than } k + 1/2 \text{ of the pairwise averages exceed } t\}\\ &= \inf\{t; a_{(k+1)} \le t\} = a_{(k+1)},\\ \hat{\theta} &= a_{(k+1)}. \end{aligned}$$

Next, suppose $M = 2k$ so that $\mu = k$. Then, arguing similarly as above, $\hat\theta_1 = a_{(k)}, \hat\theta_2 = a_{(k+1)}$ and $\hat\theta = \big(a_{(k)} + a_{(k+1)}\big)/2$. Thus the estimator obtained from the Wilcoxon signed rank statistic is the Hodges-Lehmann estimator

$$\hat\theta_{HL} = \text{median}\Big\{\frac{X_i + X_j}{2}, 1 \le i \le j \le n\Big\}. \tag{6.2.16}$$

Direct calculations show that F being symmetric about zero implies $\widehat\theta_{HL}$ is symmetrically distributed about θ and the continuity of F implies that of the distribution of $\widehat\theta_{HL}$.

Asymptotic distribution of $\hat\theta_{HL}$. To describe the asymptotic distribution of $\hat\theta_{HL}$ we shall use (6.2.12) and (6.2.13) with $T(\boldsymbol{X}_n) = V(\boldsymbol{X}_n)/\binom{n}{2}$.

The identity $E_0\big(V(\boldsymbol{X}_n)\big) = n(n+1)/4$ gives

$$\mu = E_0\big(T(\boldsymbol{X}_n)\big) = \frac{n(n+1)}{4\binom{n}{2}} = \frac{(n+1)}{2(n-1)}.$$

The estimator $\hat\theta_{HL}$ is the same when we replace V and $n(n+1)/4$ by T and μ in (6.2.15).

Fix $x \in \mathbb{R}$ and let $x_n = a_n^{-1}x$. First, we need to determine the sequence $0 < a_n \to \infty$ and the limiting distribution of $a_n(T - \mu)$ when the parameter value is $x_n := a_n^{-1}x$, $x \in \mathbb{R}$. For this purpose we shall use (5.5.13). Recall the definition of $U^{(1)}$ from (5.2.2) and write

$$T = \frac{2}{n(n-1)}V = \frac{2}{n(n-1)}\sum_{i=1}^{n} I(X_i > 0) + U^{(1)},$$

$$\begin{aligned}\sqrt{n}(T - \mu) &= \frac{2}{\sqrt{n}(n-1)}\sum_{i=1}^{n} I(X_i > 0) + \sqrt{n}\Big[U^{(1)} - \frac{(n+1)}{2(n-1)}\Big] \\ &= \frac{2}{\sqrt{n}(n-1)}\sum_{i=1}^{n} I(X_i > 0) + \sqrt{n}\Big(U^{(1)} - \frac{1}{2}\Big) - \frac{\sqrt{n}}{(n-1)}.\end{aligned}$$

Hence,

$$\Big|\sqrt{n}(T - \mu) - \sqrt{n}\Big(U^{(1)} - \frac{1}{2}\Big)\Big| \le \frac{3\sqrt{n}}{n-1} \to 0.$$

By Theorem 2.10.5, it thus suffices to determine the limiting distribution of $n^{1/2}\big(U^{(1)} - \frac{1}{2}\big)$, under P_{-x_n}, where $a_n = n^{1/2}$. For this we assume that F has a square integrable density f and set $I_2(f) = \int f^2(y)\,dy$. Let Φ denote the d.f. of a $\mathcal{N}(0,1)$ r.v. Then, by (5.5.13),

$$P_{-x_n}\Big(n^{1/2}(U^{(1)} - 1/2) \le t\Big) = P_{-x_n}\Big(\frac{n^{1/2}(U^{(1)} - 1/2)}{(1/3)^{1/2}} \le \frac{t}{(1/3)^{1/2}}\Big) \to \Phi\Big(\frac{t + 2xI_2(f)}{(1/3)^{1/2}}\Big).$$

Thus (6.2.12) is satisfied with $H = \Phi$, $b = 2I_2(f)$ and $\tau = (1/3)^{1/2}$. Hence, by (6.2.13),

$$P_\theta\Big(n^{1/2}(\hat\theta_{HL} - \theta) \le x\Big) \to \Phi\Big(x\sqrt{12}I_2(f)\Big).$$

In other words,

$$n^{1/2}(\hat\theta_{HL} - \theta) \to_D \mathcal{N}\Big(0, \frac{1}{12\big(\int f^2(x)\,dx\big)^2}\Big). \tag{6.2.17}$$

Two sample location model. Now consider the two sample location model, where X, Y are two independent r.v.'s satisfying $X \sim_D F$, $Y - \theta \sim_D F$, for some real number θ and some continuous d.f. F. Let $\boldsymbol{X}_{n_1} := (X_1, \cdots, X_{n_1})$ and $\boldsymbol{Y}_{n_2} := (Y_1, \cdots, Y_{n_2})$ be two independent random samples from this two sample location model. Here our estimating score will be a function of $\boldsymbol{X}_{n_1}, \boldsymbol{Y}_{n_2}$ and θ. An estimator is said to be an R estimator if its defining estimating score function depends on $\boldsymbol{X}_{n_1}, \boldsymbol{Y}_{n_2}$ only through their ranks. One such estimator is obtained by basing the defining estimating score on the Mann-Whitney-Wilcoxon rank sum statistic.

As argued in Hodges and Lehmann (1963), a way to estimate θ is to choose a value of t that will somehow align $Y_1 - t, \cdots, Y_{n_2} - t$ with $X_1, \cdots, X_{n_1}$. How one aligns these variables depends on the estimating score statistic being used in this process. One possible choice of estimating score based on ranks is the two sample Mann-Whitney-Wilcoxon statistic

$$W(t) \equiv W(\boldsymbol{X}_{n_1}, \boldsymbol{Y}_{n_2} - t\boldsymbol{1}_{n_2}) := \sum_{i=1}^{n_1}\sum_{j=1}^{n_2} I(Y_j - X_i > t).$$

Clearly, $W(t)$ is nonincreasing in t and $E_0\big(W(0)\big) = n_1 n_2/2$. Let $m = n_1 n_2$. Thus (6.2.1) and (6.2.2) with $T = W$ are satisfied. Let

$$\tilde{\theta}_1 := \sup\Big\{t : W(t) > \frac{m}{2}\Big\}, \qquad \tilde{\theta}_2 := \inf\Big\{t : W(t) < \frac{m}{2}\Big\}, \qquad \hat{\theta}_{HLS} := \frac{\tilde{\theta}_1 + \tilde{\theta}_2}{2}.$$

Arguing as for (6.2.16) in the one sample case above, one obtains

$$\hat{\theta}_{HLS} = \text{median}\{Y_i - X_j, 1 \le i \le n_1, 1 \le j \le n_2\}. \tag{6.2.18}$$

The estimator $\hat{\theta}_{HLS}$ was developed by Hodges and Lehmann (1963) and Sen (1963a), independently. For that reason we call $\hat{\theta}_{HLS}$ the Hodges-Lehmann-Sen estimator for the two sample location parameter. If either F is symmetric about zero or $n_1 = n_2$, then $\hat{\theta}_{HLS}$ is symmetrically distributed about θ and median unbiased for θ.

Arguing as for (6.2.17), one can verify that if $n_1 \wedge n_2 \to \infty$, $n_1/(n_1 + n_2) \to \lambda \in (0, 1)$ and F has a density f satisfying $I_2(f) = \int f^2(y)\,dy < \infty$, then $\sqrt{\frac{n_1 n_2}{n_1+n_2}}\big(\hat{\theta}_{HLS} - \theta\big)$ has the same limiting distribution as given at (6.2.17).

6.2.3 M Estimators in the One Sample Location Model

Let $X_1, \cdots, X_n$ be a random sample from the location model (4.3.2) for some $\theta \in \mathbb{R}$ and for some d.f. F. The classical maximum likelihood estimator of the location parameter θ, in the case F is known and has an absolutely continuous density f with a.e. derivative $\dot{f}$, is defined to be a minimizer of $\sum_{i=1}^n \big(-\log f(X_i - t)\big)$ with respect to t or a solution t of the likelihood equation $-\sum_{i=1}^n \dot{f}(X_i - t)/f(X_i - t) = 0$. The estimators thus obtained are not necessarily robust against gross errors. For example if F is $N(0, \sigma^2)$ d.f., then the MLE of θ is the sample mean $\bar{X}$ obtained by minimizing $\sum_{i=1}^n (X_i - t)^2$ with respect to t. This dispersion and the corresponding minimizer is heavily influenced by outliers. One large observation makes both of these entities large. As argued in Huber (1964), one way to overcome this is to have a dispersion that tampers the influence of the extreme observations. Accordingly, let ρ be a function from $\mathbb{R}$ to $\mathbb{R}$ and define the M dispersion

$$M(t) := \sum_{i=1}^n \rho(X_i - t), \quad t \in \mathbb{R}.$$

An M estimator corresponding to the given ρ is a minimizer $\hat{\theta}_M$ of M:

$$M(\hat{\theta}_M) := \inf_{t\in\mathbb{R}} M(t). \tag{6.2.19}$$

Suppose

$$\rho \text{ is absolutely continuous with a.e. derivative } \psi. \tag{6.2.20}$$

A closely related estimator is given by a solution of the equation

$$S(t) := \sum_{i=1}^{n} \psi(X_i - t) = 0. \tag{6.2.21}$$

We denote such a solution by $\hat{\theta}_\psi$. It serves as a candidate for an M estimator. It is well known that not every solution of (6.2.21) need to be a minimizer of $M(t)$.

Note that if $\rho = -\log f$, then the M estimator of (6.2.19) corresponds to the maximum likelihood estimator obtained by minimizing $-\sum_{i=1}^{n} \ln f(X_i - t)$ w.r.t. t and a solution of (6.2.21) is a solution t of the likelihood equation $-\sum_{i=1}^{n} \dot{f}(X_i - t)/f(X_i - t) = 0$. If $\rho(x) = x^2$, then the M estimator is the least squares estimator $\bar{X}$ and if $\rho(x) = |x|$ then an M estimator is the least absolute deviation estimator $\tilde{X} := \text{median}\{X_i, 1 \le i \le n\}$. Note that while $\bar{X}$ is extremely sensitive to gross errors, the sample median $\tilde{X}$ is robust against gross errors. A class of M estimators that combines these two scores is obtained as follows. Let $k > 0$ be given. Let

$$\begin{aligned} \rho(x) &:= \frac{x^2}{2} I(|x| \le k) + \big(k|x| - \frac{k^2}{2}\big) I(|x| > k), \\ \psi(x) &:= x I(|x| \le k) + k\text{sgn}(x). \end{aligned} \tag{6.2.22}$$

The score functions ψ, one corresponding to each k, are called Huber score functions, and the corresponding M estimators are called Huber M estimators. Observe that for those X_i's for which $|X_i - t| \le k$, the score function $S(t)$ behaves like that of the squared error score function while for those X_i's for which $|X_i - t| > k$, it is like the absolute deviations score function.

In all of the above examples the function ρ is convex and ψ is nondecreasing. In general suppose

$$\psi(x) \text{ is nondecreasing in } x \text{ and } \int \psi(x)\, dF(x) = 0. \tag{6.2.23}$$

Then the score function $S(t)$ of (6.2.21) is nonincreasing in t and satisfies $E_\theta\big(S(\theta)\big) = 0$, for all $\theta \in \mathbb{R}$. Moreover, (6.2.23) implies that the integral $\lambda(t) := \int \psi(x - t)\, dF(x)$ is well defined for each $t \in \mathbb{R}$. This is true for $t = 0$ by assumption. With $\psi^+(x) = \psi(x)I(\psi(x) > 0)$ and $\psi^-(x) = -\psi(x)I(\psi(x) < 0)$, we rewrite

$$\begin{aligned} \lambda(t) &= \int \psi(x - t)\, dF(x) \\ &= \int \psi^+(x - t)\, dF(x) - \int \psi^-(x - t)\, dF(x) = \lambda^+(t) - \lambda^-(t), \quad \text{say.} \end{aligned}$$

By (6.2.23), ψ is nondecreasing and so are ψ^+ and $-\psi^-$. Thus $t \ge 0$ implies $0 \le \lambda^+(t) \le \lambda^+(0)$ and $t \le 0$ yields $0 \le \lambda^-(t) \le \lambda^-(0)$. As $\int \psi\, dF = 0$, both $\lambda^+(0) = \int \psi^+\, dF$ and $\lambda^-(0) = \int \psi^-\, dF$ are finite. Thus at least one of the terms $\lambda^+(t)$ and $\lambda^-(t)$ is finite, and $\lambda(t)$ is well defined but may take the value ∞ or $-\infty$. Since ψ is nondecreasing, we have $\lambda(s) \ge \lambda(t)$ for $s < t$.

Write $S(\boldsymbol{X}_n - t\mathbf{1}_n)$ for $S(t)$. Analogous to R estimators of the Subsection 6.2.2, if the equation $S(\boldsymbol{X}_n - t\mathbf{1}_n) = 0$ does not have a unique solution then a solution is defined as follows. Let

$$\hat{\theta}_{1\psi} := \sup\{t : S(\boldsymbol{X}_n - t\mathbf{1}_n) > 0\}, \qquad \hat{\theta}_{2\psi} := \inf\{t; S(\boldsymbol{X}_n - t\mathbf{1}_n) < 0\}, \tag{6.2.24}$$
$$\hat{\theta}_{\psi} := \frac{\hat{\theta}_{1\psi} + \hat{\theta}_{2\psi}}{2}.$$

We have already seen examples of computing estimators like $\hat{\theta}_\psi$ when talking about R estimator in the one and two sample location models.

We shall now discuss consistency and asymptotic normality of $\hat{\theta}_\psi$. Argue as for (6.2.4) to conclude that $\hat{\theta}_\psi$ is location invariant and like (6.2.6), if $\hat{\theta}_\psi$ has a continuous distribution, then

$$P_\theta\big(S(\boldsymbol{X}_n - b\mathbf{1}_n) < 0\big) \le P_\theta\big(\hat{\theta}_\psi \le b\big) \le P_\theta\big(S(\boldsymbol{X}_n - b\mathbf{1}_n) \le 0\big), \quad \forall\, b \in \mathbb{R}. \tag{6.2.25}$$

Recall $\varepsilon = X - \theta$ and $\varepsilon_1 = X_1 - \theta, \cdots, \varepsilon_n = X_n - \theta$. Note that $\hat{\theta}_\psi - \theta$ is a solution T_n of the equation $\sum_{i=1}^n \psi(\varepsilon_i - T_n) = 0$.

Lemma 6.2.3. *Suppose* (6.2.23) *holds and that $\lambda(t) > 0$, for all $t < 0$ and $\lambda(t) < 0$ for all $t > 0$. Then, $T_n \to_p 0$ or equivalently, $\hat{\theta}_\psi \to_p \theta$.*

Proof. Fix an $\eta > 0$. Let $\boldsymbol{\varepsilon}_n := (\varepsilon_1, \cdots, \varepsilon_n)'$. Note that $S(\boldsymbol{X}_n - (\theta \pm \eta)\mathbf{1}_n) = S(\boldsymbol{\varepsilon}_n \mp \eta\mathbf{1}_n)$. Moreover, $\hat{\theta}_\psi < \theta - \eta$ implies $n^{-1}S(\boldsymbol{\varepsilon}_n + \eta\mathbf{1}_n) \le 0$ and $\hat{\theta}_\psi > \theta + \eta$ implies $n^{-1}S(\boldsymbol{\varepsilon}_n - \eta\mathbf{1}_n) \ge 0$. By Theorem 2.10.6,

$$n^{-1}S(\boldsymbol{\varepsilon}_n - \eta\mathbf{1}_n) \to_p \lambda(\eta) < 0, \quad n^{-1}S(\boldsymbol{\varepsilon}_n + \eta\mathbf{1}_n) \to_p \lambda(-\eta) > 0.$$

Hence, the above set inclusions imply

$$P_\theta\big(\hat{\theta}_\psi > \theta + \eta\big) \le P_0\big(n^{-1}S(\boldsymbol{\varepsilon}_n - \eta\mathbf{1}_n) \ge 0\big) \to 0,$$
$$P_\theta\big(\hat{\theta}_\psi < \theta - \eta\big) \le P_0\big(n^{-1}S(\boldsymbol{\varepsilon}_n + \eta\mathbf{1}_n) \le 0\big) \to 0,$$

thereby completing the proof of the lemma. □

Next, we shall prove the asymptotic normality of $\hat{\theta}_\psi$ under the following additional conditions.

$$\lambda \text{ is differentiable at } t = 0 \text{ with } \dot{\lambda}(0) < 0. \tag{6.2.26}$$
$$\sigma_\psi^2 := E(\psi^2(\varepsilon)) < \infty \text{ and } \lim_{t\to 0} E\big(|\psi(\varepsilon - t) - \psi(\varepsilon)|^2\big) = 0. \tag{6.2.27}$$

Lemma 6.2.4. *Under* (6.2.23), (6.2.26) *and* (6.2.27), *the following holds.*

$$n^{1/2}(\hat{\theta}_\psi - \theta) \to_D \mathcal{N}(0, \tau^2), \qquad \tau^2 = \sigma_\psi^2/\big(\dot{\lambda}(0)\big)^2.$$

Proof. By the definition (6.2.24) of $\hat{\theta}_\psi$ and analogous to (6.2.11), for any $x \in \mathbb{R}$,

$$P\Big(S(\boldsymbol{\varepsilon}_n - n^{-1/2}x\mathbf{1}_n) < 0\Big) \le P_\theta\Big(n^{1/2}(\hat{\theta}_\psi - \theta) \le x\Big) \tag{6.2.28}$$
$$\le P\Big(S(\boldsymbol{\varepsilon}_n - n^{-1/2}x\mathbf{1}_n) \le 0\Big).$$

The r.v.'s

$$Y_{ni} := n^{-1/2}\Big[\psi(\varepsilon_i - n^{-1/2}x) - \psi(\varepsilon_i) - \lambda\big(n^{-1/2}x\big)\Big], \quad i = 1, \cdots, n.$$

are i.i.d. with mean zero and by (6.2.27),

$$\mathrm{Var}\Big(\sum_{i=1}^{n} Y_{ni}\Big) = \sum_{i=1}^{n} \mathrm{Var}(Y_{ni}) \le E\Big(\big|\psi(\varepsilon - n^{-1/2}x) - \psi(\varepsilon)\big|^2\Big) \to 0.$$

Thus Corollary 2.10.2 implies that, for every $x \in \mathbb{R}$,

$$\sum_{i=1}^{n} Y_{ni} = n^{-1/2}\Big(S(\boldsymbol{\varepsilon}_n - n^{-1/2}x\mathbf{1}_n) - S(\boldsymbol{\varepsilon}_n) - n\lambda(n^{-1/2}x)\Big) \to_p 0.$$

Moreover, by (6.2.26),

$$n^{1/2}\lambda(n^{-1/2}x) = n^{1/2}\big[\lambda(n^{-1/2}x) - \lambda(0)\big] \to x\dot{\lambda}(0).$$

The last two statements together yield

$$n^{-1/2}S(\boldsymbol{\varepsilon}_n - n^{-1/2}x\mathbf{1}_n) = n^{-1/2}S(\boldsymbol{\varepsilon}_n) + x\dot{\lambda}(0) + o_p(1).$$

Thus by Theorem 2.10.9 and Slutsky's Theorem (2.10.5), we conclude

$$n^{-1/2}S(\boldsymbol{\varepsilon}_n - n^{-1/2}x\mathbf{1}_n) \to_D \mathcal{N}(x\dot{\lambda}(0), \sigma_\psi^2).$$

From this we derive $P\big(S(\boldsymbol{\varepsilon}_n - n^{-1/2}x\mathbf{1}_n) < 0\big)$ and $P\big(S(\boldsymbol{\varepsilon}_n - n^{-1/2}x\mathbf{1}_n) \le 0\big)$ converge to the same limit $\Phi\big(-x\dot{\lambda}(0)/\sigma_\psi\big)$, where Φ is the d.f. of a $\mathcal{N}(0,1)$ r.v.. Hence, in view of (6.2.28), we obtain that

$$P_\theta\big(n^{1/2}(\hat{\theta}_\psi - \theta) \le x\big) \to \Phi\big(-x\dot{\lambda}(0)/\sigma_\psi\big) = \Phi\big(x|\dot{\lambda}(0)|/\sigma_\psi\big).$$

But this limit is the d.f. of a $\mathcal{N}\big(0, \sigma_\psi^2/(\dot{\lambda}(0))^2\big)$ r.v. at x. □

The following lemmas give sufficient conditions for the differentiability of λ at 0.

Lemma 6.2.5. *Suppose ψ is absolutely continuous with a.e. derivative $\dot{\psi}$, $H(s) := \int \dot{\psi}(x - s)\,dF(x)$ is finite for s close to 0 and $H(s) \to 0$ as $s \to 0$. Then λ is differentiable at 0 with derivative $\dot{\lambda}(0) = -\int \dot{\psi}(x)\,dF(x)$.*

Proof. Using the absolute continuity of ψ and then Theorem 2.3.2, one derives

$$\begin{aligned}\lambda(t) - \lambda(0) &= \int [\psi(x - t) - \psi(x)]\,dF(x) = -t\int\int_0^1 \dot{\psi}(x - st)\,ds\,dF(x)\\ &= -t\int_0^1\int \dot{\psi}(x - st)\,dF(x)\,ds, \quad \forall\, t \in \mathbb{R}.\end{aligned}$$

By assumptions, $H(s) := \int \dot{\psi}(x - s)\,F(x)$ is continuous at $s = 0$ and thus bounded for s close to 0. Hence, by Theorem 2.10.10,

$$\dot{\lambda}(0) := \lim_{t\to 0}\frac{\lambda(t) - \lambda(0)}{t} = \lim_{t\to 0} -\int_0^1 H(st)\,ds = -\int_0^1 H(0)\,ds = -\int \dot{\psi}(x)\,dF(x)$$

as asserted. □

Lemma 6.2.6. *Suppose* (6.2.27) *holds and F has a density with finite Fisher information for location, see Definition 2.11.1. Then λ is differentiable at* 0 *with derivative $\dot{\lambda}(0) = \int \dot{f}(x)\psi(x)\,dx$.*

Proof. For any measurable function h with $\int h^2(y)f(y)\,dy$ finite, (2.11.1) and (2.4.4) give the inequality

$$\Big(\int \dot f(y)h(y)dy\Big)^2 = \Big(\int \frac{\dot f(y)}{f(y)}\psi(y)f(y)dy\Big)^2 \le I_L(f)\int h^2(y)f(y)\,dy < \infty. \tag{6.2.29}$$

This inequality shows that $\int \dot f(y) psi(y)\,dy$ is finite and that $\int \dot f(y)(\psi(y-t)-\psi(y))\,du \to 0$ as $t \to 0$ in view of (6.2.27). To see this, we have

$$\lambda(t) = \int \psi(x-t)f(x)dx = \int f(y+t)\psi(y)\,dy$$

and arguing as n the previous proof we now obtain

$$\begin{aligned}\lambda(t)-\lambda(0)-t\int \dot f(y)\psi(y)\,dy &= \int [f(y+t)-f(y)-t\dot f(y)]\psi(y)\,dy\\ &= t\int_0^1\int \dot f(y+st)-\dot f(y)\psi(y)\,dy\,ds\\ &= t\int_0^1\int \dot f(y)(p(y-st)-\psi(y))\,dy\,ds = o(t).\end{aligned}$$

This shows that λ has derivative $\int \dot f(y)\psi(y)\,dy$ at 0. □

Asymptotic minimaxity. We shall now talk about an asymptotic minimaxity property of M estimators. Suppose the assumptions of Lemma 6.2.6 hold and $\dot\lambda(0) = \int \dot f(y)\psi(y)\,dy$ is negative. Then, by (6.2.29), we have the inequality $(\dot\alpha(0))^2 \le I_L(f)\sigma_\psi^2 < \infty$. Since $\big(\dot\lambda(0)\big)^2$ is positive, this inequality says that the asymptotic variance $\tau^2(\psi,F)$ of $\hat\theta_\psi$ satisfies

$$\tau^2 \equiv \tau^2(\psi,F) = \frac{E(\psi^2(\varepsilon))}{\big(\dot\lambda(0)\big)^2} \ge \frac{1}{I_L(f)},$$

with equality attained if and only if ψ is proportional to $\dot f/f$. Recall that the asymptotic variance of $n^{1/2}(\hat\theta_{MLE}-\theta)$ is $1/I_L(f)$. The above inequality says that the lower bound on the asymptotic variance is achieved by the maximum likelihood estimator.

Next, fix $0 \le \epsilon < 1$ and let G be a d.f. on $\mathbb{R}$ and let $\mathcal{F}_\epsilon := \{F = (1-\epsilon)G + \epsilon H;\ H \text{ d.f. on } \mathbb{R}\}$. Assume that the support of G is $\mathbb{R}$ and G has twice continuously differentiable density g with $-\log g$ convex on $\mathbb{R}$, which in particular implies that $-\dot g/g$ is nondecreasing on $\mathbb{R}$.

Now imagine playing a game where nature gets to pick an $F \in \mathcal{F}_\epsilon$ to maximize your loss $\tau^2(\psi,F)$, and you get to pick a ψ so as to minimize this maximum loss. Huber (1964) proved that this game has a saddle point. We state his Theorem 1, without proof, as the following theorem.

Theorem 6.2.1. *Under the above setup there is an $F_0 = (1-\epsilon)G + \epsilon H_0 \in \mathcal{F}_\epsilon$ and a ψ_0 such that*

$$\sup_{F\in\mathcal{F}_\epsilon} \tau^2(\psi_0,F) = \tau^2(\psi_0,F_0) = \inf_\psi \tau^2(\psi,F_0), \tag{6.2.30}$$

where $F_0 \in \mathcal{F}_\epsilon$ such that $\int \psi_0(x)\,dF_0(x) = 0$ and where ψ in $\inf_\psi$ is a real-valued nondecreasing bounded function with $\int \dot f_0(x)\psi(x)\,dx < 0$. Let $y_0 < y_1$ be the endpoints of the interval $\{y \in \mathbb{R}; |\dot g(y)/g(y)| \le k\}$, where k is related to ϵ by the relation

$$\frac{1}{1-\epsilon} = \int_{y_0}^{y_1} g(y)dy + \frac{g(y_0)+g(y_1)}{k}.$$

Then the density f_0 *of the least favorable d.f.* F_0 *is given by*

$$\begin{aligned} f_0(y) &= (1-\epsilon)g(y_0)e^{k(y-y_0)}, & y \le y_0, \\ &= (1-\epsilon)g(y), & y_0 < y < y_1, \\ &= (1-\epsilon)g(y_1)e^{-k(y-y_1)}, & y \ge y_1. \end{aligned}$$

The optimal score function $\psi_0 = -\dot{f}_0/f_0$ *given by*

$$\psi_0(y) = -kI(y \le y_0) - (1-\epsilon)\frac{\dot{g}(y)}{g(y)}I(y_0 < y < y_1) + kI(y \ge y_1)$$

is monotone and bounded. It corresponds to the MLE of the parameter θ *when* F_0 *is the error d.f.*

As noted by Huber, it is a bit surprising that in the above game the nature is not very cruel, i.e., the tails of the least favorable error d.f. F_0 are not very heavy. They are exponential tails. If G is the d.f. of a $\mathcal{N}(0,1)$ r.v., then we need to protect only against exponential distribution contamination and not against heavy tail distributions like Cauchy. Also, in this case the optimal score function is $\psi_0(y) = -kI(y \le y_0) + (1-\epsilon)yI(y_0 < y < y_1) + kI(y \ge y_1)$. It behaves like the score function for the normal error distribution in the middle and for the Laplace distribution in the tails.

6.2.4 Minimum Distance Estimators

In this subsection we shall discuss some minimum distance estimators in the one and two sample location model. Some common estimators can be viewed as minimum distance estimators. For example, the sample mean minimizes the squared distances

$$\sum_{i=1}^{n}(X_i - t)^2, \quad t \in \mathbb{R},$$

and is thus a least squares estimator. The sample median is a minimizer of the sum of the absolute deviations

$$\sum_{i=1}^{n}|X_i - t|, \quad t \in \mathbb{R},$$

and is thus a least absolute deviation (LAD) estimator. In this section we will be concerned with minimum distance estimators that are based on distances involving residual empirical d.f.'s.

Consider first the one sample symmetric location model with location parameter θ, error d.f. $F \in \mathcal{F}_s$ and observations $X_1, \cdots, X_n$, see (4.3.11). Since $F \in \mathcal{F}_s$ is symmetric about 0 and continuous, θ is the point of symmetry of the distribution of X. Hence, the random vectors $(X_1 - \theta, \cdots, X_n - \theta)'$ and $(-(X_1 - \theta), \cdots, -(X_n - \theta))'$ have the same distribution. Therefore, an estimator of θ can be defined as a value of t which minimizes some distance between the empirical d.f. of the residuals $X_1 - t, \cdots, X_n - t$ with that of their symmetrized versions $-(X_1 - t), \cdots, -(X_n - t)$. The empirical d.f.'s of $X_1 - t, \cdots, X_n - t$ and $-(X_1 - t), \cdots, -(X_n - t)$, respectively, are

$$n^{-1}\sum_{i=1}^{n} I(X_i - t \le x), \qquad n^{-1}\sum_{i=1}^{n} I(-X_i + t \le x), \quad x \in \mathbb{R}.$$

Let G be a nondecreasing right continuous function from $\mathbb{R}$ to $\mathbb{R}$. A useful distance between the above two empirical d.f.'s is their L_2 distance with respect to G, defined as

$$K_G(t) := n^{-2}\int\Big[\sum_{i=1}^{n}\Big\{I(X_i - t \le x) - I(-X_i + t \le x)\Big\}\Big]^2 dG(x), \quad t \in \mathbb{R}.$$

A minimum distance (m.d.) estimator $\widehat{\theta}_G$ is then a minimizer of the criterion function K_G, i.e.,

$$K_G(\widehat{\theta}_G) = \inf_{t \in \mathbb{R}} K_G(t). \tag{6.2.31}$$

If we take $G(x) = I(x \geq 0)$, $x \in \mathbb{R}$, then we have

$$K_G(t) = n^{-2}\Big(\sum_{i=1}^{n} \text{sgn}(X_i - t)\Big)^2,$$

and the sample median $\tilde{X} = \text{median}\{X_i, 1 \leq i \leq n\}$ is an m.d. estimator. If we take $G(x) = x$, $x \in \mathbb{R}$, then we have

$$K_G(t) = n^{-2} \sum_{i,j=1}^{n} |X_i + X_j - 2t| - n^{-2} \sum_{i,j=1} |X_i - X_j| \tag{6.2.32}$$

as will be shown below and an m.d. estimator is given by the Hodges-Lehmann estimator $\widehat{\theta}_{HL} = \text{median}\big\{(X_i + X_j)/2; 1 \leq i \leq j \leq n\big\}$.

We shall now derive a computational formula for $K_G(t)$ which gives the formula (6.2.32) as a special case. For this we set $H(x) = G(x_-)$, $r_i = X_i - t$ and $B = \max_{1 \leq i \leq n} |r_i|$. Note that for $x < -B$ or $x \geq B$, the integrand in $K_G(t)$ vanishes. Hence

$$K_G(t) = n^{-2} \int_{-B \leq x < B} \Big[\sum_{i=1}^{n} I(r_i \leq x) - \sum_{i=1}^{n} I(-r_i \leq x)\Big]^2 dG(x).$$

Recall, for any two numbers a, b,

$$a \vee b := \max(a, b) = \frac{a + b + |a - b|}{2}.$$

Since H is nondecreasing, for any $x, y \in \mathbb{R}$, we obtain

$$H(x \vee y) = H(x) \vee H(y) = \frac{1}{2}\big[H(x) + H(y) + |H(x) - H(y)|\big].$$

Now expand the quadratic in $K_G(t)$, carry out the term-by-term integration, and use the above properties to obtain the following identity.

$$\begin{aligned} n^2 K_G(t) &= \sum_{i,j=1}^{n} \Big\{\big[H(B) - H(r_i \vee r_j)\big] - 2\big[H(B) - H(r_i \vee -r_j)\big] \\ &\qquad + \big[H(B) - H(-r_i \vee -r_j)\big]\Big\} \\ &= \sum_{i,j=1}^{n} \Big\{2\big[H(r_i) \vee H(-r_j)\big] - H(r_i) \vee H(r_j) - H(-r_i) \vee H(-r_j)\Big\} \\ &= \sum_{i,j=1}^{n} \Big\{2\frac{H(r_i) + H(-r_j) + \big|H(r_i) - H(-r_j)\big|}{2} \\ &\qquad - \frac{H(r_i) + H(r_j) + \big|H(r_i) - H(r_j)\big|}{2} \\ &\qquad - \frac{H(-r_i) + H(-r_j) + \big|H(-r_i) - H(-r_j)\big|}{2}\Big\} \end{aligned} \tag{6.2.33}$$

$$= \sum_{i,j=1}^{n} \Big\{ \big|H(r_i) - H(-r_j)\big| - \frac{1}{2}\big|H(-r_i) - H(-r_j)\big| - \frac{1}{2}\big|H(r_i) - H(r_j)\big| \Big\}.$$

Suppose G also satisfies $|G(a) - G(b)| = |G(-a) - G(-b)|$, for all $a, b \in \mathbb{R}$. Then G is continuous and symmetric about zero and the above formula reduces to

$$K_G(t) = n^{-2} \sum_{i,j=1}^{n} \Big\{ \big|G(r_i) - G(-r_j)\big| - \big|G(r_i) - G(r_j)\big| \Big\} \tag{6.2.34}$$

$$= n^{-2} \sum_{i,j=1}^{n} \Big\{ \big|G(X_i - t) - G(-X_j + t)\big| - \big|G(X_i - t) - G(X_j - t)\big| \Big\}.$$

In particular if $G(x) \equiv x$, then this formula reduces to the formula (6.2.32).

The next theorem describes the asymptotic distribution of $\widehat{\theta}_G$ of (6.2.31) for a general G. Its proof can be found in Chapter 5 of Koul (2002).

Theorem 6.2.2. *Suppose the one sample location model (4.3.2) holds for some $\theta \in \mathbb{R}$ and some error d.f. F. Suppose G is symmetric about 0, i.e., $dG(x) = -dG(-x)$, for all $x \in \mathbb{R}$, and F has a density f that is symmetric about 0 and satisfies the following conditions.*

$$0 < \int f^j(x)\, dG(x) < \infty, \quad j = 1, 2. \tag{6.2.35}$$

$$\lim_{s \to 0} \int \big[f(x+s) - f(x)\big]^j\, dG(x) = 0, \quad j = 1, 2. \tag{6.2.36}$$

$$\int I(x \geq 0)(1 - F(x))\, dG(x) < \infty. \tag{6.2.37}$$

Let $\psi(x) := \int I(y \leq x) f(y)\, dG(y)$, $x \in \mathbb{R}$. Then, $n^{1/2}(\widehat{\theta}_G - \theta) \to_D \mathcal{N}(0, \sigma_G^2)$, where

$$\sigma_G^2 := \frac{\mathrm{Var}\big(\psi(\varepsilon)\big)}{\big(\int f^2(x)\, dG(x)\big)^2} = \frac{\int \big[F(x \wedge y) - F(x)F(y)\big] f(x) f(y)\, dG(x)\, dG(y)}{\big(\int f^2(x)\, dG(x)\big)^2}.$$

Remark 6.2.1. If G is a d.f., then the assumptions (6.2.35) to (6.2.37) are implied by

$$\int f\, dG > 0, \quad \int f^2\, dG < \infty, \quad \lim_{s \to 0} \int \big(f(x+s) - f(x)\big)^2\, dG(x) = 0.$$

Indeed, in view of the moment inequality (2.4.7), the first two properties give (6.2.35), the third property implies (6.2.36), and (6.2.37) is satisfied as $1 - F$ is bounded. Note that (6.2.37) implies that $E\big(K_G(\theta)\big) < \infty$, for all $n \geq 1$. Hence, by (2.4.8), $K_G(\theta) = O_p(1)$.

We shall now discuss the assumptions of the above theorem and describe the asymptotic variance σ_G^2 in three cases.

Case 1: Let $G(x) = I(x \geq 0)$, i.e, G is degenerate at zero. Then $\hat{\theta}_G$ is the sample median $\tilde{X}$. Now assume f is symmetric about 0, and continuous and positive at 0. Then the assumptions of the theorem are met. As G is a d.f. which is symmetric about 0, we only need to verify the three conditions in Remark 6.2.1. They follow from the identities $\int f^j\, dG = (f(0))^j$, $j = 1, 2$ and $f(0) > 0$ and from

$$\lim_{s \to 0} \int \big[f(y+s) - f(y)\big]^2\, dG(y) = \lim_{s \to 0} \big[f(s) - f(0)\big]^2 = 0,$$

in view of the continuity of f at 0. Moreover, we calculate

$$\psi(\varepsilon) = \int I(x \le \varepsilon) f(x)\, dG(x) = f(0) I(\varepsilon \ge 0), \qquad \mathrm{Var}\big(\psi(\varepsilon)\big) = \frac{f^2(0)}{4},$$
$$\int f^2(x)\, dG(x) = f^2(0), \qquad \sigma_G^2 = \sigma_0^2 = \frac{1}{4f^2(0)}.$$

Case 2: Let $G(x) = x$, $x \in \mathbb{R}$. This G is clearly continuous and symmetric about 0. Assume f is symmetric about 0 and the integrals $\int |x| f(x)\, dx$ and $\int f^2(x)\, dx$ are finite. Since f is a square integrable density, (6.2.35) and (6.2.36) are met, for the latter see Lemma 2.10.10. Finally (6.2.37) holds, because f has a finite mean. Here $\psi(\varepsilon) = F(\varepsilon)$, $\mathrm{Var}(\psi(\varepsilon)) = 1/12$ and

$$\sigma_G^2 = \frac{1}{12\Big(\int f^2(x)\, dx\Big)^2}.$$

Case 3: Suppose F is known and $G = F$. In this case it suffices to assume f to be symmetric about 0 and $\int f^3(x)\, dx < \infty$. Then, all conditions of Remark 6.2.1 are satisfied, and hence those of the above theorem hold. We then have $\psi(\varepsilon) = \int I(x \le \varepsilon) f^2(x)\, dx$,

$$\sigma_G^2 = \sigma_F^2 = \frac{\iint \big[F(x \wedge y) - F(x)F(y)\big] f^2(x) f^2(y)\, dx\, dy}{\big(\int f^3(x)\, dx\big)^2}.$$

Some numerical values of these asymptotic variances for some distributions F appear in Koul (1992, 2002). From Table 1 on p. 207 of Koul (2002), one sees that the estimator $\hat{\theta}_F$ is asymptotically more efficient than the LAD at logistic error distribution while it is asymptotically more efficient than the HL estimator at the double exponential and Cauchy error distributions.

Next, consider the **two sample location model** where for some $\theta \in \mathbb{R}$, $X_1, \cdots, X_{n_1}$ and $Y_1 - \theta, \cdots, Y_{n_2} - \theta$ are i.i.d. F, an unknown distribution function. Let

$$F_{n_1}(x) := \frac{1}{n_1}\sum_{i=1}^{n_1} I(X_i \le x), \qquad G_{n_2}(x) := \frac{1}{n_2}\sum_{j=1}^{n_2} I(Y_j \le x).$$

Then the empirical d.f. of $Y_1 - t, \cdots, Y_n - t$ is

$$G_{n_2}(x + t) := \frac{1}{n_2}\sum_{j=1}^{n_2} I(Y_j \le x + t).$$

Here we align the two data sets $X_1, \cdots, X_{n_1}$ and $Y_1 - t, \cdots, Y_{n_2} - t$ by choosing a t that brings the following L_2 distance of their empirical d.f.'s close to each other. With G as above, define

$$M_G(t) := \int \big[F_{n_1}(x) - G_{n_2}(x + t)\big]^2\, dG(x), \qquad \widetilde{\theta}_G := \operatorname*{argmin}_t M_G(t). \tag{6.2.38}$$

We shall now give a computational formula for $M_G(t)$. As in the one sample setup, one can derive a formula for a general G that may not be continuous. But, for simplicity, we shall assume G to be continuous. Let $A = \min\{X_1, \cdots, X_{n_1}, Y_1 - t, \cdots, Y_{n_2} - t\}$, and $B = \max\{X_1, \cdots, X_{n_1}, Y_1 - t, \cdots, Y_{n_2} - t\}$. Observe that if $x < A$ or $x \ge B$, then the integrand $F_{n_1}(x) - G_{n_2}(x + t)$ is zero. Argue as in the one sample case to obtain that

$$M_G(t) = \int_A^B \Big[\frac{1}{n_1}\sum_{i=1}^{n_1} I(X_i \le x) - \frac{1}{n_2}\sum_{j=1}^{n_2} I(Y_j - t \le x)\Big]^2\, dG(x)$$

$$= \frac{1}{n_1 n_2}\sum_{i=1}^{n_1}\sum_{j=1}^{n_2}\big|G(Y_j - t) - G(X_i)\big| - \frac{1}{2n_2^2}\sum_{i=1}^{n_2}\sum_{j=1}^{n_2}\big|G(Y_i - t) - G(Y_j - t)\big|$$
$$- \frac{1}{2n_1^2}\sum_{i=1}^{n_1}\sum_{j=1}^{n_1}\big|G(X_i) - G(X_j)\big|.$$

In the case $G(x) = x, x \in \mathbb{R}$, we write M_I for M_G, and the above formula becomes

$$M_I(t) = \frac{1}{n_1 n_2}\sum_{i=1}^{n_1}\sum_{j=1}^{n_2}\big|Y_j - X_i - t\big| - \frac{1}{n_2^2}\sum_{i=1}^{n_2}\sum_{j=1}^{n_2}\frac{|Y_i - Y_j|}{2} - \frac{1}{n_1^2}\sum_{i=1}^{n_1}\sum_{j=1}^{n_1}\frac{|X_i - X_j|}{2}.$$

Therefore, $\widetilde{\theta}_I = \text{med}\{Y_j - X_i; 1 \le i \le n_1,\ 1 \le j \le n_2\} = \hat{\theta}_{HLS}$, where $\hat{\theta}_{HLS}$, as in (6.2.18), is the HLS two sample location parameter estimator obtained via the Mann-Whitney-Wilcoxon statistic. Fine (1966) was the first author to point out this fact.

6.3 Influence function

Consider the one sample location setup with error d.f. F. An important concept for describing the robustness (or lack of it) of an estimator is to see how much an extreme observation influences the given estimator. To make this precise, we need to view estimators and parameters as functionals of d.f.'s. For a given $x \in \mathbb{R}$, let $\delta_x(\cdot)$ be the d.f. of the r.v. that is degenerate at x, i.e., $\delta_x(t) := I(t \ge x)$. Given $0 \le \epsilon \le 1$, consider the ϵ-contaminated d.f.

$$F_\epsilon = (1-\epsilon)F + \epsilon\delta_x = F + \epsilon(\delta_x - F) = F - \epsilon(F - \delta_x),$$

of F. Note that $\epsilon = 0$ gives $F_0 = F$.

Let T be a functional from a class of d.f.'s to $\mathbb{R}$. Sometimes we shall write T_ϵ for $T(F_\epsilon)$. The rate of change in the functional T at F (that results from adding an infinitesimal point mass at x) is called the 'influence function' of T at F and is defined as

$$\text{IF}(x) = \lim_{\epsilon\downarrow 0}\frac{T(F_\epsilon) - T(F_0)}{\epsilon}.$$

Example 6.3.1. Mean Functional. Let $T(F) = \mu_F = \int t\,dF(t)$. Then

$$T(F_\epsilon) = \int t\,dF_\epsilon(t) = (1-\epsilon)\int t\,dF(t) + \epsilon\int t d\delta_x(t)$$
$$= (1-\epsilon)\mu_F + \epsilon x = \mu_F + \epsilon(x - \mu_F).$$

Hence

$$\text{IF}_1(x) = \lim_{\epsilon\downarrow 0}\frac{T(F_\epsilon) - T(F)}{\epsilon} = \lim_{\epsilon\downarrow 0}\frac{(1-\epsilon)\mu_F + \epsilon x - \mu_F}{\epsilon} = \lim_{\epsilon\downarrow 0}\frac{\epsilon(x - \mu_F)}{\epsilon} = x - \mu_F,$$

which is an unbounded function of x. This shows that the mean functional is very sensitive to an extreme observation, and the sample mean is not robust against heavy tails.

Example 6.3.2. Median Functional. Assume that F has a positive derivative $f(\theta)$ at its unique median θ. The median functional $\theta = \theta(F) = T(F)$ satisfies

$$\int \operatorname{sgn}(t - T(F))\, dF(t) = 0.$$

Hence $T_\epsilon \equiv T(F_\epsilon)$ satisfies

$$(1-\epsilon)\int \operatorname{sgn}(t - T_\epsilon)\, dF(t) + \epsilon \operatorname{sgn}(x - T_\epsilon) = 0.$$

Or, equivalently

$$(1-\epsilon)(1 - 2F(T_\epsilon)) + \epsilon \operatorname{sgn}(x - T_\epsilon) = 0.$$

Differentiating now wrt ϵ at $\epsilon = 0$ yields the equation

$$-(1 - 2F(\theta)) - 2f(\theta)\dot{T}_0 + \operatorname{sgn}(x - \theta) = 0.$$

Because of $F(\theta) = 1/2$, this gives

$$\mathrm{IF}_2(x) = \frac{\operatorname{sgn}(x-\theta)}{2f(\theta)}.$$

Since this influence function is bounded in x, the sample median is not sensitive to gross errors.

Example 6.3.3. M Functional. The M functional is a solution $T(F)$ of the equation

$$E_F(\psi(X - T(F))) = \int \psi(x - T(F))\, dF(x) = 0. \tag{6.3.1}$$

Assume that the function $\lambda(t) = \int \psi(x - t)\, dF(x)$, $t \in \mathbb{R}$, has a nonzero derivative $\dot{\lambda}(\theta)$ at $\theta = T(F)$. Also, the M functional corresponding to F_ϵ is a solution T_ϵ of

$$\lambda(T_\epsilon) = \int \psi(t - T_\epsilon)\, dF_\epsilon(t) = 0.$$

Or equivalently, upon replacing dF_ϵ by $(1-\epsilon)dF + \epsilon d\delta_x$, it is a solution of the equation

$$(1-\epsilon)\int \psi(t - T_\epsilon)\, dF(t) + \epsilon\, \psi(x - T_\epsilon) = 0.$$

Differentiate the above expression w.r.t. ϵ at $\epsilon = 0$ to obtain the equation

$$-\lambda(\theta) + \dot{T}_0\dot{\lambda}(\theta) + \psi(x - \theta) = 0.$$

In view of $\lambda(\theta) = 0$, this yields the influence function

$$\mathrm{IF}(x) = \mathrm{IF}_{\psi,F}(x) = \frac{\psi(x-\theta)}{\dot{\lambda}(\theta)}.$$

Under additional smoothness conditions on ψ, $\dot{\lambda}(\theta)$ will equal $-\int \dot{\psi}(t-\theta)\, dF(t)$. Note that the shape of $\mathrm{IF}_{\psi,F}$ depends on that of ψ. The median functional is the special case $\psi = \operatorname{sgn}$ for which $\lambda(t) = \int \operatorname{sgn}(x - t)\, dF(x)$ and $\dot{\lambda}(\theta) = -2f(\theta)$.

6.4 Robust Estimators in Linear Regression

This section contains classes of R, M and minimum distance estimators of the underlying parameters in the multiple linear regression model (4.6.7).

For the time being consider the linear regression model with non-random predictors. Accordingly, let $\boldsymbol{x}_i \in \mathbb{R}^p$ be non-random design vectors and Y_i be the response variables such that

$$Y_i = \beta_0 + \boldsymbol{x}_i'\boldsymbol{\beta} + \varepsilon_i, \quad \text{for some } \beta_0 \in \mathbb{R}, \boldsymbol{\beta} \in \mathbb{R}^p, \quad 1 \le i \le n, \tag{6.4.1}$$

where the errors $\varepsilon_1, \cdots, \varepsilon_n$ are i.i.d. r.v.'s with continuous d.f. F. Also, the vectors $\boldsymbol{x}_i$'s are allowed to depend on n, but we do not exhibit this dependence. For later use, we set $\boldsymbol{\beta}^* = (\beta_0, \boldsymbol{\beta}')'$. We are interested in estimating $\boldsymbol{\beta}$ and $\boldsymbol{\beta}^*$.

6.4.1 R Estimators in Linear Regression

The R estimators in this model are those that depend on the responses only through the ranks or the signed ranks of the residuals. In the multiple linear regression model (6.4.1), R estimators were developed by Jurečková (1969, 1971) and Jaeckel (1972).

Let R_{it} denote the rank of $Y_i - \boldsymbol{x}_i't$ among $Y_1 - \boldsymbol{x}_1't, \cdots, Y_n - \boldsymbol{x}_n't$, $t \in \mathbb{R}^p$, $\bar{\boldsymbol{x}} := n^{-1}\sum_{i=1}^n \boldsymbol{x}_i$, J be a nondecreasing function from $(0,1)$ to $\mathbb{R}$ and define

$$T(t) := \sum_{i=1}^n (\boldsymbol{x}_i - \bar{\boldsymbol{x}}) J\Big(\frac{R_{it}}{n+1}\Big), \qquad \mathcal{J}(t) := \sum_{i=1}^n J\Big(\frac{R_{it}}{n+1}\Big)(Y_i - \boldsymbol{x}_i'\boldsymbol{t}), \quad t \in \mathbb{R}^p.$$

Note that $R_{1\boldsymbol{\beta}}, \cdots, R_{n\boldsymbol{\beta}}$ is the vector of ranks of the i.i.d. r.v.'s $\varepsilon_1, \cdots, \varepsilon_n$, and hence is uniformly distributed over the set of permutations of $1, \cdots, n$, see Lemma 3.2.1. Thus,

$$E\Big(J\Big(\frac{R_{i\boldsymbol{\beta}}}{n+1}\Big)\Big) = \frac{1}{n}\sum_{j=1}^n J\Big(\frac{j}{n+1}\Big), \qquad E\Big(T(\boldsymbol{\beta})\Big) = \frac{1}{n}\sum_{i=1}^n (\boldsymbol{x}_i - \bar{\boldsymbol{x}}) \sum_{j=1}^n J\Big(\frac{j}{n+1}\Big) = 0.$$

For any vector $y = (y_1, \cdots, y_p)' \in \mathbb{R}^p$, let $|y| := \sum_{k=1}^p |y_k|$ and define

$$\widehat{\boldsymbol{\beta}}_R = \operatorname*{argmin}_{t \in \mathbb{R}^p} |T(t)|, \qquad \widetilde{\boldsymbol{\beta}} := \operatorname*{argmin}_{t \in \mathbb{R}^p} \mathcal{J}(t).$$

The estimator $\widehat{\boldsymbol{\beta}}_R$ was proposed by Jurečková (1969, 1971) and the estimator $\widetilde{\boldsymbol{\beta}}$ by Jaeckel (1972). They are known as R estimators of the regression parameter $\boldsymbol{\beta}$.

Consider the condition

$$\sum_{i=1}^n J\Big(\frac{i}{n+1}\Big) = 0. \tag{6.4.2}$$

Jaeckel (1972) showed that under (6.4.2) and the continuity of the error d.f., $\mathcal{J}(t)$ is convex in t with its a.e. vector of derivatives equal to $-T(t)$. The estimator $\widetilde{\boldsymbol{\beta}}$ exists and is also a solution of the equation $T(t) = 0$, i.e., $\widehat{\boldsymbol{\beta}}_R = \widetilde{\boldsymbol{\beta}}$. This result continues to hold when the predicting variables are random, and errors may be dependent provided their joint distribution is continuous so that their ranks have no ties, almost surely.

Since R_{it} is also the rank of $Y_i - s - t'\boldsymbol{x}_i$ among $Y_1 - s - t'\boldsymbol{x}_1, \cdots, Y_n - s - t'\boldsymbol{x}_n$, for any real number s, these ranks cannot be used to estimate β_0 of the above regression model.

Recall that any nonzero vector t can be written as $t = su$, where $u \in \mathbb{R}^p$ is a unit vector, i.e., $\|u\| = 1$ and $s \in \mathbb{R}$. The fact that is useful in the computation of the above R estimators is that if J is nondecreasing then $e'T(se)$ is nonincreasing in s, for every unit vector e. This is a consequence of Theorem II.7E, p. 35 of Hájek (1969). For the sake of completeness and its importance in the computation of R estimators, we reproduce this theorem as the following lemma.

Lemma 6.4.1. *Let $c_1, \cdots, c_n$, $d_1, \cdots, d_n$ and $v_1, \cdots, v_n$ be real numbers such that not all c_i and d_i are the same and no two v_i's are the same. For $s \in \mathbb{R}$, let r_{is} denote the rank of $v_i - sd_i$ among $v_1 - sd_1, \cdots, v_n - sd_n$. Let $b(1) \le \cdots \le b(n)$ be a set of nondecreasing real numbers. Set*

$$\mathcal{T}(s) := \sum_{i=1}^{n} c_i b(r_{is}), \quad s \in \mathbb{R}.$$

Assume

$$(c_i - c_j)(d_i - d_j) \ge 0, \quad \forall 1 \le i \le j \le n. \tag{6.4.3}$$

Then the step function $\mathcal{T}(s)$ is nonincreasing for all those $s \in \mathbb{R}$ for which there are no ties among $v_1 - sd_1, \cdots, v_n - sd_n$, i.e., $v_i - sd_i \ne v_j - sd_j$, for all $i \ne j$.

Proof. The following proof is as in Hájek (1969). We reproduce it for the sake of convenience. Let $s_1 < s_2$ be such that there are no ties among $v_1 - s_1 d_j, \cdots, v_n - sd_n$ and among $v_1 - s_2 d_1, \cdots, v_n - s_2 d_n$. We wish to show that $\mathcal{T}(s_1) \ge \mathcal{T}(s_2)$. The proof is by induction.

Let $R_1 = (r_{1s_1}, \cdots, r_{ns_1})$ and $R_2 = (r_{1s_2}, \cdots, r_{ns_2})$. Suppose there is only one change between the rank sets R_1 and R_2, i.e., suppose for some $i, k = 1, \cdots, n$, $r_{ks_1} = r_{is_2}$, $r_{is_1} = r_{ks_2}$, and $r_{js_1} = r_{js_2}$, $j \ne i, k$. Then,

$$\begin{aligned}
\mathcal{T}(s_1) - \mathcal{T}(s_2) &= c_i b(r_{is_1}) + c_k\, b(r_{ks_1}) - c_i b(r_{is_2}) - c_k b(r_{ks_2}) \\
&= c_i[b(r_{is_1}) - b(r_{is_2})] + c_k\,[b(r_{ks_1}) - b(r_{ks_2})] \\
&= (c_i - c_k)[b(r_{ks_2}) - b(r_{ks_1})].
\end{aligned}$$

Now suppose without loss of any generality $r_{ks_2} > r_{is_2}$. Then $r_{is_1} > r_{ks_1}$ and

$$\begin{aligned}
& v_k - s_2 d_k > v_i - s_2 d_i, \quad v_i - s_1 d_i > v_k - s_1 d_k \\
& \text{implies } v_k - v_i > s_2(d_k - d_i), \quad v_i - v_k > s_1(d_i - d_k) \\
& \text{implies } (s_2 - s_1)(d_k - d_i) < 0, \\
& \text{implies } d_i > d_k, \text{ since } s_2 > s_1, \\
& \text{implies } c_i > c_k, \quad \text{by (6.4.3).}
\end{aligned}$$

Hence, $\mathcal{T}(s_1) \ge \mathcal{T}(s_2)$, because $b(i)$ is non-decreasing in i.

Now suppose there are m changes in the rank sets R_1 and R_2. Then write $s_2 - s_1 = \sum_{k=1}^{m} \theta_k$, such that for each $\theta_k > 0$ and each pair of sets of ranks of the following variables

$$\begin{aligned}
&\{v_j - (s_1 + \theta_1)\, d_j\}_{j=1}^{n} && \& \ \{v_j - s_1\, d_j\}_{j=1}^{n}, \\
&\{v_j - (s_1 + \theta_1 + \theta_2)\, d_j\}_{j=1}^{n} && \& \ \{v_j - (s_1 + \theta_1) d_j\}_{j=1}^{n}, \\
&\cdots && \\
&\{v_j - s_2\, d_j\}_{j=1}^{n} && \& \ \{v_j - (s_1 + \delta_{m-1})\, d_j\}_{j=1}^{n}
\end{aligned}$$

will have exactly one change between them, where $\delta_{m-1} := \sum_{k=1}^{m-1} \theta_k$. Then,

$$\mathcal{T}(s_2) - \mathcal{T}(s_1) = [\mathcal{T}(s_1 + \theta_1) - \mathcal{T}(s_1)] + [\mathcal{T}(s_1 + \theta_1 + \theta_2) - \mathcal{T}(s_1 + \theta_1)] + \cdots + [\mathcal{T}(s_2) - \mathcal{T}(s_1 + \theta_1 + \cdots + \theta_{m-1})].$$

By the first step of the above proof, each of these summands is non-positive. □

In the case $p = 1$, $T(t)$ is a non-increasing step function of t. Also, $E\big(T(\beta)\big) = 0$. Hence, analogous to (6.2.3), the estimator $\widehat{\beta}_R$ is obtained as follows.

$$\hat{\beta}_1 := \sup\{t : T(t) > 0\}, \quad \hat{\beta}_2 := \inf\{t : T(t) < 0\}, \quad \widehat{\beta}_R := (\hat{\beta}_1 + \hat{\beta}_2)/2.$$

There are programs in the R language for the computation of the above R estimators $\widehat{\boldsymbol{\beta}}_R$ or $\widetilde{\boldsymbol{\beta}}$. See Kloke and McKean (2012, 2014) and the references therein. The most popular R estimator is the one for which $J(u) = u - 1/2$, $0 \le u \le 1$. The corresponding estimator is called the Wilcoxon R estimator, because when the model (6.4.1) is specialized to the two sample location model by choosing $p = 1$, $x_i = 0$, for $i = 1, \cdots, n_1$ and $x_i = 1$ for $i = n_1 + 1, \cdots, n$, where $n_1 < n$ is the sample size from the first population, then $T(t)$ reduces to the two sample Wilcoxon rank sum test statistic. Since in this case the R estimator based on the Wilcoxon rank sum statistic is known as the Hodges-Lehmann-Sen estimator, the estimator $\widehat{\boldsymbol{\beta}}_R$ is also known as an analog of the Hodges-Lehmann-Sen estimator in the linear regression model.

We shall now describe the limiting distribution of $\widehat{\boldsymbol{\beta}}_R$. First, we note that the estimator $\widehat{\boldsymbol{\beta}}_R$ is translation invariant. To make this precise, let $\boldsymbol{X}$ denote the $n \times p$ matrix whose ith row is $\boldsymbol{x}_i'$, $i = 1, \cdots, n$ and $\boldsymbol{Y} := (Y_1, \cdots, Y_n)'$. Write $T(\boldsymbol{X}, \boldsymbol{Y} - \boldsymbol{X}t)$ for $T(t)$ and $\widehat{\boldsymbol{\beta}}_R(\boldsymbol{X}, \boldsymbol{Y})$ for $\widehat{\boldsymbol{\beta}}_R$. Then, we readily see from the definition that $\widehat{\boldsymbol{\beta}}_R(\boldsymbol{X}, \boldsymbol{Y} + \boldsymbol{X}b) = b + \widehat{\boldsymbol{\beta}}_R(\boldsymbol{X}, \boldsymbol{Y})$, for all $b \in \mathbb{R}^p$. Hence, the distribution of $\widehat{\boldsymbol{\beta}}_R - \boldsymbol{\beta}$ does not depend on $\boldsymbol{\beta}$.

Now, to describe the asymptotic distribution, we need some more notation. Let $\boldsymbol{X}_c$ denote the $n \times p$ matrix whose ith row is $(\boldsymbol{x}_i - \bar{\boldsymbol{x}})'$, $i = 1, \cdots, n$. Recall $\bar{\boldsymbol{x}} := n^{-1} \sum_{i=1}^n \boldsymbol{x}_i$. Assume

$$\boldsymbol{X}_c'\boldsymbol{X}_c \text{ is nonsingular and } \max_{1 \le i \le n} \left\| \left(\boldsymbol{X}_c'\boldsymbol{X}_c\right)^{-1/2} (\boldsymbol{x}_i - \bar{\boldsymbol{x}}) \right\| \to 0. \tag{6.4.4}$$

About the error distribution consider the following conditions.

F has a density f with finite Fisher information for location, see Definition 2.11.1. (6.4.5)

F has a uniformly continuous density f on $\mathbb{R}$. (6.4.6)

The following theorem gives two sets of conditions for the asymptotic normality of $\widehat{\boldsymbol{\beta}}_R$, where $\mathcal{I}_p$ is the $p \times p$ identity matrix and

$$C(J, f) := \int J(F(x)) \dot{f}(x)\, dx, \quad D(J, f) := \int f(x)\, dJ(F(x)), \tag{6.4.7}$$

$$\sigma_J^2 := \int_0^1 J^2(u)\, du - \Big(\int_0^1 J(u)\, du\Big)^2, \quad I_L(f) = \int \Big(\frac{\dot{f}(x)}{f(x)}\Big)^2 f(x)\, dx.$$

The Cauchy-Schwarz inequality (2.4.4) shows that σ_J^2 finite and (6.4.5) together imply $C^2(J, F) \le \int_0^1 J^2(u)\, du\, I_L(f) < \infty$. Under (6.4.6), f is bounded, see Lemma 2.11.2, which in turn implies that for nondecreasing and bounded J, $D^2(J, f) < \infty$. Integration by parts shows that if (6.4.5) holds and J is nondecreasing and bounded then $C^2(J, f) = D^2(J, f)$. We are now ready to state the following theorem.

Theorem 6.4.1. *Suppose* (6.4.1) *and* (6.4.4) *hold and* J *is nondecreasing right continuous function with* σ_J^2 *finite and positive. Then the following are true.*
(a) *If* (6.4.5) *holds and* $C(J,f)$ *is not zero, then*

$$\big(\boldsymbol{X}_c'\boldsymbol{X}_c\big)^{1/2}\big(\widehat{\boldsymbol{\beta}}_R-\boldsymbol{\beta}\big)\to_D \mathcal{N}_p(0,\tau^2\mathcal{I}_p),\qquad \tau^2:=\frac{\sigma_J^2}{C^2(J,f)}. \tag{6.4.8}$$

(b) *If* (6.4.6) *holds,* J *is bounded and* $D(J,f)$ *is not zero, then* (6.4.8) *holds with* $\tau^2=\sigma_J^2/D^2(J,f)$.

The proof of part (a) of the above theorem is available in Jurečková (1969, 1971) while that of part (b) is in Koul (2002).

The condition (6.4.5) is stronger than (6.4.6). An example of a uniformly continuous density f for which $I_L(f)=\infty$ is given by $f(x)=\big(1-|x|\big)I\big(|x|\le 1\big)$. For other examples, see Exercises 11 and 12 at the end of this chapter.

Part (a) of the above theorem is applicable to an unbounded but square integrable J like $J=\Phi^{-1}$, where Φ is the d.f. of a $\mathcal{N}(0,1)$ r.v.. The corresponding R estimator $\widehat{\boldsymbol{\beta}}_R$ is called the normal scores R estimator. To balance against the unboundedness of the score function J one needs the stronger condition (6.4.5) on F. Part (b) of the above theorem is applicable to bounded scores under the weaker condition (6.4.6) on F. Thus, for example, the asymptotic normality of the Wilcoxon R estimator, which corresponds to $J(u)=u-1/2$, $0<u<1$, holds under (6.4.6) while that of the normal scores R estimator holds under (6.4.5).

Signed Rank Estimators. As mentioned earlier, the above ranks cannot be used to estimate the intercept parameter β_0. Recall that in the one sample location model the location parameter is only identiable under additional assumptions on F such as symmetry of F. Thus in order to estimate β_0 in the model (6.4.1) using some kind of ranks, we shall assume that the error d.f. is symmetric about zero so that $F(-x)=1-F(x)$ holds for all $x\in\mathbb{R}$ by the continuity of F. In this case we can use signed ranks of the residuals to define R estimators. Proceeding a bit more precisely, set $\boldsymbol{z}_i:=(1,\boldsymbol{x}_i')'$, and for $\boldsymbol{b}:=(b_0,b_1,\cdots,b_p)'\in\mathbb{R}^{p+1}$, let $R_{i\boldsymbol{b}}^+$ denote the absolute rank of the ith residual, i.e., $R_{i\boldsymbol{b}}^+$ is the rank of $|Y_i-\boldsymbol{b}'\boldsymbol{z}_i|$ among $|Y_1-\boldsymbol{b}'\boldsymbol{z}_1|,\cdots,|Y_n-\boldsymbol{b}'\boldsymbol{z}_n|$. Let $\text{sgn}(x):=I(x>0)-I(x<0)$ be the sign function and φ be a nonnegative and nondecreasing function defined on $(0,1)$. Let

$$\boldsymbol{S}(\boldsymbol{b}):=\sum_{i=1}^n \boldsymbol{z}_i\varphi\Big(\frac{R_{i\boldsymbol{b}}^+}{n+1}\Big)\text{sgn}(Y_i-\boldsymbol{b}'\boldsymbol{z}_i),\qquad \widehat{\boldsymbol{\beta}^*}:=\underset{\boldsymbol{b}\in\mathbb{R}^{p+1}}{\text{argmin}}\,\big|\boldsymbol{S}(\boldsymbol{b})\big|.$$

If there are no slope parameters in the regression model (6.4.1), i.e., if all $\boldsymbol{x}_i$'s are zero vectors then the model (6.4.1) reduces to the one sample location model. In this case the only parameter in the model is β_0. Let $\tilde{\boldsymbol{b}}:=(t,0,\cdots,0)'\in\mathbb{R}^{p+1}$. Then $\boldsymbol{S}(\tilde{\boldsymbol{b}})$ is equivalent to the statistic $S\big(\boldsymbol{X}-t\boldsymbol{1}_n\big)$ of (6.2.14). And if we take $\varphi(u)=u$, then $\boldsymbol{S}(\tilde{\boldsymbol{b}})$ coincides with the Wilcoxon signed rank statistics and $\widehat{\boldsymbol{\beta}^*}$ coincides with the estimator $\hat{\theta}_{HL}$ of (6.2.16). For this reason, in the general linear regression model (6.4.1), the estimator $\widehat{\boldsymbol{\beta}^*}$ corresponding to $\varphi(u)\equiv u$ is called the Wilcoxon R estimator or an analog of the Hodges-Lehmann estimator of $\boldsymbol{\beta}^*$.

Note that $R_{i\boldsymbol{\beta}^*}$ is the rank of $|\varepsilon_i|$ among $|\varepsilon_1|,\cdots,|\varepsilon_n|$. Since the error r.v.'s are continuous and symmetrically distributed about zero, by Lemma 4.3.2, $|\varepsilon_1|,\cdots,|\varepsilon_n|$ are independent of $\text{sgn}(\varepsilon_1),\cdots,\text{sgn}(\varepsilon_n)$. Hence the absolute ranks $R_{1\boldsymbol{\beta}^*}^+,\cdots,R_{n\boldsymbol{\beta}^*}^+$ are independent of $\text{sgn}(\varepsilon_1),\cdots,\text{sgn}(\varepsilon_n)$. This fact and the identity $E\big(\text{sgn}(\varepsilon_1)\big)=0$ give

$$E\big(S(\boldsymbol{\beta}^*)\big) = \sum_{i=1}^{n} \boldsymbol{z}_i E\Big(\varphi\Big(\frac{R^+_{i\boldsymbol{\beta}^*}}{n+1}\Big)\Big) E\big(\text{sgn}(\varepsilon_i)\big) = 0.$$

Thus the estimators $\widehat{\boldsymbol{\beta}^*}$ will not have any asymptotic bias.

An analog of Lemma 6.4.1 describing certain monotonicity property of $S(\boldsymbol{b})$ is proved in van Eeden (1972). Using this one can compute these estimators. Computation aspects of these estimators are also discussed in Kloke and McKean (2012, 2014).

An analog of Theorem 6.4.1 is available for these estimators. Let $\boldsymbol{Z}$ denote the $n \times (p+1)$ matrix whose i-th row is $\boldsymbol{z}_i'$, $i = 1, \cdots, n$. Assume the following.

$$\boldsymbol{Z}'\boldsymbol{Z} \text{ is nonsingular} \quad \text{and} \quad \max_{1 \le i \le n} \big\| \big(\boldsymbol{Z}'\boldsymbol{Z}\big)^{-1/2} \boldsymbol{z}_i \big\| \to 0. \tag{6.4.9}$$

The role of J will now be played by the function ψ defined by

$$\psi(u) := \varphi(|2u-1|)\text{sgn}(2u-1), \quad 0 < u < 1.$$

As φ is nonnegative and nondecreasing, ψ is nondecreasing. Moreover, ψ is skew symmetric about 1/2, i.e., $\psi(u) = -\psi(1-u)$, $0 < u < 1$. Note also the identity $\varphi(u) = \psi((u+1)/2)$, $0 < u < 1$. We shall require that $\nu_\psi^2 := \int \psi^2(u)\, du$ is positive and finite.

Theorem 6.4.2. *Suppose* (6.4.1) *and* (6.4.9) *hold and F is continuous and symmetric about zero. Assume also that φ is nonnegative and nondecreasing and that $\nu_\psi^2 := \int_0^1 \psi^2(u)\, du$ is positive and finite. Then the following hold.*

(a) *If* (6.4.5) *holds and $C(\psi, F)$ is not zero, then $\big(\boldsymbol{Z}'\boldsymbol{Z}\big)^{1/2}\big(\widehat{\boldsymbol{\beta}^*} - \boldsymbol{\beta}^*\big) \to_D \mathcal{N}_{p+1}\big(0, \gamma^2 \mathcal{I}_{p+1}\big)$, where $\gamma^2 := \nu_\psi^2 / C^2(\psi, f)$.*

(b) *If ψ is bounded,* (6.4.6) *holds and $D_1(\psi, f)$ is not zero, then the conclusion of part* (a) *holds with $\gamma^2 := \nu_\psi^2 / D^2(\psi, f)$.*

The proof of part (a) can be deduced from the arguments in van Eeden (1972) while that of part (b) is available in Koul (1996, 2002).

6.4.2 M Estimators

Let ε denote a copy of ε_1 of (6.4.1), $\boldsymbol{z}_i$ and $\boldsymbol{Z}$ be as in the previous subsection. Let ψ be a nondecreasing function from $\mathbb{R}$ to $\mathbb{R}$ such that $E\big(\psi(\varepsilon)\big) = 0$. An M estimator corresponding to the given ψ is defined as follows:

$$M(t) := \sum_{i=1}^{n} \boldsymbol{z}_i \psi(Y_i - \boldsymbol{z}_i' t), \quad t \in \mathbb{R}^{p+1}; \qquad \widehat{\boldsymbol{\beta}^*_\psi} : M\big(\widehat{\boldsymbol{\beta}^*_\psi}\big) = 0.$$

The computational issues for general $p > 1$ of these estimators have been discussed in Huber (1981). As in the case of R estimators, $e'M(se)$ is nonincreasing function of $s \in \mathbb{R}$, for every $e \in \mathbb{R}^{p+1}$, $\|e\| = 1$. In particular, for $p = 0$, $M(t)$ is nonincreasing function of t and the estimator $\widehat{\beta^*_\psi}$ can be defined as in the one sample location model, see (6.2.24).

In addition to the above assumptions, suppose $\tau_\psi^2 := E\big(\psi^2(\varepsilon)\big)$ is positive and finite, $\int f d\psi$ is positive and (6.4.9) holds. Then, $\big(\boldsymbol{Z}'\boldsymbol{Z}\big)^{1/2}\big(\widehat{\boldsymbol{\beta}^*_\psi} - \boldsymbol{\beta}^*\big) \to_D \mathcal{N}_{p+1}\big(0, \tau_\psi^2 \mathcal{I}_{p+1} / \big(\int f d\psi\big)^2\big)$. For a proof see Huber (1973, 1981).

Huber (1973) investigated the asymptotic distribution of suitably standardized $\widehat{\boldsymbol{\beta}}_\psi$ when $p = p_n$ depends on n, $p_n \to \infty, p_n^2/n \to 0$. Portnoy (1984, 1985) and Mammen (1989) extend these results further that allow $p_n \to \infty$ at different rates.

6.4.3 Minimum Distance Estimators in Linear Regression

Here we shall discuss the analogs of the minimum distance estimators (6.2.31) of the one sample location parameter. Assume the model (6.4.1) holds with the error d.f. continuous and symmetric about the origin. The question is how should one define an analog of the m.d. estimator (6.2.31) of the one sample location parameter for the regression parameter vector $\boldsymbol{\beta}^* = (\beta_0, \boldsymbol{\beta}')'$. For the time being suppose $\beta_0 = 0$ so that the model is $Y_i = \boldsymbol{x}_i'\boldsymbol{\beta} + \varepsilon_i$. Since the i.i.d. errors $\varepsilon_1, \cdots, \varepsilon_n$ are symmetrically distributed about zero, the random vector $(Y_1 - \boldsymbol{x}_1'\boldsymbol{\beta}, \cdots, Y_n - \boldsymbol{x}_n'\boldsymbol{\beta})$ has the same distribution as the random vector $(-Y_1 + \boldsymbol{x}_1'\boldsymbol{\beta}, \cdots, -Y_n + \boldsymbol{x}_n'\boldsymbol{\beta})$, and similarly to (6.2.31) in the one sample location model, one is tempted to define a m.d. estimator of $\boldsymbol{\beta}$ to be a minimizer of $\tilde{K}$ given by

$$\tilde{K}(\boldsymbol{t}) := \frac{1}{n^2}\int \Big[\sum_{i=1}^n \Big\{I\big(Y_i - \boldsymbol{x}_i'\boldsymbol{t} \le x\big) - I\big(-Y_i + \boldsymbol{x}_i'\boldsymbol{t} \le x\big)\Big\}\Big]^2 dG(x), \quad \boldsymbol{t} \in \mathbb{R}^p.$$

Here and below, G is a nondecreasing right continuous function on $\mathbb{R}$. The set of minimizers of $\tilde{K}$ is not a bounded set in general, as illustrated by the following example.

Example 6.4.1. Suppose $\beta_0 = 0$, $p = 1$ and $G(y) = y, y \in \mathbb{R}$. Then, arguing as for (6.2.34), we obtain

$$n^2\tilde{K}(t) = \sum_{i=1}^n 2\big|Y_i - x_i t\big| + 2\sum_{1\le i<j\le n}\Big\{\big|Y_j + Y_i - (x_i + x_j)t\big| - \big|Y_j - Y_i - (x_j - x_i)t\big|\Big\}.$$

Suppose $n = 2$. Then

$$2\tilde{K}(t) = |Y_1 - x_1 t| + |Y_2 - x_2 t| + \big|Y_2 + Y_1 - (x_2 + x_1)t\big| - \big|Y_2 - Y_1 - (x_2 - x_1)t\big|.$$

Next, take $Y_1 = 5, Y_2 = 8, x_1 = 1$ and $x_2 = -1$ to obtain $2\tilde{K}(t) = |5-t| + |8+t| + 13 - |3+2t|$, i.e.,

$$\begin{aligned} 2\tilde{K}(t) &= 13, & t &< -8,\\ &= 13 + 16 + 2t, & -8 \le t &< -3/2\\ &= 13 + 10 - 2t, & -3/2 \le t &< 5\\ &= 13, & t &\ge 5. \end{aligned}$$

The smallest value $\tilde{K}(t)$ takes is $13/2$, obtained for $t \in (-\infty, -8] \cup [5, \infty)$. So the set of minimizers is $(-\infty, -8] \cup [5, \infty)$, an unbounded set, which clearly is undesirable.

The right analog is obtained by using a certain weighted empirical process as follows. Consider the model (6.4.1). Let $\boldsymbol{z}_i = (1, \boldsymbol{x}_i')'$ and $\boldsymbol{Z}$ be the $n \times (p+1)$ matrix with rows $\boldsymbol{z}_i'$, $i = 1, \cdots, n$. Assume $\boldsymbol{Z}'\boldsymbol{Z}$ is positive definite and set $\boldsymbol{w}_i = \big(\boldsymbol{Z}'\boldsymbol{Z}\big)^{-1/2}\boldsymbol{z}_i$, $i = 1, \cdots, n$. Now define the minimum distance estimator $\widehat{\boldsymbol{\beta}_{md}^*}$ of $\boldsymbol{\beta}^*$ to be minimizer of

$$K(\boldsymbol{t}) := \int \Big\|\sum_{i=1}^n \boldsymbol{w}_i\Big\{I\big(Y_i - \boldsymbol{z}_i'\boldsymbol{t} \le x\big) - I\big(-Y_i + \boldsymbol{z}_i'\boldsymbol{t} \le x\big)\Big\}\Big\|^2 dG(x), \quad \boldsymbol{t} \in \mathbb{R}^{p+1}. \quad (6.4.10)$$

If G is degenerate at the origin, i.e., $G(x) := \delta_0(x) = I(x \geq 0)$, then we have $K(\boldsymbol{t}) = \left\|\sum_{i=1}^n \boldsymbol{w}_i \text{sgn}(Y_i - \boldsymbol{z}_i'\boldsymbol{t})\right\|^2$ and $\widehat{\boldsymbol{\beta}_{md}^*}$ is the LAD estimator. If $|G(y) - G(x)| = |G(-y) - G(-x)|$, $\forall x, y \in \mathbb{R}$, then analogous to (6.2.34), we have

$$K(\boldsymbol{t}) = \sum_{i=1}^n \boldsymbol{w}_i'\boldsymbol{w}_i \big|G(Y_i - \boldsymbol{z}_i'\boldsymbol{t}) - G(-Y_i + \boldsymbol{z}_i'\boldsymbol{t})\big| \tag{6.4.11}$$
$$+ 2 \sum_{1 \leq i < j \leq n} \boldsymbol{w}_i'\boldsymbol{w}_j \Big\{\big|G(Y_i - \boldsymbol{z}_i'\boldsymbol{t}) - G(-Y_j + \boldsymbol{z}_j'\boldsymbol{t})\big| - \big|G(Y_i - \boldsymbol{z}_i'\boldsymbol{t}) - G(Y_j - \boldsymbol{z}_j'\boldsymbol{t})\big|\Big\}.$$

In particular, if $G(y) = y, y \in \mathbb{R}$, then

$$K(\boldsymbol{t}) = 2\sum_{i=1}^n \boldsymbol{w}_i'\boldsymbol{w}_i \big|Y_i - \boldsymbol{z}_i'\boldsymbol{t}\big|$$
$$+ 2 \sum_{1 \leq i < j \leq n} \boldsymbol{w}_i'\boldsymbol{w}_j \Big\{\big|Y_i + Y_j - (\boldsymbol{z}_i + \boldsymbol{z}_j)'\boldsymbol{t}\big| - \big|Y_i - Y_j - (\boldsymbol{z}_i - \boldsymbol{z}_j)'\boldsymbol{t}\big|\Big\}$$

and $\widehat{\boldsymbol{\beta}_{md}^*}$ is an analog of the Hodges-Lehmann estimator (6.2.16) in the regression setup.

Example 6.4.2. Suppose $p = 1$, $\beta_0 = 0$ and $G(y) = y, y \in \mathbb{R}$. Then the model is $Y_i = \beta_1 x_i + \varepsilon_i$ and the above formula becomes

$$K(t) = 2\sigma_x^{-2}\Big[\sum_{i=1}^n x_i^2 |Y_i - x_i t|$$
$$+ \sum_{1 \leq i < j \leq n} x_i x_j \Big\{\big|Y_i + Y_j - (x_i + x_j)t\big| - \big|Y_i - Y_j - (x_i - x_j)t\big|\Big\}\Big]$$

with $\sigma_x^2 := \sum_{i=1}^n x_i^2$ assumed to be positive. Suppose $n = 2$. Then

$$K(t) = 2\sigma_x^{-2}\Big\{\Big[x_1^2|Y_1 - x_1 t| + x_2^2|Y_2 - x_2 t|\Big]$$
$$+ x_1 x_2\Big[\big|Y_2 + Y_1 - (x_2 + x_1)t\big| - \big|Y_2 - Y_1 - (x_2 - x_1)t\big|\Big]\Big\}.$$

Now reconsider the data of Example 6.4.1, where $Y_1 = 5$, $Y_2 = 8$, $x_1 = 1$, $x_2 = -1$. Then we have $\sigma_x^2 = 2$ and obtain $K(t) = |5 - t| + |8 + t| - 13 + |3 + 2t|$, i.e.,

$$\begin{aligned} K(t) &= -19 - 4t, & t &< -8, \\ &= -3 - 2t, & -8 \leq t &< -3/2 \\ &= 3 + 2t, & -3/2 \leq t &< 5 \\ &= -7 + 4t, & t &\geq 5. \end{aligned}$$

Now $K(-3/2) = 0$ and the minimizer equals $-3/2$.

Dhar (1992) discusses the existence and some computational aspects of $\widehat{\boldsymbol{\beta}_{md}^*}$ for general G while Dhar (1993) does the same for the analogs of these estimators in the p-th order linear autoregressive time series model. Kim (2018) gives a fast algorithm for the computation of these minimum distance estimators. His package KoulLrMde() in R pertains to the computation of these estimators in the linear regression model while the package KoulArMde() in R is used for the computation of their analogs in linear autoregressive time series model.

The following theorem gives some sufficient conditions for the asymptotic normality of this class of estimators. It is an extension of Theorem 6.2.2 and proved in Chapter 5 of Koul (2002). We state it here for an interested reader.

Theorem 6.4.3. *Assume the regression model* (6.4.1) *holds for some* $\boldsymbol{\beta} \in \mathbb{R}^p$ *and some error d.f.* F *with* $\boldsymbol{z}_i$*'s satisfying* (6.4.9). *Suppose* G *is symmetric about* 0, *i.e.,* $dG(x) = -dG(-x)$, *for all* $x \in \mathbb{R}$, F *is symmetric about* 0 *and has density* f *satisfying* (6.2.35), (6.2.36) *and* (6.2.37). *Let* $\psi(x) := \int_{-\infty}^{x} f(y)\, dG(y)$, $x \in \mathbb{R}$. *Then,* $\boldsymbol{A}\big(\widehat{\boldsymbol{\beta}}_{md} - \boldsymbol{\beta}\big) \to_D \mathcal{N}_p\big(0, \sigma_G^2 \mathcal{I}_p\big)$ *with*

$$\sigma_G^2 := \frac{\mathrm{Var}\big(\psi(\varepsilon)\big)}{\big(\int f^2(x)\, dG(x)\big)^2} = \frac{\int\int \big[F(x \wedge y) - F(x)F(y)\big] f(x) f(y)\, dG(x)\, dG(y)}{\big(\int f^2(x)\, dG(x)\big)^2}. \tag{6.4.12}$$

For G degenerate at zero, $\sigma_G^2 = 1/4f^2(0)$ and for $G(y) = y$, $\sigma_G^2 = 1/12(\int f^2(x)\, dx)^2$. The efficiency comparison discussed following Theorem 6.2.2 in Subsection 6.2.4 also applies in the regression setup.

An analog of the above theorem in the linear autoregressive time series of order p is proved in Ch. 7 of Koul (2002).

Remark 6.4.1. Random Designs. All of the above results about R, M and m.d. estimators continue to hold under appropriate conditions when predictors are random. We shall state analogs of the asymptotic normality results for these estimators in this case. More precisely, consider the linear regression model

$$Y_i = \beta_0 + \boldsymbol{X}_i' \boldsymbol{\beta} + \varepsilon_i, \quad \text{for some } \beta_0 \in \mathbb{R}, \boldsymbol{\beta} \in \mathbb{R}^p, \tag{6.4.13}$$

where $Y_1, \cdots, Y_n$ are the response variables, the p-dimensional random vectors $\boldsymbol{X}_1, \cdots, \boldsymbol{X}_n$ are i.i.d. with positive definite variance-covariance matrix $\boldsymbol{\Sigma}$ and the r.v.'s $\varepsilon_1, \cdots, \varepsilon_n$ are i.i.d. with continuous d.f. F and independent of $\boldsymbol{X}_1, \cdots, \boldsymbol{X}_n$.

The above estimators $\widehat{\boldsymbol{\beta}}_R$ and $\widehat{\boldsymbol{\beta}}_\psi$ continue to be well defined when in their definition the non-random predictors $\boldsymbol{x}_i$ and $\boldsymbol{z}_i$ are, respectively, replaced by the random predictors $\boldsymbol{X}_i$ and $\boldsymbol{\mathcal{Z}}_i$, where $\boldsymbol{\mathcal{Z}}_i' := \big(1, \boldsymbol{X}_i'\big)$. Similarly, the analogs of $\widehat{\boldsymbol{\beta}^*}$ and $\widehat{\boldsymbol{\beta}_{md}^*}$ where $\boldsymbol{z}_i$'s are replaced by $\boldsymbol{\mathcal{Z}}_i$'s are well defined provided F is also symmetric about zero. Their asymptotic normality results also continue to hold provided the conditions (6.4.4) and (6.4.9) are replaced by the positive definiteness of the variance-covariance matrix $\boldsymbol{\Sigma}$. For an interested reader we state these results explicitly for the analogs of these estimators. First, consider $\widehat{\boldsymbol{\beta}}_R$, which only estimates $\boldsymbol{\beta}$. Analogous to Theorem 6.4.1, we have the following theorem, where $C(J, f)$ and $D(J, f)$ are as in (6.4.7).

Theorem 6.4.4. *Suppose* (6.4.13) *holds,* $\boldsymbol{\Sigma}$ *is positive definite and* J *is nondecreasing right continuous function with* σ_J^2 *positive and finite. Then the following hold.*

(a) *If* (6.4.5) *holds and* $C(J, f)$ *is not zero, then* $n^{1/2}\big(\widehat{\boldsymbol{\beta}}_R - \boldsymbol{\beta}\big) \to_D \mathcal{N}_p\big(0, \tau^2 \boldsymbol{\Sigma}^{-1}\big)$ *holds with* $\tau^2 := \sigma_J^2 / C^2(J, f)$.

(b) *If* J *is bounded,* (6.4.6) *holds and* $D(J, f)$ *is not zero, then* $n^{1/2}\big(\widehat{\boldsymbol{\beta}}_R - \boldsymbol{\beta}\big) \to_D \mathcal{N}_p\big(0, \tau_1^2 \boldsymbol{\Sigma}^{-1}\big)$ *holds with* $\tau_1^2 = \sigma_J^2 / D^2(J, f)$.

Similarly, the analog of Theorem 6.4.2 is given in the next theorem. Let $\boldsymbol{\Sigma}_1 := E\big(\boldsymbol{\mathcal{Z}}_1 \boldsymbol{\mathcal{Z}}_1'\big)$.

Theorem 6.4.5. *Suppose* (6.4.13) *holds,* $\boldsymbol{\Sigma}_1$ *is positive definite, and the error d.f.* F *is symmetric about zero. Let* ψ *be as in Theorem 6.4.2 with* $\nu_\psi^2 := \int_0^1 \psi^2(u)\, du$ *positive and finite.*

(a) *If* (6.4.5) *holds and* $C(\psi, f) \neq 0$, *then* $n^{1/2}\big(\widehat{\boldsymbol{\beta}^*} - \boldsymbol{\beta}^*\big) \to_D \mathcal{N}_{p+1}\big(0, \gamma^2 \boldsymbol{\Sigma}_1^{-1}\big)$, *where* $\gamma^2 := \nu_\psi^2 / C^2(\psi, f)$.

(b) *If* ψ *is bounded,* (6.4.6) *holds, and* $D(\psi, f) \neq 0$, *then* $n^{1/2}\big(\widehat{\boldsymbol{\beta}^*} - \boldsymbol{\beta}^*\big) \to_D \mathcal{N}_{p+1}\big(0, \gamma_1^2 \boldsymbol{\Sigma}_1^{-1}\big)$, *where* $\gamma_1^2 := \nu_\psi^2 / D^2(\psi, f)$.

The proofs of the (b) parts of both of these theorems are available in Koul (1996, 2002), while those of the (a) parts can be deduced from the work of Jurečková (1969, 1971).

For the analogs of M estimators we have the following asymptotic normality result. Suppose $\mathbf{\Sigma}_1$ is positive definite, $E\big(\psi(\varepsilon)\big) = 0$, $0 < \sigma_\psi^2 := E\big(\psi^2(\varepsilon)\big) < \infty$, F satisfies (6.4.6) and $0 < \int f\,d\psi < \infty$. Then, $n^{1/2}\big(\widehat{\boldsymbol{\beta}}_\psi - \boldsymbol{\beta}^*\big) \to_D \mathcal{N}_{p+1}\big(0, \sigma_\psi^2 \mathbf{\Sigma}_1^{-1}/(\int f\,d\psi)^2\big)$.

For the analog of the m.d. estimator $\widehat{\boldsymbol{\beta}^*_{md}}$, we have the following asymptotic normality result. Suppose the model (6.4.13) holds, $\mathbf{\Sigma}_1$ is positive definite and F and G satisfy the conditions of Theorem 6.4.3. Then, $n^{1/2}\big(\widehat{\boldsymbol{\beta}^*_{md}} - \boldsymbol{\beta}^*\big) \to_D \mathcal{N}_{p+1}\big(0, \sigma_G^2 \mathbf{\Sigma}_1^{-1}\big)$, where σ_G^2 is as in (6.4.12). A proof of this result appears in Koul (1996).

Remark 6.4.2. Other minimum dispersion estimators. Consider the one sample parametric model, where $X_1, \cdots, X_n$ is a random sample from the probability model P_θ, $\theta \in \Theta \subseteq \mathbb{R}^p$, p a known positive integer. Let $\rho(x, \theta)$ be a function from $\mathbb{R} \times \Theta$ to $\mathbb{R}$, measurable in x, for each θ. A general M estimator of θ is defined to be $\widetilde{\theta} := \operatorname{argmin}_{t\in\Theta} \sum_{i=1}^n \rho(X_i, t)$. Properties of such estimators have been extensively discussed in the literature. In particular, conditions under which these estimators are consistent and asymptotically normal have been discussed by Huber (1967), Pfanzagl (1994), van der Vaart and Wellner (1996) and references therein. A method of deriving asymptotic distributions of a class of minimum dispersion estimators that are applicable to more general models including regression and autoregressive models is discussed in Koul (2002, Ch. 5).

6.4.4 Regression Quantiles and Rank Scores

As in Chapter 3, order statistics and ranks are very useful in developing some broadly applicable inference procedures. Regression quantiles of Koenker and Bassett (1978) (KB) and the corresponding rank scores of Gutenbrunner and Jurečková (1992) provide, respectively, an appropriate extension of the one sample order statistics and ranks to the linear regression model (6.4.1). They are also useful in providing an extension of the class of linear combinations of order statistics in the one sample location model to the linear regression model. In this section we shall briefly discuss regression quantiles and the related regression rank scores. Recall that the error d.f. F is assumed to be continuous.

As in Subsection 6.4.1, let $\boldsymbol{z}_i' := (1, \boldsymbol{x}_i')$, and for $0 \le \alpha \le 1$, $\boldsymbol{t} \in \mathbb{R}^{p+1}$, define

$$\rho_\alpha(u) := \alpha\, u\, I(u > 0) - (1-\alpha)\, u\, I(u \le 0) = u\big[\alpha - I(u \le 0)\big],$$

$$\boldsymbol{S}_\alpha(\boldsymbol{t}) := \sum_{i=1}^n \boldsymbol{z}_i\big\{I(Y_i - \boldsymbol{z}_i'\boldsymbol{t} \le 0) - \alpha\big\}, \qquad \widehat{\boldsymbol{\beta}^*_S}(\alpha) := \operatorname{argmin}_{\boldsymbol{t}} \|\boldsymbol{S}_\alpha(\boldsymbol{t})\|^2,$$

$$M_\alpha(\boldsymbol{t}) := \sum_{i=1}^n \rho_\alpha(Y_i - \boldsymbol{z}_i'\boldsymbol{t}), \qquad \widehat{\boldsymbol{\beta}^*_M}(\alpha) := \operatorname{argmin}_{\boldsymbol{t}} M_\alpha(\boldsymbol{t}).$$

Note that $-\boldsymbol{S}_\alpha(\boldsymbol{t})$ is an almost everywhere derivative of $M_\alpha(\boldsymbol{t})$. The estimator $\widehat{\boldsymbol{\beta}^*_M}(\alpha)$ is called an *αth regression quantile*. Both estimators $\widehat{\boldsymbol{\beta}^*_M}(\alpha)$ and $\widehat{\boldsymbol{\beta}^*_S}(\alpha)$ provide extension of the one sample order statistics to the linear regression model (6.4.1). For example, take $\alpha = 0.5$. Then,

$$\rho_{0.5}(u) = \frac{1}{2}\,|u|, \quad M_{0.5}(\boldsymbol{t}) := \frac{1}{2}\sum_{i=1}^n \big|Y_i - \boldsymbol{z}_i'\boldsymbol{t}\big|, \quad S_{0.5}(\boldsymbol{t}) = \sum_{i=1}^n \boldsymbol{z}_i\big[I(Y_i - \boldsymbol{z}_i'\boldsymbol{t} \le 0) - 0.5\big].$$

Hence, $\widehat{\boldsymbol{\beta}^*_M}(0.5)$ and $\widehat{\boldsymbol{\beta}^*_S}(.5)$ are equal to the LAD estimator of the parameter vector $\boldsymbol{\beta}^*$. Further, suppose there are no covariates in (6.4.1), i.e., the model is the one sample location model $Y_i =$

$\beta_0 + \varepsilon_i$. As we have seen in Section 6.2.3 on M estimators in the one sample location model, the sample median is a LAD estimator. Hence, $\widehat{\boldsymbol{\beta}^*_M}(0.5)$ and $\hat{\boldsymbol{\beta}}^*_S(.5)$ provide an extension of the sample median in the one sample location model to the regression model (6.4.1).

As another example, let $\alpha = 0.75$. Then,

$$M_{0.75}(\boldsymbol{t}) := \sum_{i=1}^n (Y_i - \boldsymbol{z}_i'\boldsymbol{t})\left[0.75 - I(Y_i - \boldsymbol{z}_i'\boldsymbol{t} \le 0)\right],$$

$$S_{0.75}(t) = \sum_{i=1}^n \boldsymbol{z}_i\left[I(Y_i - \boldsymbol{z}_i'\boldsymbol{t} \le 0) - 0.75\right].$$

If we further specialize (6.4.1) to the one sample location model, then

$$M_{0.75}(t) = \sum_{i=1}^n (Y_i - t)\left[0.75 - I(Y_i - t \le 0)\right],$$

$$S_{0.75}(t) = \sum_{i=1}^n \left[I(Y_i - t \le 0) - 0.75\right] = \sum_{i=1}^n \left[I(Y_i \le t) - 0.75\right], \quad t \in \mathbb{R}.$$

It readily thus follows that both estimators are seventy-fifth sample quantile. In general, if $(i - 1)/n \le \alpha < i/n$, then $\widehat{\beta^*_M}(\alpha) = Y_{(i)}$, the ith order statistic.

In the one sample location model, the statistic $L := \sum_{i=1}^n w_{ni}Y_{(i)}$, where $w_{n1}, \cdots, w_{nn}$ are some real numbers, is known as a linear combination of order statistics. See Gastwirth (1966) and Stigler (1967, 1974) for more on this topic. For each set of weights $\{w_{ni}\}$, we have one such statistic. This class of statistics also contains some robust estimators of the location parameter. For example, if $w_{n1} = 0 = w_{nn}$ and $w_{ni} = 1/(n-2)$, for $i = 2, \cdots, n-1$, then $L = \sum_{i=2}^{n-1} Y_{(i)}/(n-2)$ is a trimmed mean. If n is odd, say $n = 2m+1$, m a positive integer, and we choose w_{ni} equal to 1 for $i = m+1$ and zero for all other i's, then $L = Y_{(m+1)}$, the sample median. Similarly, if n is even, say $n = 2m$ and we choose $w_{ni} = 1/2$, for $i = m, m+1$ and zero for all other i's, then $L = \left[Y(m) + Y_{(m+1)}\right]/2$ is again the sample median. Both, trimmed mean and median are known to be robust against some gross errors. The above class also contains the sample range obtained by taking $w_{nn} = 1, w_{n1} = -1$ and the rest of w_{ni}'s equal to zero.

An extension of the class of statistics L to the regression model (6.4.1) is given by $\mathcal{L} := \int_0^1 \widehat{\boldsymbol{\beta}^*_M}(s)g(s)\,ds$, where g is a nonnegative integrable function on $(0,1)$. Let $G(t) := \int_0^t g(s)\,ds$. Then in the one sample location model, the statistic L with $w_{ni} = G(i/n) - G((i-1)/n)$ is the same as $\mathcal{L}$. Thus regression quantiles are useful in providing a natural extension of the class of linear combination of order statistics from the one sample location model to the linear regression model.

Now we shall describe some properties of the regression quantiles in general. Recall $\mathbf{1}_n := (1, \cdots, 1)'$ denotes the vector of 1's of dimension n. Set $\boldsymbol{Y} := (Y_1, \cdots, Y_n)'$ and let $\boldsymbol{Z}$ be the $n \times (p+1)$ matrix with rows $\boldsymbol{z}_i'$, $i = 1, \cdots, n$. Assume $\boldsymbol{Z}$ has full rank $p+1$. For a subset π_n of $\{1, 2, \cdots, n\}$ of size $p+1$, we write $\mathcal{Y}_{\pi_n}$ for the vector of Y_i's with $i \in \pi_n$ and $\mathcal{Z}_{\pi_n}$ for the $(p+1) \times (p+1)$ matrix with rows $\boldsymbol{z}_i'$, $i \in \pi_n$. Let $A_n := \{\pi_n; \text{rank of } \mathcal{Z}_{\pi_n} = p+1\}$.

Now, let

$$\mathscr{B}_n(\alpha) := \left\{\boldsymbol{t} \in \mathbb{R}^{p+1}; M_\alpha(\boldsymbol{t}) = \text{minimum}\right\},$$

and consider the following linear programming problem.

$$\text{minimize } \alpha\mathbf{1}_n'\mathbf{r}^+ + (1-\alpha)\mathbf{1}_n'\mathbf{r}^-, \text{ subject to } \boldsymbol{Y} - \boldsymbol{Z}\boldsymbol{t} = \mathbf{r}^+ - \mathbf{r}^-, \qquad (6.4.14)$$
$$\text{w.r.t. } (\boldsymbol{t}, \mathbf{r}^+, \mathbf{r}^-) \in \mathbb{R}^{p+1} \times (0,\infty)^n \times (0,\infty)^n.$$

Note that $\widehat{\boldsymbol{\beta}^*_M}(\alpha) \in \mathscr{B}_n(\alpha)$. Moreover, as proved in Theorem 3.1 of KB, the set $\mathscr{B}_n(\alpha)$ is the convex hull of one or more basic solutions of the form

$$\boldsymbol{b}_{\pi_n} := \mathcal{Z}_{\pi_n}^{-1}\mathcal{Y}_{\pi_n}, \quad \text{for some } \pi_n \in A_n. \tag{6.4.15}$$

Recall from (3.3.5), in the one sample setup when the sample size is even, say $n = 2m$, then sample median is in the interval $[X_{(m)}, X_{(m+1)}]$ but to make it unique we define it as the average of these two order statistics. The above result (6.4.15) is an extension of this notion to the current regression setup.

Closely related entities to the regression quantiles are the so called *regression rank scores* as developed in Gutenbrunner and Jurečková (1992) (GJ). Consider the following dual of the above linear programming problem.

$$\text{Maximize } \boldsymbol{Y}'\boldsymbol{a}, \text{ w.r.t. } \boldsymbol{a}, \text{ subject to } \boldsymbol{Y}'\boldsymbol{a} = (1-\alpha)\boldsymbol{Y}'\mathbf{1}_n, \quad \boldsymbol{a} \in [0,1]^n. \tag{6.4.16}$$

By the linear programming theory, the optimal solution $\hat{\boldsymbol{a}}_n(\alpha) = (\hat{a}_{n1}(\alpha), \cdots, \hat{a}_{nn}(\alpha))'$ of this problem can be computed in terms of $\widehat{\boldsymbol{\beta}^*_M}(\alpha)$ as follows: If $\widehat{\boldsymbol{\beta}^*_M}(\alpha) = \mathcal{Z}_{\pi_n(\alpha)}^{-1}\mathcal{Y}_{\pi_n(\alpha)}$, for some $(p+1)$-dimensional subset $\pi_n(\alpha)$ of $\{1, \cdots, n\}$, then, for $i \notin \pi_n(\alpha)$,

$$\begin{aligned} \hat{a}_{ni}(\alpha) &= 1, \quad Y_i > \boldsymbol{z}_i'\widehat{\boldsymbol{\beta}^*_M}(\alpha), \\ &= 0, \quad Y_i < \boldsymbol{z}_i'\widehat{\boldsymbol{\beta}^*_M}(\alpha), \end{aligned} \tag{6.4.17}$$

and, for $i \in \pi_n(\alpha)$, $\hat{a}_{ni}(\alpha)$ is the solution of the $p+1$ linear equations

$$\sum_{j \in \pi_n(\alpha)} \boldsymbol{z}_j \hat{a}_{nj}(\alpha) = (1-\alpha)\sum_{j=1}^n \boldsymbol{z}_j - \sum_{j=1}^n \boldsymbol{z}_j I(Y_j > \boldsymbol{z}_j'\widehat{\boldsymbol{\beta}^*_M}(\alpha)).$$

As shown in GJ , the continuity of F implies that the regression rank scores $\hat{\boldsymbol{a}}_n(\alpha)$ are unique for all $0 < \alpha < 1$, wp1. Each component of the process $\hat{\boldsymbol{a}}_n \in [0,1]^n$ has piecewise linear paths in $\mathcal{C}[0,1]$, the space of real-valued continuous functions define on $[0,1]$, and $\hat{\boldsymbol{a}}_n(0) = \mathbf{1}_n = \mathbf{1}_n - \hat{\boldsymbol{a}}_n(1)$. It is invariant in the sense that $\hat{\boldsymbol{a}}_n(\alpha)$ based on the vector $\boldsymbol{Y} + \boldsymbol{Z}\boldsymbol{t}$ is the same as the $\hat{\boldsymbol{a}}_n(\alpha)$ based on $\boldsymbol{Y}$, for all $\boldsymbol{t} \in \mathbb{R}^{p+1}$, $0 < \alpha < 1$.

Koenker and d'Odrey (1987, 1993), Koenker (1997), Portnoy (1997) and Portnoy and Koenker (1997) discuss some computational aspects of regression quantiles and regression rank scores.

We shall now state some of the asymptotic results about $\widehat{\boldsymbol{\beta}^*_M}(\alpha)$ and $\hat{\boldsymbol{a}}_n(\alpha)$ without proofs. Most of the proofs use the methodology developed in the previous sections of this chapter. For these results we need the following assumption.

$$\text{The error d.f. } F \text{ has a positive and continuous density } f \text{ on } \mathbb{R}. \tag{6.4.18}$$

First, we focus on the regression quantiles $\widehat{\boldsymbol{\beta}^*_M}(\alpha)$. To proceed further, we need some additional notation. Set $\boldsymbol{A} := (\boldsymbol{Z}'\boldsymbol{Z})^{1/2}$ and let

$$\boldsymbol{\beta}(\alpha) := (\beta_0 + F^{-1}(\alpha), \beta_1, \cdots, \beta_p)', \quad q(\alpha) := f(F^{-1}(\alpha)), \quad 0 < \alpha < 1.$$

For $\mathbb{R}^{p+1}$-valued random vectors $W_n(\boldsymbol{s}, \alpha)$ indexed by $0 < \alpha < 1$ and $\boldsymbol{s} \in \mathbb{R}^{p+1}$, the statement $W_n(\boldsymbol{s}, \alpha) = o_p^*(1)$ means that for every $0 < a \le 1/2$, $0 < b < \infty$, $\sup\{\|W_n(\boldsymbol{s}, \alpha)\|; \|\boldsymbol{s}\| \le b, a \le \alpha \le 1-a\} = o_p(1)$. The proof of the following lemma is similar to that which deals with M estimators in linear regression.

Lemma 6.4.2. *Suppose* (6.4.1), (6.4.9) *and* (6.4.18) *hold. Then,* $\forall\, 0 < a \leq 1/2,\ 0 < b < \infty$,

$$\left\|\boldsymbol{A}^{-1}\{\boldsymbol{S}_\alpha(\boldsymbol{\beta}(\alpha)+\boldsymbol{A}^{-1}\boldsymbol{s})-\boldsymbol{S}_\alpha(\boldsymbol{\beta}(\alpha)\}-\boldsymbol{s}q(\alpha)\right\| = o_p^*(1). \tag{6.4.19}$$

Moreover,

$$\boldsymbol{A}(\widehat{\boldsymbol{\beta}_S^*}(\alpha)-\boldsymbol{\beta}(\alpha)) = -\{q(\alpha)\}^{-1}\boldsymbol{A}^{-1}\boldsymbol{S}_\alpha(\boldsymbol{\beta}(\alpha))+o_p^*(1), \tag{6.4.20}$$

$$n^{1/2}(\widehat{\boldsymbol{\beta}_S^*}(\alpha)-\widehat{\boldsymbol{\beta}_M^*}(\alpha)) = o_p^*(1). \tag{6.4.21}$$

If F *satisfies* (6.4.6), *then for every* $0 < b < \infty$,

$$\sup\left\|\boldsymbol{A}^{-1}\{\boldsymbol{S}_\alpha(\boldsymbol{\beta}(\alpha)+\boldsymbol{A}^{-1}\boldsymbol{s})-\boldsymbol{S}_\alpha(\boldsymbol{\beta}(\alpha))\}-\boldsymbol{s}q(\alpha)\right\| = o_p(1), \tag{6.4.22}$$

where the supremum is taken over $(\alpha, \boldsymbol{s}) \in [0,1]\times\{\boldsymbol{s}\in\mathbb{R}^{p+1}; \|\boldsymbol{s}\| \leq b\}$.

To state the next result we recall the definition of Kronecker matrix product $\otimes$. For an $m\times n$ matrix $A = ((a_{ij}))_{i=1,\cdots,m,j=1,\cdots,n}$ and $p\times q$ matrix B, their Kronecker matrix product is the $pm\times qn$ matrix

$$A\otimes B = \begin{pmatrix} a_{11}B, & \cdots & a_{1n}B \\ \vdots, & \vdots & \vdots \\ a_{m1}B, & \cdots & a_{mn}B \end{pmatrix}.$$

The following corollary follows from the previous lemma.

Corollary 6.4.1. *Under the assumptions* (6.4.9) *and* (6.4.18), *the following hold.*

$$\boldsymbol{A}\left(\widehat{\boldsymbol{\beta}_M^*}(\alpha)-\boldsymbol{\beta}(\alpha)\right) = -\{q(\alpha)\}^{-1}\boldsymbol{A}^{-1}\boldsymbol{S}_\alpha(\boldsymbol{\beta}(\alpha))+o_p^*(1). \tag{6.4.23}$$

Moreover, for every $0 < \alpha_1 < \cdots < \alpha_k < 1$, *the asymptotic joint distribution of the vector* $\boldsymbol{A}\left(\widehat{\boldsymbol{\beta}_M^*}(\alpha_1)-\boldsymbol{\beta}(\alpha_1),\cdots,\widehat{\boldsymbol{\beta}_M^*}(\alpha_k)-\boldsymbol{\beta}(\alpha_k)\right)$ *is* $(p+1)\times k$ *normal distribution with the mean matrix* **0** *and the variance-covariance matrix the Kronecker product* $\mathcal{I}_{p+1}\otimes\mathbf{B}$, *where*

$$\mathbf{B} = \left(\frac{\alpha_i\wedge\alpha_j-\alpha_i\alpha_j}{q(\alpha_i)q(\alpha_j)}\right)_{1\leq i,j\leq k}.$$

Now we shall discuss the asymptotic behaviour of the *regression rank scores* defined at (6.4.17). To that effect we need to introduce some more notation. Let $c_{n1},\cdots,c_{nn}$ be real numbers so that $\sum_{i=1}^n c_{ni}^2 = 1$. For $0<\alpha<1$, define the processes

$$\hat{U}_n(\alpha) := \sum_{i=1}^n c_{ni}\{\hat{a}_{ni}(\alpha)-(1-\alpha)\}, \qquad U_n(\alpha) := \sum_{i=1}^n c_{ni}\{I(\varepsilon_i > F^{-1}(\alpha))-(1-\alpha)\}.$$

Let $\boldsymbol{c}_n := (c_{n1},\cdots,c_{nn})'$, $\boldsymbol{k}_n := (\boldsymbol{c}_n'\boldsymbol{Z}\boldsymbol{A}^{-1})'$ and $\hat{\boldsymbol{\Delta}}_n(\alpha) := \boldsymbol{A}\left(\widehat{\boldsymbol{\beta}_M^*}(\alpha)-\boldsymbol{\beta}(\alpha)\right)$. Note that $\boldsymbol{k}_n$ is a $(p+1)$-dimensional vector satisfying $\left\|\boldsymbol{k}_n\right\| \leq 1$. This follows from $\boldsymbol{k}_n'\boldsymbol{k}_n = \boldsymbol{c}_n'\boldsymbol{Z}(\boldsymbol{Z}'\boldsymbol{Z})^{-1}\boldsymbol{Z}'\boldsymbol{c}_n$, $\|\boldsymbol{c}_n\| = 1$ and the fact that $\boldsymbol{Z}(\boldsymbol{Z}'\boldsymbol{Z})^{-1}\boldsymbol{Z}'$ has eigenvalues 1 and 0 only. We are now ready to state the following lemma without a proof.

Lemma 6.4.3. *In addition to the assumptions* (6.4.9) *and* (6.4.18), *suppose* $\max_{1\leq i\leq n}|c_{ni}| = o(1)$. *Then, for every* $0 < a \leq 1/2$,

$$\hat{U}_n(\alpha) = U_n(\alpha) + \boldsymbol{k}_n'\,\boldsymbol{A}^{-1}\boldsymbol{S}_\alpha(\boldsymbol{\beta}(\alpha)) + o_p^*(1). \tag{6.4.24}$$

Let ε be an independent copy of ε_i and set $U := F(\varepsilon), U_i := F(\varepsilon_i)$. Then we have

$$U_n(\alpha) := \sum_{i=1}^n c_{ni}\{I(U_i > \alpha) - (1-\alpha)\}, \qquad \boldsymbol{S}_\alpha(\boldsymbol{\beta}(\alpha)) = \sum_{i=1}^n \boldsymbol{z}_i\{I(U_i \le \alpha) - \alpha\}.$$

Because U, U_i's are i.i.d. Unif$(0,1)$ r.v.'s, we readily obtain that for any $0 \le \alpha_1 < \alpha_2 \le 1$,

$$\begin{aligned}
E\big(U_n(\alpha_1)U_n(\alpha_2)\big) &= \alpha_1(1-\alpha_2),\\
E\big(U_n(\alpha_1)\boldsymbol{S}'_\alpha(\boldsymbol{\beta}(\alpha_2))\big) &= \sum_{i=1}^n c_{ni}\boldsymbol{z}'_i\,\alpha_1(1-\alpha_2) = -\boldsymbol{c}'_n\boldsymbol{Z}\,\alpha_1(1-\alpha_2)\\
E\big(\boldsymbol{S}_\alpha(\boldsymbol{\beta}(\alpha_1))U_n(\alpha_2)\big) &= \sum_{i=1}^n c_{ni}\boldsymbol{z}_i\,\alpha_1(1-\alpha_2) = -\boldsymbol{Z}'\boldsymbol{c}_n\,\alpha_1(1-\alpha_2),\\
E\big(\boldsymbol{S}_\alpha(\boldsymbol{\beta}(\alpha_1))\boldsymbol{S}_\alpha(\boldsymbol{\beta}(\alpha_2))'\big) &= \boldsymbol{Z}'\boldsymbol{Z}\,\alpha_1(1-\alpha_2).
\end{aligned}$$

Let $T_n(\alpha)$ denote the leading r.v. on the right-hand side of (6.4.24), i.e.,

$$T_n(\alpha) := U_n(\alpha) + \boldsymbol{k}'_n\,\boldsymbol{A}^{-1}\boldsymbol{S}_\alpha(\boldsymbol{\beta}(\alpha)).$$

The above derivations, the definition of $\boldsymbol{k}_n$ and the identity $\boldsymbol{A}^{-1}\boldsymbol{Z}'\boldsymbol{Z}\boldsymbol{A}^{-1} = \mathcal{I}_{p+1}$ yield

$$\begin{aligned}
&\mathrm{Cov}\big(T_n(\alpha_1), T_n(\alpha_2)\big) = E\big(T_n(\alpha_1)T'_n(\alpha_2)\big)\\
&= E\big(U_n(\alpha_1)U_n(\alpha_2)\big) + E\big(U_n(\alpha_1)\boldsymbol{S}'_{\alpha_2}(\boldsymbol{\beta}(\alpha_2))\boldsymbol{A}^{-1}\boldsymbol{k}_n\\
&\quad + \boldsymbol{k}'_n\boldsymbol{A}^{-1}E\big(\boldsymbol{S}_{\alpha_1}(\boldsymbol{\beta}(\alpha_1))U_n(\alpha_2)\big) + \boldsymbol{k}'_n\boldsymbol{A}^{-1}E\big(\boldsymbol{S}_{\alpha_1}(\boldsymbol{\beta}(\alpha_1))\boldsymbol{S}'_{\alpha_2}(\boldsymbol{\beta}(\alpha_2))\big)\boldsymbol{A}^{-1}\boldsymbol{k}_n\\
&= \alpha_1(1-\alpha_2)\Big\{1 - \boldsymbol{c}'_n\boldsymbol{Z}\boldsymbol{A}^{-1}\boldsymbol{k}_n - \boldsymbol{k}'_n\boldsymbol{A}^{-1}\boldsymbol{Z}'\boldsymbol{c}_n + \boldsymbol{k}'_n\boldsymbol{A}^{-1}\boldsymbol{Z}'\boldsymbol{Z}\boldsymbol{A}^{-1}\boldsymbol{k}_n\Big\}\\
&= \alpha_1(1-\alpha_2)\Big\{1 - 2\|\boldsymbol{k}_n\|^2 + \|\boldsymbol{k}_n\|^2\Big\} = \alpha_1(1-\alpha_2)\Big\{1 - \|\boldsymbol{k}_n\|^2\Big\}.
\end{aligned}$$

Now let m be a fixed positive integer and $0 < \alpha_1 < \cdots < \alpha_m < 1$ be given numbers. Set $\boldsymbol{W}_n := \big(T_n(\alpha_1), \cdots, T_n(\alpha_m)\big)'$ and let $\boldsymbol{C}$ denote the $m \times m$ covariance matrix whose $(i,j)^{\text{th}}$ entry is $\alpha_i(1-\alpha_j), 1 \le i \le j \le m$. Then $\boldsymbol{\Sigma}_n := E\big(\boldsymbol{W}_n\boldsymbol{W}'_n\big) = \boldsymbol{C}\big(1 - \|\boldsymbol{k}_n\|^2\big)$.

Recall from Chapter 3 that in the one sample setup when the given d.f. has a density, ranks and order statistics are independent. The following corollary asserts a similar relationship for large samples between regression quantiles and regression rank scores. Its proof can be deduced from results in GJ.

Corollary 6.4.2. *Under the assumptions of Lemma 6.4.3, the regression quantile and regression rank score processes are asymptotically independent. If, in addition, we assume that* $\liminf_n \big(1 - \|\boldsymbol{k}_n\|^2\big) > 0$, *then for every* $m \ge 1$ *and for every* $0 < \alpha_1 < \cdots < \alpha_m < 1$, $\boldsymbol{\Sigma}_n^{-1}$ *exists for all sufficiently large* n *and* $\boldsymbol{\Sigma}_n^{-1}\big(\hat{U}_n(\alpha_1), \cdots, \hat{U}_n(\alpha_m)\big) \Rightarrow \mathcal{N}_m\big(0_m, \mathcal{I}_m\big)$.

6.5 Asymptotic Relative Efficiency of Estimators

Suppose $\widehat{\theta}_1$ and $\widehat{\theta}_2$ are two estimators of the p-dimensional parameter such that $n^{1/2}\big(\widehat{\theta}_j - \theta\big) \to_D \mathcal{N}_p\big(0, \Sigma_{j\theta}\big)$, where $\Sigma_{j\theta}$ is a positive definite matrix, $j = 1, 2$. Then the asymptotic relative efficiency of $\widehat{\theta}_2$, relative to $\widehat{\theta}_1$, is defined to be

$$\text{ARE}(\widehat{\theta}_2, \widehat{\theta}_1) = \Big(\frac{\det(\Sigma_{1\theta})}{\det(\Sigma_{2\theta})}\Big)^{1/p}, \tag{6.5.1}$$

where $\det(A)$ denotes the determinant of a square matrix A.

As an example consider the one sample location model with location parameter θ and symmetric error density f. Consider the three estimators $\widehat{\theta}_1 = \bar{X}$, $\widehat{\theta}_2 = \widetilde{X}$ and $\widehat{\theta}_3 = \widehat{\theta}_{HLS}$ of θ. Suppose f is uniformly continuous, $f(0) > 0$ and has mean zero and finite variance σ^2. By the classical CLT, $n^{1/2}(\bar{X} - \theta) \to_D \mathcal{N}(0, \sigma^2)$. By Theorem 3.3.2, $n^{1/2}(\widetilde{X} - \theta) \to_D \mathcal{N}(0, 1/4f^2(0))$ and by (6.2.17), $n^{1/2}(\hat{\theta}_{HLS} - \theta) \to_D \mathcal{N}\big(0, 1/\big[12\big(\int f^2(x)\,dx\big)^2\big]\big)$. Note also that $p = 1$. Hence,

$$\text{ARE}(\widetilde{X}, \bar{X}) = 4\sigma^2 f^2(0), \quad \text{ARE}(\widehat{\theta}_{HLS}, \bar{X}) = 12\sigma^2\Big(\int f^2(x)\,dx\Big)^2.$$

These are the same as the ARE's of the sign test and Wilcoxon sign rank test, relative to the one sample t test, respectively. See (5.10.2).

Next, consider the linear regression model (6.4.1). Suppose, in addition to the assumptions of Theorem 6.4.1, $n^{-1}\boldsymbol{X}_c'\boldsymbol{X}_c$ tends to some positive definite $p \times p$ symmetric matrix $\boldsymbol{\Sigma}$. Then, by Theorem 6.4.1, $n^{1/2}(\widehat{\boldsymbol{\beta}}_R - \boldsymbol{\beta}) \to_D \mathcal{N}_p(0, \tau_J^2\boldsymbol{\Sigma}^{-1})$, where $\tau_J^2 := \sigma_J^2/\big(\int J(F(x))\dot{f}(x)\,dx\big)^2$ under the assumptions of Part (a) and $\tau_J^2 = \sigma_J^2/\big(\int f(x)dJ(F(x))\big)^2$ under those of Part (b). Also, under the assumption (6.4.4) and finite error variance σ^2, the least squares estimator $\widehat{\boldsymbol{\beta}}_{LS} := (\boldsymbol{X}_c'\boldsymbol{X}_c)^{-1}\boldsymbol{X}_c'\boldsymbol{Y}$ of $\boldsymbol{\beta}$ satisfies $n^{1/2}(\widehat{\boldsymbol{\beta}}_{LS} - \boldsymbol{\beta}) \to_D \mathcal{N}_p(0, \sigma^2\boldsymbol{\Sigma}^{-1})$. Hence,

$$\text{ARE}(\widehat{\boldsymbol{\beta}}_R, \widehat{\boldsymbol{\beta}}_{LS}) = \left(\frac{\det(\sigma^2\boldsymbol{\Sigma}^{-1})}{\det(\tau_J^2\boldsymbol{\Sigma}^{-1})}\right)^{1/p} = \frac{\sigma^2}{\tau_J^2}. \tag{6.5.2}$$

We shall now consider some special choices of J.

Example 6.5.1. Normal scores R estimator. Suppose $J(u) = \Phi^{-1}(u)$, where Φ denotes the d.f. of $\mathcal{N}(0,1)$ r.v. and φ its density. The R estimator $\widehat{\boldsymbol{\beta}}_R$ corresponding to this J is called the normal scores rank estimator and will be denoted by $\widehat{\boldsymbol{\beta}}_{NS}$. Then,

$$E(J(U)) = \int_0^1 \Phi^{-1}(u)du = \int x d\Phi(x) = 0, \quad \sigma^2_{\Phi^{-1}} = \int_0^1 \big(\Phi^{-1}(u)\big)^2 du = \int x^2\,d\Phi(x) = 1,$$

$$\tau^2_{\Phi^{-1}} = \frac{1}{\big(\int \Phi^{-1}(F(x))\dot{f}(x)\,dx\big)^2}, \quad \text{ARE}(\widehat{\boldsymbol{\beta}}_{NS}, \widehat{\boldsymbol{\beta}}_{LS}) = \sigma^2\Big(\int \Phi^{-1}(F(x))\dot{f}(x)\,dx\Big)^2.$$

Integration by parts and by the change of variable $F(x) = u$ we obtain

$$\int \Phi^{-1}(F(x))\dot{f}(x)\,dx = -\int_0^1 \frac{f(F^{-1}(u))}{\varphi(\Phi^{-1}(u))}du. \tag{6.5.3}$$

Therefore,

$$\text{ARE}(\widehat{\boldsymbol{\beta}}_{NS}, \widehat{\boldsymbol{\beta}}_{LS}) = \sigma^2\Big(\int_0^1 \frac{f(F^{-1}(u))}{\varphi(\Phi^{-1}(u))}du\Big)^2.$$

This is the same expression as in (5.10.3), and the discussion following (5.10.3) equally applies when comparing the normal scores estimator with the least squares estimator.

Suppose further that $F(x) \equiv \Phi(x/\sigma)$. Then,

$$F^{-1}(u) = \sigma\Phi^{-1}(u), \quad f(F^{-1}(u)) = \frac{\varphi(\Phi^{-1}(u))}{\sigma}, \quad \forall 0 < u < 1,$$

and hence

$$\text{ARE}(\widehat{\beta}_{NS}, \widehat{\beta}_{LS}) = 1.$$

This is a remarkable result. We know that the least squares estimator is the most efficient estimator at the Gaussian errors, for all sample sizes. The above ARE result says that the normal scores rank estimator is asymptotically a most efficient estimator of the regression parameter vector at the Gaussian errors. At the same time this rank estimator is not sensitive to the lack of normality of the errors. It is a robust estimator against gross errors. Moreover, by (5.10.4), the estimator $\widehat{\beta}_{Nsc}$ is asymptotically more efficient than the least squares estimator, for all non-Gaussian error distributions with zero mean and finite variance.

Example 6.5.2. Wilcoxon rank estimator. The choice $J(u) = u - 1/2$, $0 < u < 1$, gives the Wilcoxon rank estimator. This J is bounded with $\sigma_J^2 = 1/12$ and $\int f(x)\,dJ(F(x)) = \int f(x)\,dF(x) = \int f^2(x)\,dx$. Hence, the asymptotic relative efficiency of the Wilcoxon rank estimator relative to the least squares estimator is

$$\text{ARE}(\widehat{\beta}_W, \widehat{\beta}_{LS}) = 12\sigma^2\Big(\int f^2(x)\,dx\Big)^2.$$

This is the same expression as when computing the ARE of the Wilcoxon signed rank test relative to the t test, see (5.10.2), so the discussion that follows (5.10.2) equally applies here in terms of Wilcoxon rank and the least squares estimators.

6.6 Estimation of $\int f^2(x)\,dx$

In this section we shall discuss the estimation of $I_2(f) := \int f^2(x)\,dx$ for a square integrable density f on $\mathbb{R}$. As seen in (5.10.1), the quantity $I_2(f)$ appears in the asymptotic relative efficiency comparisons of the Wilcoxon signed ranks and two sample rank sum tests. It also appears in the asymptotic variances of the Hodges-Lehmann and Hodges-Lehmann-Sen estimators of the one and two sample location parameters as in (6.2.17) and in the asymptotic variance of a certain minimum distance estimator of the regression parameter vector as in Theorem 6.4.3. Hettmansperger (1984) argues that the rank analysis of variance using Wilcoxon type linear rank statistics is incomplete without having a consistent estimator of $I_2(f)$.

We shall discuss consistent estimators of $I_2(f)$ in the one sample location model and in the linear regression model. Consider first the one sample location model $Y_i = \theta + \varepsilon_i$, where ε_i are i.i.d. f. Let K be a density kernel symmetric about zero. Then an estimator of $I_2(f)$ is provided by the U-statistic

$$\hat{Q}_1 := \frac{2}{n(n-1)h_n} \sum_{1\le i<j\le n} K\Big(\frac{Y_i - Y_j}{h_n}\Big),$$

with $h_n > 0$. This estimator can be written in terms of the unobservable errors as

$$\hat{Q}_1 = \frac{2}{n(n-1)h_n} \sum_{1\le i<j\le n} K\Big(\frac{\varepsilon_i - \varepsilon_j}{h_n}\Big),$$

irrespective of any identifiability assumptions for the location parameter θ such as symmetry of f. Clearly, $\hat{Q}_1 = U_n^{(3)}$ of (5.2.3) with $X_i = \varepsilon_i$, $i = 1, \cdots, n$. The following lemma gives the consistency of this estimator, see Exercise 10 at the end of this Chapter 5.

Lemma 6.6.1. *If f and K are square integrable, $h_n \to 0$ and $nh_n \to \infty$, then $\hat{Q}_1 \to_p I_2(f)$.*

Next, we shall discuss the consistent estimation of $I_2(f)$ in the linear regression model

$$Y_{ni} = \beta_0 + \boldsymbol{x}_{ni}'\boldsymbol{\beta} + \varepsilon_i, \quad i = 1, \cdots, n,$$

where $\boldsymbol{x}_{n1}, \cdots, \boldsymbol{x}_{nn}$ are known p-dimensional vectors, β_0 is an unknown intercept parameter and $\boldsymbol{\beta}$ is a p-dimensional slope parameter, and the error variables $\varepsilon_1, \cdots, \varepsilon_n$ are i.i.d. f. We set $\tilde{\boldsymbol{x}}_{ni} = \boldsymbol{x}_{ni} - \bar{\boldsymbol{x}}_n$, $i = 1, \cdots, n$, with $\bar{\boldsymbol{x}}_n := (1/n)\sum_{i=1}^n \boldsymbol{x}_{ni}$. Let $\hat{\boldsymbol{\beta}}$ be an estimator of $\boldsymbol{\beta}$. We assume the following.

Assumption 6.6.1. The matrix $\boldsymbol{A}_n = \sum_{i=1}^n \tilde{\boldsymbol{x}}_{ni}\tilde{\boldsymbol{x}}_{ni}'$ is positive definite, $\max_{1\le i\le n} \|\boldsymbol{A}_n^{-1/2}\tilde{\boldsymbol{x}}_{ni}\| \to 0$ and $\|\boldsymbol{A}_n^{1/2}(\hat{\boldsymbol{\beta}} - \boldsymbol{\beta})\| = O_p(1)$.

Note that

$$\hat{\varepsilon}_i := Y_i - \tilde{\boldsymbol{x}}_{ni}'\hat{\boldsymbol{\beta}} = \varepsilon_i + \beta_0 + \bar{\boldsymbol{x}}_n'\boldsymbol{\beta} - \tilde{\boldsymbol{x}}_{ni}'(\hat{\boldsymbol{\beta}} - \boldsymbol{\beta}), \quad i = 1, \cdots, n.$$

Thus the differences $\hat{\varepsilon}_i - \hat{\varepsilon}_j = \varepsilon_i - \varepsilon_j - (\tilde{\boldsymbol{x}}_{ni} - \tilde{\boldsymbol{x}}_{nj})'(\hat{\boldsymbol{\beta}} - \boldsymbol{\beta})$, $1 \le i < j \le n$, do not depend on the intercept parameter. In view of this we now consider the analog of $\hat{Q}_1$ where Y_i's are replaced by $\hat{\varepsilon}_i$'s, i.e.,

$$\hat{Q}_2 := \frac{2}{n(n-1)h_n} \sum_{1\le i<j\le n} K\Big(\frac{\hat{\varepsilon}_i - \hat{\varepsilon}_j}{h_n}\Big).$$

To describe the consistency of this estimator in the next lemma, we recall that a real-valued function ψ defined on $\mathbb{R}$ is said to be Lipschitz continuous if there is a constant L such that

$$|\psi(y) - \psi(x)| \le L\,|y - x|, \quad \forall x, y \in \mathbb{R}. \tag{6.6.1}$$

Lemma 6.6.2. *Suppose f is square-integrable, K is Lipschitz continuous, $h_n \to 0$ and $nh_n^4 \to \infty$. Then $\hat{Q}_2 \to_p I_2(f)$.*

Proof. With L a Lipschitz constant of K and $c_{ni} := \|\boldsymbol{A}_n^{-1/2}\tilde{\boldsymbol{x}}_{ni}\|$, we derive the inequality

$$\begin{aligned}
|\hat{Q}_2 - \hat{Q}_1| &\le \frac{2}{n(n-1)h_n^2} \sum_{1\le i<j\le n} L|(\tilde{\boldsymbol{x}}_{ni} - \tilde{\boldsymbol{x}}_{nj})'(\hat{\boldsymbol{\beta}} - \boldsymbol{\beta})| \\
&\le \frac{2L}{n(n-1)h_n^2} \sum_{1\le i<j\le n} (c_{ni} + c_{nj})\,\|\boldsymbol{A}_n^{1/2}(\hat{\boldsymbol{\beta}} - \boldsymbol{\beta})\| \\
&= \frac{2L}{nh_n^2} \sum_{i=1}^n c_{ni}\|\boldsymbol{A}_n^{1/2}(\hat{\boldsymbol{\beta}} - \boldsymbol{\beta})\|.
\end{aligned}$$

Furthermore, by (2.4.4),

$$\Big(\sum_{i=1}^n c_{ni}\Big)^2 \le n\sum_{i=1}^n c_{ni}^2 = n\sum_{i=1}^n \tilde{\boldsymbol{x}}_{ni}'\boldsymbol{A}_n^{-1}\tilde{\boldsymbol{x}}_{ni} = n\,\mathrm{trace}\Big(\boldsymbol{A}_n^{-1}\sum_{i=1}^n \tilde{\boldsymbol{x}}_{ni}\tilde{\boldsymbol{x}}_{ni}'\Big) = n\,\mathrm{trace}(\mathcal{I}_p) = np.$$

From the above inequalities and by Theorem 2.10.5, we conclude $\hat{Q}_2 = \hat{Q}_1 + O_p(n^{-1/2}h_n^{-2})$. The desired result then follows from the assumptions on h_n and Lemma 6.6.1. □

We shall next describe an estimator of $I_2(f)$ where the choice of bandwidth is partly data dependent. For this we shall assume f is positive a.e. and uniformly continuous. Let

$$\Psi(y) := \int [F(y+x) - F(-y+x)]dF(x) = P(|\varepsilon_1 - \varepsilon_2| \le y), \quad y \ge 0.$$

The uniform continuity of f implies $\Psi(y)/(2y) \to I_2(f)$ as $y \to 0$. This suggests that an estimator of $I_2(f)$ can be obtained by estimating $\Psi(y)/(2y)$ for small positive y. A natural estimator of $I_2(f)$ is $\hat{\psi}_n := \hat{\Psi}_n(h_n)/2h_n$, where $0 < h_n \to 0$ and

$$\hat{\Psi}_n(y) := \frac{2}{n(n-1)} \sum_{1 \le i < j \le n} I(|\hat{\varepsilon}_i - \hat{\varepsilon}_j| \le y), \quad y \ge 0.$$

By the continuity of F, $P(\hat{\Psi}_n(0) = 0) = 1$.

There is an arbitrariness in this estimator due to the choice of the window width h_n. Koul, Sievers and McKean (1987) recommended that h_n be determined from the spread of the data as follows. Let t_n^α be the α quantile of $\hat{\Psi}_n$. Note that t_n^α is an estimator of the α quantile t^α of the d.f. Ψ. Taking $h_n = n^{-1/2} t_n^\alpha$ results in the estimator

$$\psi_n^\alpha := \hat{\Psi}_n(n^{-1/2} t_n^\alpha)/(2n^{-1/2} t_n^\alpha).$$

In ψ_n^α, the window width depends on t_n^α, which is data dependent and known as soon as α is known. Thus, the estimator ψ_n^α shifts the burden of choosing the window width to the choice of α. In an empirical study that investigated level and power of some rank tests in some linear regression models, $\alpha = 0.8$ was found to be the most desirable, see Koul, McKean and Severs (1987). These authors also show that if f is positive a.e. and uniformly continuous and Assumption 6.6.1 holds, then $\psi_n^\alpha \to_p I_2(f)$. A similar result holds for general classes of estimators and parameters that include the estimator ψ_n^α and the parameter $I_2(f)$, cf., Theorem 4.5.1 of Koul (2002).

Asymptotic normality of $n^{1/2}(\hat{Q}_1 - I_2(f))$ in the one sample location model can be deduced from Theorem 5.6.2 while that of $n^{1/2}(\psi_n^\alpha - I_2(f))$ in the linear regression setup is available in Thewarapperuma (1987). Both of these results require some additional conditions on f.

6.7 Exercises

1. **Least Absolute Deviation Estimator.** Consider the one sample location model where $X_1, \cdots, X_n$ are i.i.d. $F(\cdot - \theta_0)$, for some continuous d.f. F on $\mathbb{R}$ and some $\theta_0 \in \mathbb{R}$. Let $M_n(t) := \sum_{i=1}^n |X_i - t|$ and $\widehat{\theta}_n = \text{arginf}_t M_n(t)$. Let $\widetilde{X}_n$ denote the sample median.

   (a) Show that $\widehat{\theta}_n = \widetilde{X}_n$.

   (b) Let $f(x) := e^{-|x|}/2$, $x \in \mathbb{R}$ and F denote the corresponding d.f.. Obtain the three random samples of sizes $n = 5, 10, 15$ from this F and graph $(t, M_n(t))$, $t \in \mathbb{R}$ for $n = 5, 8, 10$ and compute $\widehat{\theta}_n$ in each case. Does $M_n(t)$ look like a quadratic form in t with increasing n?

   (c) Let $\varepsilon_i := X_i - \theta_0$. Show that for every $0 < b < \infty$,

$$\sup_{|t| \le b} \left| \sum_{i=1}^n \Big[|\varepsilon_i + n^{-1/2} t| - |\varepsilon_i| - t n^{-1/2} \text{sgn}(\varepsilon_i) - t^2 f(0) \Big] \right| \to_p 0.$$

2. Obtain an expression for $M(t)$ of (6.2.38) for a general nondecreasing right continuous function, not necessarily continuous, analogous to the expression for $K(t)$ given at (6.2.33).
3. In the two sample location model, prove that if either F is continuous and symmetric about zero or $n_1 = n_2$, then $\hat{\theta}_{HLS}$ of (6.2.18) is symmetrically distributed about θ and median unbiased for θ.
4. Prove the identity (6.5.3).
5. Find $\text{IF}(x)$ for the U statistics of degree 2.
6. Find $\text{IF}(x)$ corresponding to HLS estimator.
7. Consider the simple linear regression model $Y_i = \beta c_i + \varepsilon_i$, $i = 1, \cdots, n$. Let $\widehat{\beta}_1 := \text{argmin}_t K_1(t)$, $\widehat{\beta}_2 := \text{argmin}_t K_2(t)$, where

$$K_1(t) := \sum_{i=1}^{n}\sum_{j=1}^{n}\Big\{\big|Y_i + Y_j - t(c_i + c_j)\big| - \big|Y_i - Y_j - t(c_i - c_j)\big|\Big\},$$

$$K_2(t) := \sum_{i=1}^{n}\sum_{j=1}^{n} c_i c_j\Big\{\big|Y_i + Y_j - t(c_i + c_j)\big| - \big|Y_i - Y_j - t(c_i - c_j)\big|\Big\}.$$

(A) For the following data, draw the graphs of $K_1(t)$, $K_2(t)$ for $t \in \mathbb{R}$ and show that $\widehat{\beta}_1 := -3.427$, $\widehat{\beta}_2 := -0.274 :\ n = 7,\ Y_i = 1.5, -2.3, 6, 5.5, 9, 4.6, 15;\ c_i = -1, 2, -3, 5, -10, 6, -4$.

(B) Consider the following two data-generating processes:

(a) The errors $\varepsilon_i, i = 1, \ldots, 10$ are generated from standard Cauchy distribution, and $\beta = 2$: ε_i: $0.1488, -0.7444, 0.9672, -1.4819, 0.6649, -1.5380, -0.4384, 0.4165, 1.2106, -3.8383$.

(b) The errors $\varepsilon_i, i = 1, \ldots, 10$ are generated from standard double exponential distribution, and $\beta = 2$: $\varepsilon_i : 0.0987, -0.5232, 0.6721, -0.9731, 0.4677, -1.0023, -0.3052, 0.2893, 0.8220, -1.8186$.

For (a) and (b) consider the following 3 cases of c_i:

Case1:$c_i = i,\ 1 \le i \le 10$

Case2:$c_i = 1, 2, -3, -5, -10, 8, 6, -12, 14, -22$.

Case3:$c_i = 0, 1 \le i \le 5;\ c_i = 1,\ 6 \le i \le 10$.

In each of the above 6 cases some one computed the estimators $\widehat{\beta}_1, \widehat{\beta}_2$ given in Table 6.1, where 'dblexp' denotes the double exponential distribution. Do you get the same values for these estimates in these 6 cases?

Table 6.1 $\widehat{\beta}_1$ and $\widehat{\beta}_2$

	B: $\varepsilon_i \sim$ Cauchy		C: $\varepsilon_i \sim$ dblexp	
	$\widehat{\beta}_1$	$\widehat{\beta}_2$	$\widehat{\beta}_1$	$\widehat{\beta}_2$
Case 1	1.926	1.919	1.956	1.949
Case 2	2.280	2.076	2.117	2.040
Case 3	1.011	1.439	1.501	1.643

8. Let σ_G^2 be as in (6.4.12). Show that $\sigma_G^2 = 1/4f^2(0)$ when G is degenerate at zero and $\sigma_G^2 = 1/12(\int f^2(x)\,dx)^2$ when $G(x) \equiv x$.
9. Let G be a real-valued function defined on $\mathbb{R}$. Suppose G is nondecreasing right continuous, has left limits and such that $|G(x) - G(y)| = |G(-x) - G(-y)|$, for all $x, y \in \mathbb{R}$. Show that G is continuous on $\mathbb{R}$.
10. Let G be given by the identity $dG(x) \equiv F(x)\big(1 - F(x)\big)\,dx$. Compute σ_G^2 of (6.4.12) in the following two cases: 1) $f(x) = F(x)\big(1 - F(x)\big)$, $x \in \mathbb{R}$, 2) $f(x) = (1/2)\exp\{-|x|\}$, $x \in \mathbb{R}$.
11. Let

$$\begin{aligned} f(x) &:= (1-x)/2, \ \ 0 \le x \le 1, \\ &:= (x - 2j + 1)/2^{j+2}, \quad 2j-1 \le x \le 2j, \\ &:= (2j + 1 - x)/2^{j+2}, \quad 2j \le x \le 2j+1, \ \ j \ge 1; \end{aligned}$$

For $x \le 0$, define $f(x) = f(-x)$. Show that for this f, $I_L(f) = \infty$ while it satisfies (6.4.6).
12. Let $f_\delta(x) = 2^{-1}(\delta + 1)\big(1 - |x|\big)^\delta I\big(|x| < 1\big)$, $\delta > 0$. Show that the family of densities f_δ satisfies (6.4.5) only if $\delta > 1$.
13. Let $\hat\theta_\psi$ be an M estimator of the one sample location parameter corresponding to the given score function ψ. By Lemma 6.2.4, $n^{1/2}(\hat\theta_\psi - \theta) \to_D \mathcal{N}\big(0, \sigma_\psi^2\big)$ where $\sigma_\psi^2 = E\psi^2(\varepsilon)/\big(\dot f(x)\psi(x)\,dx\big)^2$. Then,

$$\mathrm{ARE}\big(\hat\theta_\psi, \hat\theta_{HLS}\big) = \frac{\Big(\int \dot f(x)\psi(x)\,dx\Big)^2}{12\Big(\int f^2(x)\,dx\Big)^2 E(\psi^2(\varepsilon))}.$$

Compute this efficiency when $\psi(x) := |x| I(|x| \le k) + k\,\mathrm{sgn}(x)$, for each $k \ge 0$ and when $f(x) = e^{-|x|}/2, x \in \mathbb{R}$ and f is the density of $\mathcal{N}(0,1)$. Obtain the limits of these ARE's as $k \to 0$ and $k \to \infty$.
14. Prove the identity (6.4.11).
15. Let k and $1 \le n_j \le n, j = 1, \cdots, k$ be positive integers such that $\sum_{j=1}^k n_j = n$. In the linear regression model (6.4.1), take $p = k$ and let x_{ij} denote the jth coordinate of the vector $\boldsymbol{x}_i$, $1 \le i \le n, 1 \le j \le k$. Let $m_0 := 0, m_j := n_1 + \cdots + n_j$ and $\mathcal{I}_j := \{\ell; m_{j-1} + 1 \le \ell \le m_j\}$. Suppose $x_{ij} := 1$, for $i \in \mathcal{I}_j$ and $x_{ij} := 0$, for $i \notin \mathcal{I}_j$, $j = 1, \cdots, k$. Then the model (4.6.7) is called the k sample location model with the location parameters $\beta_1, \cdots, \beta_k$. Show that $\boldsymbol{X}'\boldsymbol{X}$ = diagonal matrix$(n_1, n_2, \cdots, n_k)$.
16. Let k and $1 \le n_j \le n, j = 1, \cdots, k$ and x_{ij} be as in the previous exercise. Let Y_{ij} denote the ith response from jth sample, $1 \le i \le n_j, 1 \le j \le k$. Show that the m.d. estimator $\widehat{\boldsymbol{\beta}}_{md}$ of (6.4.10) with $G(y) = y, y \in \mathbb{R}$ is equal to $(\tilde M_1, \cdots, \tilde M_k)'$, where

$$\tilde M_j := \mathrm{median}\big\{(Y_{ij} + Y_{\ell j})/2; 1 \le i \le \ell \le n_j\big\}, \ 1 \le j \le k.$$

17. Determine $\widehat{\boldsymbol{\beta}}_{md} := \mathrm{arginf}_{\boldsymbol{t}} \big\|\boldsymbol{A}^{-1}\sum_{i=1}^n \boldsymbol{x}_i \mathrm{sgn}(Y_i - \boldsymbol{x}_i'\boldsymbol{t})\big\|^2$ for the k sample location model, i.e., when $\boldsymbol{x}_i$'s are as in the Exercise 15.

18. Let X, Y be r.v.'s such that $E(|Y|) < \infty$ and let $m(x)$ denote the median of the conditional distribution of Y, given $X = x$. Let $\mathcal{L}_1$ denote the class of integrable functions of X. Show that

$$\inf_{g \in \mathcal{L}_1} E|Y - g(X)| = E|Y - m(X)|.$$

19. Let $Y_1, \cdots, Y_n$ be a random sample from the one sample location model with the location parameter θ and for $0 < \alpha < 1$, define

$$Q_\alpha(t) := \sum_{i=1}^{n} \alpha(Y_i - t)I(Y_i - t > 0) - \sum_{i=1}^{n}(1 - \alpha)(Y_i - t)I(Y_i - t \leq 0), \quad t \in \mathbb{R},$$
$$\tilde{\theta}(\alpha) := \text{arginf}_{t \in \mathbb{R}} Q_\alpha(t).$$

Show that $\tilde{\theta}(0.5)$ is the sample median and $\tilde{\theta}(\alpha)$ is α-th sample quantile.

Notes: Formal definitions of robustness and influence function were developed by Hampel (1968, 1971). For some recent developments in the robust rank inference procedures, see the collection of papers by Liu and McKean (2016). The idea of using distances based on empirical distribution function to define minimum distance estimators in the one sample setting was first proposed in a series of papers by Wolfowitz (1953, 1955, 1957). Beran (1977, 1978) and Donoho and Liu (1988a,b) convincingly argued that the minimum distance estimators of the one sample location parameter based on L_2 distances involving either nonparametric density estimators or empirical distribution function have some desirable finite sample properties, tend to be robust against some contaminated models, and are also asymptotically efficient at some error distributions. The minimum distance estimators using empirical or weighted empirical of the residuals discussed in this chapter in the one and two sample location models and in the linear regression model advances this methodology convincingly.

7

Density Estimation

7.1 Introduction

In classical parametric inference the underlying population density is assumed to be a known parametric function depending on some unknown Euclidean parameters. In this case the problem of estimating the underlying density is equivalent to that of estimating an Euclidean parameter. In nonparametric inference, the population density itself is an infinite-dimensional parameter of interest.

The main problem of interest here is the estimation of the univariate population density based on a random sample from this density. A common approach is the kernel density estimation method. This method needs two entities for its implementation, a density kernel and a window width. The former is a density on the real line and the latter is a (small) positive number which serves as a smoothing parameter.

This chapter contains some asymptotic analysis of the bias, the mean squared error, the mean integrated squared error, the asymptotic distributions of a class of kernel density estimators, and some methods for choosing the bandwidth. As applications of density estimation to statistical inference, this chapter discusses the estimation of the convolutions $\int f(x \pm y)f(y)\,dy$ and $\int f^2(x)\,dx$, where f is a density, and minimum Hellinger distance estimators of the one sample location parameter.

7.2 Histogram Estimator

One of the earliest density estimation or kernel smoothing methods is that of the histogram. To describe this method, let $X_1, \cdots, X_n$ be a random sample from some density f on $\mathbb{R}$. The problem is to estimate the density f at a given point x, i.e., to estimate the real parameter $f(x)$.

Let F denote the continuous d.f. corresponding to the density f. Suppose f is continuous at x. Then $f(x)$ is the derivative of F at x. Thus $f(x)$ is well approximated by $P(x - h < X_1 < x + h)/(2h) = [F(x+h) - F(x-h)]/(2h)$ for small positive h. This expression can be estimated by the proportion of observations that fall into the open interval $(x-h, x+h)$ divided by the length $2h$ of the interval. The resulting estimator is

$$\tilde{f}_n(x) := \frac{1}{2nh} \sum_{i=1}^{n} I\big(x - h < X_i < x + h\big). \tag{7.2.1}$$

The parameter h is known as the window width or bandwidth. Typically, the window width depends on the sample size n and tends to 0 as $n \to \infty$. For convenience in notation, we do not exhibit the dependence of h on n. All limits in this chapter are taken as $n \to \infty, h \to 0$, unless

DOI: 10.1201/9781003585770-7

mentioned otherwise. The estimator $\tilde{f}_n(x)$ is known as the histogram density estimator. We shall now compute its bias and variance and then address consistency.

Throughout this chapter we shall adopt the convention that for any real-valued integrable function g defined on $\mathbb{R}$, the range of integration in $\int g(x)\,dx$ is $\mathbb{R}$, unless specified otherwise.

7.2.1 The Bias of the Histogram Estimator

The bias of an estimator is the difference between its expected value and the target value. It is easy to check that $\tilde{f}_n(x)$ has mean

$$E\big(\tilde{f}_n(x)\big) = \frac{1}{2h}\big[F(x+h) - F(x-h)\big]$$

so that the bias of $\tilde{f}_n(x)$ is given by

$$\begin{aligned}\text{Bias}\big(\tilde{f}_n(x)\big) = E\big(\tilde{f}_n(x)\big) - f(x) &= \frac{1}{2h}\big[F(x+h) - F(x-h) - 2hf(x)\big]\\ &= \frac{1}{2h}\int_{-h}^{h}\big[f(x+s) - f(x)\big]\,ds.\end{aligned}$$

Thus the absolute value of the bias is bounded from above by

$$w_f(x,h) := \sup_{|s|<h} |f(x+s) - f(x)|.$$

This bound converges to zero if f is continuous at x. In other words, $\tilde{f}_n(x)$ is asymptotically unbiased at all continuity points x of f:

$$\text{Bias}\big(\tilde{f}_n(x)\big) \to 0. \tag{7.2.2}$$

If f is uniformly continuous, then $\sup_{x\in\mathbb{R}} w_f(x,h)$ converges to zero and $\tilde{f}_n$ is uniformly asymptotically unbiased, i.e.,

$$\big\|\text{Bias}\big(\tilde{f}_n\big)\big\|_\infty \to 0. \tag{7.2.3}$$

7.2.2 The Variance of the Histogram Estimator

Since $X_1,\cdots,X_n$ are i.i.d. F and F is continuous, the distribution of the r.v. $2nh\tilde{f}_n(x) = \sum_{i=1}^n I\big(x-h < X_i < x+h\big)$ is Bin(n,p) with $p = F(x+h) - F(x-h)$. Hence,

$$\text{Var}\big(\tilde{f}_n(x)\big) = \frac{1}{4nh^2}\big(F(x+h) - F(x-h)\big)\big(1 - [F(x+h) - F(x-h)]\big).$$

If f is continuous at x, then we have

$$2nh\text{Var}\big(\tilde{f}_n(x)\big) \to f(x). \tag{7.2.4}$$

We also have the bound

$$2nh\text{Var}\big(\tilde{f}_n(x)\big) \le \frac{1}{2h}\big(F(x+h) - F(x-h)\big) = \frac{1}{2h}\int_{-h}^{h} f(x+s)\,ds, \quad \forall\, x\in\mathbb{R}. \tag{7.2.5}$$

For bounded f we thus have the inequality

$$2nh\big\|\text{Var}\big(\tilde{f}_n\big)\big\|_\infty \le \|f\|_\infty. \tag{7.2.6}$$

7.2.3 Consistency of the Histogram Estimator

For an estimator $\widehat{f}_n$ of f and $x \in \mathbb{R}$ such that $E(\widehat{f}_n^{\,2}(x))$ is finite, the mean squared error (MSE) of $\widehat{f}_n$ at x is defined as

$$\mathrm{MSE}(\widehat{f}_n(x)) := E(\widehat{f}_n(x) - f(x))^2.$$

If f is square integrable, the *mean integrated squared error* (MISE) of $\widehat{f}_n$ is defined as

$$\mathrm{MISE}(\hat{f}_n) := E\int (\widehat{f}_n(y) - f(y))^2\,dy = \int E(\widehat{f}_n(y) - f(y))^2\,dy.$$

Definition 7.2.1. Let $\widehat{f}_n$ be a sequence of estimators for f. It is said to be *consistent at* x if $\widehat{f}_n(x) \to_p f(x)$. It is said to be *uniformly consistent* if $\|\widehat{f}_n - f\|_\infty \to_p 0$. It is said to be *MSE consistent* at a point x if $\mathrm{MSE}(\widehat{f}_n(x)) \to 0$. It is said to be *uniformly MSE consistent* if $\|\mathrm{MSE}(\widehat{f}_n)\|_\infty \to 0$. It is said to be *mean integrated squared error* (MISE) consistent if $\mathrm{MISE}(\widehat{f}_n) \to 0$.

In the sequel we shall often use the following well-known identities.

$$\mathrm{MSE}(\widehat{f}_n(x)) = \mathrm{Var}(\widehat{f}_n(x)) + \mathrm{Bias}^2(\widehat{f}_n(x)), \quad \forall x \in \mathbb{R}. \tag{7.2.7}$$

$$\mathrm{MISE}(\widehat{f}_n) = \int \mathrm{Var}(\widehat{f}_n(x))\,dx + \int \mathrm{Bias}^2(\widehat{f}_n(x))\,dx. \tag{7.2.8}$$

The following lemma proves the consistency of the above types for $\tilde{f}_n$. Often we shall assume the following about h.

$$h \to 0, \quad nh \to \infty. \tag{7.2.9}$$

Lemma 7.2.1. *Suppose* (7.2.9) *holds. Then the following are true.*

(a) *The estimator $\tilde{f}_n$ is MSE consistent, and hence consistent, at each continuity point x of f.*

(b) *If f is uniformly continuous, then $\tilde{f}_n$ is uniformly MSE consistent.*

(c) *If f is square-integrable, then $\tilde{f}_n$ is MISE consistent.*

Proof. If f is continuous at x, then the MSE consistency at x follows from (7.2.2), (7.2.4), the identity (7.2.7) and (7.2.9). In view of (2.4.8), MSE consistency at x implies consistency at x. If f is uniformly continuous, then f is bounded by Lemma 2.11.2 and the uniform MSE consistency follows from (7.2.3), (7.2.6), (7.2.7) and (7.2.9). This proves parts (a) and (b). Now assume f is square-integrable. By (7.2.5) and Fubini's theorem,

$$2nh\int \mathrm{Var}(\widehat{f}_n(x))\,dx \le \frac{1}{2h}\int_{-h}^{h}\int f(x+s)\,dx\,ds = 1. \tag{7.2.10}$$

By (2.4.6), the substitution $s = hu$ and by a change of order of integration we obtain

$$\begin{aligned}\int \mathrm{Bias}^2(\tilde{f}_n(x))\,dx &= \int \Big(\frac{1}{2h}\int_{-h}^{h} (f(x+s) - f(x))\,ds\Big)^2\,dx \\ &\le \int \frac{1}{2h}\int_{-h}^{h} (f(x+s) - f(x))^2\,ds\,dx \\ &= \frac{1}{2}\int_{-1}^{1}\int (f(x+hu) - f(x))^2\,dx\,du.\end{aligned}$$

Since f is square-integrable, Corollary 2.10.8, applied with $g = f$ and $v(u) = I(-1 < u < 1)/2$, implies that the last double integral converges to zero as $h \to 0$. This, (7.2.8), (7.2.9) and (7.2.10) complete the proof of part (c) and hence of the lemma. □

7.3 Kernel Density Estimators

The above histogram estimator is an example of the class of kernel density estimators introduced by Rosenblatt (1956) and Parzen (1962). These estimators are defined as follows. Given a nonnegative function K defined on $\mathbb{R}$, such that $\int K(u)\,du = 1$, i.e., given a density kernel K, and a bandwidth $h > 0$, the corresponding kernel density estimator is defined to be

$$\hat{f}_n(x) = \frac{1}{nh}\sum_{i=1}^{n} K\Big(\frac{x - X_i}{h}\Big), \quad x \in \mathbb{R}. \tag{7.3.1}$$

The estimator $\hat{f}_n$ corresponding to $K(s) = \frac{1}{2}I_{(-1,1)}(s)$ is the histogram density estimator $\tilde{f}_n$. This kernel gives uniform weights to all X_i's around x. Intuitively, the observations closer to x ought to have more information about $f(x)$, and those farther away from x have less information about $f(x)$. Thus, it is natural to use kernels that give more weight to the observations closer to x and less weight to the observations further away from x in either direction. This goal is achieved by choosing a density kernel K that is symmetric about zero and unimodal. Some commonly used density kernels of this type are as follows.

Gaussian kernel:	$K(y) = (2\pi)^{-1/2}e^{-y^2/2}$,	$y \in \mathbb{R}$,
Logistic kernel:	$K(y) = e^{-y}/(1+e^{-y})^2$,	$y \in \mathbb{R}$,
Epanechnikov kernel:	$K(y) = \frac{3}{4}(1-y^2)$,	$\lvert y\rvert \le 1$,
Biweight kernel:	$K(y) = \frac{15}{16}(1-y^2)^2$,	$\lvert y\rvert \le 1$.

7.3.1 Some Preliminaries

In this subsection we shall collect some preliminary results that are useful in analyzing the bias, MSE and MISE of $\hat{f}_n$.

Lemma 7.3.1. *Let g and γ be real-valued integrable functions on $\mathbb{R}$ such that $|y\gamma(y)| \to 0$ as $|y| \to \infty$. Then the following hold.*
(a) *If g is continuous at $x \in \mathbb{R}$, then*

$$\lim_{h\to 0}\int |g(x-ht) - g(x)|\,|\gamma(t)|\,dt = 0.$$

(b) *If g is uniformly continuous on $\mathbb{R}$, then*

$$\lim_{h\to 0}\sup_{x\in\mathbb{R}}\int |g(x-ht) - g(x)|\,|\gamma(t)|\,dt = 0.$$

Proof. Fix $x \in \mathbb{R}$ and set $w(x,\delta) = \sup_{|y|\le\delta}|g(x-y) - g(x)|$, $\delta \ge 0$. Then, for each positive δ, the integral in the lemma is bounded by the sum of the following three terms.

$$A_\delta(x,h) := \int_{|ht|\le\delta}|g(x-ht) - g(x)|\,|\gamma(t)|\,dt, \qquad B_\delta(x,h) := \int_{|ht|>\delta}|g(x-ht)|\,|\gamma(t)|\,dt$$

$$C_\delta(x,h) := |g(x)|\int_{|ht|>\delta}|\gamma(t)|\,dt.$$

Clearly, $A_\delta(x,h) \le w(x,\delta)\int |\gamma(t)|\,dt$, and use $x - ht = y$ in the second bound to obtain

$$B_\delta(x,h) \le \sup_{|th|>\delta} |t\gamma(t)| \int_{|ht|\ge\delta} \frac{g(x-ht)}{ht} h\,dt \le \sup_{|t|\ge\delta/h} |t\gamma(t)| \frac{1}{\delta}\int |g(y)|\,dy.$$

The integrability of γ implies $\lim_{h\to0} C_\delta(x,h) = 0$, the decay assumption $|y\gamma(y)| \to 0$ as $|y| \to \infty$ implies $\lim_{h\to0} B_\delta(x,h) = 0$, and the continuity of g at x implies $\lim_{\delta\to0} w(x,\delta) \to 0$. Thus part (a) follows by first letting h and then δ tend to zero.

If g is uniformly continuous, then g is bounded and $\sup_{x\in\mathbb{R}} w(x,\delta) \to 0$ as $\delta \to 0$. In this case, the above convergences are uniform in x. This yields part (b). □

Remark 7.3.1. Note, part (a) of the above lemma implies the convergence

$$\lim_{h\to0} \int g(x-ht)\gamma(t)\,dt = g(x)\int \gamma(t)\,dt,$$

while part (b) yields the uniform version

$$\lim_{h\to0} \sup_{x\in\mathbb{R}} \left| \int g(x-ht)\gamma(t)\,dt - g(x)\int \gamma(t)\,dt \right| = 0.$$

7.3.2 The Bias of the Kernel Estimator

We shall now use the above preliminaries to analyse the bias of the kernel density estimators $\hat{f}_n(x)$. We shall often use the following decay condition on K.

$$\lim_{|y|\to\infty} yK(y) = 0. \tag{7.3.2}$$

Because $X_1, \cdots, X_n$ are i.i.d. f,

$$E(\hat{f}_n(x)) = \frac{1}{h} E\Big(K\Big(\frac{x-X_1}{h}\Big)\Big) = \frac{1}{h}\int K\Big(\frac{x-t}{h}\Big) f(t)\,dt.$$

Put $y = (x-t)/h$. Then $t = x - hy$, $dt = hdy$ and

$$f_h(x) := E(\hat{f}_n(x)) = \int f(x-hy)K(y)\,dy. \tag{7.3.3}$$

Thus $\hat{f}_n(x)$ is not an unbiased estimator of $f(x)$. Its bias is

$$\text{Bias}(\hat{f}_n(x)) := f_h(x) - f(x) = \int \big[f(x-hy) - f(x)\big]K(y)\,dy.$$

Lemma 7.3.1 applied with $g = f$ and $\gamma = K$ yields the following results.

Lemma 7.3.2. *Under* (7.3.2) *the following hold.*
(a) *If f is continuous at x, then $\hat{f}_n(x)$ is asymptotically unbiased for $f(x)$, i.e.,* $\text{Bias}(\hat{f}_n(x)) \to 0$.
(b) *If f is uniformly continuous, then $\|\text{Bias}(\hat{f}_n)\|_\infty \to 0$, i.e., $\hat{f}_n(x)$ is asymptotically uniformly unbiased for $f(x)$.*

It is of interest to know the rate at which the above bias converges to zero. For this purpose we shall assume that K has mean zero and finite variance, i.e.,

$$\int yK(y)\,dy = 0, \qquad \kappa_2 := \int y^2 K(y)\,dy < \infty. \tag{7.3.4}$$

Every symmetric density K with a finite second moment satisfies these requirements. In particular, symmetric densities with bounded support meet these conditions. The kernel K has bounded support if $K(y) = 0$ holds for all $|y| > B$ and for some finite $B > 0$.

Lemma 7.3.3. *Suppose K satisfies (7.3.4) and $y^3K(y) \to 0$ as $|y| \to \infty$. Suppose f is twice differentiable on $\mathbb{R}$ and its second derivative $\ddot{f}$ is continuous and integrable. Then, for each x in $\mathbb{R}$,*

$$\operatorname{Bias}(\hat{f}_n(x)) = \frac{h^2}{2}\ddot{f}(x)\kappa_2 + o(h^2). \tag{7.3.5}$$

If $\ddot{f}$ is uniformly continuous, then

$$\sup_{x\in\mathbb{R}} \left|\operatorname{Bias}(\hat{f}_n(x)) - \frac{h^2}{2}\ddot{f}(x)\kappa_2\right| = o(h^2). \tag{7.3.6}$$

Proof. Set

$$D(x, b) := \int (\ddot{f}(x - by) - \ddot{f}(x))y^2K(y)\,dy, \quad x, b \in \mathbb{R}, \tag{7.3.7}$$

$$R(x, \delta) := f(x + \delta) - f(x) - \delta\dot{f}(x) - \frac{\delta^2}{2}\ddot{f}(x), \quad x, \delta \in \mathbb{R}. \tag{7.3.8}$$

Since K has mean zero, we have the identity

$$\operatorname{Bias}(\hat{f}_n(x)) - \frac{h^2\ddot{f}(x)\kappa_2}{2} = \int R(x, -hy)K(y)\,dy, \quad x \in \mathbb{R}. \tag{7.3.9}$$

Since f is twice continuously differentiable, the integral formula for the remainder term in the Taylor expansion yields

$$R(x, \delta) = \frac{\delta^2}{2}\int_0^1 (\ddot{f}(x + s\delta) - \ddot{f}(x))2(1 - s)\,ds, \quad x, \delta \in \mathbb{R}.$$

Using this with $\delta = -hy$ and then changing the order of integration yields the identity

$$\operatorname{Bias}(\hat{f}_n(x)) - \frac{h^2\ddot{f}(x)\kappa_2}{2} = \frac{h^2}{2}\int_0^1 D(x, sh)\,2(1 - s)\,ds. \tag{7.3.10}$$

The absolute value of this quantity is bounded by h^2 multiplied by the supremum of $|D(x, hs)|$ over $0 < s < 1$. Thus (7.3.5) follows from part (a) and (7.3.6) from part (b) of Lemma 7.3.1 applied with $g = \ddot{f}$ and $\gamma(y) = y^2K(y)$. □

The conclusion (7.3.5) for a fixed x can be obtained under weaker assumptions if the density kernel K has bounded support, as shown in the next lemma.

Lemma 7.3.4. *Suppose K is symmetric about zero and has bounded support, f is continuously differentiable in a neighborhood of x and $\dot{f}$ has derivative $\ddot{f}(x)$ at x. Then (7.3.5) holds.*

Proof. Fix $x \in \mathbb{R}$. Let $R(x, \delta)$ be as in (7.3.8) and define

$$H(\delta) := \begin{cases} \dfrac{\dot{f}(x + \delta) - \dot{f}(x)}{\delta} - \ddot{f}(x), & \delta \neq 0, \\ 0, & \delta = 0. \end{cases}$$

Since $\dot{f}$ is differentiable at x, $H(\delta)$ is continuous at $\delta = 0$. Since the integrals $\int_0^1 1\,ds$ and $\int_0^1 2s\,ds$ equal one and f is continuously differentiable in a neighborhood of x, we obtain the identity

$$R(x,\delta) = \frac{\delta^2}{2}\int_0^1 \Big\{\frac{\dot{f}(x+\delta s) - \dot{f}(x)}{\delta\, s} - \ddot{f}(x)\Big\} 2s\,ds = \delta^2 \int_0^1 H(\delta s) s\,ds, \quad \text{for } \delta \text{ close to } 0.$$

This and (7.3.9) yield

$$\mathrm{Bias}\big(\hat{f}_n(x)\big) - \frac{h^2}{2}\ddot{f}(x)\kappa_2 = h^2 \int\int_0^1 H(-hys)s\,ds\, y^2 K(y)\,dy.$$

The absolute value of this quantity is bounded by $h^2\kappa_2 \sup_{|u|\le Bh}|H(u)|$, with B chosen such that $K(y) = 0$ for $|y| > B$. By the continuity of H at 0, the error term has the desired order $o(h^2)$. □

7.3.3 The Variance of the Kernel Estimator

We shall now evaluate the variance of the kernel density estimator. To proceed further, assume that K is also square-integrable:

$$I_2(K) := \int K^2(y)\,dy < \infty. \tag{7.3.11}$$

Every bounded K has this property. Note also that $\int K(y)\,dy = 1$ implies $I_2(K) > 0$. We calculate

$$\begin{aligned} nh\mathrm{Var}\big(\hat{f}_n(x)\big) &= \frac{1}{h}\mathrm{Var}\Big(K\Big(\frac{x - X_1}{h}\Big)\Big) \\ &= \frac{1}{h}\Big[\int K^2\Big(\frac{x-t}{h}\Big) f(t)\,dt - \Big(\int K\Big(\frac{x-t}{h}\Big) f(t)\,dt\Big)^2\Big]. \end{aligned}$$

Substitute $x = t + hy$ in both integrals to obtain the formula

$$nh\mathrm{Var}\big(\hat{f}_n(x)\big) = \int K^2(y) f(x - hy)\,dy - h\Big(\int K(y) f(x-hy)\,dy\Big)^2. \tag{7.3.12}$$

Lemma 7.3.5. *Suppose* (7.2.9), (7.3.2) *and* (7.3.11) *are satisfied. Then the following hold.*
(a) *If f is continuous at x, then*

$$nh\mathrm{Var}\big(\hat{f}_n(x)\big) = f(x)I_2(K) + o(1).$$

(b) *If f is uniformly continuous, then*

$$\sup_{x\in\mathbb{R}}\big|nh\mathrm{Var}\big(\hat{f}_n(x)\big) - f(x)I_2(K)\big| = o(1).$$

Proof. The assumption (7.3.2) implies $yK^2(y) \to 0$ as $|y| \to \infty$. In view of the formula (7.3.12), the desired results now follow from Lemma 7.3.1 applied with $g = f$ and $\gamma = K^2$ and $\gamma = K$. □

7.3.4 The MSE of the Kernel Estimator

We are now ready to consider the MSE of $\hat{f}_n(x)$. Lemmas 7.3.2 and 7.3.5 and formula (7.2.7) give the following results.

Corollary 7.3.1. *Suppose* (7.2.9) *holds and* K *satisfies* (7.3.2) *and* (7.3.11). *Then the following are true.*
(a) *If* f *is continuous at* x, *then* $\mathrm{MSE}(\hat{f}_n(x)) \to 0$, *i.e.,* $\hat{f}_n(x)$ *is* MSE *consistent, and hence consistent, for* $f(x)$.
(b) *If* f *is uniformly continuous, then* $\|\mathrm{MSE}(\hat{f}_n)\|_\infty \to 0$, *i.e.,* $\hat{f}_n$ *is uniformly* MSE *consistent for* f.

Lemmas 7.3.3 and 7.3.5 and formula (7.2.7) yield the following rates of convergence for the MSE.

Corollary 7.3.2. *Suppose* (7.2.9) *holds,* K *satisfies* (7.3.4), (7.3.11) *and* $y^3K(y) \to 0$ *as* $|y| \to \infty$. *Suppose* f *is twice differentiable and its second derivative* $\ddot{f}$ *is continuous and integrable. Then, for each* x *in* $\mathbb{R}$,

$$\mathrm{MSE}\big(\hat{f}_n(x)\big) = \frac{f(x)}{nh} I_2(K) + h^4\Big(\frac{\ddot{f}(x)}{2}\kappa_2\Big)^2 + o\Big(\frac{1}{nh}\Big) + o(h^4).$$

If $\ddot{f}$ *is also uniformly continuous, then the expansion is uniform,*

$$\sup_{x\in\mathbb{R}} \Big|\mathrm{MSE}\big(\hat{f}_n(x)\big) - \frac{f(x)}{nh} I_2(K) - h^4\Big(\frac{\ddot{f}(x)}{2}\kappa_2\Big)^2\Big| = o\Big(\frac{1}{nh}\Big) + o(h^4).$$

The following result is obtained using Lemmas 7.3.4 and 7.3.5.

Corollary 7.3.3. *If* (7.2.9) *holds,* K *has bounded support and satisfies* (7.3.4) *and* (7.3.11), f *is continuously differentiable in a neighborhood of* x *and* $\dot{f}$ *has derivative* $\ddot{f}(x)$ *at* x, *then*

$$\mathrm{MSE}\big(\hat{f}_n(x)\big) = \frac{f(x)}{nh} I_2(K) + h^4\Big(\frac{\ddot{f}(x)}{2}\kappa_2\Big)^2 + o\Big(\frac{1}{nh}\Big) + o(h^4).$$

7.3.5 The MISE of the Kernel Estimator

We are now ready to study the mean integrated square error of the kernel estimator and derive its asymptotic behavior. For this we assume that f is square integrable. Let $I_2(f) := \int f^2(x)\,dx$. Integrating the formula (7.3.12) with respect to x yields

$$\int nh\mathrm{Var}\big(\hat{f}_n(x)\big)\,dx = \iint K^2(y)f(x-hy)\,dy\,dx - h\int\Big(\int K(y)f(x-hy)dy\Big)^2\,dx.$$

Changing the order of integration and then using the substitution $u = x - hy$ shows that the first integral on the right-hand side equals $I_2(K)$. Because K is a density, an application of (2.4.4) shows that the second integral on the right-hand side is bounded by $\iint f^2(x-hy)K(y)\,dy\,dx = I_2(f)$. This establishes the approximation

$$\int \mathrm{Var}\big(\hat{f}_n(x)\big)\,dx = \frac{I_2(K)}{nh} + O\Big(\frac{1}{n}\Big). \tag{7.3.13}$$

Since f is square integrable and K is a density, Corollary 2.10.8 yields

$$\int \mathrm{Bias}^2\big(\hat{f}_n(x)\big)\,dx = \int\Big(\int \big(f(x-hy) - f(x)\big)K(y)\,dy\Big)^2\,dx \to 0.$$

This fact, (7.3.13) and (7.2.8) together yield the following corollary.

Corollary 7.3.4. *If* (7.2.9) *and* (7.3.11) *hold and f is square integrable, then $\hat{f}_n$ is* MISE-*consistent.*

The next lemma gives rates of convergence for the MISE.

Lemma 7.3.6. *Suppose* (7.2.9) *holds, K satisfies* (7.3.4) *and* (7.3.11), *f is square integrable, two times continuously differentiable with $\ddot{f}$ square integrable. Then,*

$$\operatorname{MISE}(\hat{f}_n) = \frac{I_2(K)}{nh} + \frac{h^4\kappa_2^2}{4}\int \big(\ddot{f}(x)\big)^2\,dx + O\Big(\frac{1}{n}\Big) + o(h^4).$$

Proof. Recall the definition of $D(x,b)$ from (7.3.7). By (7.2.8) and (7.3.13), it suffices to show

$$\int \operatorname{Bias}^2(\hat{f}_n(x))\,dx = \frac{h^4\kappa_2^2}{4}\int \big(\ddot{f}(x)\big)^2\,dx + o(h^4). \tag{7.3.14}$$

In view of (7.3.10), the desired result will follow if we show that

$$\int \Big(\int_0^1 D(x,hs)2(1-s)\,ds\Big)^2\,dx \to 0.$$

To prove this, apply (2.4.6) and Theorem 2.3.2, to conclude that the above integral is bounded by

$$\int y^2K(y)\,dy \iiint (\ddot{f}(x-hst)-\ddot{f}(x))^2\,dx\,t^2K(t)\,dt\,2(1-s)\,ds$$

and thus converges to zero by (2.10.8) applied with $p=2$, $g(x)=\ddot{f}(x)$, $v(s)=2(1-s)I(0<s<1)$ and $w(t)=t^2K(t)$. □

7.3.6 Consistency of the Kernel Estimator

Corollary 7.3.1 gives sufficient conditions for pointwise and uniform MSE consistency of $\hat{f}_n$. Clearly, pointwise MSE consistency implies consistency: $\hat{f}_n(x) \to_p f(x)$. MISE consistency is given in Corollary 7.3.4, which in turn implies consistency in the integrated squared error.

We shall now address the question of uniform and L_1 consistency of $\hat{f}_n$. The following theorem discusses uniform consistency.

Theorem 7.3.1. *Assume f is uniformly continuous, K satisfies* (7.3.2), *and there exists an integrable function $\dot{K}$ from $\mathbb{R}$ to $\mathbb{R}$ such that*

$$K(y) = \int I(s\le y)\dot{K}(s)\,ds, \quad y\in\mathbb{R}.$$

If, in addition, $h\to 0, nh^2\to\infty$, then $\hat{f}_n$ is uniformly consistent: $\|\hat{f}_n - f\|_\infty \to_p 0$.

Proof. Recall the definition of f_h from (7.3.3). Let $F_n(x) := n^{-1}\sum_{i=1}^n I(X_i\le x)$ and $\mathcal{D}_n := n^{1/2}\|F_n - F\|_\infty$. By (2.12.2), we have $\mathcal{D}_n = O_p(1)$ and calculate

$$\hat{f}_n(x) = \frac{1}{nh}\sum_{i=1}^n \int I(hs\le x-X_i)\dot{K}(s)\,ds = \frac{1}{h}\int F_n(x-sh)\dot{K}(s)\,ds,$$

$$\hat{f}_n(x) - f_h(x) = \frac{1}{h}\int \big[F_n(x-hs) - F(x-hs)\big]\dot{K}(s)\,ds,$$

$$\big\|\hat{f}_n - f_h\big\|_\infty \le \frac{\mathcal{D}_n}{n^{1/2}h}\int |\dot{K}(s)|\,ds = o_p(1),$$

because of $nh^2\to\infty$. The triangle inequality and Lemma 7.3.2(b) now give $\|\hat{f}_n - f\|_\infty \le \|\hat{f}_n - f_h\|_\infty + \|\operatorname{Bias}(\hat{f}_n)\|_\infty = o_p(1) + o(1)$. □

Since $f \in L_1$, the set of all integrable functions on $\mathbb{R}$, it is natural to seek conditions under which $\hat{f}_n$ is consistent for f in the L_1 norm, i.e., $\|\hat{f}_n - f\|_1 := \int |\hat{f}_n(x) - f(x)|\, dx \to_p 0$. The following theorem proves the L_1 consistency of $\hat{f}_n$.

Theorem 7.3.2. *Assume* (7.2.9) *and, for some* $\alpha > 1$,

$$\int (1 + |x|)^\alpha (f(x) + K^2(x))\, dx < \infty. \tag{7.3.15}$$

Then $\hat{f}_n$ *is* L_1 *consistent, i.e.,* $\|\hat{f}_n - f\|_1 \to_p 0$.

Proof. Recall the definition of f_h from (7.3.3). Then, by the triangle inequality, $\|\hat{f}_n - f\|_1 \leq \|\hat{f}_n - f_h\|_1 + \|f_h - f\|_1$. By (2.10.7) applied with $p = 1, g = f$ and $v = K$, we have $\|f_h - f\|_1 \to 0$. Let us now verify $\|\hat{f}_n - f_h\|_1 = o_p(1)$. Recall, for any $\alpha > 1$, the integral $\mathcal{I}_\alpha := \int (1 + |x|)^{-\alpha}\, dx$ is finite. By the Hölder inequality (2.4.6),

$$\begin{aligned}
\|\hat{f}_n - f_h\|_1^2 &= \Big(\int (1 + |x|)^{\alpha/2} |\hat{f}_n(x) - f_h(x)| (1 + |x|)^{-\alpha/2}\, dx\Big)^2 \\
&\leq \mathcal{I}_\alpha \int (1 + |x|)^\alpha (\hat{f}_n(x) - f_h(x))^2\, dx.
\end{aligned}$$

Hence, in view of (7.3.12), the substitution $x = z + hy$ and the inequality $1 + |z + hy| \leq 1 + |z| + |hy| \leq (1 + |z|)(1 + |hy|)$ give

$$\begin{aligned}
E(\|\hat{f}_n - f_h\|_1^2) &\leq \mathcal{I}_\alpha \int (1 + |x|)^\alpha \mathrm{Var}(\hat{f}_n(x))\, dx \\
&\leq \frac{\mathcal{I}_\alpha}{nh} \iint (1 + |x|)^\alpha f(x - hy) K^2(y)\, dx\, dy \\
&\leq \frac{\mathcal{I}_\alpha}{nh} \int (1 + |z|)^\alpha f(z)\, dz \int (1 + |hy|)^\alpha K^2(y)\, dy.
\end{aligned}$$

Thus (7.2.9) and (7.3.15) imply $E(\|\hat{f}_n - f_h\|_1^2) \to 0$. This proves $\|\hat{f}_n - f_h\|_1 = o_p(1)$. □

7.3.7 Bandwidth Selection

One of the most challenging tasks in the application of the kernel density estimation method is the choice of the bandwidth h. In choosing a bandwidth, one has to be careful in not having it too small or too large. A bandwidth that is too small can result in under-smoothing. An extreme case of small bandwidth can yield a density estimate consisting of peaks around each observation. A large bandwidth can result in over-smoothing. We illustrate this in the following figures, for the bandwidths $h = 0.5$, $h = 0.3$, $h = 0.15$, $h = 0.05$. The figures are for the bimodal density

$$f(x) = .7 \frac{3}{\sqrt{2\pi}} \exp\Big(-\frac{9(x-1)^2}{2}\Big) + .3 \frac{4}{\sqrt{2\pi}} \exp\Big(-\frac{16(x-2)^2}{2}\Big)$$

and based on a random sample of size $n = 30$ from this density. Note that f is the density of a mixture of two normal distributions, one with mean 1 and variance $1/9$ and the other with mean 2 and variance $1/16$ and mixture coefficients .7 and .3. We used as kernel the standard normal density. In each figure, the true density and the kernel density estimator are shown over the interval $[0, 3]$. The kernel density estimator is the dotted curve. For the bandwidths $h = .5$ and $h = .3$, the kernel density estimator does not recover the bimodality of the density f. For the bandwidth $h = .15$ the kernel density estimator recovers the bimodal nature and provides a good approximation to the true density. For the bandwidth $h = .05$, the kernel density estimator indicates too many modes.

Figure 1. Bandwidth=.5

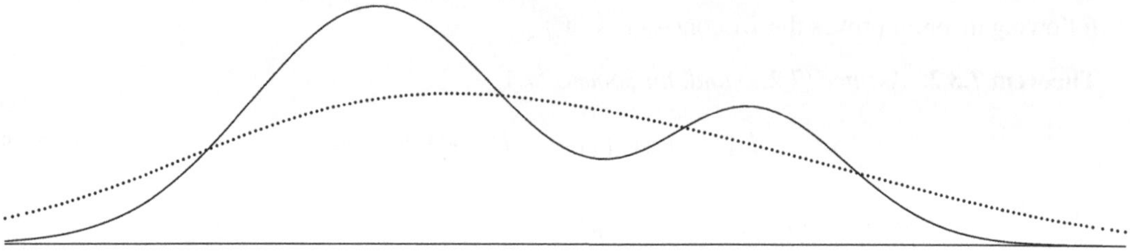

Figure 2. Bandwidth=.3

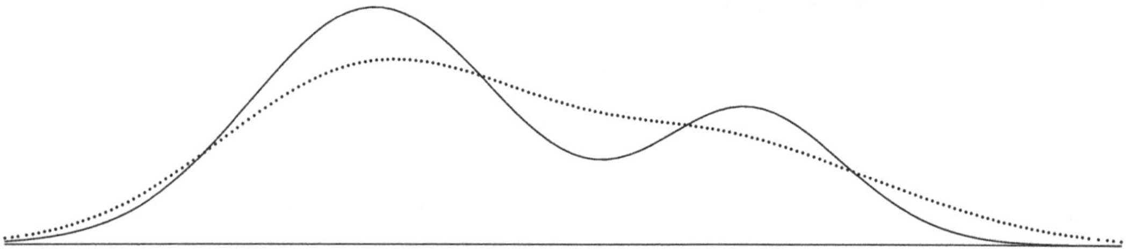

Figure 3. Bandwidth=.15

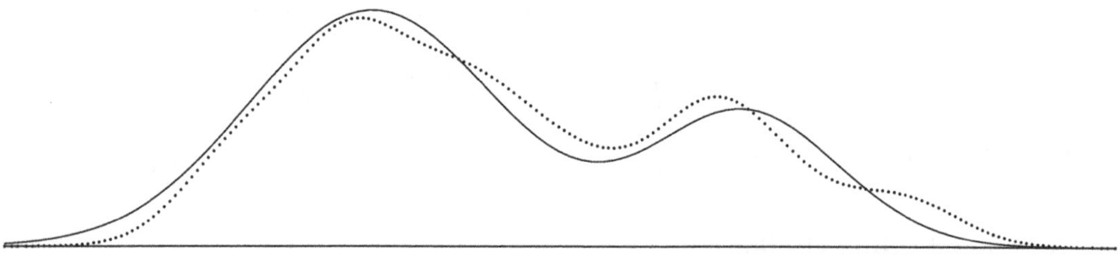

Figure 4. Bandwidth=.05

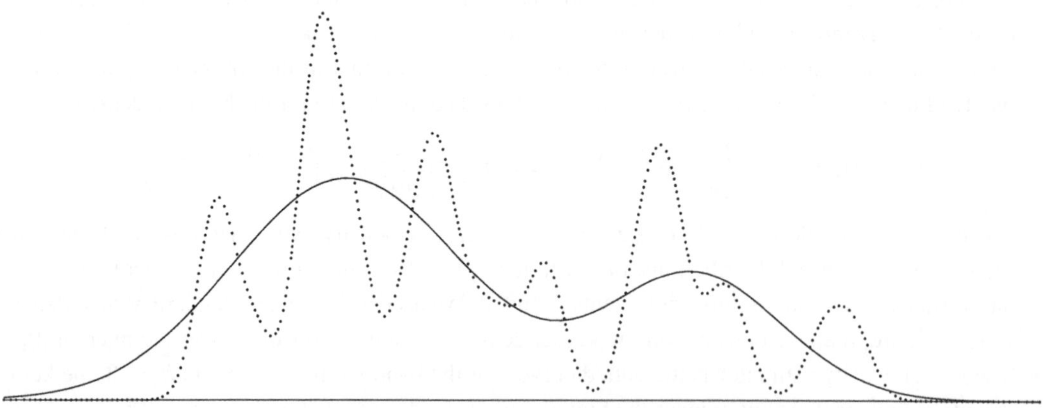

The following R code was used to compute the kernel density estimator for a grid of x values.

```
kde <- function(xgrid,sample,bandwidth)
{  y <- outer(xgrid,sample,FUN="-")
   fhat <- apply(dnorm(y,0,sd=bw),1,mean)
   fhat    }
```

The graphs can be produced in R using the command

```
plot(xgrid, f(xgrid),type="l")
fhat <- Kde(xgrid,sample,bandwidth)
points(xgrid,fhat,type="p",pch=".")
```

For our figures we used the grid `xgrid <- seq(0,3,.03)`. We encourage the reader to plot these curves for various choices of bandwidth by either generating their own sample or using the sample we used:

0.8870307	0.9235737	0.9123626	1.2659351	1.1685819
1.0164643	0.9448395	1.4558921	2.3726990	1.2384722
1.0294911	2.0598917	1.2878430	1.9109843	1.8413046
1.2278612	1.3678292	2.4596729	2.1189634	1.8562973
0.6035644	1.5683156	0.6313649	1.9303797	1.9405705
0.7317141	1.5811666	0.9082981	0.9623889	0.6274249

There are several methods in the literature for choosing the bandwidth. Some are ad hoc while others use some objective functions to define them. We shall describe two of these methods. One of them is based on the asymptotic MSE and MISE, and the other is based on cross-validation, which also uses MISE. For more on this topic, and more generally on the density estimation, see the monographs of Wand and Jones (1995) and Fan and Gijbels (1996) and the references therein.

The first method we shall discuss is the one that minimizes the asymptotic MSE and the MISE. For any square integrable function g defined on $\mathbb{R}$, let $I_2(g) := \int g^2(x)\,dx$. Suppose f is twice continuously differentiable with square integrable derivatives $\dot{f}$ and $\ddot{f}$, and K is a symmetric density that satisfies (7.3.4), (7.3.11) and the decay condition $y^3K(y) \to 0$ as $|y| \to \infty$. Recall $\kappa_2 := \int y^2K(y)\,dy$. Then, by Corollary 7.3.2,

$$\mathrm{MSE}\big(\hat{f}_n(x)\big) \approx A(h) := \frac{1}{nh} f(x)I_2(K) + \frac{h^4}{4}\kappa_2^2\big(\ddot{f}(x)\big)^2.$$

One way to choose h is to minimize the asymptotic MSE $A(h)$. For this assume $\ddot{f}(x) \neq 0$. Differentiate $A(h)$ twice w.r.t. h to obtain

$$\dot{A}(h) = -\frac{1}{nh^2} f(x)I_2(K) + h^3\kappa_2^2\big(\ddot{f}(x)\big)^2, \quad \ddot{A}(h) = \frac{2}{nh^3} f(x)I_2(K) + 3h^2\kappa_2^2\big(\ddot{f}(x)\big)^2 > 0.$$

Hence the solution of the equation $\dot{A}(h) = 0$ gives the minimizer of $A(h)$:

$$h_{\mathrm{opt}}(x) = \left(\frac{f(x)I_2(K)}{\kappa_2^2\big(\ddot{f}(x)\big)^2}\right)^{1/5} n^{-1/5}.$$

Besides depending on the unknown entities $f(x)$ and $\ddot{f}(x)$, this optimal h also depends on x, so it is only locally optimal. A globally optimal h is obtained by minimizing the asymptotic MISE. By Lemma 7.3.6,

$$\mathrm{MISE}(\hat{f}_n) \approx B(h) := \frac{1}{nh} I_2(K) + \frac{h^4}{4}\kappa_2^2 I_2(\ddot{f}). \tag{7.3.16}$$

Since f is a density, we have $I_2(\ddot{f}) > 0$. Minimizing $B(h)$ w.r.t. $h > 0$ gives

$$h_{\text{opt}} = \left(\frac{I_2(K)}{\kappa_2^2 I_2(\ddot{f})}\right)^{1/5} n^{-1/5}. \tag{7.3.17}$$

Note that the larger the integral $I_2(\ddot{f})$, which is a measure of the roughness of f, the smaller the optimal window width. Note also that this optimal window width satisfies (7.2.9). Upon substituting h_{opt} in the right hand side of (7.3.16), we obtain the smallest asymptotic MISE to be equal to

$$\frac{5}{4}\left(\kappa_2 I_2^4(K) I_2(\ddot{f})\right)^{1/5} n^{-4/5}.$$

If in (7.3.16), $h = cn^{-1/5}$ for some $c > 0$, then (7.3.16) is equivalent to saying that $n^{4/5}\text{MISE}(\hat{f}_n) \to \frac{I_2(K)}{c} + \frac{c^4}{4}\kappa_2^2 I_2(\ddot{f})$. This limit is minimized over c by the choice

$$c_{opt} = \left(\frac{I_2(K)}{\kappa_2^2 I_2(\ddot{f})}\right)^{1/5}.$$

Suppose f is the density of a $\mathcal{N}(0, \sigma^2)$ r.v.. Then,

$$\ddot{f}(x) = \frac{1}{\sqrt{2\pi}\sigma}\left(\frac{x^2}{\sigma^4} - \frac{1}{\sigma^2}\right)e^{-x^2/2\sigma^2}, \qquad I_2(\ddot{f}) = \frac{3}{8\pi^{1/2}\sigma^5},$$

$$h_{\text{opt}} = \left(\frac{8\pi^{1/2} I_2(K)}{3\kappa_2^2}\right)^{1/5} \sigma n^{-1/5}, \qquad c_{opt} = \left(\frac{8\pi^{1/2} I_2(K)}{3\kappa_2^2}\right)^{1/5} \sigma.$$

If K is the standard normal kernel, then $c_{opt} \approx 1.06\sigma$. Scott's (1992) easy to implement bandwidth recommends the use of a standard normal kernel and the data-driven bandwidth $h = 1.06\hat{\sigma}n^{-1/5}$, where $\hat{\sigma}$ is a consistent estimate of σ. This usually works well when f is a normal density and not so well when data is not from a normal distribution. Silverman (1986) proposed instead $h = 0.9\min(\hat{\sigma}, \text{IQR}/1.34)n^{-1/5}$, where IQR is the sample interquartile range. This works well for normal densities and densities close to normal. These are two ad hoc rules of thumb for choosing bandwidths.

A data-driven bandwidth method. There are several data-driven methods of bandwidth selection. These methods produce a random bandwidth $\hat{h}$ and use the kernel estimator with h replaced by $\hat{h}$. One possibility is to replace $I_2(\ddot{f})$ in (7.3.17) by an estimator and use the resulting random bandwidth $\hat{h}$ instead of h_{opt}. But estimation of $I_2(\ddot{f})$ poses its own challenges that may require its own bandwith selection. Other approaches avoid estimating this quantity by minimizing some criterion. Here we shall present such a method based on minimizing an estimator of the integrated squared error of a kernel density estimator.

The integrated squared error (ISE) of the kernel density estimator $\hat{f}_n$ is

$$\int \left(\hat{f}_n(y) - f(y)\right)^2 dy = \int \hat{f}_n^2(y)dy - 2\int \hat{f}_n(y)f(y)\,dy + \int f^2(y)\,dy.$$

Since the last term does not depend on h, instead of minimizing the above expression, one minimizes an estimator of

$$Q(h) := \int \hat{f}_n^2(y)dy - 2\int \hat{f}_n(y)f(y)\,dy$$

w.r.t. h, where the only unknown term is the product term

$$J_n(F) := \int \hat{f}_n(y)f(y)\,dy = \int \hat{f}_n(y)dF(y).$$

A natural estimator of this term is obtained by replacing F by the empirical d.f. F_n, thereby giving $J_n(F_n) := \frac{1}{n}\sum_{i=1}^n \hat{f}_n(X_i)$ as an estimate of the cross product term. But X_i also appears in $\hat{f}_n$. In order to avoid double counting, one replaces $\hat{f}_n(X_i)$ by the leave-one-out estimator

$$\hat{f}_{n,-i}(X_i) := \frac{1}{(n-1)h}\sum_{j=1}^{n} I(j \neq i) K\Big(\frac{X_i - X_j}{h}\Big).$$

Assume K is symmetric about zero. Replacing $f_n(X_i)$ in $J(F_n)$ by $f_{n,-i}(X_i)$ yields the estimator

$$\tilde{J}_n(F_n) := \frac{1}{n}\sum_{i=1}^{n} \hat{f}_{n,-i}(X_i) = \frac{2}{n(n-1)h}\sum_{1\le i<j\le n} K\Big(\frac{X_i - X_j}{h}\Big).$$

Plugging this in $Q(h)$ and using the symmetry of K yield

$$\begin{aligned} Q(h) = \frac{I_2(K)}{nh} &+ \frac{2}{n^2 h}\sum_{1\le i<j\le n} \int K\Big(\frac{X_i - X_j}{h} - y\Big) K(y)\, dy \\ &- \frac{4}{n(n-1)h}\sum_{1\le i<j\le n} K\Big(\frac{X_i - X_j}{h}\Big). \end{aligned}$$

Replace n^2 by $n(n-1)$ in the second summand to obtain the estimator

$$\hat{Q}(h) = \frac{I_2(K)}{nh} - \frac{2}{n(n-1)h}\sum_{1\le i<j\le n} \tilde{K}\Big(\frac{X_i - X_j}{h}\Big)$$

of $Q(h)$, where $\tilde{K}(t) := 2K(t) - \int K(t-s)K(s)\, ds$, $t \in \mathbb{R}$. The minimizer

$$\hat{h}_{CV} := \text{argmin}_{h>0} \hat{Q}(h) \tag{7.3.18}$$

of $\hat{Q}(h)$ is called the *least-squares cross-validation window width estimator.* The term cross-validation refers to the fact that a leave-one-out density estimator is used in estimating $Q(h)$.

In view of the fact that the optimal bandwidth is proportional to $n^{-1/5}$, see (7.3.17), sometimes practitioners choose $\hat{h} = c_n^* n^{-1/5}$, where $c_n^* := \text{argmin}_{c>0} \hat{Q}(cn^{-1/5})$. Theorem 5.6.3 gives the expansion

$$n^{4/5}\hat{Q}(cn^{-1/5}) = \frac{I_2(K)}{c} + \frac{\kappa_2^2 c^4 I_2(\ddot{f})}{4} - n^{4/5} S_n^{(3)} + o_p(1).$$

This suggests that the minimizer c_n^* converges in probability to c_{opt}. This can be made precise by showing that the above expansion is uniform in c on subintervals $[a, b]$ of $(0, \infty)$.

The above definitions use MISE. For that reason it is also called the least squares cross validation estimator of h, initially proposed by Rudemo (1982). The use of different measures of the error of estimation can lead to different cross-validation estimators of h. Bowman (1984) used Kullback-Leibler distance to propose cross-validation estimate of h while Bowman (1985) made a finite sample comparison of some bandwidth methods.

7.4 More Uniformity

In this section we shall revisit uniform consistency of the kernel density estimators. Our main goal is to prove the following theorem. Recall that Lipschitz continuity is defined in (6.6.1).

Theorem 7.4.1. *Suppose $K(t) = 0$ for $|t| > 1$, K is Lipschitz continuous and satisfies* (7.3.4) *and the bandwidth h satisfies $h \to 0$ and $\log(n)/(nh) = O(1)$. Suppose there exist $\gamma > 0$ and $0 < C_f < \infty$ such that*

$$f(x) \le \frac{C_f}{(1+|x|)^{1+\gamma}}, \quad x \in \mathbb{R}. \tag{7.4.1}$$

Then, the following hold.
(a) *If f is Lipschitz continuous, then*

$$\|\hat{f}_n - f\|_\infty = O(h) + O_p\Big(\Big(\frac{\log(n)}{nh}\Big)^{1/2}\Big).$$

(b) *If f has a Lipschitz continuous derivative, then*

$$\|\hat{f}_n - f\|_\infty = O(h^2) + O_p\Big(\Big(\frac{\log(n)}{nh}\Big)^{1/2}\Big).$$

Proof. With f_h as in (7.3.3), by the triangle inequality, we obtain

$$\|\hat{f}_n - f\|_\infty \le \|\hat{f}_n - f_h\|_\infty + \|f_h - f\|_\infty.$$

If f is Lipschitz continuous with constant L_f, we have

$$\big\|f_h - f\big\|_\infty \le \sup_x \int |f(x-ht) - f(x)|K(t)\,dt \le L_f \int |ht|K(t)\,dt = O(h).$$

If f has a Lipschitz continuous derivative $\dot{f}$, then using the assumption $\int tK(t)\,dt = 0$, we have

$$\begin{aligned}\|f_h - f\|_\infty &\le \sup_x \int |f(x-ht) - f(x) + ht\dot{f}(x)|K(t)\,dt \\ &= \sup_x \int \Big|ht\int_0^1 (\dot{f}(x-sht) - \dot{f}(x))\,ds\Big|K(t)\,dt \\ &\le L_{\dot{f}}h^2 \int t^2K(t)\,dt = O(h^2).\end{aligned}$$

Thus, the theorem will follow if we verify

$$\|\hat{f}_n - f_h\|_\infty = O_p\Big(\Big(\frac{\log(n)}{nh}\Big)^{1/2}\Big). \tag{7.4.2}$$

Fix an $\alpha > 1/\gamma$ and assume without loss of generality $h < 1$. Then, using (7.4.1) and $K(t) = 0$ for $|t| > 1$, we obtain

$$\begin{aligned}\sup_{|x|>n^\alpha} f_h(x) &= \sup_{|x|>n^\alpha} \int f(x-ht)K(t)\,dt \le \int \sup_{|x|>n^\alpha} \frac{C_f}{(1+|x-ht|)^{1+\gamma}}K(t)\,dt \\ &\le \frac{C_f}{n^{\alpha(1+\gamma)}} = o\Big(\frac{1}{n}\Big).\end{aligned}$$

Moreover,

$$\begin{aligned}P\Big(\sup_{|x|>n^\alpha} \hat{f}_n(x) > 0\Big) &\le P\Big(\max_{1\le j\le n} |X_j| > n^\alpha - h\Big) \le nP(|X_1| > n^\alpha - h) \\ &\le 2n\int_{x>n^\alpha-h} \frac{C_f}{(1+x)^{1+\gamma}}\,dx = \frac{2nC_f}{\gamma(1-h+n^\alpha)^\gamma} \to 0.\end{aligned}$$

Thus, we have

$$\sup_{|x|>n^{\alpha}} |\hat{f}_n(x) - f_h(x)| \le \sup_{|x|>n^{\alpha}} \hat{f}_n(x) + \sup_{|x|>n^{\alpha}} f_h(x) = o_p\Big(\frac{1}{n}\Big).$$

Consequently, the theorem will follow if we prove

$$\sup_{|x|\le n^{\alpha}} |\hat{f}_n(x) - f_h(x)| = O_p\Big(\Big(\frac{\log(n)}{nh}\Big)^{1/2}\Big). \tag{7.4.3}$$

Since this will be established in Lemma 7.4.3 below, this completes the proof. □

Remark 7.4.1. Condition (7.4.1) was only used in deriving (7.4.2). The latter can be established without this condition and under weaker conditions on K, using more sophisticated tools. This follows from Giné and Guillou (2002) whose work implies (7.4.2) for bounded f. These authors treat the multivariate case and obtain strong consistency rates.

We shall now present a general theory that yields (7.4.3) and some related results that are needed in the next chapter. We begin by stating the Bernstein inequality. For its proof, see Bennett (1962) or pp. 191–193 in Pollard (1984).

Lemma 7.4.1. Bernstein Inequality. *Suppose $\xi_1, \cdots, \xi_n$ are independent r.v.'s having mean zero and finite positive variance σ^2 and are bounded by M. Then,*

$$P\Big(\Big|\frac{1}{n}\sum_{j=1}^{n}\xi_j\Big| \ge t\Big) \le 2\exp\Big(-\frac{nt^2}{2\sigma^2 + (2/3)Mt}\Big), \qquad \forall\, t > 0.$$

Let $Z, Z_1, Z_2, \cdots$ be i.i.d. p-dimensional random vectors and for $t \in \mathbb{R}$, let $h_{n,t}$ be a bounded measurable function from $\mathbb{R}^p$ into $\mathbb{R}$. Set

$$H_n(t) = \frac{1}{n}\sum_{i=1}^{n}\big[h_{n,t}(Z_i) - E\big(h_{n,t}(Z_i)\big)\big], \quad t \in \mathbb{R}.$$

Lemma 7.4.2. *Let $a_n < b_n$ be sequences of numbers such that $(b_n - a_n) = O(n^{\alpha})$ for some $\alpha > 0$. Suppose the following three conditions hold for some positive constants κ and C.*

$$\sup_{a_n\le t\le b_n}\sup_{z}|h_{n,t}(z)| = O\Big(\frac{n}{\log(n)}\Big). \tag{7.4.4}$$

$$\sup_{a_n\le t\le b_n} E[h_{n,t}^2(Z)] = O\Big(\frac{n}{\log(n)}\Big). \tag{7.4.5}$$

$$\sup_{z\in\mathbb{R}^p}\big|h_{n,s}(z) - h_{n,t}(z)\big| \le Cn^{\kappa}|s-t|, \ \forall\, s,t \in [a_n, b_n], \ |s-t| \le 1. \tag{7.4.6}$$

Then,

$$\sup_{a_n\le t\le b_n}|H_n(t)| = O_p(1). \tag{7.4.7}$$

Proof. This lemma is a special case of Proposition 4.1 in Müller, Schick and Wefelmeyer (2007). Applying the Bernstein inequality with $\xi_j = h_{n,t}(Z_j) - E[h_{n,t}(Z)]$ and using the fact that the function $x \mapsto \exp(-c/x)$ is increasing for each positive c, we obtain

$$P\big(|H_n(t)| \ge s\big) \le 2\exp\Big(-\frac{ns^2}{2E\big(h_{n,t}^2(Z)\big) + 2s\sup_z|h_{n,t}(z)|}\Big), \qquad \forall\, s > 0.$$

Thus, in view of the conditions (7.4.4) and (7.4.5), there is a positive number a such that, for all large enough n,

$$\sup_{a_n \le t \le b_n} P(|H_n(t)| \ge s) \le 2\exp\big(-a\, s\log(n)\big), \quad s \ge 1.$$

Now let $t_{n,k} = a_n + k(b_n - a_n)n^{-m}$ for $k = 0, 1, \cdots, n^m$, with m an integer greater than $\alpha+\kappa$. Let $t^*_{n,k} = (t_{n,k-1} + t_{n,k})/2$ be the midpoint of the interval $I_{n,k} := [t_{n,k-1}, t_{n,k}]$, for $k = 1, \cdots, n^m$. The above yields for large enough $s \ge 1$,

$$P\Big(\max_{k=1,\cdots,n^m} |H_n(t^*_{n,k})| > s\Big) \le \sum_{k=1}^{n^m} P(|H_n(t^*_{n,k})| > s) = o(1).$$

This shows that

$$H_{n,1} := \max_{k=1,\cdots,n^m} |H_n(t^*_{n,k})| = O_p(1).$$

It follows from (7.4.6) and the choice of m that

$$H_{n,2} := \max_{k=1,\cdots,n^m} \sup_{t\in I_{n,k}} |H_n(t) - H_n(t^*_{n,k})| = O_p(n^\kappa n^\alpha n^{-m}) = o_p(1).$$

In view of the inequality $\sup_{a_n\le t\le b_n} |H_n(t)| \le H_{n,1} + H_{n,2}$, we have the desired result (7.4.7). □

We are now ready to give a proof of (7.4.3).

Lemma 7.4.3. *Suppose the kernel K is Lipschitz continuous, the bandwidth h satisfies $h \to 0$ and $\log(n)/(nh) = O(1)$, and f is bounded. Then (7.4.3) holds for every $\alpha > 0$.*

Proof. Assume n is large enough so that $\log(n) > 1$ and $1/h \le Bn/\log(n) \le Bn$ for some finite B. We apply Lemma 7.4.2 with $Z = X$ and

$$h_{n,t}(x) = \Big(\frac{n}{h\log(n)}\Big)^{1/2} K\Big(\frac{t-x}{h}\Big).$$

The assumption on K implies that K is bounded and hence square-integrable. Then, for all s, t, x in $\mathbb{R}$, we have, with L_K the Lipschitz constant of K,

$$\frac{\log(n)}{n}|h_{n,t}(x)| \le \|K\|_\infty \Big(\frac{\log(n)}{nh}\Big)^{1/2} = O(1),$$
$$\frac{\log(n)}{n}E(h^2_{n,t}(X)) = \int \frac{1}{h}K^2\Big(\frac{t-x}{h}\Big)f(x)\,dx = \int f(t-hu)K^2(u)\,du \le \|f\|_\infty \int K^2(u)\,du,$$
$$|h_{n,s}(x) - h_{n,t}(x)| \le \Big(\frac{n}{h\log(n)}\Big)^{1/2} L_K\frac{|s-t|}{h} \le B^{3/2}L_K n^2|s-t|.$$

Thus, the assumptions of Lemma 7.4.2 are met with $\kappa = 2$ and $C = B^{3/2}L_K$ and we obtain

$$\sup_{|t|\le n^\alpha}\Big|\frac{1}{n}\sum_{i=1}^n \Big(\frac{n}{h\log(n)}\Big)^{1/2}\Big\{K\Big(\frac{t-X_i}{h}\Big) - E\Big(K\Big(\frac{t-X_i}{h}\Big)\Big)\Big\}\Big| = O_p(1).$$

The claim (7.4.3) follows from this result because the left-hand side of the above equation equals $(nh/\log(n))^{1/2}\sup_{|t|\le n^\alpha}|\hat f_n(t) - f_h(t)|$. □

Now fix a closed interval $[a,b]$ with $a<b$. Our next goal is to derive uniform convergence on this interval for some quantities that will play a key role in Chapter 8. In what follows we set $w=gK$ for some polynomial g. It is easy to check that w is Lipschitz continuous if K is Lipschitz continuous and $K(t)=0$ for $|t|>1$. We first study the uniform behavior of

$$W_n(t) := \frac{1}{nh}\sum_{i=1}^{n} w\Big(\frac{X_i-t}{h}\Big), \quad t\in\mathbb{R},$$

on the interval $[a,b]$.

Lemma 7.4.4. *Assume K is Lipschitz continuous and satisfies $K(t)=0$ for $|t|>1$, and the bandwidth h satisfies $h\to 0$ and $\log(n)/(nh)=O(1)$. Assume that f is bounded on $I_\delta=(a-\delta,b+\delta)$ for some $\delta>0$. Then we have*

$$\sup_{a\le t\le b}\big|W_n(t)-E(W_n(t))\big| = O_p\Big(\Big(\frac{\log(n)}{nh}\Big)^{1/2}\Big). \tag{7.4.8}$$

Proof. Assume n is large enough so that $\log(n)>1$, $h<\delta$, and $1/h\le Bn/\log(n)\le Bn$ for some finite B. We apply Lemma 7.4.2 with $Z=X$ and

$$h_{n,t}(x)=\Big(\frac{n}{h\log(n)}\Big)^{1/2} w\Big(\frac{x-t}{h}\Big).$$

The assumptions on K imply that w is bounded, integrable and Lipschitz continuous. Then, for all s,t in $[a,b]$ and x in $\mathbb{R}$, we have with $B_f=\sup\{f(x): x\in I_\delta\}$ and L_w a Lipschitz constant for w,

$$\begin{aligned}
\frac{\log(n)}{n}\big|h_{n,t}(x)\big| &\le \|w\|_\infty\Big(\frac{\log(n)}{hn}\Big)^{1/2}=O(1),\\
\frac{\log(n)}{n}E\big(h_{n,t}^2(X)\big) &= \int \frac{1}{h}w^2\Big(\frac{x-t}{h}\Big)f(x)\,dx=\int f(t+hu)w^2(u)\,du\le B_f\int w^2(u)\,du,\\
\big|h_{n,s}(x)-h_{n,t}(x)\big| &\le \Big(\frac{n}{h\log(n)}\Big)^{1/2} L_w\frac{|s-t|}{h}\le B^{3/2}L_w n^2|s-t|.
\end{aligned}$$

These bounds show that the assumptions of Lemma 7.4.2 are met with $\kappa=2$ and $C=B^{3/2}L_w$, and hence (7.4.8) follows. □

Next, we study the uniform behavior of

$$U_n(t) := \frac{1}{nh}\sum_{i=1}^{n}\varepsilon_i w\Big(\frac{X_i-t}{h}\Big), \quad t\in\mathbb{R},$$

on the interval $[a,b]$, where $(\varepsilon_1,X_1),\cdots,(\varepsilon_n,X_n)$ are independent copies of a pair (ε,X) with X having density f.

Lemma 7.4.5. *Assume that $E(\varepsilon|X)=0$ and $E(|\varepsilon|^\gamma)$ is finite for some $\gamma>2$ and that the function v_f defined by $v_f(x)=f(x)E(\varepsilon^2|X=x)$, $x\in\mathbb{R}$, is bounded on $I_\delta=(a-\delta,b+\delta)$ for some $\delta>0$. Let K be Lipschitz continuous and satisfy $K(t)=0$ for $|t|>1$ and h satisfy*

$$h\to 0, \qquad \frac{\log(n)}{hn^{1-2/\gamma}}=O(1). \tag{7.4.9}$$

Then,

$$\sup_{a\le t\le b}\big|U_n(x)\big| = O_p\Big(\Big(\frac{\log(n)}{nh}\Big)^{1/2}\Big). \tag{7.4.10}$$

Proof. Assume n is large enough so that $\log(n) > 1$, $h < \delta$, and $1/h \leq Bn^{1-2/\gamma}/\log(n) \leq Bn^{1-2/\gamma}$ for some finite B. For $t \in \mathbb{R}$, set

$$\widetilde{U}_n(t) := \Big(\frac{nh}{\log(n)}\Big)^{1/2} U_n(t) = \Big(\frac{n}{h\log(n)}\Big)^{1/2} \frac{1}{n}\sum_{i=1}^{n} \varepsilon_i w\Big(\frac{X_i - t}{h}\Big).$$

Then (7.4.10) is equivalent to

$$\sup_{a \leq t \leq b} \big|\widetilde{U}_n(t)\big| = O_p(1).$$

It follows from $E(\varepsilon|X) = 0$ that $E\big(\widetilde{U}_n(t)\big) = 0$.

Let $M := E(|\varepsilon|^\gamma)$. Define

$$R_{ni}(t) := \Big(\frac{n}{h\log(n)}\Big)^{1/2} \varepsilon_i I\big[|\varepsilon_i| \leq Mn^{1/\gamma}\big] w\Big(\frac{X_i - t}{h}\Big)$$
$$S_{ni}(t) := \Big(\frac{n}{h\log(n)}\Big)^{1/2} \varepsilon_i I\big[|\varepsilon_i| > Mn^{1/\gamma}\big] w\Big(\frac{X_i - t}{h}\Big).$$

Then, using the identity $\varepsilon_i = \varepsilon_i I\big[|\varepsilon_i| \leq Mn^{1/\gamma}\big] + \varepsilon_i I\big[|\varepsilon_i| > Mn^{1/\gamma}\big]$, we obtain that $\forall\, t \in \mathbb{R}$,

$$\big(\widetilde{U}_n(t) - E(\widetilde{U}_n(t)\big) = \frac{1}{n}\sum_{i=1}^{n} \Big[R_{ni}(t) - E\big(R_{ni}(t)\big)\Big] + \frac{1}{n}\sum_{i=1}^{n} \Big[S_{ni}(t) - E\big(S_{ni}(t)\big)\Big].$$

It thus suffices to prove the following three results.

$$T_{n1} := \sup_{a \leq t \leq b} \Big|\frac{1}{n}\sum_{i=1}^{n} \big\{R_{ni}(t) - E\big(R_{ni}(t)\big)\big\}\Big| = O_p(1),$$

$$T_{n2} := \sup_{a \leq t \leq b} \Big|\frac{1}{n}\sum_{i=1}^{n} S_{ni}(t)\Big| = o_p(1), \qquad T_{n3} := \sup_{a \leq t \leq b} \Big|\frac{1}{n}\sum_{i=1}^{n} E\big(S_{ni}(t)\big)\Big| = o(1).$$

Because of $M = E\big(|\varepsilon|^\gamma\big) < \infty$, we obtain from Lemma 2.10.8

$$P\Big(T_{n2} > 0\Big) \leq P\Big(\max_{1 \leq i \leq n} |\varepsilon_i| > Mn^{1/\gamma}\Big) \to 0.$$

Hence, $T_{n2} = o_p(1)$. Moreover,

$$T_{n3} \leq \Big(\frac{n}{h\log(n)}\Big)^{1/2} \|w\|_\infty E\Big(|\varepsilon| I\big(|\varepsilon| > Mn^{1/\gamma}\big)\Big) \leq \Big(\frac{n}{h\log(n)}\Big)^{1/2} \frac{\|w\|_\infty E\big(|\varepsilon|^\gamma\big)}{(Mn^{1/\gamma})^{\gamma-1}}$$
$$= O\Big(\Big(\frac{1}{n^{1-2/\gamma} h\log(n)}\Big)^{1/2}\Big) = O\Big(\frac{1}{\log(n)}\Big),$$

by the assumption (7.4.9). Therefore, $T_{n3} = o(1)$.

It remains to prove $T_{n1} = O_p(1)$. For this we apply Lemma 7.4.2 with $Z = (X, \varepsilon)$ and

$$h_{n,t}(x, y) = \Big(\frac{n}{h\log(n)}\Big)^{1/2} y I\big(|y| \leq Mn^{1/\gamma}\big) w\Big(\frac{x - t}{h}\Big).$$

For $a \leq s, t \leq b$ and x, y in $\mathbb{R}$, we compute

$$\frac{\log(n)}{n}\big|h_{n,t}(x, y)\big| \leq \Big(\frac{\log(n)}{hn}\Big)^{1/2} Mn^{1/\gamma}\|w\|_\infty \leq M\|w\|_\infty B^{1/2},$$
$$\big|h_{n,s}(x, y) - h_{n,t}(x, y)\big| \leq \Big(\frac{n}{h\log(n)}\Big)^{1/2} Mn^{1/\gamma} L_w \frac{|s - t|}{h} \leq Cn^{2-\frac{2}{\gamma}}|s - t|,$$
$$\frac{\log(n)}{n} E\big(h_{n,t}^2(X, \varepsilon)\big) \leq \int \frac{1}{h} w^2\Big(\frac{x - t}{h}\Big) v_f(x)\, dx \leq B_* \int w^2(u)\, du,$$

with $B_* = \sup_{u \in I_\delta} v_f(u)$ and $C = B^{3/2} M L_w$. Thus, the assumptions of Lemma 7.4.2 hold, and we obtain $T_{n1} = O_p(1)$. □

7.5 Asymptotic Distribution of $\hat{f}_n(x)$

This section addresses the question of the asymptotic normality of $\hat{f}_n(x)$. Recall $f_h(x) := E(\hat{f}_n(x))$. By Lemma 7.3.5, under some conditions, $\mathrm{Var}(\sqrt{nh}\hat{f}_n(x)) \to f(x) I_2(K)$. This suggests a scale factor $\sqrt{nh}$ and asymptotic normality of the quantity $\sqrt{nh}(\hat{f}_n(x) - f_h(x))$ with mean zero and variance $f(x)I_2(K)$. Having obtained this asymptotic normality result, we then use the representation

$$\sqrt{nh}(\hat{f}_n(x)) - f(x)) = \sqrt{nh}(\hat{f}_n(x)) - f_h(x)) + \sqrt{nh}\,\mathrm{Bias}(\hat{f}_n(x))$$

and properties of the bias to obtain the asymptotic normality of $\sqrt{nh}(\hat{f}_n(x) - f(x))$. Indeed, under the conditions of Lemma 7.3.3,

$$\mathrm{Bias}(\hat{f}_n(x)) = \frac{1}{2} h^2 \ddot{f}(x) \kappa_2 + o(h^2). \tag{7.5.1}$$

If the bandwidth is chosen such that $nh^5 \to c$ for some $c \geq 0$, then, $\sqrt{nh}\,\mathrm{Bias}(\hat{f}_n(x)) \to \sqrt{c}\,\ddot{f}(x)\kappa_2/2$ and $\sqrt{nh}(\hat{f}_n(x) - f(x))$ is asymptotically normal with mean $\sqrt{c}\,\ddot{f}(x)\kappa_2/2$ and variance $f(x)I_2(K)$. The precise conditions are given in the following theorem.

Theorem 7.5.1. *Suppose the density f is continuous at x and positive at x, the kernel K satisfies* (7.3.2) *and* (7.3.4), *and the bandwidth h satisfies* (7.2.9). *Then,*

$$\sqrt{nh}(\hat{f}_n(x) - f_h(x)) \to_D \mathcal{N}(0, f(x) I_2(K)). \tag{7.5.2}$$

Suppose, in addition, $y^3 K(y) \to 0$ as $|y| \to \infty$, f is twice differentiable and its second derivative $\ddot{f}$ is continuous and integrable and h satisfies $nh^5 \to c$, for some $c \geq 0$. Then, (7.5.1) *holds and*

$$\sqrt{nh}(\hat{f}_n(x) - f(x)) \to_D \mathcal{N}(\sqrt{c}\,\ddot{f}(x)\kappa_2/2,\ f(x) I_2(K)). \tag{7.5.3}$$

Consequently, if $c = 0$, then

$$\sqrt{nh}(\hat{f}_n(x) - f(x)) \to_D \mathcal{N}(0,\ f(x) I_2(K)). \tag{7.5.4}$$

Proof. In view of the above comments, we only need to prove (7.5.2). Its left-hand side equals $\sum_{i=1}^n [Z_{ni}(x) - E(Z_{ni}(x))]$ with

$$Z_{ni}(x) := \frac{1}{\sqrt{nh}} K\Big(\frac{x - X_i}{h}\Big), \quad i = 1, \ldots, n.$$

The r.v.'s $Z_{n1}(x), \cdots, Z_{nn}(x)$ are i.i.d. and satisfy, by Lemma 7.3.1 and Remark 7.3.1,

$$\sum_{i=1}^n \Big(E(Z_{ni}(x))\Big)^2 = h\Big(\int K(y) f(x - hy)\, dy\Big)^2 \to 0,$$
$$\sum_{i=1}^n E(Z_{ni}^2(x)) = \int K^2(y) f(x - hy)\, dy \to I_2(K) f(x).$$

Moreover, by the square integrability of K and Lemma 7.3.1, we obtain that $\forall\, \epsilon > 0$,

$$L_n(\varepsilon) := \sum_{i=1}^{n} E\Big(Z_{ni}^2(x) I(|Z_{ni}(x)| > \epsilon)\Big) = \int K^2(y) I(|K(y)| > \epsilon\sqrt{nh}) f(x - hy)\, dy$$

$$\le f(x) \int K^2(y) I(|K(y)| > \epsilon\sqrt{nh})\, dy + \int K^2(y) |f(x - hy) - f(x)|\, dy \to 0.$$

Hence, (7.5.2) follows from Corollary 2.10.7. □

If $h = bn^{-1/5}$ for some b >0, then $nh^5 = b^5$ and $\sqrt{nh} = \sqrt{b} n^{2/5}$ and the conclusion (7.5.3) becomes

$$n^{2/5}\big(\hat{f}_n(x) - f(x)\big) \to_D \mathcal{N}\big(b^2 \ddot{f}(x)\kappa_2/2, f(x) I_2(K)/b\big).$$

Sometimes one is interested in the joint distribution of $\hat{f}_n(x_1)$ and $\hat{f}_n(x_2))$ for two real numbers $x_1 < x_2$. With $Z_{ni}(x)$ as in the previous proof, we have $T(x_j) := \sqrt{nh}\big(\hat{f}_n(x_j) - E\hat{f}_n(x_j)\big) = \sum_{i=1}^{n} \big(Z_{ni}(x_j) - E(Z_{ni}(x_j)\big)$, $j = 1, 2$. Because $T(x_1)$ and $T(x_2)$ are sums of i.i.d. r.v.'s, their covariance equals $n\mathrm{Cov}\big(Z_{n1}(x_1), Z_{n1}(x_2)\big) = C_1(h) - C_2(h)$, where

$$C_1(h) := h^{-1} E\Big\{K\Big(\frac{x_1 - X}{h}\Big) K\Big(\frac{x_2 - X}{h}\Big)\Big\} = \int K\Big(\frac{x_2 - x_1}{h} - t\Big) K(t) f(x_2 - ht)\, dt,$$

$$C_2(h) := h^{-1} E\Big\{K\Big(\frac{x_1 - X}{h}\Big)\Big\} E\Big\{K\Big(\frac{x_2 - X}{h}\Big)\Big\}$$

$$= h \int K(t) f(x_1 - ht)\, dt \int K(s) f(x_2 - hs)\, ds.$$

Assume that K is bounded and satisfies (7.3.2) and f is continuous at x_1 and x_2. Then $C_1(h)$ and $C_2(h)$ converge to zero as $h \to 0$. Indeed, Lemma 7.3.1(a) applied with $\gamma = K$ and $g = f$ yields $C_2(h) \to 0$, while $C_1(h) \to 0$ follows from the following two assertions,

$$f(x_2) \int K\Big(\frac{x_2 - x_1}{h} - t\Big) K(t)\, dt \to 0,$$

$$\int K\Big(\frac{x_2 - x_1}{h} - t\Big) K(t) (f(x_2 - ht) - f(x_2))\, dt \to 0.$$

As K is bounded and $K\big(h^{-1}(x_2 - x_1) - t\big) \to 0$ for every t in view of (7.3.2), the first assertion follows from Theorem 2.10.10. The second assertion follows because the absolute value of its left-hand side is bounded by

$$\|K\|_\infty \int |f(x - h_2 t) - f(x_2)| K(t)\, dt$$

which tends to zero by part (a) of Lemma 7.3.1.

Theorem 7.5.2. *Suppose the density f is twice differentiable with a continuous and integrable second derivative, the kernel K is bounded and satisfies* (7.3.4) *and $y^3 K(y) \to 0$ as $|y| \to \infty$, and the bandwidth satisfies $h \to 0$ and $nh^5 \to 0$. Then, for all $x_1 < x_2$ with $f(x_1)$ and $f(x_2)$ positive, we have*

$$(nh)^{1/2} \begin{pmatrix} \hat{f}_n(x_1) - f(x_1) \\ \hat{f}_n(x_2) - f(x_2) \end{pmatrix} \to_D \mathcal{N}_2(0, I_2(K)\Sigma), \qquad \Sigma = \begin{bmatrix} f(x_1) & 0 \\ 0 & f(x_2) \end{bmatrix}.$$

Because Σ is a diagonal matrix, the above theorem implies the asymptotic independence of the r.v.'s $(nh)^{1/2}\big(\hat{f}_n(x_1) - f(x_1)\big)$ and $(nh)^{1/2}\big(\hat{f}_n(x_2) - f(x_2)\big)$. The theorem easily extends to the case of p points $x_1 < \cdots < x_p$. The smoothness assumptions on f stem from Corollary 7.3.2. If a kernel with bounded support is used, we can work with the weaker smoothness assumptions of Corollary 7.3.3.

7.6 Estimation of Convolution Densities

While the estimation of a density f is of interest in its own right, sometimes quantities of interest are functions of the density. In this case a natural estimator of the unknown quantity is the plug-in estimator which replaces in the formula the unknown density f by a kernel estimator. We shall illustrate this with some important examples, namely, the estimation of the densities of $X_1 + X_2$ and $X_1 - X_2$. To begin with consider the convolution $g := f * f$ of f with itself, i.e., the density of $X_1 + X_2$. Recall

$$g(x) := f * f(x) = \int f(x-y)f(y)\,dy, \quad x \in \mathbb{R}.$$

The plug-in estimator of g is $\hat{g}_n := \hat{f}_n * \hat{f}_n$, where

$$\hat{f}_n(x) := \frac{1}{nh}\sum_{i=1}^{n} K\Big(\frac{x-X_i}{h}\Big).$$

In this section we shall establish the stochastic expansion

$$\hat{g}_n(x) = g(x) + \frac{2}{n}\sum_{i=1}^{n}\big(f(x-X_i) - E(f(x-X_i))\big) + o_p(n^{-1/2}), \tag{7.6.1}$$

under the following two different sets of assumptions on f which were already used in the study of $U_n^{(3)}$ in Chapter 5.

(F1) There is a square-integrable function $\dot{f}$ such that $f(y) - f(x) = \int_x^y \dot{f}(t)\,dt, \forall\, x < y$.

(F2) The density f is bounded and L_1-Lipschitz, see (5.6.4).

By Lemma 5.6.3, (F1) implies that f is bounded. In view of the CLT and Slutsky's Theorem 2.10.5, if $\tau^2(x) := \mathrm{Var}\big(f(x - X_1)\big)$ is finite and positive, then (7.6.1) implies the asymptotic normality result:

$$n^{1/2}\big(\hat{g}_n(x) - g(x)\big) \to_D \mathcal{N}\big(0, 4\tau^2(x)\big). \tag{7.6.2}$$

The finiteness of $\tau^2(x)$ is guaranteed under (F1) or (F2), because under either condition f is bounded.

To proceed further, let $K_2 := K * K$ and set

$$T_n(x) := \frac{1}{nh}\sum_{i=1}^{n} K_2\Big(\frac{x-2X_i}{h}\Big), \quad U_n(x) := \frac{2}{n(n-1)}\sum_{1\le i<j\le n}\frac{1}{h}K_2\Big(\frac{x-X_i-X_j}{h}\Big).$$

Use the substitution $y = X_i + hu$ below to conclude the representation

$$\begin{aligned}
\hat{f}_n * \hat{f}_n(x) &= \frac{1}{n^2h^2}\sum_{i=1}^{n}\sum_{j=1}^{n}\int K\Big(\frac{x-y-X_j}{h}\Big)K\Big(\frac{y-X_i}{h}\Big)\,dy \\
&= \frac{1}{n^2h}\sum_{i=1}^{n}\sum_{j=1}^{n}\int K\Big(\frac{x-X_i-X_j}{h} - u\Big)K(u)\,du \\
&= \frac{1}{n^2h}\sum_{i=1}^{n}\sum_{j=1}^{n} K_2\Big(\frac{x-X_i-X_j}{h}\Big) = \frac{T_n(x)}{n} + \frac{n-1}{n}U_n(x).
\end{aligned}$$

Since $T_n(x)$ is a kernel density estimator of the density of $2X_1$, we have

$$E(T_n(x)) = \int \frac{1}{2} f\Big(\frac{x-hu}{2}\Big) K_2(u)\,du, \quad \mathrm{Var}(T_n(x)) \le \frac{1}{2nh}\int f\Big(\frac{x-hu}{2}\Big) K_2^2(u)\,du.$$

Thus, if f is bounded, then $T_n(x) = O_p(1)$ holds and (7.6.1) follows from the expansion

$$U_n(x) = g(x) + \frac{1}{n}\sum_{i=1}^{n}\big[f(x-X_i) - E(f(x-X_i))\big] + o_p(n^{-1/2}). \tag{7.6.3}$$

Recall Theorem 5.5.1 and the notation used there. Note that $U_n(x)$ is a U-statistic with kernel $\varphi_n(y,z) := K_2\big((x-y-z)/h\big)/h$ depending on n via h. We calculate

$$\begin{aligned}
E(\varphi_n(X_1,X_2)) &= \int g(x-ht)K_2(t)\,dt,\\
E(\varphi_n^2(X_1,X_2)) &\le (1/h)\int g(x-ht)K_2^2(t)\,dt,\\
\varphi_{n1}(X_1) &= \int \varphi_n(X_1,z)f(z)\,dz = \int f(x-X_1-ht)K_2(t)\,dt.
\end{aligned}$$

The expansion (7.6.3) can now be derived from Theorem 5.5.1 by requiring $nh \to \infty$ and by verifying the following three conditions.

$$\int g(x-ht)K_2(t)\,dt - g(x) = o(n^{-1/2}), \tag{7.6.4}$$

$$\int g(x-ht)K_2^2(t)\,dt = O(1), \tag{7.6.5}$$

$$\int\Big(\int f(x-y-ht)K_2(t)\,dt - f(x-y)\Big)^2 f(y)\,dy \to 0. \tag{7.6.6}$$

Recall K_2 is a density. For bounded f, the integral in (7.6.6) is bounded by

$$\|f\|_\infty \iint \big(f(u-ht)-f(u)\big)^2\,du K_2(t)\,dt \le 2\|f\|_\infty^2 \iint |f(u-ht)-f(u)|\,du\,K_2(t)\,dt.$$

The first bound follows from (2.4.4), the Fubini Theorem 2.3.2 and the substitution $u = x-y$, while the second is an immediate consequence of the first bound. Both bounds tend to zero by Corollary 2.10.8. Thus, (7.6.6) holds under either (F1) or (F2). Under these conditions we can even derive rates of convergence. Indeed, if K_2 has a finite mean, then (F2) implies the rate

$$2\|f\|_\infty^2 \iint |f(u-ht)-f(u)|\,du\,K_2(t)\,dt = O(h), \tag{7.6.7}$$

and if K_2 has a finite second moment, then (F1) yields the rate

$$\|f\|_\infty \iint \big(f(u-ht)-f(u)\big)^2\,du\,K_2(t)\,dt = o(h^2). \tag{7.6.8}$$

Note that K_2 has a finite kth moment if K has a finite kth moment.

If f is bounded, so is g, and (7.6.5) follows if K_2 is square integrable. The latter is guaranteed by (7.3.11). Indeed, (2.4.4) and Theorem 2.3.2 give

$$\begin{aligned}
\int K_2^2(t)\,dt &= \int\Big(\int K(t-x)K(x)\,dx\Big)^2 dt \qquad (7.6.9)\\
&\le \iint K^2(t-x)K(x)\,dx\,dt = I_2(K) < \infty.
\end{aligned}$$

Theorem 7.6.1. *Suppose* (F1) *holds, the kernel K satisfies* (7.3.4) *and* (7.3.11)*, and the bandwidth satisfies $nh \to \infty$ and $nh^4 \to 0$. Then* (7.6.1) *holds for all x in $\mathbb{R}$.*

Proof. By the above discussion, we only need to verify (7.6.4). The density g is twice differentiable with first derivative $\dot{g} := \dot{f} * f$ and second derivative $\ddot{g} := \dot{f} * \dot{f}$. The Cauchy-Schwarz inequality and a substitution yield $\|\ddot{g}\|_\infty \le I_2(\dot{f})$. The above imply $|g(x+\delta) - g(x) - \delta\dot{g}(x)| \le \delta^2 I_2(\dot{f})/2$. Since K_2 has mean zero and finite variance $2\kappa_2$, the error term in (7.6.4) is bounded by

$$\Big|\int (g(x-ht) - g(x) + ht\dot{g}(x))K_2(t)\,dt\Big| \le h^2\kappa_2 I_2(\dot{f}).$$

This establishes (7.6.4) in view of $nh^4 \to 0$. □

A more detailed analysis using (7.6.8) shows that the error in the approximation is of order $O_p(1/n + h^2 + hn^{-1/2} + n^{-1}h^{-1/2})$. The choice $h = n^{-1/3}$ makes the error of order $O_p(n^{-2/3})$, the choice $h = n^{-1/2}$ makes this error of order $O_p(n^{-3/4})$.

Theorem 7.6.2. *Suppose* (F2) *holds, the kernel K has finite mean and satisfies* (7.3.11)*, and the bandwidth satisfies $nh \to \infty$ and $nh^2 \to 0$. Then* (7.6.1) *holds for all x in $\mathbb{R}$.*

Proof. Again, we only need to verify (7.6.4). The assumption (7.3.11) implies that $I_2(K_2) < \infty$, see (7.6.9). Because $\int |t|K_2(t)\,dt \le 2\int |t|K(t)\,dt$, K having finite mean implies the same about K_2. The assumption (F2) implies that g is Lipschitz continuous with Lipschitz constant $C = L\|f\|_\infty$. This follows from the inequality

$$\begin{aligned}
|g(x_1) - g(x_2)| &\le \Big|\int (f(x_1 - y) - f(x_2 - y))f(y)\,dy\Big| \\
&\le \|f\|_\infty \int |f(x_1 - y) - f(x_2 - y)|\,dy \\
&= \|f\|_\infty \int |f(u - (x_2 - x_1)) - f(u)|\,du \\
&\le \|f\|_\infty L|x_1 - x_2|, \quad x_1, x_2 \in \mathbb{R}.
\end{aligned}$$

This property of g yields that the error term in (7.6.4) is bounded by

$$\int |g(x-ht) - g(x)|K_2(t)\,dt \le Ch\int |t|K_2(t)\,dt.$$

This proves (7.6.4) in view of $nh^2 \to 0$. □

A more detailed analysis using (7.6.7) shows that the error in the approximation is of order $O_p(1/n + h + (h/n)^{1/2} + n^{-1}h^{-1/2})$. The choice $h = n^{-2/3}$ makes the error of order $O_p(n^{-2/3})$.

Similar results can be obtained about the estimation of the density γ of $X_1 - X_2$ given by

$$\gamma(x) := f \star f(x) := \int f(x+y)f(y)\,dy, \quad x \in \mathbb{R}.$$

The plug-in estimator is $\hat{\gamma}_n = \hat{f}_n \star \hat{f}_n$. We now have the representation

$$\hat{\gamma}_n(x) = \frac{1}{nh}\tilde{K}_2\Big(\frac{x}{h}\Big) + \frac{n-1}{n}\bar{U}_n(x)$$

where $\tilde{K}_2 := K \star K$ and $\bar{U}_n(x)$ is the U-statistic

$$\frac{1}{n(n-1)h}\sum_{1\le i<j\le n}\Big[\tilde K_2\Big(\frac{x-X_i+X_j}{h}\Big)+\tilde K_2\Big(\frac{x+X_i-X_j}{h}\Big)\Big].$$

Set $Y_i(x) = f(X_i+x)+f(X_i-x)$, $x \in \mathbb{R}$, $i = 1,\cdots,n$. We then have the following analogs to Theorems 7.6.1 and 7.6.2. Their proofs are similar to the proofs of these theorems, respectively, and are left as exercises.

Theorem 7.6.3. *Suppose* (F1) *holds, the kernel K satisfies* (7.3.4) *and* (7.3.11), *and the bandwidth satisfies $nh \to \infty$ and $nh^4 \to 0$. Then, for each x in $\mathbb{R}$, we have the following expansion.*

$$\bar U_n(x) = \gamma(x) + \frac{1}{n}\sum_{i=1}^{n}\big(Y_i(x) - E(Y_i(x))\big) + o_p(n^{-1/2}). \tag{7.6.10}$$

Theorem 7.6.4. *Suppose* (F2) *holds, the kernel K satisfies* (7.3.4) *and* (7.3.11), *and the bandwidth satisfies $nh \to \infty$ and $nh^2 \to 0$. Then* (7.6.10) *holds for all $x \in \mathbb{R}$.*

If $\mathrm{Var}\big(Y_1(x)\big)$ is positive and finite, then (7.6.10) implies

$$n^{1/2}\big(\hat\gamma_n(x)) - \gamma(x)\big) \to_D \mathcal{N}\big(0, \mathrm{Var}(Y_1(x))\big).$$

The finiteness of this variance is guaranteed under (F1) and (F2).

Of particular interest is the parameter $\gamma(0) = \int f^2(y)dy = I_2(f)$ and its plugin estimator $\hat\gamma_n(0)$. The following corollary summarizes the asymptotic normality of this estimator obtained from the above two theorems.

Corollary 7.6.1. *Suppose the density K is square-integrable symmetric about zero having finite second moment and $nh \to \infty$. Then the following holds.*
(a) *If, in addition, (F1) holds and $nh^4 \to 0$, then $n^{1/2}\big(\hat\gamma_n(0) - \gamma(0)\big) \to_D \mathcal{N}\big(0, 4\sigma^2\big)$, where $\sigma^2 := \mathrm{Var}\big(f(-X)\big)$.*
(b) *If, in addition (F2) holds and $nh^2 \to 0$, then $n^{1/2}\big(\hat\gamma_n(0) - \gamma(0)\big) \to_D \mathcal{N}\big(0, 4\tau^2\big)$, where $\tau^2 := \mathrm{Var}\big(f(X)\big)$.*

Recall some other estimators of $I_2(f)$ in the presence of some nuisance parameters were discussed in Section 6.6.

Remark 7.6.1. The $\sqrt{n}$ consistency rate of $\hat g_n$ given at (7.6.2) is surprising as the kernel density estimator generally has a much slower consistency rate. Frees (1994) was the first to recognize that densities of some functions of several independent r.v.'s can be estimated at this faster rate. Schick and Wefelmeyer (2004) showed that (7.6.1) can be generalized to hold for the supremum norm and the L_1-norm and obtained functional central limit theorems in associated normed spaces generalizing (7.6.2). Schick and Wefelmeyer (2007) obtained generalizations to weighted L_1-norms. Giné and Mason (2007) generalized the results of Schick and Wefelmeyer to the setting of Frees (1994) and to L_p-spaces.

7.7 Minimum Hellinger Distance Estimators

As a further application of the kernel density estimation in statistical inference, we shall discuss a sequence of minimum distance estimators of Beran (1978) for the location parameter in the symmetric location model. This sequence of estimators is asymptotically efficient at every error density with finite Fisher information for location, see (2.11.1), and has some robustness properties. They are based on smoothly truncated versions of kernel density estimators. Assume that the error d.f. F has finite Fisher information for location, see (6.4.5). Recall from (2.11.1) that $I_L(f) = \int \big(\dot f(x)/f(x)\big)^2 f(x)\,dx$.

Let $X_1, \cdots, X_n$ be a random sample from the location model (4.3.2) for some $\theta \in \mathbb{R}$ and some error density f symmetric about zero. Let $\hat\theta_n$ be a preliminary estimator of θ satisfying

$$n^{1/2}|\hat\theta_n - \theta| = O_p(1). \tag{7.7.1}$$

Let $\hat f_n$ be the kernel density estimator defined at (7.3.1) with kernel $K(x) = e^{-|x|}/2$, $x \in \mathbb{R}$. Let $b(x) := (1 - x^2)^2 I(|x| \le 1)$, and a_n, $n \ge 1$, be a sequence of positive numbers and define the sequence of smooth truncation functions

$$\begin{aligned} b_n(x) &:= 1, & |x| &\le a_n, \\ &:= b(x - a_n), & a_n \le x &\le a_n + 1, \\ &:= b(x + a_n), & -a_n - 1 \le x &\le -a_n, \\ &:= 0, & &\text{otherwise.} \end{aligned}$$

Let $\widehat{\kappa}_n$ denote the smoothly truncated version of the estimator $\hat f_n$ defined as follows.

$$\ell_n(x) := \hat f_n^{1/2}(x) b_n(x - \hat\theta_n), \quad B_n := \int \ell_n^2(x\,dx, \quad \widehat{\kappa}_n(x) := B_n^{-1}\ell_n^2(x), \quad x \in \mathbb{R}.$$

The Hellinger distance between any two densities ℓ_1, ℓ_2 on $\mathbb{R}$ is defined to be $\mathcal{H}^{1/2}(\ell_1, \ell_2)$, where

$$\mathcal{H}(\ell_1, \ell_2) := \int \big(\ell_1^{1/2}(x) - \ell_2^{1/2}(x)\big)^2 dx = 2 - 2\int \ell_1^{1/2}(x)\ell_2^{1/2}(x)\,dx.$$

The last equality is obtained by expanding the quadratic and using the fact: $\int \ell_j(x)\,dx = 1, j = 1, 2$.

A motivation for the minimum Hellinger distance estimator of θ is as follows. By the symmetry of f around the origin, $X_1 - \theta, \cdots, X_n - \theta$ have the same distribution as $-X_1 + \theta, \cdots, -X_n + \theta$. It is thus natural to define an estimator of θ to be that value of t which will align the data set $X_1 - t, \cdots, X_n - t$ with the data set $-X_1 + t, \cdots, -X_n + t$ in some sense. One way to do that is via minimizing the Hellinger distance between the density estimates based on these two data sets, as done by Beran (1978).

The density estimate of $f(x)$ based on $X_1 - t, \cdots, X_n - t$ is $\widehat{\kappa}_n(t + x)$. By the symmetry of b and K about the origin, $\widehat{\kappa}_n(x)$ based on $-X_1 + t, \cdots, -X_n + t$ is $\widehat{\kappa}_n(t - x)$. The square of the Hellinger distance between these two estimates is

$$\begin{aligned} \mathcal{H}_n(t) &:= \int \Big(\widehat{\kappa}_n^{1/2}(x + t) - \widehat{\kappa}_n^{1/2}(t - x)\Big)^2 dx = \int \Big(\widehat{\kappa}_n^{1/2}(2t - y) - \widehat{\kappa}_n^{1/2}(y)\Big)^2 dy \\ &= 2 - 2\int \widehat{\kappa}_n^{1/2}(2t - y)\widehat{\kappa}_n^{1/2}(y)\,dy. \end{aligned}$$

The corresponding minimum Hellinger distance estimator of θ is defined to be

$$\widehat{\theta}_{n,\mathcal{H}} := \operatorname*{argmin}_{t\in\mathbb{R}} \mathcal{H}_n(t) = \operatorname*{argmax}_{t\in\mathbb{R}} \int \widehat{\kappa}_n^{1/2}(2t - y)\widehat{\kappa}_n^{1/2}(y)\,dy. \tag{7.7.2}$$

Assume also that $h_n \to 0$ and $a_n \to \infty$. Beran has shown that the above minimizers exist but may not be unique. Theorem 1 of Beran (1978) shows that under the additional requirement $n^{1/2}h_n \to \infty$, every solution $\widehat{\theta}_{n,\mathcal{H}}$ of (7.7.2) tends to θ in probability, i.e., is consistent for θ.

From Theorem 3 of Beran it follows that if f satisfies (2.11.1) and $\left(nh_n^4\right)^{-1}a_n^2 \to \infty$ holds, then $n^{1/2}\left(\widehat{\theta}_{n,\mathcal{H}} - \theta\right) \to_D \mathcal{N}\left(0, 1/I_L(f)\right)$.

Recall, say from Rao (1973), that under some regularity conditions, the asymptotic distribution of $n^{1/2}(\tilde{\theta}_n - \theta)$ is $\mathcal{N}\left(0, 1/I_L(f)\right)$, where $\tilde{\theta}_n$ is a consistent root of the likelihood equation, when f is the error density. The entity $1/I_L(f)$ is also the lower bound on the asymptotic variance of $n^{1/2}(\theta_n - \theta)$, where θ_n is any location invariant consistent estimator of θ such that $n^{1/2}(\theta_n - \theta)$ is asymptotically normally distributed. Because the asymptotic variance of $n^{1/2}(\widehat{\theta}_{n,\mathcal{H}} - \theta)$ achieves this lower bound, we conclude that this sequence of estimators is asymptotically efficient at f.

We shall next describe a robustness property of these estimators. Recall we assume (6.4.5). Let γ_n be a sequence of densities with finite Fisher information for location such that $I_L(\gamma_n) \to I_L(f)$. Let φ be a square integrable function defined on $\mathbb{R}$ satisfying

$$\int \left(\dot{f}(x)/f(x)\right)\varphi(x) f^{1/2}(x)\, dx = 0, \tag{7.7.3}$$

$$D_n := \int \left(n^{1/2}\left[\gamma_n^{1/2}(x) - f^{1/2}(x)\right] - \varphi(x)\right)^2 dx \to 0.$$

We shall show that these two conditions imply $\mathcal{H}(\gamma_n, f) = O(1/n)$. Let $g_n(x) := n^{1/2}\left[\gamma_n^{1/2}(x) - f^{1/2}(x)\right]$. Then,

$$\begin{aligned} n\mathcal{H}(\gamma_n, f) &= \int g_n^2(x)\, dx = \int \left(g_n(x) - \varphi(x) + \varphi(x)\right)^2 dx \\ &= D_n + 2\int \varphi(x)\left(g_n(x) - \varphi(x)\right) dx + \int \varphi^2(x)\, dx. \end{aligned}$$

By the Cauchy-Schwarz inequality (2.4.4) and by (7.7.3),

$$\left| n\mathcal{H}(\gamma_n, f) - \int \varphi^2(x)\, dx \right| \le D_n + 2D_n^{1/2}\left(\int \varphi^2(x)\, dx\right)^{1/2} \to 0.$$

This is the desired result.

Beran (1978) showed that, under the above assumptions, the distribution of $n^{1/2}\left(\widehat{\theta}_{n,\mathcal{H}} - \theta\right)$ when ε_i's in the location model (4.3.4) are i.i.d. γ_n, converges weakly to the same limit as when they are i.i.d. f, i.e., to $\mathcal{N}\left(0, 1/I_L(f)\right)$. In other words, the asymptotic distribution of this sequence of estimators is robust against small deviations (in the sense of Hellinger metric) in the error distributions from the true error distribution.

Actually, Beran proves these results for a larger class of minimum Hellinger distance estimators corresponding to a larger class of K's and smoothing windows b_n. The statements of the results for these larger classes of estimators are complicated because of the complicated nature of the assumptions needed.

7.8 Exercises

1. We say f is locally Lipschitz at x, if there is a constant L_x such that

$$|f(y) - f(x)| \le L_x|y - x|, \quad \forall\, y \text{ in a neighborhood of } x. \tag{7.8.1}$$

With $\tilde{f}_n$ as in (7.2.1), prove the following.

(a) If f is locally Lipschitz at x, then $\text{Bias}(\tilde{f}_n(x)) = O(h)$.

(b) If f is Lipschitz continuous, then $\|\text{Bias}(\tilde{f}_n)\|_\infty = O(h)$.

2. Let $\hat{f}_n$ be as in (7.3.1). Assume K has bounded support. Prove the following.

(a) If f is locally Lipschitz at x, then $\text{Bias}(\hat{f}_n(x)) = O(h)$.

(b) If f is Lipschitz continuous, then $\|\text{Bias}(\hat{f}_n)\|_\infty = O(h)$.

3. Let $\hat{f}_n$ be as in (7.3.1). Suppose K satisfies $K(x) = 0$, for all $|x| > 1$. Also assume f has derivative $\dot{f}(x)$ at x. Show that $\text{Bias}(\hat{f}_n(x)) = h\dot{f}(x)\int tK(t)\,dt + o(h)$.

4. Let the random variables $X_1, \cdots, X_n$ be i.i.d. with density f. Assume that K and f satisfy the conditions of Theorem 7.3.1. Use Theorem 2.12.2 to show that if $n^\alpha h_n^2 \to \infty$ for some $0 < \alpha < 1$, then $\|\hat{f}_n - f\|_\infty \to_{a.s.} 0$.

In the next three problems, K is real-valued function defined on $\mathbb{R}$ such that $\int K(s)\,ds = 1$. The function K is said to be *of the order* r if r is a positive integer, $\int |s|^r|K(s)|\,ds < \infty$ and

$$\nu_j := \int s^j K(s)\,ds = 0, \qquad j = 1, 2, \ldots, r-1,$$
$$\ne 0, \qquad j = r.$$

5. Prove the following. Let r be a positive integer. Suppose that K is of order r and $y^{r+1}K(y) \to 0$ as $|y| \to \infty$. Suppose f is r-times differentiable and its r-th derivative $f^{(r)}$ is continuous and integrable. Then, for each x in $\mathbb{R}$,

$$\text{Bias}(\hat{f}_n(x)) = h^r \frac{f^{(r)}(x)}{r!}\nu_r + o(h^r).$$

If $f^{(r)}$ is also uniformly continuous, then the expansion is uniform,

$$\sup_{x\in\mathbb{R}} \left|\text{Bias}(\hat{f}_n(x)) - h^r \frac{f^{(r)}(x)}{r!}\nu_r\right| = o(h^r).$$

6. Let r be a positive integer. Suppose (7.2.9) holds, K is square-integrable and of order r, and $y^{r+1}K(y) \to 0$ as $|y| \to \infty$. Suppose f is r-times differentiable and its r-th derivative $f^{(r)}$ is continuous and integrable. Then, for each x in $\mathbb{R}$,

$$\text{MSE}(\hat{f}_n(x)) = \frac{f(x)}{nh}I_2(K) + h^{2r}\Big(\frac{f^{(r)}(x)}{r!}\nu_r\Big)^2 + o\Big(\frac{1}{nh}\Big) + o(h^{2r}).$$

If $f^{(r)}$ is also uniformly continuous, then the expansion is uniform,

$$\sup_{x\in\mathbb{R}}\left|\text{MSE}(\hat{f}_n(x)) - \frac{f(x)}{nh}I_2(K) - h^{2r}\Big(\frac{f^{(r)}(x)}{r!}\nu_r\Big)^2\right| = o\Big(\frac{1}{nh}\Big) + o(h^{2r}).$$

7. Let r be a positive integer. Suppose (7.2.9) holds, K is square integrable and of order r, f is square integrable, r times continuously differentiable with $f^{(r)}$ square integrable. Then, prove that

$$\mathrm{MISE}(\hat{f}_n) = \frac{I_2(K)}{nh} + \frac{h^{2r}\nu_r^2}{(r!)^2}\int \left(f^{(r)}(x)\right)^2 dx + o(1/n) + o(h^{2r}).$$

8. If f_1 and f_2 are integrable functions, then show that $f_1 * f_2(x) := \int f_1(x-y)f_2(y)\,dy$ is integrable and $f_1 * f_2 = f_2 * f_1$.
9. If g_1 and g_2 are square integrable, then show that $g_1 * g_2$ is bounded and uniformly continuous.
10. If f_1 and f_2 are integrable functions, then show that $f_1 \star f_2(x) := \int f_1(x+y)f_2(y)\,dy$ is integrable and symmetric about zero: $f_1 \star f_2(x) = f_2 \star f_1(-x)$, for all $x \in \mathbb{R}$.
11. Show that $f_1 \star f_2 = f_1 * \check{f}_2$, where $\check{f}_2(x) = f_2(-x)$.
12. Let f be a density of bounded variation and set $g = f * f$.

(a) Show that there is a function $\dot{g}$ such that

$$g(y) - g(x) = \int_x^y \dot{g}(t)\,dt, \quad x < y,$$

and for some finite constant C,

$$\int |\dot{g}(x+t) - \dot{g}(x)|\,dx \le C|t|, \quad t \in \mathbb{R}.$$

(b) Let K be a symmetric density with finite second moment and set $K_2 = K * K$. Use part (a) to derive the inequality

$$\left|\int g(x-hy)K_2(y)\,dy - g(x)\right| \le 2Ch^2 \int t^2 K(t)\,dt$$

(c) Let f_h be as defined in (7.3.3). Verify $\|f_h * f_h - g\|_1 = O(h^2)$.

13. Let $\hat{f}_n$ be the kernel density estimator based on the kernel K and the bandwidth h and let f_h be as defined in (7.3.3). Set $K_2 = K * K$ and

$$\bar{H}_n(x) = \frac{1}{n}\sum_{i=1}^n \Big[f(x - X_i) - E(f(x - X_i))\Big], \quad x \in \mathbb{R}.$$

Assume (7.2.9) and, for some $\alpha > 1$,

$$\int (1+|x|)^\alpha \Big(f(x) + f^2(x) + K^2(x)\Big)\,dx < \infty.$$

(a) Verify the identity $\hat{f}_n * \hat{f}_n - f_h * f_h = 2f_h * (\hat{f}_n - f_h) + (\hat{f}_n - f_h) * (\hat{f}_n - f_h)$.

(b) Show that $\|(\hat{f}_n - f_h) * (\hat{f}_n - f_h)\|_1 \le \|\hat{f}_n - f_h\|_1^2 = O(1/(nh)$.

(c) Use the inequalities in the proof of Theorem 7.3.2 to show that $nE\big(\|\bar{H}_n\|_1^2\big) = O(1)$.

(d) Verify the identity $f_h * (\hat{f}_n - f)(x) = \int \bar{H}_n(x - hy)K_2(y)\,dy$, $x \in \mathbb{R}$.

(e) Show that $\|f_h * (\hat{f}_n - f) - H_n\|_1 = o_p(n^{-1/2})$.

(f) Conclude $\|\hat{f}_n * \hat{f}_n - f_h * f_h - 2\bar{H}_n\|_1 = o_p(n^{-1/2})$.

14. Assume (7.2.9), K is a bounded symmetric kernel with finite variance and the density f is of bounded variation and satisfies $\int |x|^\alpha f(x)\,dx < \infty$ for some $\alpha > 1$. Show that the corresponding kernel density estimator $\hat{f}_n$ satisfies

$$\left\| \hat{f}_n * \hat{f}_n - f * f - \bar{H}_n \right\|_1 = o_p(n^{-1/2}).$$

15. Prove Theorem 7.6.3

16. Prove Theorem 7.6.4

Notes: The literature on nonparametric density estimation is vast. Several monographs have been written on the subject including those by Prakasa Rao (1983) and Devroye and Györfi (1985). The monographs of Silverman (1986), Scott (1992), Wand and Jones (1995) and Simonoff (1996) discuss numerous properties of kernel estimators, methods of choosing bandwidth sequences and some applications in univariate and multivariate settings. Bosq (1998) discusses similar topics for the density estimators of the stationary density for some stochastic processes, while Rüschendorf (1977) proves the a.s. uniform consistency for a class of kernel estimators of multivariate density functions in the independence case and under a mixing condition.

Lehmann (1963) and Sen (1966) proposed estimators of $I_2(f)$ in the one and two sample location models. These estimators are given in terms of lengths of confidence intervals based on Wilcoxon type linear rank and signed rank statistics. Schweder (1975) discusses some estimators of $I_2(f)$ in the one and two sample location models. Cheng and Serfling (1981) discuss several estimators of $I_2(f)$ when observations are i.i.d. F, i.e., when there are no nuisance parameters.

8

Nonparametric Regression

8.1 Introduction

A classical problem in statistics is to use the given information X to predict a response Y subject to some loss. Suppose the second moment of Y is finite and one chooses to use a function of X, say $g(X)$ with finite second moment, to predict Y, using the mean squared error loss, i.e., incurring the loss $E\big(|Y - g(X)|^2\big)$. In Example 5.4.1 we have seen that this loss is minimized by the conditional mean $\mu(X) := E\big(Y\big|X\big)$. The function μ is known as the *regression function.*

We write the regression model that regresses Y on X as

$$Y = \mu(X) + \varepsilon,$$

where the error r.v. ε has a finite second moment σ^2 and satisfies $E(\varepsilon|X = x) = 0$, for all x. Let $v(x) := E(\varepsilon^2|X = x)$ denote the conditional variance of ε, given $X = x$, for $x \in \mathbb{R}$. The regression model is called homoscedastic if $v(x)$ does not depend on x, i.e., $v(x) \equiv \sigma^2$. This is the case when ε and X are independent. The regression model is called a parametric regression model if μ is of a known parametric form and a nonparametric regression model if μ is an arbitrary function.

Let $(X_1, Y_1), \cdots, (X_n, Y_n)$ be a random sample from the above nonparametric regression model so that

$$Y_i = \mu(X_i) + \varepsilon_i, \qquad i = 1, \cdots, n, \tag{8.1.1}$$

where $\varepsilon_1, \cdots, \varepsilon_n$ are i.i.d. with the same distribution as ε. The problem of interest here is the estimation of the regression function μ based on the given data $(X_1, Y_1), \cdots, (X_n, Y_n)$ under minimal assumptions on μ. The problem of fitting a parametric model to the regression function will be addressed in Chapter 9.

There are two major estimation methods. One is a local method and the other is a global method. In the local approach, μ is estimated at a point x, and so the problem is to estimate a real parameter $\mu(x)$. Assume μ is smooth. This assumption suggests that observations (X_i, Y_i) with X_i close to x carry more information about $\mu(x)$ than those with X_i farther away. Thus a good estimator of $\mu(x)$ ought to give more weight to the observations closest to x. Typically such estimators are weighted averages

$$\sum_{i=1}^{n} w_i(x) Y_i$$

of the responses, with larger weights $w_i(x)$ for X_i close to x. A simple such estimator is the local average

$$\frac{\sum_{i=1}^{n} Y_i I\big(|X_i - x| \le h\big)}{\sum_{i=1}^{n} I\big(|X_i - x| \le h\big)},$$

DOI: 10.1201/9781003585770-8

for some positive h. If we replace $I(|X_i - x| \le h)$ by $K((X_i - x)/h)$ with K a density, we arrive at the Nadaraya–Watson estimator

$$\frac{\sum_{i=1}^{n} Y_i K((X_i - x)/h)}{\sum_{i=1}^{n} K((X_i - x)/h)}. \tag{8.1.2}$$

This estimator was independently developed by Nadaraya (1964) and Watson (1964). It is a special case of a larger class, the so-called local polynomial smoothers that will be studied in this chapter.

In the global approach, a coordinate system in the space of functions is introduced so as to reduce the problem of estimating μ to that of estimating a set of real-valued parameters. Suppose p is a known positive integer, W is a p-dimensional vector space, and $\{w_1, \cdots, w_p\}$ is a basis of W. Then any $w \in W$ can be uniquely expressed as $w = \sum_{j=1}^{p} \theta_j w_j$, where $\{\theta_j\}_{j=1}^{p}$ is the representation of w using the coordinate system. Likewise, using an appropriate set of linearly independent functions (basis) in the space of all r.v.'s $a(X)$ with $E(a^2(X))$ finite, any μ with $E(\mu^2(X))$ finite can be uniquely expressed by the set of coordinates. Indeed, given a set of basis functions $\{w_j\}_{j=1}^{\infty}$, the function μ has a set of unique coefficients $\{\theta_j\}_{j=1}^{\infty}$ such that $\mu(X) = \sum_{j=1}^{\infty} \theta_j w_j(X)$, a.s., where $\{\theta_j\}_{j=1}^{\infty}$ is the representation of "μ" using the coordinate system. This approach reduces the problem of estimating μ into a problem of estimating a set $\{\theta_j\}_{j=1}^{\infty}$ of real numbers. Polynomial series, Fourier series (which are infinitely differentiable everywhere), polynomial spline bases and Wavelet bases are some of the well-known coordinate systems/bases that are commonly used. See Loève (1963), Green and Silverman (1994) and Hernández and Weiss (1996) for more on this approach.

We shall focus on the method of local polynomial smoothers. For this we require a kernel K and a bandwidth h and set

$$K_h(t) = K(t/h)/h, \quad t \in \mathbb{R}.$$

To motivate the local method consider the problem of estimating the population mean. Let $Z_1, \cdots, Z_n$ be a random sample from the given population. Then an estimator of the population mean is the sample mean which minimizes the average squared error $n^{-1} \sum_{i=1}^{n} (Z_j - a)^2$ with respect to $a \in \mathbb{R}$.

Assume μ to be continuous at x. Then, for X_i close to x, $\mu(X_i)$ is close to $\mu(x)$ and this suggests to estimate $\mu(x)$ by a minimizer, with respect to β_0, of the local version of the above average squared error, viz,

$$\frac{1}{n}\sum_{i=1}^{n} (Y_i - \beta_0)^2 K_h(X_i - x), \quad \beta_0 \in \mathbb{R}. \tag{8.1.3}$$

The Nadaraya–Watson estimator is a minimizer of this function, see Section 8.3 below.

If μ is differentiable at x, then a better approximation of $\mu(X_i)$ for X_i close to x is $\mu(x) + (X_i - x)\dot{\mu}(x)$ and this suggests to estimate $\mu(x)$ by the first entry $\hat{\beta}_{1,0}(x)$ of the minimizer $(\hat{\beta}_{1,0}(x), \hat{\beta}_{1,1}(x))$ of

$$\frac{1}{n}\sum_{i=1}^{n} \Big(Y_i - \beta_0 - \beta_1(X_i - x)\Big)^2 K_h(X_i - x), \quad \beta_0, \beta_1 \in \mathbb{R}. \tag{8.1.4}$$

If μ has a second derivative at x, then an even better approximation of $\mu(X_i)$ for X_i close to x is $\mu(x) + (X_i - x)\dot{\mu}(x) + (X - x)^2\ddot{\mu}(x)/2$ and this suggests to estimate $\mu(x)$ by the first entry $\hat{\beta}_{2,0}(x)$ of the minimizer $(\hat{\beta}_{2,0}(x), \hat{\beta}_{2,1}(x), \hat{\beta}_{2,2}(x))$ of the criterion function

$$\frac{1}{n}\sum_{i=1}^{n} \Big(Y_i - \beta_0 - \beta_1(X_i - x) - \beta_2(X_i - x)^2\Big)^2 K_h(X_i - x), \quad \beta_0, \beta_1, \beta_2 \in \mathbb{R}. \tag{8.1.5}$$

The above are all special cases of the local polynomial smoothing method.

A local polynomial smoother of order k for the regression function $\mu(x)$ is the first component $\hat{\beta}_{k,0}(x)$ of the minimizer $\big(\hat{\beta}_{k,0}(x), \cdots, \hat{\beta}_{k,k}(x)\big)$ of the criterion function

$$\frac{1}{n}\sum_{i=1}^{n}\Big(Y_i - \sum_{j=0}^{k}\beta_j(X_i-x)^j\Big)^2 K_h(X_i-x), \quad \beta_0,\cdots,\beta_k \in \mathbb{R}. \tag{8.1.6}$$

For $k=0$, this criterion function reduces to (8.1.3), and a local smoother of order 0 is also called a local constant smoother. For $k=1$, the criterion function reduces to (8.1.4), and a local smoother of order 1 is also called a local linear smoother. For $k=2$, the criterion function is that in (8.1.5), and a local polynomial smoother of order 2 is also called a local quadratic smoother.

We shall study these smoothers under the following standing assumption.

Assumption 8.1.1. The covariate X has a density f, the kernel K is a bounded symmetric density with $K(t) > 0$ for $|t| < 1$ and $K(t) = 0$ for $|t| \ge 1$ and the bandwidth $h = h_n$ satisfies (7.2.9). i.e., $h_n \to 0$ and $nh_n \to \infty$.

Such a kernel K is square-integrable and has finite moments of all orders with all odd moments equal to zero and satisfies

$$|t|^j K(t) \le K(t), \quad \forall t \in \mathbb{R}. \tag{8.1.7}$$

We set

$$\kappa_j := \int |t|^j K(t)\,dt, \qquad \kappa_j^+ := \int_0^1 t^j K(t)\,dt, \qquad j = 0,1,2,\cdots. \tag{8.1.8}$$

By the symmetry of K about zero, $2\kappa_j^+ = \kappa_j$ and $\kappa_j^+/\kappa_0^+ = \kappa_j$ as $\kappa_0^+ = 1/2$. Also, by (8.1.7), $\kappa_j^+ \le \kappa_0^+$.

In the next section we shall provide some preparatory results needed to study local polynomial smoothers. Detailed discussions of local constant, local linear and local quadratic smoothers are then discussed in the subsequent sections.

8.2 Preparatory Results

For each nonnegative integer j and each $x \in \mathbb{R}$, we set

$$\hat{p}_{n,j}(x) := \frac{1}{n}\sum_{i=1}^{n}\Big(\frac{X_i-x}{h}\Big)^j K_h(X_i-x), \qquad \hat{q}_{n,j}(x) := \frac{1}{n}\sum_{i=1}^{n} Y_i\Big(\frac{X_i-x}{h}\Big)^j K_h(X_i-x),$$

$$\hat{s}_{n,j}(x) := \frac{1}{n}\sum_{i=1}^{n}\varepsilon_i\Big(\frac{X_i-x}{h}\Big)^j K_h(X_i-x).$$

The behavior of local polynomial smoothers is closely linked to the behavior of the above quantities. For this reason, we collect properties of these quantities in this section. We first derive properties for a fixed x and then obtain uniform versions on intervals. A reader may skip the rest of this section at a first reading and return to it later when properties of these quantities are needed.

In the next section we shall discuss pointwise consistency of the local constant smoother at some point x and provide associated rates of convergence. This does not require a consistent estimator of f at x. Thus we do not have to impose that f is continuous at x. We only require f to have one-sided limits at x and one of these limits to be positive. This will guarantee that the denominator of the local constant estimator converges in probability to a positive constant.

We note that $\hat{p}_{n,j}(x)$ has mean

$$p_j(x,h) := E(\hat{p}_{n,j}(x)) = \int f(x+ht)t^j K(t)\,dt, \quad j = 0, 1, \cdots.$$

Also recall the definition of κ_j^+, see (8.1.8). Let $I_2(K) = \int K^2(t)\,dt$. If f has finite one-sided limits $f(x_+) = \lim_{t\to 0+} f(x+t)$ and $f(x_-) = \lim_{t\to 0+} f(x-t)$ at x, then we set

$$p_j(x) := \big[f(x_+) + (-1)^j f(x_-)\big]\kappa_j^+. \tag{8.2.1}$$

In particular,

$$\begin{aligned} p_0(x) &= \big(f(x_+) + f(x_-)\big)/2, \qquad p_1(x) = \big(f(x_+) - f(x_-)\big)\kappa_1^+, \\ p_2(x) &= \big(f(x_+) + f(x_-)\big)\kappa_2^+ = p_0(x)\kappa_2. \end{aligned} \tag{8.2.2}$$

If f is continuous at x, then $p_0(x) = f(x)$ and $p_1(x) = 0$.

Lemma 8.2.1. *Suppose f has finite one-sided limits $f(x_+)$ and $f(x_-)$ at x. Then, for each nonnegative integer j, we have the expansion*

$$\hat{p}_{n,j}(x) = p_j(x,h) + O_p((nh)^{-1/2}) = p_j(x) + o_p(1). \tag{8.2.3}$$

Proof. Fix a nonnegative integer j. By the symmetry of K about 0, we have

$$\int f(x+ht)t^j K(t) = \int_0^1 \big[f(x+ht) + (-1)^j f(x-ht)\big]t^j K(t)dt. \tag{8.2.4}$$

Since the left hand side equals $p_j(x,h)$, we conclude

$$p_j(x,h) = \int_0^1 \big[f(x+ht) + (-1)^j f(x-ht)\big]t^j K(t)\,dt \to p_j(x). \tag{8.2.5}$$

Moreover, by the fact $t^{2j} \le 1$ for $|t| \le 1$, we obtain

$$\begin{aligned} nh\mathrm{Var}\big(\hat{p}_{n,j}(x)\big) &\le \int f(x+ht)t^{2j}K^2(t)\,dt \\ &\le \int_0^1 [f(x+ht) + f(x-ht)]K^2(t)\,dt \to p_0(x)I_2(K). \end{aligned}$$

This establishes $\mathrm{Var}\big(\hat{p}_{n,j}(x)\big) = O\big((nh)^{-1}\big)$ and gives the first part of (8.2.3). The second part follows from $p_j(x,h) \to p_j(x)$. □

Properties of $\hat{s}_{n,j}(x)$ follow from the following lemma by taking $g(t) = t^j$. Recall $v(x) := E(\epsilon^2|X=x)$, $x \in \mathbb{R}$. To simplify notation we set $v_f = vf$ so that

$$v_f(t) = v(t)f(t) = E(\varepsilon^2|X=t)f(t), \quad t \in \mathbb{R},$$

and introduce

$$v_f^*(x) := \limsup_{s\to x} v_f(s). \tag{8.2.6}$$

Lemma 8.2.2. *Set $w = gK$ for some polynomial g. Then the following hold.*
(a) *The condition $v_f^*(x) < \infty$ implies*

$$\hat{s}_n(x) := \frac{1}{nh}\sum_{i=1}^{n} \varepsilon_i w\Big(\frac{X_i - x}{h}\Big) = O_p\big((nh)^{-1/2}\big).$$

(b) *If v_f has finite one-sided limits $v_f(x_+)$ and $v_f(x_-)$ at x, their sum $v_f(x_+)+v_f(x_-)$ is positive, g is not identically zero, and the condition*

$$E\Big\{\varepsilon^2 K_h(X-x) I\big(|\varepsilon| > a\sqrt{nh}\big)\Big\} \to 0, \quad \forall a > 0, \tag{8.2.7}$$

holds, then

$$\sqrt{nh}\,\hat{s}_n(x) \to_D \mathcal{N}\big(0, \sigma^2(x)\big)$$

with

$$\sigma^2(x) = v_f(x_+)\int_0^1 w^2(t)\,dt + v_f(x_-)\int_0^1 w^2(-t)\,dt.$$

Proof. Express $\sqrt{nh}\,\hat{s}_n(x)$ as the sum $\sum_{i=1}^n \xi_{ni}$, where the r.v.'s

$$\xi_{ni} := \frac{1}{\sqrt{nh}}\varepsilon_i w\Big(\frac{X_i - x}{h}\Big), \quad i = 1, \cdots, n,$$

are i.i.d. with mean zero and variance

$$\frac{1}{nh}E\Big(\varepsilon^2 w^2\Big(\frac{X-x}{h}\Big)\Big) = \frac{1}{nh}E\Big(v(X) w^2\Big(\frac{X-x}{h}\Big)\Big) = \frac{1}{n}\int \frac{1}{h} v_f(y) w^2\Big(\frac{y-x}{h}\Big)\,dy.$$

Thus $\sqrt{nh}\,\hat{s}_n(x)$ has mean zero and, in view of the substitution $y = x + ht$, variance

$$\sigma_n^2(x) := \int v_f(x+ht) w^2(t)\,dt \le \sup_{|s|\le h} v_f(x+s)\int w^2(t)\,dt.$$

Since $\sup_{|s|\le h} v_f(x+s)$ converges to $v_f^*(x)$, we conclude

$$\limsup_n \operatorname{Var}\big(\sqrt{nh}\,\hat{s}_n(x)\big) \le v_f^*(x)\int w^2(t)\,dt.$$

Thus, $v_f^*(x) < \infty$ implies part (a).

To prove part (b), analogous to (8.2.4), we have

$$\sigma_n^2(x) = \int_0^1 v_f(x+ht) w^2(t)\,dt + \int_0^1 v_f(x-ht) w^2(-t)\,dt.$$

Hence, under the assumptions of part (b), this identity yields $\sigma_n^2(x) \to \sigma^2(x)$ and $\sigma^2(x) > 0$.

We shall apply Corollary 2.10.4 to achieve the desired asymptotic normality. We thus need to verify the Lindeberg condition:

$$L_n(\eta) := \sum_{i=1}^{n} E\big(\xi_{ni}^2 I(|\xi_{ni}| > \eta)\big) \to 0, \quad \forall \eta > 0.$$

Fix $\eta > 0$ and set $B_w := \|w\|_\infty$, $a := \eta/B_w$ and $B_g := \sup_{|t|\le 1}|g(t)|$. Then (8.2.7) and the inequality

$$\begin{aligned} L_n(\eta) &= nE\Big\{\frac{1}{nh}\varepsilon^2 w^2\Big(\frac{X-x}{h}\Big)I\Big(\Big|\varepsilon w\Big(\frac{X-x}{h}\Big)\Big| > \eta\sqrt{nh}\Big)\Big\} \\ &\le B_w B_g E\Big\{\varepsilon^2 K_h(X-x)I\big(|\varepsilon| > a\sqrt{nh}\big)\Big\} \end{aligned}$$

imply the Lindeberg condition. Thus Corollary 2.10.4 applies and gives the required asymptotic normality. □

Part (b) of the above lemma applied with $g(t) = t^j$ gives

$$\sqrt{nh}\,\hat{s}_{n,j}(x) \to_D \mathcal{N}\big(0, \gamma_{2j}(x)\big), \quad j = 0, 1, \cdots, \tag{8.2.8}$$

with

$$\gamma_j(x) := \big(v_f(x_+) + (-1)^j v_f(x_-)\big)\int_0^1 t^j K^2(t)\,dt, \quad j = 0, 1, 2, \cdots. \tag{8.2.9}$$

With these quantities associate the $(k+1)\times(k+1)$ matrix

$$\Gamma_k(x) = \begin{bmatrix} \gamma_0(x) & \cdots & \gamma_k(x) \\ \vdots & \ddots & \vdots \\ \gamma_k(x) & \cdots & \gamma_{2k}(x) \end{bmatrix}, \quad k = 1, 2, \cdots. \tag{8.2.10}$$

Then, by Lemma 8.2.2 and the Cramér-Wold device, see Chapter 2, we obtain the following result.

Corollary 8.2.1. *Suppose v_f has finite one-sided limits $v_f(x_+)$ and $v_f(x_-)$ at x, their sum $v_f(x_+) + v_f(x_-)$ is positive, and the condition (8.2.7) holds. Let k be a positive integer. Then $S_{n,k}(x) = (\hat{s}_{n,0}(x), \cdots, \hat{s}_{n,k}(x))'$ satisfies $\sqrt{nh}\,S_{n,k}(x) \to_D \mathcal{N}_k(0, \Gamma_k(x))$.*

Proof. In view of the Cramér-Wold device, it suffices to show

$$\sqrt{nh}\,u'S_{n,k}(x) = \sqrt{nh}\big(u_0\hat{s}_{n,0}(x) + \cdots + u_k\hat{s}_{n,k}(x)\big) \to_D \mathcal{N}(0, u'\Gamma_k(x)u)$$

for all unit vectors $u = (u_0, \cdots, u_k)'$ in $\mathbb{R}^{k+1}$. Fix such a vector u. Then $u'S_{n,k}(x)$ equals $\hat{s}_n(x)$ of Lemma 8.2.2 with $g(t) = u_0 + u_1 t + \cdots + u_k t^k = u'a(t)$ with $a(t) = (1, t, \cdots, t^k)'$. Thus Lemma 8.2.2 yields the asymptotic normality $\sqrt{nh}\,u'S_{n,k}(x) \to_D N(0, \sigma^2(x))$ with

$$\begin{aligned} \sigma^2(x) &= v_f(x_+)\int_0^1 \big(u'a(t)\big)^2 K^2(t)\,dt + v_f(x_-)\int_0^1 \big(u'a(-t)\big)^2 K^2(t)\,dt \\ &= u'\Big(v_f(x_+)\int_0^1 a(t)(a(t))'K^2(t)\,dt + v_f(x_-)\int_0^1 a(-t)(a(-t))'K^2(t)\,dt\Big)u. \end{aligned}$$

This is the desired result because $\sigma^2(x)$ equals $u'\Gamma_k(x)u$ as is easily checked. □

A byproduct of the proof is that $\Gamma_k(x)$ is positive definite as $u'\Gamma_k(x)u$ is positive for all unit vectors u. Sufficient conditions for (8.2.7) will be given in a later section. Finally, consider $\hat{q}_{n,j}(x)$.

Lemma 8.2.3. *Suppose f has finite one-sided limits $f(x_+)$ and $f(x_-)$ at x. Then the following hold for every nonnegative integer j.*

(a) *If μ is continuous at x, then $\hat{q}_{n,j}(x) = \mu(x)\hat{p}_{n,j}(x) + \hat{s}_{n,j}(x) + o_p(1)$.*

(b) *If μ is k times differentiable at x for some positive integer k, then*

$$\hat{q}_{n,j}(x) = \sum_{r=0}^{k} \frac{h^r \mu^{(r)}(x)}{r!} \hat{p}_{n,j+r}(x) + \hat{s}_{n,j}(x) + o_p(h^k),$$

where $\mu^{(r)}(x)$ is the r-th derivative of μ at x.

Proof. Since $Y_i = \mu(x) + \varepsilon_i + \mu(X_i) - \mu(x)$, we have the identity

$$\hat{q}_{n,j}(x) = \mu(x)\hat{p}_{n,j} + \hat{s}_{n,j}(x) + \hat{r}_{n,j,0}(x) \tag{8.2.11}$$

with

$$\hat{r}_{n,j,0}(x) := \frac{1}{n} \sum_{i=1}^{n} \Big[\mu(X_i) - \mu(x)\Big] \Big(\frac{X_i - x}{h}\Big)^j K_h(X_i - x).$$

Using (8.1.7), we derive the inequality $|\hat{r}_{n,j,0}(x)| \le \omega_0(x,h)\hat{p}_{n,0}(x)$ with

$$\omega_0(x,h) = \sup_{|t| \le h} |\mu(x+t) - \mu(x)|.$$

Note that $\omega_0(x,h)$ tends to zero if μ is continuous at x. Thus part (a) follows from this and $\hat{p}_{n,0}(x) \to_p p_0(x)$.

Now assume that μ is k-times differentiable at x and let

$$R_i(x) := \mu(X_i) - \sum_{r=0}^{k} \frac{h^r \mu^{(r)}(x)}{r!} \Big(\frac{X_i - x}{h}\Big)^r.$$

Then the error term in the expansion of part (b) is

$$\hat{r}_{n,j,k}(x) = \frac{1}{n} \sum_{i=1}^{n} R_i(x) \Big(\frac{X_i - x}{h}\Big)^j K_h(X_i - x).$$

Again using (8.1.7), we derive the inequality $|\hat{r}_{n,j,k}(x)| \le \omega_k(x,h)\hat{p}_{n,0}(x)$ with

$$\omega_k(x,h) := \sup_{|t| \le h} \Big|\mu(x+t) - \sum_{r=0}^{k} \frac{t^r}{r!} \mu^{(r)}(x)\Big|. \tag{8.2.12}$$

The result in part (b) then follows as $\omega_k(x,h) = o(h^k)$. □

We shall now study uniform versions of some of the above pointwise results. For this we fix an interval $[a,b]$ with $a < b$ and set $\mathcal{I}_\delta = (a - \delta, b + \delta)$ for $\delta > 0$. The function g is called zero-smooth if its domain contains I_δ for some $\delta > 0$ and g is continuous at each point in $[a,b]$. If g is zero-smooth, then

$$\sup_{|t| \le h} \sup_{a \le x \le b} \big|g(x+t) - g(x)\big| \to 0. \tag{8.2.13}$$

This follows from the uniform continuity of the restriction of g to $[a,b]$ and the continuity of g at the endpoints. The function g is called one-smooth if there is a zero-smooth function $\dot{g}$ satisfying

$$g(y) - g(x) = \int_x^y \dot{g}(t)\, dt, \quad a - \delta < x < y < b + \delta,$$

for some $\delta > 0$. Let k be an integer greater than 1. Then the function g is called k-smooth if, for some $\delta > 0$, g is $k-1$ times differentiable on I_δ and its derivative of order $k-1$ is one-smooth. If g is k-smooth for some nonnegative integer k, then, with $g^{(k)}$ the $\dot{g}$ of $g^{(k-1)}$,

$$\sup_{|t|\le h}\sup_{a\le x\le b}\Big|g(x+t)-\sum_{j=0}^{k}\frac{t^k g^{(j)}(x)}{j!}\Big| = o(h^k). \tag{8.2.14}$$

For $k=0$, this follows from (8.2.13). For $k>0$, use the Taylor expansion

$$g(x+t)-\sum_{j=0}^{k}\frac{t^k g^{(j)}(x)}{j!} = \frac{t^k}{k!}\int_0^1 (g^{(k)}(x+st)-g^{(k)}(x))k(1-s)^{k-1}\,ds$$

to bound the left hand side of (8.2.14) by

$$\frac{h^k}{k!}\sup_{|t|\le h}\sup_{a\le x\le b}\big|g^{(k)}(x+t)-g^{(k)}(x)\big|.$$

This bound is of order $o(h^k)$ as $g^{(k)}$ is zero-smooth. Note, if g is k times continuously differentiable on I_δ for some $\delta>0$, then g is k-smooth.

We first study $\hat{p}_{n,j}(x)$ and then its mean $p_j(x,h)$ on $[a,b]$. Lemma 7.4.4 applied with $g(t)=t^j$ gives the following result.

Lemma 8.2.4. *Assume K is Lipschitz continuous, $\log(n)/(nh) = O(1)$, f is bounded on $\mathcal{I}_\delta$ for some $\delta>0$, and j is a nonnegative integer. Then we have*

$$\sup_{a\le x\le b}\big|\hat{p}_{n,j}(x)-p_j(x,h)\big| = O_p\Big(\Big(\frac{\log(n)}{nh}\Big)^{1/2}\Big). \tag{8.2.15}$$

Let

$$\nu_j = \int t^j K(t)\,dt, \quad j=0,1,\cdots.$$

Lemma 8.2.5. *The following hold for each nonnegative integer j.*

(a) *If f is bounded on $\mathcal{I}_\delta$ for some $\delta>0$, then*

$$\sup_{a\le x\le b}\big|p_j(x,h)\big| = O(1).$$

(b) *If f is zero-smooth, then*

$$\sup_{a\le x\le b}\big|p_j(x,h)-f(x)\nu_j\big| \to 0.$$

(c) *If f is one-smooth, then*

$$\sup_{a\le x\le b}\big|p_j(x,h)-f(x)\nu_j-h\dot{f}(x)\nu_{j+1}\big| = o(h).$$

Proof. Since f is bounded on $\mathcal{I}_\delta$ and by (8.1.7), we have for $h<\delta$,

$$\big|p_j(x,h)\big| \le \int f(x+ht)K(t)\,dt \le \sup_{s\in\mathcal{I}_\delta} f(s), \quad \forall x\in[a,b].$$

This proves part (a). If f is zero-smooth, then

$$g_0(h) := \sup_{a\le y\le b}\sup_{|s|\le h}\big|f(y+s)-f(y)\big| \to 0,$$

and therefore part (b) follows from the inequality

$$|p_j(x,h) - f(x)\nu_j| \leq \int |f(x+ht) - f(x)|K(t)\,dt \leq g_0(h), \quad \forall\, x \in [a,b].$$

If f is one-smooth, then

$$g_1(h) := \sup_{a\leq y\leq b}\sup_{|s|\leq h} |\dot f(y+s) - \dot f(y)| \to 0,$$

and part (c) follows from the inequality

$$\left|p_j(x,h) - f(x)\nu_j - h\dot f(x)\nu_{j+1}\right| \leq \int |f(x+ht) - f(x) - ht\dot f(x)|K(t)\,dt \leq hg_1(h),$$

which holds for all $x \in [a,b]$. □

Lemma 7.4.5 applied with $g(t) = t^j$ gives the following result.

Lemma 8.2.6. *Assume that $E(|\varepsilon|^\gamma)$ is finite for some $\gamma > 2$ and that v_f is bounded on $\mathcal{I}_\delta$ for some $\delta > 0$. Additionally, assume that K is Lipschitz continuous and that the bandwidth satisfies*

$$\frac{\log(n)}{hn^{1-2/\gamma}} = O(1). \tag{8.2.16}$$

Then, for each nonnegative integer j,

$$\sup_{a\leq t\leq b}\left|\hat s_{n,j}(t)\right| = O_p\Big(\Big(\frac{\log(n)}{nh}\Big)^{1/2}\Big). \tag{8.2.17}$$

Lemma 8.2.7. *Assume K is Lipschitz continuous, the bandwidth satisfies $\log(n)/(nh) = O(1)$, and f is bounded on $\mathcal{I}_\delta$ for some $\delta > 0$. Then the following holds for nonnegative integers j and k. If μ is k-smooth, then*

$$\sup_{a\leq x\leq b}\left|\hat q_{n,j}(x) - \sum_{r=0}^{k}\frac{h^r\mu^{(r)}(x)}{r!}\hat p_{n,j+r}(x) - \hat s_{n,j}(x)\right| = o_p(h^k). \tag{8.2.18}$$

Proof. Lemmas 8.2.4 and 8.2.5 give $B_{n,0} := \sup_{a\leq x\leq b}\hat p_{n,0}(x) = O_p(1)$. Fix j and k. It follows from the proof of Lemma 8.2.3 that the supremum in (8.2.18) is bounded by $\omega_k(h)B_{n,0}$, where

$$\omega_k(h) = \sup_{a\leq x\leq b}\omega_k(x,h) = o(h^k),$$

as μ is k-smooth. □

8.3 Local Constant Smoothers

Here we shall study the local constant smoother. This estimator is a minimizer of the criterion function

$$\frac{1}{n}\sum_{i=1}^{n}(Y_i - \beta_0)^2 K_h(X_i - x), \quad \beta_0 \in \mathbb{R}.$$

This function is the quadratic polynomial

$$\frac{1}{n}\sum_{i=1}^{n} Y_i^2 K_h(X_i - x) - 2\beta_0 \hat{q}_{n,0}(x) + \beta_0^2 \hat{p}_{n,o}(x)$$

in β_0 and is easily seen to be minimized by

$$\hat{\mu}_n(x) := \frac{\hat{q}_{n,0}(x)}{\hat{p}_{n,0}(x)}.$$

Since $\hat{p}_{n,0}(x) = 0$ implies $\hat{q}_{n,0}(x) = 0$, we interpret $\mu_n(x)$ as zero for $\hat{p}_{n,0}(x) = 0$. This estimator is a weighted average $\sum_{i=1}^{n} Y_i w_i(x)$ with weights

$$w_i(x) = \frac{K_h(X_i - x)}{\sum_{j=1}^{n} K_h(X_j - x)}.$$

This is precisely the Nadaraya–Watson estimator given at (8.1.2).

One can also derive it as a plug-in estimator. For this assume that the pair (X, Y) has joint density g. Then, for $x \in \mathbb{R}$ with $f(x) > 0$, we have

$$\mu(x) = \frac{\int y g(x,y)\,dy}{f(x)} = \frac{\int y g(x,y)\,dy}{\int g(x,y)\,dy}.$$

Suppose we estimate g by the kernel density estimator

$$\hat{g}_n(x,y) = \frac{1}{n}\sum_{i=1}^{n} K_h(x - X_i) K_h(y - Y_i), \quad x, y \in \mathbb{R}.$$

Then the plug-in estimator of $\mu(x)$ is

$$\widetilde{\mu}_n(x) = \frac{\int y \hat{g}(x,y)\,dy}{\int \hat{g}(x,y)\,dy} = \frac{\frac{1}{n}\sum_{i=1}^{n} K_h(x - X_i) \int y K_h(y - Y_i)\,dy}{\frac{1}{n}\sum_{i=1}^{n} K_h(x - X_i) \int K_h(y - Y_i)\,dy}.$$

Since the symmetric kernel K satisfies $\int t K(t)\,dt = 0$, we calculate $\int K_h(y - Y_i)\,dy = 1$ and $\int y K_h(y - Y_i)\,dy = Y_i$ and obtain

$$\widetilde{\mu}_n(x) = \frac{\frac{1}{n}\sum_{i=1}^{n} Y_i K_h(x - X_i)}{\frac{1}{n}\sum_{i=1}^{n} K_h(x - X_i)}.$$

By the symmetry of K, $\widetilde{\mu}_n(x)$ equals $\hat{\mu}_n(x)$.

In the following subsections we shall study consistency, asymptotic normalty and uniform convergence of the local constant smoother.

8.3.1 Consistency

In this subsection we shall discuss the pointwise consistency of the local constant smoother $\hat{\mu}_n$ at some point x and provide associated rates of convergence. For this we shall require that the density f has finite one-sided limits at x. Thus our results are valid for densities with discontinuities such as uniform densities and exponential densities. Recall the definitions of $p_j(x)$ and $v_f^*(x)$ from (8.2.1) and (8.2.6), respectively.

Theorem 8.3.1. *Suppose f has finite one-sided limits $f(x_+)$ and $f(x_-)$ at x, the sum $f(x_+) + f(x_-)$ is positive, and $v_f^*(x)$ is finite. Then the following hold.*
(a) *If μ is continuous at x, then $\hat{\mu}_n(x) \to_p \mu(x)$.*
(b) *If μ is differentiable at x, then*

$$\hat{\mu}_n(x) - \mu(x) = hm_1(x) + \frac{\hat{s}_{n,0}(x)}{p_0(x)} + o_p(h) + o_p((nh)^{-1/2})$$

holds with

$$m_1(x) := \dot{\mu}(x)\frac{p_1(x)}{p_0(x)} = \dot{\mu}(x)\frac{f(x_+) - f(x_-)}{f(x_+) + f(x_-)}\kappa_1. \tag{8.3.1}$$

(c) *If μ has a second derivative at x and f is differentiable at x, then*

$$\hat{\mu}_n(x) - \mu(x) = h^2 m_2(x) + \frac{\hat{s}_{n,0}(x)}{p_0(x)} + o_p(h^2) + o_p((nh)^{-1/2})$$

holds with

$$m_2(x) = \Big[\frac{\ddot{\mu}(x)}{2} + \dot{\mu}(x)\frac{\dot{f}(x)}{f(x)}\Big]\kappa_2. \tag{8.3.2}$$

Proof. Recall $\kappa_j = \kappa_j^+/\kappa_0^+$, see (8.1.8). By (8.2.3), applied with $j = 0, 1, 2$, we obtain

$$\hat{p}_{n,0}(x) \to_p p_0(x) = \big(f(x_+) + f(x_-)\big)\kappa_0^+ > 0 \tag{8.3.3}$$

$$\hat{p}_{n,1}(x) \to_p p_1(x) = \big(f(x_+) - f(x_-)\big)\kappa_1^+, \tag{8.3.4}$$

$$\hat{p}_{n,2}(x) \to_p p_2(x) = \big(f(x_+) + f(x_-)\big)\kappa_2^+ = p_0(x)\kappa_2. \tag{8.3.5}$$

If f is differentiable at x, then we have the stronger conclusion

$$\hat{p}_{n,1}(x) = h\dot{f}(x)\kappa_2 + o_p(h) + O_p\big((nh)^{-1/2}\big). \tag{8.3.6}$$

This follows as $d_n(x) := p_1(x,h) - h\dot{f}(x)\kappa_2 = o(h)$. Indeed, in view of $\int tK(t)\,dt = 0$, $d_n(x)$ equals $\int \{f(x+ht) - f(x) - ht\dot{f}(x)\}tK(t)\,dt$ and thus satisfies

$$|d_n(x)| \le \sup_{|s|\le h} \big|f(x+s) - f(x) - s\dot{f}(x)\big| \int |t|K(t)\,dt = o(h).$$

Lemma 8.2.2 applied with $g(t) = 1$ yields

$$\hat{s}_{n,0}(x) = O_p((nh)^{-1/2}) = o_p(1). \tag{8.3.7}$$

If μ is continuous at x, then Lemma 8.2.3 gives

$$\hat{q}_{n,0}(x) = \mu(x)\hat{p}_{n,0}(x) + \hat{s}_{n,0}(x) + o_p(1)$$

and part (a) follows from this, (8.3.3) and (8.3.7).

If μ is differentiable at x, then Lemma 8.2.3 with $k = 1$ gives

$$\hat{q}_{n,0}(x) = \mu(x)\hat{p}_{n,0}(x) + h\dot{\mu}(x)\hat{p}_{n,1}(x) + \hat{s}_{n,0}(x) + o_p(h)$$

and part (b) follows from this, (8.3.3), (8.3.4) and (8.3.7).

If μ has second derivative at x and f is differentiable at x, then Lemma 8.2.3 with $k = 2$ gives

$$\hat{q}_{n,0}(x) = \mu(x)\hat{p}_{n,0}(x) + h\dot{\mu}(x)\hat{p}_{n,1}(x) + (h^2/2)\ddot{\mu}(x)\hat{p}_{n,2}(x) + \hat{s}_{n,0}(x) + o_p(h^2)$$

and part (c) follows from this, (8.3.3), (8.3.5), (8.3.6) and (8.3.7). □

8.3.2 Asymptotic Normality

The following theorem establishes asymptotic normality of $\hat{\mu}_n(x)$ under the smoothness assumptions of parts (a) and (b) of Theorem 8.3.1. Below $m_1(x)$ and $m_2(x)$ are as in (8.3.1) and (8.3.2) and $\gamma_0(x)$ is defined in (8.2.9), i.e.,

$$\gamma_0(x) = \big(v_f(x_+) + v_f(x_-)\big) \int_0^1 K^2(t)\, dt = \frac{1}{2}\big(v_f(x_+) + v_f(x_-)\big) I_2(K).$$

Theorem 8.3.2. *Suppose f and v_f have finite one-sided limits at x, $f(x_+) + f(x_-)$ and $v_f(x_+) + v_f(x_-)$ are positive, and* (8.2.7) *is met. Then the following hold with $\tau(x) = \gamma_0(x)/(p_0(x))^2$.*
(a) *If μ is differentiable at x and there exists a $c \geq 0$ such that $n^{1/2}h^{3/2} \to c$, then*

$$\sqrt{nh}\,\big(\hat{\mu}_n(x) - \mu(x)\big) \to_D \mathcal{N}\big(c\, m_1(x), \tau(x)\big).$$

(b) *If μ has a second derivative $\ddot{\mu}(x)$ at x, f is differentiable at x, and there exists a $c \geq 0$ such that $n^{1/2}h^{5/2} \to c$, then*

$$\sqrt{nh}\,\big(\hat{\mu}_n(x) - \mu(x)\big) \to_D \mathcal{N}\big(c\, m_2(x)), \tau(x)\big).$$

Proof. Under the assumptions of (a), part (b) of Theorem 8.3.1 gives

$$\sqrt{nh}\,\big(\hat{\mu}_n(x) - \mu(x)\big) = c m_1(x) + \sqrt{nh}\frac{\hat{s}_{n,0}(x)}{p_0(x)} + o_p(1),$$

while under the assumptions of (b), part (c) of Theorem 8.3.1 implies

$$\sqrt{nh}\,\big(\hat{\mu}_n(x) - \mu(x)\big) = c m_2(x) + \sqrt{nh}\frac{\hat{s}_{n,0}(x)}{p_0(x)} + o_p(1).$$

Thus, the desired result follows from

$$\sqrt{nh}\,\hat{s}_{n,0}(x) \to_D \mathcal{N}\big(0, \gamma_0(x)\big). \tag{8.3.8}$$

But this is a consequence of Lemma 8.2.2 applied with $g(t) = 1$, see (8.2.8). □

Remark 8.3.1. We shall now provide some sufficient conditions for (8.2.7). For this we let (A1) refer to the assumption that f has finite one-sided limits at x. For $a > 0$, set

$$M_n(a, x) := E\Big\{\varepsilon^2 K_h(X - x) I\big(|\varepsilon| > a\sqrt{nh}\big)\Big\},$$
$$V_n(a, x) := \sup_{|t| \leq h} E\Big\{\varepsilon^2 I\big(|\varepsilon| > a\sqrt{nh}\big) \big| X = x + t\Big\}.$$

Then (8.2.7) is equivalent to $M_n(a, x) \to 0$, $\forall\, a > 0$. We provide five sets of sufficient conditions for (8.2.7) as follows.

(1) If ε and X are independent, then (A1) implies (8.2.7). By (8.2.5), (A1) implies $p_0(x, h) \to p_0(x)$. Hence, by the assumed independence of X and ε and the square integrability of ε, we have

$$M_n(a, x) = E\{\varepsilon^2 I\big(|\varepsilon| > a\sqrt{nh}\big)\} p_0(x, h) \to 0.$$

(2) Suppose ε has a finite third moment and $\liminf_n nh^3$ is positive. Then (8.2.7) follows from the following computation with $B = \|K\|_\infty$ and the finiteness of $E(|\varepsilon|^3)$.

$$\begin{aligned} M_n(a,x) &\le (B/h)E\big(\varepsilon^2 I(|\varepsilon| > a\sqrt{nh})\big) \\ &\le (B/h)E\Big(\varepsilon^2 \frac{|\varepsilon|}{a\sqrt{nh}} I(|\varepsilon| > a\sqrt{nh})\Big) \\ &= \frac{B}{a\sqrt{nh^3}} E(|\varepsilon|^3 I(|\varepsilon| > a\sqrt{nh}). \end{aligned}$$

(3) Suppose $E(|\varepsilon|^{5/2})$ is finite and $\liminf_n nh^5$ is positive. Then (8.2.7) follows again by a similar argument as in (2) above.

(4) If (A1) holds and

$$V_n(a,x) \to 0, \quad \forall a > 0, \tag{8.3.9}$$

then (8.2.7) holds. This follows from (8.2.5) and the following inequality:

$$\begin{aligned} M_n(a,x) &= \int E\Big\{\varepsilon^2 I\Big(|\varepsilon| > a\sqrt{nh}\Big)\big|X = y\Big\}K_h(y-x)f(y)\,dy \\ &= \int E\Big\{\varepsilon^2 I\Big(|\varepsilon| > a\sqrt{nh}\Big)\big|X = x+ht\Big\}K(t)f(x+ht)\,dt \\ &\le V_n(a,x)\int f(x+ht)K(t)\,dt = V_n(a,x)p_0(x,h). \end{aligned}$$

(5) A sufficient condition for (8.3.9) is that for some $\delta > 0$,

$$\limsup_{t\to 0} E\big(|\varepsilon|^{2+\delta}|X = x+t\big) < \infty.$$

This follows from the inequality

$$V_n(a,x) \le \sup_{|t|\le h}\Big(\frac{1}{(a\sqrt{nh})^\delta}E\big(|\varepsilon|^{2+\delta}|X = x+t\big)\Big).$$

By the above comment (4), this sufficient condition for (8.3.9) and (A1) imply (8.2.7).

Remark 8.3.2. If f and v are continuous and positive at x, then $m_1(x) = 0$ and $\tau(x)$ simplifies to $I_2(K)v(x)/f(x)$. In this case, the conclusion of part (a) of the theorem becomes

$$\sqrt{nh}\big(\hat{\mu}_n(x) - \mu(x)\big) \to_D \mathcal{N}\Big(0, \frac{v(x)}{f(x)}I_2(K)\Big). \tag{8.3.10}$$

Remark 8.3.3. Suppose μ is differentiable at x, f and v are continuous and positive at x, and ε has a finite fourth moment. Then we have (8.3.10) if $\sqrt{nh}\,h$ converges to some positive c. A consistent estimator of $v(x)$ is given by

$$\hat{v}_n(x) = \frac{\sum_{1\le i<j\le n}(Y_i - Y_j)^2 K_h(x-X_i)K_h(x-X_j)}{2\sum_{1\le i<j\le n}K_h(x-X_i)K_h(x-X_j)}.$$

We leave the verification of the claim as an exercise. Thus a consistent estimator of the asymptotic variance $\tau(x) = I_2(K)v(x)/f(x)$ is given by $\hat{\tau}_n(x) := I_2(K)\hat{v}_n(x)/\hat{f}_n(x)$ and an asymptotic $1-\alpha$ confidence interval for $\mu(x)$ is given by the interval with endpoints $\hat{\mu}_n(x) \pm z_{\alpha/2}\sqrt{\hat{\tau}_n(x)/nh}$.

Example 8.3.1. Assume $f(x) = e^{-x}I(x > 0)$, v is continuous and positive on $[0, \infty)$, (8.2.7) holds and μ is twice differentiable. Then $f(0_-) = v_f(0-)) = 0$. If $n^{1/2}h^{3/2} \to c$, then by part (a) of Theorem 8.3.2,

$$\sqrt{nh}\big(\hat{\mu}_n(0) - \mu(0)\big) \to_D \mathcal{N}\big(c\dot{\mu}(0)\kappa_1, 2v(0)I_2(K)\big),$$
$$\sqrt{nh}\big(\hat{\mu}_n(x) - \mu(x)\big) \to_D \mathcal{N}\big(0, v(x)e^x I_2(K)\big), \quad x > 0.$$

If $nh^{5/2} \to c$, then part (b) gives

$$\sqrt{nh}\big(\hat{\mu}_n(x) - \mu(x)\big) \to_D \mathcal{N}\big(c(\ddot{\mu}(x)/2 - \dot{\mu}(x))\kappa_2, e^x v(x)I_2(K)\big), \quad x > 0.$$

8.3.3 Uniform Consistency

We now address uniform consistency on a closed interval $[a, b]$ with $a < b$. For this we need to impose higher moment assumptions on ε, require smoothness of K and need to restrict the decay rate of the bandwidth. We also need to assume

$$c_f := \inf_{a<x<b} f(x) > 0. \tag{8.3.11}$$

Theorem 8.3.3. *Suppose K is Lipschitz continuous, $E(|\varepsilon|^\gamma) < \infty$, for some $\gamma > 2$, (8.3.11) holds and $\log(n)/(n^{1-2/\gamma}h_n) = O(1)$. In addition, assume that f and v_f are bounded on an open interval $\mathcal{I}_\delta = (a - \delta, b + \delta)$, for some $\delta > 0$. Then the following hold.*
(a) *If μ is zero-smooth, then*

$$\sup_{a\le x\le b} \big|\hat{\mu}_n(x) - \mu(x)\big| = o_p(1).$$

(b) *If μ is one-smooth, then*

$$\sup_{a\le x\le b} \big|\hat{\mu}_n(x) - \mu(x)\big| = O_p\Big(\Big(\frac{\log(n)}{nh}\Big)^{1/2}\Big) + O(h).$$

If, in addition, f is zero-smooth, then

$$\sup_{a\le x\le b} \big|\hat{\mu}_n(x) - \mu(x)\big| = O_p\Big(\Big(\frac{\log(n)}{nh}\Big)^{1/2}\Big) + o(h).$$

(c) *If μ is two-smooth and f is one-smooth, then*

$$\sup_{a\le x\le b} \big|\hat{\mu}_n(x) - \mu(x)\big| = O_p\Big(\Big(\frac{\log(n)}{nh}\Big)^{1/2}\Big) + O(h^2).$$

We shall discuss some examples before giving the proof.

Example 8.3.2. If $\gamma = 3$ in the above theorem, then we can take $h = h_n \sim (\log(n)/n)^{1/3}$ and part (b) yields

$$\sup_{a\le x\le b} \big|\hat{\mu}_n(x) - \mu(x)\big| = O_p\Big(\Big(\frac{\log(n)}{n}\Big)^{1/3}\Big).$$

If $\gamma = 5/2$ in the above theorem, then we can take $h = h_n \sim (\log(n)/n)^{1/5}$ and part (c) yields

$$\sup_{a\le x\le b} \big|\hat{\mu}_n(x) - \mu(x)\big| = O_p\Big(\Big(\frac{\log(n)}{n}\Big)^{2/5}\Big).$$

Example 8.3.3. Assume $f(x) = e^{-x}I(x \geq 0)$, v_f is bounded, μ is twice continuously differentiable, and $E(|\varepsilon|^3)$ is finite. Then, the following hold. If $a = 0$, then we take $h = h_n \sim (\log(n)/n)^{1/3}$ and obtain

$$\sup_{0 \leq x \leq b} |\hat{\mu}_n(x) - \mu(x)| = O_p\Big(\Big(\frac{\log(n)}{n}\Big)^{1/3}\Big).$$

If $a > 0$, then we take $h = h_n \sim (\log(n)/n)^{1/5}$ and obtain

$$\sup_{a \leq x \leq b} |\hat{\mu}_n(x) - \mu(x)| = O_p\Big(\Big(\frac{\log(n)}{n}\Big)^{2/5}\Big).$$

Example 8.3.4. Let g be a function such that $0 < \inf_{a \leq x \leq b} g(x) \leq \sup_{a \leq x \leq b} g(x) < \infty$ and $\int_a^b g(x)\,dx = 1$. Suppose $f(x) = g(x)I(a \leq x \leq b)$ and v satisfies $\sup_{a \leq x \leq b} v(x) < \infty$. Assume μ is continuously differentiable, $E(|\varepsilon|^3) < \infty$ and $h = h_n \sim (\log(n)/n)^{1/3}$. Then we obtain

$$\sup_{a \leq x \leq b} |\hat{\mu}_n(x) - \mu(x)| = O_p\Big(\Big(\frac{\log(n)}{n}\Big)^{1/3}\Big).$$

Proof. *Proof of Theorem 8.3.3.* For any function g defined on $I := [a,b]$, let $\|g\|_I := \sup_{a \leq x \leq b} |g(x)|$. Since v_f is bounded on $\mathcal{I}_\delta$, Lemma 8.2.6 gives

$$\|\hat{s}_{n,0}\|_I = O_p\Big(\Big(\frac{\log(n)}{nh}\Big)^{1/2}\Big). \tag{8.3.12}$$

With c_f as in (8.3.11), we calculate

$$p_0(x,h) \geq c_f \int I(a < x + ht < b) K(t)\,dt = c_f \int_{(a-x)/h}^{(b-x)/h} K(t)\,dt.$$

Since $h \to 0$, we may and do assume $h < (b-a)/2$. For $a \leq x \leq (a+b)/2$, we then have $(a-x)/h \leq 0$ and $(b-x)/h \geq 1$ and the range of integration includes the interval $[0,1]$. For $(a+b)/2 < x \leq b$, we have $(a-x)/h < -1$ and $(b-x)/h \geq 0$ and the range of integration includes the interval $[-1,0]$. This establishes

$$p_0(x,h) \geq c_f/2, \quad \forall x \in [a,b].$$

Since f is bounded on $\mathcal{I}_\delta$, we derive from Lemmas 8.2.4 and 8.2.5 that $\|\hat{p}_{n,j}\|_I = O_p(1)$, for $j = 0,1,2$, and

$$P\Big(\inf_{a \leq x \leq b} \hat{p}_{n,0}(x) > c_f/4\Big) \to 1. \tag{8.3.13}$$

If f is zero-smooth, then we have $\|\hat{p}_{n,1}\|_I = o_p(1)$. If f is one-smooth, then we have the faster rate

$$\|\hat{p}_{n,1}\|_I = O_p(h) + O_p\Big(\Big(\frac{\log(n)}{nh}\Big)^{1/2}\Big). \tag{8.3.14}$$

We are now ready to treat the various cases.

(a) If μ is zero-smooth, then Lemma 8.2.7 applied with $k = 0$ yields

$$\sup_{a \leq x \leq b} |\hat{q}_{n,0}(x) - \mu(x)\hat{p}_{n,0}(x) - \hat{s}_{n,0}(x)| = o_p(1)$$

and the conclusion of part (a) follows from this, (8.3.12) and (8.3.13).

(b) If μ is one-smooth, then Lemma 8.2.7 applied with $k = 1$ yields

$$\sup_{a \le x \le b} \left|\hat{q}_{n,0}(x) - \mu(x)\hat{p}_{n,0}(x) - h\dot{\mu}(x)\hat{p}_{n,1}(x) - \hat{s}_{n,0}(x)\right| = o_p(h).$$

Thus, the first conclusion of part (b) follows from this, (8.3.13), (8.3.12) and $\left\|\hat{p}_{n,1}\right\|_I = O_p(1)$. The second conclusion follows as $\left\|\hat{p}_{n,1}\right\|_I = o_p(1)$ holds if f is continuous on $\mathcal{I}_\delta$.

(c) If μ is two-smooth, then Lemma 8.2.7 applied with $k = 2$ yields

$$\sup_{a \le x \le b} \left|\hat{q}_{n,0}(x) - \mu(x)\hat{p}_{n,0}(x) - h\dot{\mu}(x)\hat{p}_{n,1}(x) - \frac{h^2}{2}\hat{p}_{n,2}(x) - \hat{s}_{n,0}(x)\right| = o_p(h^2).$$

Thus, part (c) follows from this, (8.3.13), (8.3.12), (8.3.14) and $\left\|\hat{p}_{n,2}\right\|_I = O_p(1)$. □

8.4 Local Linear Smoothers

Recall, the local linear smoother estimates $\mu(x)$ by $\hat{\beta}_{1,0}(x)$, where the pair $\left(\hat{\beta}_{1,0}(x), \hat{\beta}_{1,1}(x)\right)$ is a minimizer of the criterion function

$$\frac{1}{n}\sum_{i=1}^{n}\left(Y_i - \beta_0 - \beta_1\frac{X_i - x}{h}\right)^2 K_h(X_i - x), \quad \beta_0, \beta_1 \in \mathbb{R}.$$

With $w_i = n^{-1/2}\left(K_h(X_i - x)\right)^{1/2}$, the criterion function becomes

$$\sum_{i=1}^{n}\left(Y_i w_i - \beta_0 w_i - \beta_1\frac{X_i - x}{h}w_i\right)^2, \quad \beta_0, \beta_1 \in \mathbb{R}.$$

Thus the above minimization problem is a least squares problem where the i-th covariate vector is $\widetilde{X}_i = \left(w_i, w_i(X_i - x)/h\right)'$ and the ith response variable is $\widetilde{Y}_i = Y_i w_i$. The associated normal equations for this problem are

$$\begin{bmatrix}\hat{p}_{n,0}(x) & \hat{p}_{n,1}(x)\\ \hat{p}_{n,1}(x) & \hat{p}_{n,2}(x)\end{bmatrix}\begin{bmatrix}\beta_0\\ \beta_1\end{bmatrix} = \begin{bmatrix}\hat{q}_{n,0}(x)\\ \hat{q}_{n,1}(x)\end{bmatrix}.$$

If $\hat{p}_{n,2}(x)\hat{p}_{n,0}(x) - \hat{p}^2_{n,1}(x)$ is positive, then the minimizers are

$$\hat{\beta}_{1,0}(x) = \frac{\hat{p}_{n,2}(x)\hat{q}_{n,0}(x) - \hat{p}_{n,1}(x)\hat{q}_{n,1}(x)}{\hat{p}_{n,2}(x)\hat{p}_{n,0}(x) - \hat{p}^2_{n,1}(x)},$$

$$\hat{\beta}_{1,1}(x) = \frac{\hat{p}_{n,0}(x)\hat{q}_{n,1}(x) - \hat{p}_{n,1}(x)\hat{q}_{n,0}(x)}{\hat{p}_{n,2}(x)\hat{p}_{n,0}(x) - \hat{p}^2_{n,1}(x)}.$$

Let

$$\hat{\Lambda}_{n,1}(x) := \begin{bmatrix}\hat{p}_{n,0}(x) & \hat{p}_{n,1}(x)\\ \hat{p}_{n,1}(x) & \hat{p}_{n,2}(x)\end{bmatrix}, \qquad \Lambda_1(x) := \begin{bmatrix}p_0(x) & p_1(x)\\ p_1(x) & p_2(x)\end{bmatrix}.$$

Assume that f has finite one-sided limits at x. Then, by (8.2.3),

$$\hat{\Lambda}_{n,1}(x) \to \Lambda_1(x). \tag{8.4.1}$$

Rewrite

$$\Lambda_1(x) = f(x_+)\begin{bmatrix}\kappa_0^+ & \kappa_1^+\\ \kappa_1^+ & \kappa_2^+\end{bmatrix} + f(x_-)\begin{bmatrix}\kappa_0^+ & -\kappa_1^+\\ -\kappa_1^+ & \kappa_2^+\end{bmatrix}.$$

We note that this matrix is positive definite if $f(x_+)+f(x_-)$ is positive. This follows from the fact that the matrices

$$M_a = \begin{bmatrix}\kappa_0^+ & a\kappa_1^+\\ a\kappa_1^+ & \kappa_2^+\end{bmatrix} = \int_0^1 \begin{bmatrix}1 & at\\ at & t^2\end{bmatrix} K(t)\,dt, \quad a=-1,1,$$

are positive definite as is easily checked, see the exercises for details. We shall write $\Lambda_1^{-1}(x)$ for $(\Lambda_1(x))^{-1}$.

To describe our results, we set

$$S_{n,1}(x) := \begin{bmatrix}\hat{s}_{n,0}(x)\\ \hat{s}_{n,1}(x)\end{bmatrix}, \qquad D_n(x) := \begin{bmatrix}\hat\beta_{1,0}(x)-\mu(x)\\ \hat\beta_{1,1}(x)-h\dot\mu(x)\end{bmatrix}.$$

In the next subsections we shall address consistency and asymptotic normality at a fixed point x and uniform consistency on an interval.

8.4.1 Consistency

In this subsection we shall establish consistency of the local linear smoother and provide rates of convergence. Recall the definition of v_f^* from (8.2.6).

Theorem 8.4.1. *Suppose f has finite one-sided limits $f(x_+)$ and $f(x_-)$ at x, their sum $f(x_+)+f(x_-)$ is positive, and $v_f^*(x)$ is finite. Then the following hold.*
(a) *If μ is continuous at x, then*

$$\begin{bmatrix}\hat\beta_{1,0}(x)\\ \hat\beta_{1,1}(x)\end{bmatrix} \to_p \begin{bmatrix}\mu(x)\\ 0\end{bmatrix}.$$

(b) *If μ is differentiable at x, then*

$$D_n(x) = \Lambda_1^{-1}(x)S_{n,1}(x) + o_p(h) + o_p((nh)^{-1/2}) = o_p(h) + O_p\big((nh)^{-1/2}\big).$$

(c) *If μ has a second derivative at x, then*

$$D_n(x) = \Lambda_1^{-1}(x)\Big[\frac{h^2}{2}\ddot\mu(x)m(x) + S_{n,1}(x)\Big] + o_p(h^2) + o_p\big((nh)^{-1/2}\big)$$

with

$$m(x) := \begin{bmatrix}p_2(x)\\ p_3(x)\end{bmatrix} = \begin{bmatrix}(f(x_+)+f(x_-))\kappa_2^+\\ (f(x)-f(x-))\kappa_3^+\end{bmatrix}.$$

Proof. By Lemma 8.2.2(a), we obtain the rate $S_{n,1}(x) = O_p\big((nh)^{-1/2}\big) = o_p(1)$.

If μ is continuous at x, then Lemma 8.2.3 (a) gives

$$\begin{bmatrix}\hat q_{n,0}(x)\\ \hat q_{n,1}(x)\end{bmatrix} = \begin{bmatrix}\hat p_{n,0}(x) & \hat p_{n,1}(x)\\ \hat p_{n,1}(x) & \hat p_{n,2}(x)\end{bmatrix}\begin{bmatrix}\mu(x)\\ 0\end{bmatrix} + \begin{bmatrix}\hat s_{n,0}(x)\\ \hat s_{n,1}(x)\end{bmatrix} + o_p(1).$$

Part (a) follows from this, the normal equations (8.4.1) and $S_{n,1}(x) = o_p(1)$.

If μ is differentiable at x, then Lemma 8.2.3 (b) gives

$$\begin{bmatrix}\hat{q}_{n,0}(x)\\ \hat{q}_{n,1}(x)\end{bmatrix} = \begin{bmatrix}\hat{p}_{n,0}(x) & \hat{p}_{n,1}(x)\\ \hat{p}_{n,1}(x) & \hat{p}_{n,2}(x)\end{bmatrix}\begin{bmatrix}\mu(x)\\ h\dot{\mu}(x)\end{bmatrix} + \begin{bmatrix}\hat{s}_{n,0}(x)\\ \hat{s}_{n,1}(x)\end{bmatrix} + o_p(h)$$

and therefore $\hat{\Lambda}_{n,1}(x)D_n(x) = S_{n,1}(x) + o_p(h)$. Part (b) follows from this, (8.4.1), and $S_{n,1}(x) = O_p((nh)^{-1/2})$.

If μ has a second derivative at x, then Lemma 8.2.3 gives

$$\begin{bmatrix}\hat{q}_{n,0}(x)\\ \hat{q}_{n,1}(x)\end{bmatrix} = \begin{bmatrix}\hat{p}_{n,0}(x) & \hat{p}_{n,1}(x)\\ \hat{p}_{n,1}(x) & \hat{p}_{n,2}(x)\end{bmatrix}\begin{bmatrix}\mu(x)\\ h\dot{\mu}(x)\end{bmatrix} + \frac{h^2}{2}\ddot{\mu}(x)\begin{bmatrix}\hat{p}_{n,2}(x)\\ \hat{p}_{n,3}(x)\end{bmatrix} + \begin{bmatrix}\hat{s}_{n,0}(x)\\ \hat{s}_{n,1}(x)\end{bmatrix} + o_p(h^2).$$

Therefore, by (8.2.3) applied with $j = 2, 3$, we obtain $\hat{\Lambda}_{n,1}(x)D_n(x) = (h^2/2)\ddot{\mu}(x)m(x) + S_{n,1}(x) + o_p(h^2)$. Part (c) follows from this, (8.4.1), and $S_{n,1}(x) = O_p((nh)^{-1/2})$. □

8.4.2 Asymptotic Normality

Next, we address asymptotic normality of local linear smoothers. We shall again require that f and v_f have finite one-sided limits at x. Upon putting $k = 1$ in (8.2.10), we obtain the matrix

$$\Gamma_1(x) = \begin{bmatrix}\gamma_0(x) & \gamma_1(x)\\ \gamma_1(x) & \gamma_2(x)\end{bmatrix}. \tag{8.4.2}$$

This matrix is positive definite if $v_f(x_+) + v_f(x_-)$ is positive, as observed after Corollary 8.2.1.

Theorem 8.4.2. *Suppose f and v_f have finite one-sided limits at x, $f(x_+) + f(x_-)$ and $v_f(x_+) + v_f(x_-)$ are positive, and (8.2.7) is met. Then the following hold with $\Sigma_1(x) = \Lambda_1^{-1}(x)\Gamma_1(x)\Lambda_1^{-1}(x)$.*

(a) *If μ is differentiable at x and $n^{1/2}h^{3/2}$ converges to some nonnegative c, then*

$$\sqrt{nh}\, D_n(x) \to_D \mathcal{N}_2\big(0, \Sigma_1(x)\big).$$

(b) *If μ has a second derivative $\ddot{\mu}(x)$ at x and $n^{1/2}h^{5/2}$ converges to some nonnegative c, then*

$$\sqrt{nh}\, D_n(x) \to_D \mathcal{N}_2\Big(\frac{c\ddot{\mu}(x)}{2}\Lambda_1^{-1}(x)m(x), \Sigma_1(x)\Big)$$

holds with $m(x)$ as in Theorem 8.4.1.

Proof. Under the assumptions of (a), Theorem 8.4.1(b) gives

$$\sqrt{nh}\, D_n(x) = \sqrt{nh}\Lambda_1^{-1}(x)S_{n,1}(x) + o_p(1),$$

while under the assumptions of (b), Theorem 8.4.1(c) implies

$$\sqrt{nh}\, D_n(x) = \frac{c}{2}\ddot{\mu}(x)\Lambda_1^{-1}(x))m(x) + \sqrt{nh}\Lambda_1^{-1}(x)S_{n,1}(x) + o_p(1).$$

Thus, the desired result follows from

$$\sqrt{nh}\, S_{n,1}(x) \to_D \mathcal{N}_2\big(0, \Gamma_1(x)\big),$$

which is the special case $k = 1$ of Corollary 8.2.1. □

Remark 8.4.1. Assume f and v are continuous at x. Then $p_3(x) = 0$ and the matrices $\Lambda_1(x)$, $\Gamma_1(x)$ and $\Sigma_1(x)$ are diagonal matrices. One calculates, with $\lambda_j = \int t^j K^2(t)\,dt$,

$$\Lambda_1(x) = f(x)\begin{bmatrix}1 & 0\\ 0 & \kappa_2\end{bmatrix} \qquad \Gamma_1(x) = v(x)f(x)\begin{bmatrix}\lambda_0 & 0\\ 0 & \lambda_2\end{bmatrix}$$

$$\Sigma_1(x) = \frac{v(x)}{f(x)}\begin{bmatrix}\lambda_0 & 0\\ 0 & \frac{\lambda_2}{\kappa_2^2}\end{bmatrix}, \qquad m(x) = f(x)\begin{bmatrix}\kappa_2\\ 0\end{bmatrix}.$$

We thus conclude that under the conditions of part (a) of Theorem 8.4.2,

$$\sqrt{nh}\big(\hat\beta_{1,0}(x) - \mu(x)\big) \to_D \mathcal{N}\Big(0, \frac{v(x)I_2(K)}{f(x)}\Big).$$

Recall from Theorem 8.3.2, the same conclusion holds for the Nadaraya-Watson estimator $\hat\mu_n(x)$ under exactly the same conditions.

Next, part (b) of Theorem 8.4.2 gives

$$\sqrt{nh}\big(\hat\beta_{1,0}(x) - \mu(x)\big) \to_D \mathcal{N}\Big(c\kappa_2^+\ddot\mu(x), \frac{v(x)I_2(K)}{f(x)}\Big).$$

Under the additional assumption that f is differentiable at x, Theorem 8.3.2 gives

$$\sqrt{nh}\big(\hat\mu_n(x) - \mu(x)\big) \to_D \mathcal{N}\Big(c\kappa_2^+\Big[\ddot\mu(x) + 2\dot\mu(x)\frac{\dot f(x)}{f(x)}\Big], \frac{v(x)I_2(K)}{f(x)}\Big).$$

In this case the limiting distributions are the same if $\dot f(x)\dot\mu(x) = 0$. If this fails to hold, we prefer the estimator with the smaller absolute asymptotic mean. In other words, the linear smoother is preferable if $|\ddot\mu(x)| \le |\ddot\mu(x) + 2\dot\mu(x)\dot f(x)/f(x)|$ which happens if $\ddot\mu(x)$ and $\dot\mu(x)\dot f(x)$ have the same sign. As μ and f are unknown, we do not know this. Since the asymptotic normality of $\hat\beta_{1,0}(x)$ does not need differentiability of f, the local linear smoother $\hat\beta_{1,0}(x)$ is preferable over the local constant estimator.

8.4.3 Uniform Consistency

In this subsection we prove uniform consistency of the local linear smoothers on an interval $[a, b]$, $-\infty < a < b < \infty$.

Theorem 8.4.3. *Suppose K is Lipschitz continuous, $E(|\varepsilon|^\gamma) < \infty$, for some $\gamma > 2$,* (8.3.11) *holds and $\log(n)/(n^{1-2/\gamma}h_n) = O(1)$. Let $\delta > 0$ and assume that f and v_f are bounded on the open interval $\mathcal{I}_\delta = (a - \delta, b + \delta)$. Then the following hold.*
(a) *If μ is zero-smooth, then*

$$\sup_{a\le x\le b}\left\|\begin{bmatrix}\hat\beta_{1,0}(x) - \mu(x)\\ \hat\beta_{1,1}(x)\end{bmatrix}\right\| = o_p(1).$$

(b) *If μ is one-smooth, then*

$$\sup_{a\le x\le b}\|D_n(x)\| = O_p\Big(\Big(\frac{\log(n)}{nh}\Big)^{1/2}\Big) + o(h).$$

(c) *If μ is two-smooth, then*

$$\sup_{a\le x\le b}\|D_n(x)\| = O_p\Big(\Big(\frac{\log(n)}{nh}\Big)^{1/2}\Big) + O(h^2).$$

The proof of this theorem is facilitated by the following lemma.

Lemma 8.4.1. *Let k be a positive integer and A be a $k \times k$ matrix. Then the inequality $\|Aw\| \geq \inf_{\|u\|=1} u'Au \, \|w\|$ holds for all $w \in \mathbb{R}^k$.*

Proof. We use the well-known identity $\|y\| = \sup_{\|u\|=1} u'y$ valid for all $y \in \mathbb{R}^k$ and write $w = \|w\|z$ for some unit vector z. We then calculate

$$\|Aw\| = \sup_{\|u\|=1} u'Aw = \|w\| \sup_{\|u\|=1} u'Az \geq \|w\|z'Az \geq \|w\| \inf_{\|u\|=1} u'Au,$$

which completes the proof of the lemma. □

Proof. [*Proof of Theorem 8.4.3*] Set

$$P(x,h) = \begin{bmatrix} p_0(x,h) & p_1(x,h) \\ p_1(x,h) & p_2(x,h) \end{bmatrix}$$

Let $u = (u_0, u_1)' \in \mathbb{R}^2$ be such that $\|u\| = 1$. Then we have the identity

$$u'P(x,h)u = \int f(x+ht)(u_0 + u_1 t)^2 K(t)\, dt.$$

Assume that n is large enough so that $h < (b-a)/2$. Recall the definition of c_f from (8.3.11). Arguing as in the proof of Theorem 8.3.3 we derive

$$u'P(x,h)u \geq c_f \int_0^1 (u_0 + u_1 t)^2 K(t) = c_f u' M_1 u, \quad a \leq x \leq (a+b)/2,$$

$$u'P(x,h)u \geq c_f \int_{-1}^0 (u_0 + u_1 t)^2 K(t) = c_f u' M_{-1} u, \quad (a+b)/2 < x \leq b.$$

This shows that $u'P(x,h)u \geq \eta := c_f \min(\lambda_1, \lambda_{-1})$, $a \leq x \leq b$, where $\lambda_r = \inf_{\|u\|=1} u'M_r u$, $r = -1, 1$. Since M_{-1} and M_1 are positive definite, we conclude $\eta > 0$. From this and Lemma 8.2.1, we conclude

$$P\Big(\inf_{\|u\|=1} \inf_{a \leq x \leq x} u'\hat{\Lambda}_{n,1}(x)u > \eta/2\Big) \to 0.$$

The desired results are now derived in a straightforward matter using uniform versions of the arguments used in the proof of Theorem 8.4.1 and Lemma 8.4.1. □

8.5 Local Quadratic Smoothers

In this section we present the theory for local quadratic smoothers. As the results are very similar to those of the local linear smoothers, we leave the proofs as exercises.

Recall, the local quadratic smoother estimates $\mu(x)$ by $\hat{\beta}_{2,0}(x)$, where $\hat{\beta}_{2,0}(x), \hat{\beta}_{2,1}(x), \hat{\beta}_{2,2}(x)$ are minimizers of the criterion function

$$\frac{1}{n}\sum_{i=1}^n \Big\{Y_i - \beta_0 - \beta_1 \frac{X_i - x}{h} - \beta_2\Big(\frac{X_i - x}{h}\Big)^2\Big\}^2 K_h(X_i - x), \quad \beta_0, \beta_1, \beta_2 \in \mathbb{R}.$$

This minimization problem is a least squares problem with normal equations

$$\begin{bmatrix} \hat{p}_{n,0}(x) & \hat{p}_{n,1}(x) & \hat{p}_{n,2}(x) \\ \hat{p}_{n,1}(x) & \hat{p}_{n,2}(x) & \hat{p}_{n,3}(x) \\ \hat{p}_{n,2}(x) & \hat{p}_{n,3}(x) & \hat{p}_{n,4}(x) \end{bmatrix} \begin{bmatrix} \beta_0 \\ \beta_1 \\ \beta_2 \end{bmatrix} = \begin{bmatrix} \hat{q}_{n,0}(x) \\ \hat{q}_{n,1}(x) \\ \hat{q}_{n,2}(x) \end{bmatrix}.$$

If f has finite one-sided limits, then we derive with the help of Lemma 8.2.1 the stochastic convergence

$$\hat{\Lambda}_{n,2}(x) := \begin{bmatrix} \hat{p}_{n,0}(x) & \hat{p}_{n,1}(x) & \hat{p}_{n,2}(x) \\ \hat{p}_{n,1}(x) & \hat{p}_{n,2}(x) & \hat{p}_{n,3}(x) \\ \hat{p}_{n,2}(x) & \hat{p}_{n,3}(x) & \hat{p}_{n,4}(x) \end{bmatrix} \to_p \Lambda_2(x) := \begin{bmatrix} p_0(x) & p_1(x) & p_2(x) \\ p_1(x) & p_2(x) & p_3(x) \\ p_2(x) & p_3(x) & p_4(x) \end{bmatrix}.$$

Let

$$M_a = \begin{bmatrix} \kappa_0^+ & a\kappa_1^+ & \kappa_2^+ \\ a\kappa_1^+ & \kappa_2^+ & a\kappa_3^+ \\ \kappa_2^+ & a\kappa_3^+ & \kappa_4^+ \end{bmatrix} = \int_0^1 \begin{bmatrix} 1 & at & t^2 \\ at & t^2 & at^3 \\ t^2 & at^3 & t^4 \end{bmatrix} K(t)\,dt, \quad a = -1, 1.$$

Then, $\Lambda_2(x) = f(x_+)M_1 + f(x_-)M_{-1}$. One can check that the matrices M_1 and M_{-1} are positive definite. Hence, $\Lambda_2(x)$ is positive definite if $f(x_+) + f(x_-)$ is positive. Let $\Lambda_2^{-1}(x)$ denote the inverse of $\Lambda_2(x)$.

To describe our results, we set

$$S_{n,2}(x) := \begin{bmatrix} \hat{s}_{n,0}(x) \\ \hat{s}_{n,1}(x) \\ \hat{s}_{n,2}(x) \end{bmatrix}, \qquad D_{n,0}(x) := \begin{bmatrix} \hat{\beta}_{2,0}(x) - \mu(x) \\ \hat{\beta}_{2,1}(x) \\ \hat{\beta}_{2,0}(x) \end{bmatrix},$$

$$D_{n,1}(x) := \begin{bmatrix} \hat{\beta}_{2,0}(x) - \mu(x) \\ \hat{\beta}_{2,1}(x) - h\dot{\mu}(x) \\ \hat{\beta}_{2,0}(x) \end{bmatrix}, \qquad D_{n,2}(x) := \begin{bmatrix} \hat{\beta}_{2,0}(x) - \mu(x) \\ \hat{\beta}_{2,1}(x) - h\dot{\mu}(x) \\ \hat{\beta}_{2,0}(x) - h^2/2\ddot{\mu}(x) \end{bmatrix}.$$

Our first result gives consistency at a fixed x.

Theorem 8.5.1. *Suppose f has finite one-sided limits $f(x_+)$ and $f(x_-)$ at x, their sum $f(x_+) + f(x_-)$ is positive, and $v_f^*(x)$ is finite. Then the following hold.*
(a) *If μ is continuous at x, then $D_{n,0}(x) = o_p(1)$.*
(b) *If μ is differentiable at x, then*

$$D_{n,1}(x) = \Lambda_2^{-1}(x)S_{n,2}(x) + o_p(h) + o_p((nh)^{-1/2}) = o_p(h) + O_p((nh)^{-1/2}).$$

(c) *If μ has a second derivative at x, then*

$$D_{n,2}(x) = \Lambda_2^{-1}(x)S_{n,2}(x) + o_p(h^2) + o_p((nh)^{-1/2}) = o_p(h^2) + O_p((nh)^{-1/2}).$$

Next, we address asymptotic normality at a fixed x. For this we also require that v_f has finite one-sided limits at x. Recall from (8.2.9) and (8.2.10) that

$$\Gamma_2(x) = \begin{bmatrix} \gamma_0(x) & \gamma_1(x) & \gamma_2(x) \\ \gamma_1(x) & \gamma_2(x) & \gamma_3(x) \\ \gamma_2(x) & \gamma_3(x) & \gamma_4(x) \end{bmatrix}.$$

The matrix $\Gamma_2(x)$ is positive definite if $v_f(x_+) + v_f(x_-)$ is positive.

Theorem 8.5.2. *Suppose f and v_f have finite one-sided limits at x, $f(x_+) + f(x_-)$ and $v_f(x_+) + v_f(x_-)$ are positive, and (8.2.7) is met. Then the following hold with $\Sigma_2(x) = \Lambda_2^{-1}(x)\Gamma_2(x)\Lambda_2^{-1}(x)$.*
(a) *If μ is differentiable at x and $n^{1/2}h^{3/2}$ converges to some positive c, then*

$$\sqrt{nh}\, D_{n,1}(x) \to_D \mathcal{N}_3(0, \Sigma_2(x)).$$

(b) *If μ has a second derivative $\ddot{\mu}(x)$ at x and $n^{1/2}h^{5/2}$ converges to some positive c, then*

$$\sqrt{nh}\, D_{n,2}(x) \to_D \mathcal{N}_3\big(0, \Sigma_2(x)\big).$$

Finally, we address uniform consistency on an interval $[a, b]$ with $a < b$.

Theorem 8.5.3. *Suppose K is Lipschitz continuous, $E(|\varepsilon|^\gamma) < \infty$, for some $\gamma > 2$,* (8.3.11) *holds and $\log(n)/(n^{1-2/\gamma}h_n) = O(1)$. Let $\delta > 0$ and assume that f and v_f are bounded on the open interval $\mathcal{I}_\delta = (a-\delta, b+\delta)$. Then the following hold.*

(a) *If μ is zero-smooth, then*

$$\sup_{a\le x\le b} \|D_{n,0}(x)\| = o_p(1).$$

(b) *If μ is one-smooth, then*

$$\sup_{a\le x\le b} \|D_{n,1}(x)\| = O_p\Big(\Big(\frac{\log(n)}{nh}\Big)^{1/2}\Big) + o_p(h).$$

(c) *If μ is two-smooth, then*

$$\sup_{a\le x\le b} \|D_{n,2}(x)\| = O_p\Big(\Big(\frac{\log(n)}{nh}\Big)^{1/2}\Big) + o_p(h^2).$$

8.6 Bandwidth Selection

Just as in the case of density estimation in Chapter 7, the selection of bandwidth h is also critical for local polynomial estimators. Some detailed methods are proposed in Härdle, Hall and Marron (1988), Härdle, W. (1991), Hall and Johnstone (1992), Wand and Jones (1995) and Fan and Gijbels (1996). Here we shall briefly mention a data-driven bandwidth selection method.

Let $\zeta_i := (X_i, Y_i), i = 1, \cdots, n$ be the given data and let $\hat{m}_n(x, h) = m_n(x, h, \zeta_1, \cdots, \zeta_n)$ be an estimate of $\mu(x)$ depending on a bandwidth h and the random sample $\zeta_1, \cdots, \zeta_n$. For $i = 1, \cdots, n$, let

$$\hat{m}_{n,i}(x, h) := m_{n-1}(x, h, \zeta_1, \cdots, \zeta_{i-1}, \zeta_{i+1}, \cdots, \zeta_n)$$

be the leave-one-out version of the estimator which does not use the pair ζ_i. Now consider the cross-validation criterion function

$$\mathrm{CV}(h) := \frac{1}{n}\sum_{i=1}^{n} \big(Y_i - \hat{m}_{n,i}(X_i, h)\big)^2 w(X_i), \quad h > 0,$$

where w is a nonnegative weight function. For example, if we are interested in estimating μ on some interval $\mathcal{I}$, we may choose $w(x) = I(x \in \mathcal{I})$. The weight function can also be used to eliminate observations X_i that are in the tails of the density f.

A data-driven bandwidth is then obtained by choosing h to be a minimizer $\hat{h}$ of CV. We call $\hat{h}$ the cross-validation bandwidth.

The criterion function CV has an expected value

$$E\big(\mathrm{CV}(h)\big) = \int v(x)w(x)f(x)\,dx + \int E\Big(\big|\hat{m}_{n-1}(x, h) - \mu(x)\big|^2\Big) w(x)f(x)\,dx. \tag{8.6.1}$$

Note that the first integral does not depend on h. This suggests that a minimizer of CV should be close to a minimizer of

$$\int E\Big(\big|\hat{m}_{n-1}(x,h) - \mu(x)\big|^2\Big) w(x) f(x)\,dx, \quad h > 0.$$

This can be made precise under appropriate conditions. We shall not pursue this here but refer an interested reader to the above mentioned references.

8.7 Exercises

1. Show that a symmetric $k \times k$ matrix A is positive definite if $u'Au$ is positive, for every $u \in \mathbb{R}^k$, $\|u\| = 1$.

2. Let A and B be symmetric positive definite $k \times k$ matrices and a, b be two nonnegative numbers. Show that $aA + bB$ is positive definite if $a + b > 0$.

3. Consider the matrices

$$A = \begin{bmatrix} a & b \\ b & c \end{bmatrix}, \quad B = \begin{bmatrix} a & -b \\ -b & c \end{bmatrix}.$$

Show that A is positive definite if and only if B is positive definite.

4. Consider the matrices

$$A = \begin{bmatrix} a & b & c \\ b & d & e \\ c & e & f \end{bmatrix}, \quad B = \begin{bmatrix} a & -b & c \\ -b & d & -e \\ c & -e & f \end{bmatrix}.$$

Show that A is positive definite if and only if B is positive definite.

5. Let w be an integrable function satisfying $w(t) > 0$ for $0 < t < 1$.

(a) Show that $\int_0^1 g^2(t) w(t)\,dt$ is positive for every non-zero polynomial g. *Hint: Use the fundamental theorem of algebra which says that a polynomial of degree m has at most m roots.*

(b) Let $a(t) = (1, t, \cdots, t^k)'$. Show that the following two matrices are positive definite.

$$\int_0^1 a(t)(a(t))' w(t)\,dt, \qquad \int_0^1 a(-t)(a(-t))' w(t)\,dt.$$

(c) Show that the matrices M_1 and M_{-1} of Sections 8.4 are positive definite.

(d) Show that the matrices M_1 and M_{-1} in Sections 8.5 are positive definite.

6. Suppose f and v are continuous and positive at two distinct points x and y, μ is differentiable at x and y, and $M_n(a, x) \to 0$ and $M_n(a, y) \to 0$, for all $a > 0$ with $M_n(a, x)$ as in Remark 8.3.1. Assume that $n^{1/2} h^{3/2} \to c$ for some positive c. Show that

$$\sqrt{nh} \begin{pmatrix} \hat{\mu}_n(x) - \mu(x) \\ \hat{\mu}_n(y) - \mu(y) \end{pmatrix} \to_D \mathcal{N}_2(0_2, B(x, y))$$

where

$$B(x,y) := I_2(K)\begin{bmatrix} v(x)/f(x) & 0 \\ 0 & v(y)/f(y) \end{bmatrix}.$$

Thus, like the kernel density estimator, the regression function estimators at two different points are asymptotically independent. *Hint: Some details of the proof of Theorem 7.5.2 may prove to be useful here.*

7. Verify the claim that $\hat{v}_n(x)$ is a consistent estimator of $v(x)$ under the assumptions of Remark 8.3.3.

8. Prove Theorem 8.5.1.

9. Prove Theorem 8.5.2.

10. Prove Theorem 8.5.3.

11. Consider the partially linear regression model

$$Y = \gamma Z + \mu(X) + \varepsilon$$

with $E(\varepsilon|X,Z) = 0$, $E(\varepsilon^2) < \infty$ and $0 < E((Z - E(Z|X))^2) < \infty$. Here γ is an unknown scalar and μ an unknown function. Let $(X_1, Y_1, Z_1), \cdots, (X_n, Y_n, Z_n)$ be independent copies of (X, Y, Z). If γ were known, then $Y^* = Y - \gamma Z = \mu(X) + \varepsilon$ and we could estimate $\mu(x)$ using local polynomial smoothers with the role of Y_i played by $Y_i^* = Y_i - \gamma Z_i$. As γ is unknown, we could then try local polynomial smoothers with Y_i^* replaced by $\hat{Y}_i^* = Y_i - \hat{\gamma} Z_i$, where $\hat{\gamma}$ is an estimator of γ. Assume that X has a density f, $\sqrt{n}(\hat{\gamma} - \gamma) = O_p(1)$ holds and $f(x)E(Z^2|X = x)$ is bounded in x.

(a) Show that the theorems in Section 8.3 remain true if $\hat{\mu}_n(x)$ is replaced by $\hat{\mu}_n^*(x)$, where the latter is the local constant smoother with Y_i replaced by $\hat{Y}_i^*$.

(b) Show that the theorems in Section 8.4 remain true if $(\hat{\beta}_{1,0}(x), \hat{\beta}_{1,1}(x))$ is replaced the versions $(\hat{\beta}_{1,0}^*(x), \hat{\beta}_{1,1}^*(x))$ with Y_i replaced by $\hat{Y}_i^*$.

12. Consider the random coefficient regression model where the response Y is related to covariates X and Z via the equation

$$Y = \alpha(X) + \beta(X)Z + \varepsilon,$$

where $E(\varepsilon|X,Z) = 0$, $E(\varepsilon^4) + E(Z^4) < \infty$, $E(|Z - E(Z|X)|^2) > 0$ and X has a density f. Here α and β are unknown functions. Let $(X_1, Y_1, Z_1), \cdots, (X_n, Y_n, Z_n)$ be independent copies of (X, Y, Z). We can estimate $(\alpha(x), \beta(x))'$ by the minimizer $(\hat{\alpha}(x), \hat{\beta}(x))'$ of

$$\frac{1}{n}\sum_{i=1}^{n} \big(Y_i - a - bZ_i\big)^2 K_h(X_i - x), \quad a, b \in \mathbb{R}.$$

(a) Show that a minimizer is a solution to the normal equations

$$\begin{bmatrix} z_{n,0}(x) & z_{n,1}(x) \\ z_{n,1}(x) & z_{n,2}(x) \end{bmatrix} \begin{bmatrix} a \\ b \end{bmatrix} = \begin{bmatrix} y_{n,0}(x) \\ y_{n,1}(x) \end{bmatrix},$$

with

$$z_{n,j}(x) := \frac{1}{n}\sum_{i=1}^{n} Z_i^j K_h(X_i - x), \quad j = 0, 1, 2,$$

$$y_{n,k}(x) := \frac{1}{n}\sum_{i=1}^{n} Y_i Z_i^k K_h(X_i - x), \quad k = 0, 1.$$

(b) Provide conditions that guarantee the consistency of the estimator $(\hat{\alpha}(x), \hat{b}(x))'$ at a fixed x.

13. Verify the identity (8.6.1).

Notes: The monograph of Wand and Jones (1995) discusses numerous properties of kernel estimators and methods of choosing bandwidth sequences. Fan and Gijbels (1996) contains detailed discussions about various properties of local polynomial regression estimators including some of their advantages over the Nadaraya-Watson estimator. More properties of these estimators can be also found in Fan and Yao (2003). Eubank and Speckman (1993) discuss the use of $\hat{\mu}_n$ in constructing confidence bands for $\mu(x)$ in some non-random covariates regression models. The paper of Wang and Yang (2009) contains an alternate method for constructing confidence bands for regression functions. Politis (2014) derives some bootstrap confidence intervals for the regression function in nonparametric regression setup.

Several other authors have proved the uniform consistency of $\hat{\mu}_n$ under a variety of conditions; see, e.g., Schuster and Yakowitz (1979), Mack and Silverman (1982), Einmahl and Mason (2000, 2005), Hansen (2008), Bouzebda and Nemouchi (2020), and the references therein. Several of these authors also discuss the almost sure uniform consistency of $\hat{\mu}_n$ and of the estimators of the higher-order derivatives of $\mu(x)$.

9

Model Diagnostics

9.1 Introduction

In parametric inference one often assumes that data is generated from a given parametric family of distributions and then proceeds to make inference about the underlying parameters in the assumed model. Similarly, in regression analysis, one assumes a particular parametric regression model and then develops the inference for the underlying parameters in the model. The focus of this chapter is to check whether a parametric family of distributions or regression functions fits the given data.

The problem of assessing whether a given parametric family of distributions fits the given data is known as the goodness-of-fit testing problem, while the problem of fitting a parametric family of functions to the regression or autoregressive function is called the lack of fit testing problem. In this chapter we shall study some tests for both of these problems.

The goodness-of-fit testing problem has been addressed in statistics ever since 1900 when Karl Pearson used a chi-square statistic to test for normality. Other tests for this problem are based on some distances between the empirical d.f. and the d.f. of the parametric model being fitted, while others are based on some distances between nonparametric density estimators and the given parametric family of densities. The tests based on density estimators are asymptotically distribution free but their implementation in finite samples needs a larger sample in general because of the slow rate of consistency of the estimator.

When fitting an error d.f. in location models, one uses residual empirical distribution functions. Tests based on residual empirical d.f.'s are usually not distribution free, even for the large samples. Khmaladze (1981) proposed a transformation of the residual empirical process that produces asymptotically distribution free (ADF) tests in these models. These tests for fitting an error d.f. in the one sample location-scale and linear regression models are discussed in some detail in this chapter. Sometimes the asymptotic distribution of a statistic does not provide a good approximation to the finite sample distribution. In such cases a bootstrap method may provide a better approximation. We shall discuss some of the bootstrap methods to implement some of these tests.

In the second problem, one is interested in fitting a parametric model to the regression function. A class of ADF tests based on certain integrated squared differences between the nonparametric estimators of the regression function and its conditional expectation under the parametric model being fitted will be discussed in this chapter. We also discuss some tests for comparing the two regression functions in the two independent samples setup.

For any two sequences of stochastic processes $\{\mathcal{Y}_n(x), x \in \Omega\}$ and $\{\mathcal{Z}_n(x), x \in \Omega\}$ indexed by the same set Ω, we write $\mathcal{Y}_n(x) = \mathcal{Z}_n(x) + u_p(1)$ to mean $\sup_{x\in\Omega} |\mathcal{Y}_n(x) - \mathcal{Z}_n(x)| = o_p(1)$.

DOI: 10.1201/9781003585770-9

9.2 Goodness-of-fit Testing

In this section we shall discuss goodness-of-fit tests for fitting an error d.f. in the one sample location-scale, one sample location, one sample scale, linear regression and linear autoregressive models, using residual empirical d.f.'s. In particular, we shall discuss ADF tests based on the Khmaladze transform of the underlying residual empirical process.

Let $\varepsilon_1, \cdots, \varepsilon_n$ be i.i.d. with d.f. F and let F_0 be a known continuous d.f. on $\mathbb{R}$. We are interested in testing the hypothesis $\tilde{H}_0 : F = F_0$ versus the alternatives $\tilde{H}_1 : F \neq F_0$, based on the observations $\varepsilon_1, \cdots, \varepsilon_n$. Let

$$F_n(y) := n^{-1}\sum_{i=1}^{n} I(\varepsilon_i \le y), \quad D_n(y) := F_n(y) - F_0(y), \quad y \in \mathbb{R}; \quad \mathcal{D}_n := \|D_n\|_\infty. \tag{9.2.1}$$

The Kolmogorov test rejects $\tilde{H}_0$ at level α whenever $\mathcal{D}_n \ge d_{n,\alpha}$, where $d_{n,\alpha}$ is determined from the null distribution of $\mathcal{D}_n$. Let $U_1 := F_0(\varepsilon_1), \cdots, U_n := F_0(\varepsilon_n)$ and

$$D_{n0}(u) := \frac{1}{n}\sum_{i=1}^{n}\big(I(U_i \le u) - u\big), \quad 0 \le u \le 1.$$

Because F_0 is continuous, $U_1, \cdots, U_n$ are i.i.d. Unif$(0,1)$ and $\mathcal{D}_n =_D \|D_{n0}\|_\infty$, under $\tilde{H}_0$. From this one readily sees that the null distribution of $\mathcal{D}_n$ does not depend on F_0, for all sample sizes $n \ge 1$. Thus this test is distribution free for all continuous F_0 and for all $n \ge 1$. The null distribution of $\mathcal{D}_n$ has been tabulated in Owen (1962). It can be generated using one of the programming languages, like R. Moreover, by (2.12.2), under $\tilde{H}_0$,

$$n^{1/2}\mathcal{D}_n \to_D \|\mathbb{B}^0\|_\infty, \tag{9.2.2}$$

where $\mathbb{B}^0$ is the Brownian bridge on $[0,1]$. The distribution of this limiting r.v. is also tabulated in Owen (1962). Alternatively, one may use formula (2.12.3) to compute desired asymptotic critical values of the test based on $n^{1/2}\mathcal{D}_n$.

The Kolmogorov test loses the property of being distribution free even for large samples as soon as there is a nuisance parameter present in the problem. To illustrate this, consider the problem of testing the hypothesis that the random sample is from $\mathcal{N}(\mu, 1)$, for some real number μ. More precisely, let $Y_1, \cdots, Y_n$ be a random sample from F, Φ denote the standard normal d.f. and φ its density and consider the problem of testing

$$H_0 : F(y) = \Phi(y - \mu), \quad \forall\, y \in \mathbb{R} \text{ and for some } \mu \in \mathbb{R},$$

against the alternative that H_0 is not true. The analog of the $\mathcal{D}_n$ test is the one based on the residuals $Y_1 - \bar{Y}_n, \cdots, Y_n - \bar{Y}_n$, where $\bar{Y}_n := n^{-1}\sum_{i=1}^n Y_i$. More precisely, define

$$\bar{F}_n(y) := \frac{1}{n}\sum_{i=1}^{n} I(Y_i - \bar{Y}_n \le y), \qquad \bar{D}_n(y) := \bar{F}_n(y) - \Phi(y), \quad y \in \mathbb{R}.$$

Intuitively, under H_0, the d.f. of $Y_i - \bar{Y}_n$ ought to be close to Φ. Thus, the analog of the Kolmogorov test would reject H_0 whenever $\bar{\mathcal{D}}_n := \|\bar{D}_n\|_\infty$ is large, subject to the level requirement. But the null distribution of this statistic is generally not known, even for large sample sizes. To see this, assume H_0 to be true. Let $\varepsilon_i := Y_i - \mu$ and $\bar{\varepsilon}_n := \bar{Y}_n - \mu$. Then,

$$\bar{F}_n(y) = \frac{1}{n}\sum_{i=1}^{n} I(\varepsilon_i \leq y + \bar{\varepsilon}_n) = F_n(y + \bar{\varepsilon}_n), \quad \bar{D}_n(y) = F_n(y + \bar{\varepsilon}_n) - \Phi(y).$$

By a change of variable,

$$\bar{\mathcal{D}}_n = \sup_{y \in \mathbb{R}} \left|F_n(y + \bar{\varepsilon}_n) - \Phi(y)\right| = \sup_{x \in \mathbb{R}} \left|F_n(x) - \Phi(x - \bar{\varepsilon}_n)\right|.$$

Under H_0, $n^{1/2}\bar{\varepsilon}_n \sim_D \mathcal{N}(0,1)$, for all $n \geq 1$, and by Theorem 2.10.6, $\bar{\varepsilon}_n = o_p(1)$. Hence, a Taylor expansion, using the uniform continuity of φ and Theorem 2.10.5, gives

$$\left|n^{1/2}\left[\Phi(x - \bar{\varepsilon}_n) - \Phi(x)\right] + n^{1/2}\bar{\varepsilon}_n\varphi(x)\right| \leq n^{1/2}|\bar{\varepsilon}_n| \sup_{|y-x| \leq |\bar{\varepsilon}_n|} \left|\varphi(y) - \varphi(x)\right| \to_p 0.$$

From this we readily obtain the uniform expansion

$$\begin{aligned} n^{1/2}\{F_n(x) - \Phi(x - \bar{\varepsilon}_n)\} &= n^{1/2}D_n(x) - n^{1/2}\{\Phi(x - \bar{\varepsilon}_n) - \Phi(x)\} \\ &= n^{1/2}D_n(x) + n^{1/2}\bar{\varepsilon}_n\varphi(x) + u_p(1) \\ &= Z_n(x) + u_p(1), \end{aligned} \tag{9.2.3}$$

where

$$Z_n(x) := n^{1/2}\left[D_n(x) + \bar{\varepsilon}_n\varphi(x)\right] = \frac{1}{\sqrt{n}}\sum_{i=1}^{n}\left[I(\varepsilon_i \leq x) - \Phi(x) + \varepsilon_i\varphi(x)\right], \quad x \in \mathbb{R}.$$

Note that $E\big(Z_n(x)\big) = 0$, for all $x \in \mathbb{R}$, and

$$\mathcal{C}(x,y) := \mathrm{Cov}\big(Z_n(x), Z_n(y)\big) = \Phi(x)\big(1 - \Phi(y)\big) - \varphi(x)\varphi(y), \quad x \leq y.$$

Moreover, $\|Z_n\|_\infty \to_D \|Z\|_\infty$, where $\{Z(x), x \in \mathbb{R}\}$ is a centered continuous Gaussian process with covariance function $\mathcal{C}$. The distribution of the limiting r.v. $\|Z\|_\infty$ is known in principle and depends on Φ. Even then, no one to date has computed this distribution explicitly, which would help in determining the large sample critical values of the test. In practice, one uses either the Monte Carlo method or a bootstrap approximation to implement this test.

The uniform expansion (9.2.3) is equivalent to

$$\bar{F}_n(x) = F_n(x) + \bar{\varepsilon}_n\varphi(x) + u_p(n^{-1/2}).$$

This type of expansion is true more generally. Suppose F_0 is a known d.f. having a uniformly continuous density f_0, the r.v.'s $\varepsilon_1, \cdots, \varepsilon_n$ are i.i.d. F_0 and let $X_1 = \theta + \varepsilon_1, \cdots, X_n = \theta + \varepsilon_n$, for some $\theta \in \mathbb{R}$. Let $\hat{\theta}_n$ be an estimator of θ based on the observations $X_1, \cdots, X_n$, $\Delta_n := n^{1/2}(\hat{\theta}_n - \theta)$ and define

$$\tilde{F}_n(x) := \frac{1}{n}\sum_{i=1}^{n} I(X_i - \hat{\theta}_n \leq x) = F_n(x + n^{-1/2}\Delta_n), \quad \tilde{\mathcal{D}}_n(x) := \tilde{F}_n(x) - F_0(x), \quad x \in \mathbb{R}.$$

Arguing as in the above normal distribution example, we see that if $\hat{\theta}_n$ is a $\sqrt{n}$-consistent estimator of θ, i.e., $|\Delta_n| = O_p(1)$, then the residual empirical process $\tilde{F}_n(x)$ is asymptotically uniformly linear in Δ_n, i.e., we have the uniform approximation

$$\tilde{F}_n(x) = F_n(x) + n^{-1/2}\Delta_n f_0(x) + u_p(n^{-1/2}). \tag{9.2.4}$$

Suppose, further, that $\Delta_n = n^{-1/2}\sum_{i=1}^n \psi(\varepsilon_i) + o_p(1)$, for some real-valued function ψ with $E(\psi(\varepsilon)) = 0$ and $E(\psi^2(\varepsilon)) < \infty$. Then, $\|\tilde{\mathcal{D}}_n\|_\infty \to_D \|\tilde{Z}\|_\infty$, where $\tilde{Z}$ is a centered continuous Gaussian process on $\mathbb{R}$ with covariance function

$$\mathrm{Cov}(\tilde{Z}(x), \tilde{Z}(y)) = F_0(x)(1 - F_0(y)) - \gamma(x)f_0(y) - \gamma(y)f_0(x) + E(\psi^2(\varepsilon))f_0(x)f_0(y),$$

for $x \le y$, where $\gamma(x) := E(\psi(\varepsilon)I(\varepsilon \le x))$, $x \in \mathbb{R}$. Again the distribution of the limiting r.v. $\|\tilde{Z}\|_\infty$ is not available.

Alternately, one can use the following transformation of Khmaladze (1981) to obtain ADF tests. We shall describe this in a bit more general testing problem where one fits a given distribution up to unknown location and scale parameters.

Consider the one sample location-scale model where one observes a random sample $Y_1, \cdots, Y_n$ from a univariate d.f. F on the real line $\mathbb{R}$, and the problem of interest is to test

$$H_{LS}: F(y) = F_0\Big(\frac{y-\mu}{\sigma}\Big), \quad \forall y \in \mathbb{R}, \text{ for some } \mu \in \mathbb{R}, \sigma > 0,$$

against the alternative that H_{LS} is not true. Let μ_n and σ_n denote estimators of μ and σ, respectively, under H_{LS}. Now, let $\hat{F}_n$ denote the empirical d.f. based on the residuals $(Y_i - \mu_n)/\sigma_n$, $1 \le i \le n$, and set $\hat{D}_n(y) := n^{1/2}(\hat{F}_n(y) - F_0(y))$, $y \in \mathbb{R}$.

As mentioned above, the tests based on $\hat{D}_n$ are not ADF in general. A proposal made by Rao (1972) and Durbin (1973, 1976) is to use maximum likelihood estimators μ_n, σ_n under H_{LS} of μ, σ based only on one half of the sample in constructing $\hat{F}_n$ and the entire sample to implement the test. These tests are ADF for all those F_0 that have finite and positive definite Fisher information matrix for the location and scale parameters, see Definition 2.11.3. As is obvious, different half-samples used in constructing these estimators will possibly lead to different inferences. Therefore, it is desirable to have an alternative way of obtaining ADF tests.

For the rest of this section we shall assume the following.

$$\text{The d.f. } F_0 \text{ has a density } f_0 \text{ that is positive almost everywhere.} \tag{9.2.5}$$

This assumption implies that F_0 is strictly increasing and its left inverse coincides with its inverse. Thus $F_0^{-1}(F_0(x)) = x$, for all $x \in \mathbb{R}$.

9.2.1 Tests based on Khmaladze Transform of $\hat{D}_n$

Khmaladze (1981) proposed a transformation of $\hat{D}_n$ whose asymptotic null distribution is the same as that of $\mathbb{B} \circ F_0$, where $\mathbb{B}$ is the standard Brownian motion on $[0, \infty)$, so that the tests based on this transformation are ADF for testing H_{LS}. Recall Definition 2.11.3 of finite Fisher information for location and scale. To describe this transformation, we shall assume the following.

$$\text{The density } f_0 \text{ has finite Fisher information for location and scale.} \tag{9.2.6}$$

$$n^{1/2}|\mu_n - \mu| + n^{1/2}|\sigma_n - \sigma| = O_p(1), \quad \text{under } H_{LS}. \tag{9.2.7}$$

We shall discuss the transformation of $\hat{D}_n(F_0^{-1})$. Let U be a $\mathrm{Unif}(0,1)$ r.v. and define

$$p_0(t) := f_0(F_0^{-1}(t)), \qquad \dot{p}_0(t) := \dot{f}_0(F_0^{-1}(t))/f_0(F_0^{-1}(t)),$$

$$q_0(t) := F_0^{-1}(t)p_0(t), \qquad \dot{q}_0(t) := 1 + F_0^{-1}(t)\dot{p}_0(t), \quad g(t) := (\dot{p}_0(t), \dot{q}_0(t))',$$

$$\mathcal{I}(t) := E(g(U)g(U)'I(U \ge t)), \qquad \varepsilon_i := \frac{Y_i - \mu}{\sigma}, \quad 1 \le i \le n, \ 0 \le t \le 1.$$

Then, under H_{LS}, $-g(F_0(\varepsilon_1))/\sigma$ is the score function for the observation Y_1 and $\mathcal{I}(0)/\sigma^2$ the Fisher information matrix. Assume that

$$\mathcal{I}(0) = \begin{pmatrix} E(\dot{p}_0^2(U)) & E(\dot{p}_0(U)\dot{q}_0(U)) \\ E(\dot{p}_0(U)\dot{q}_0(U)) & E(\dot{q}_0^2(U)) \end{pmatrix} \quad \text{is positive definite.}$$

Let $\mathcal{I}^{-1}(0)$ denote the inverse matrix $(\mathcal{I}(0))^{-1}$.

Suppose, under H_{LS}, the estimators μ_n, σ_n satisfy

$$\frac{\sqrt{n}}{\sigma}\begin{pmatrix} \mu_n - \mu \\ \sigma_n - \sigma \end{pmatrix} = -\mathcal{I}^{-1}(0)\frac{1}{\sqrt{n}}\sum_{i=1}^{n}\begin{pmatrix} \dot{p}_0(F_0(\varepsilon_i)) \\ \dot{q}_0(F_0(\varepsilon_i)) \end{pmatrix} + o_p(1). \tag{9.2.8}$$

Then, one can verify, as we did in the normal mean example above, that $\hat{F}_n$ admits an approximation like (9.2.4), under H_{LS}. Using this expansion one verifies that $\sup_{0\le t\le 1}|\hat{D}_n(F_0^{-1}(t))| \to_D \|\mathcal{G}\|_\infty$, where $\mathcal{G}$ is a centered continuous Gaussian process on $[0,1]$ with covariance function

$$\rho(s,t) := s\wedge t - st - (p_0(s), q_0(s))\mathcal{I}^{-1}(0)\begin{pmatrix} p_0(t) \\ q_0(t) \end{pmatrix}, \quad 0 \le s,t \le 1.$$

Some simplification of this covariance occurs if F_0 is symmetric about zero. In this case, the matrix $\mathcal{I}(0)$ is diagonal and the above covariance simplifies to

$$\rho_1(s,t) = s\wedge t - st - \frac{p_0(s)p_0(t)}{\int_0^1 \dot{p}_0^2(u)du} - \frac{q_0(s)q_0(t)}{\int_0^1 \dot{q}_0^2(u)du}, \quad 0 \le s,t \le 1.$$

The distribution of a Gaussian process with covariance function ρ or ρ_1 depends on F_0 in a complicated way, and the distribution of the supremum of the absolute value of such a Gaussian process is not available. But the following Khmaladze transformation converts such a Gaussian process to a time-transformed Brownian motion, whose distribution is known. For $0 \le t \le 1$, let $\ell_\sigma(t) := (1, -g(t)'/\sigma)' = (1, -\dot{p}_0(t)/\sigma, -\dot{q}_0(t)/\sigma)'$ and

$$A_\sigma(t) := E\Big(\ell_\sigma(U)\ell_\sigma(U)'I(U \ge t)\Big) = \begin{pmatrix} 1-t & p_0(t)/\sigma \quad q_0(t)/\sigma \\ p_0(t)/\sigma & \\ q_0(t)/\sigma & \mathcal{I}(t)/\sigma^2 \end{pmatrix}.$$

Clearly $A_\sigma(t)$ is non-negative definite for all t. We assume that

$$A_\sigma(t) \text{ is positive definite for all } 0 \le t < 1,\ \sigma > 0. \tag{9.2.9}$$

Let $A_\sigma^{-1}(t)$ stand for $(A_\sigma(t))^{-1}$ and set

$$\mathcal{T}_t(u) := I(u \le t) - \int_0^{t\wedge u} \ell_\sigma(s)'A_\sigma^{-1}(s)\,ds\,\ell_\sigma(u), \quad u \in [0,1]. \tag{9.2.10}$$

The Khmaladze transform of interest based on $\varepsilon_1, \cdots, \varepsilon_n$ is

$$\mathcal{U}_n(t,\theta) := n^{-1/2}\sum_{i=1}^{n}\mathcal{T}_t(F_0(\varepsilon_i)), \quad 0 \le t \le 1,\ \theta := (\mu,\sigma). \tag{9.2.11}$$

Since $E(\ell_\sigma(U)I(U \ge s))$ equals the first column of $A_\sigma(s)$, we obtain

$$\begin{aligned} E(\mathcal{T}_t(U)) &= t - \int_0^t \ell_\sigma(s)'A_\sigma^{-1}(s)E\big(\ell_\sigma(U)I(U \ge s)\big)\,ds \\ &= t - \int_0^t 1\,ds = 0, \qquad \forall 0 \le t \le 1. \end{aligned}$$

Hence, $E(\mathcal{U}_n(t,\theta)) = 0$, $0 \le t \le 1$. Moreover, direct calculations and Theorem 2.3.2 yield that $\text{Cov}(\mathcal{U}_n(s,\theta), \mathcal{U}_n(t,\theta)) = s \wedge t \equiv \text{Cov}(\mathbb{B}(s), \mathbb{B}(t))$. Hence, $\{\mathcal{U}_n(t,\theta),\ 0 \le t \le 1\}$ is a zero-mean process with covariance function equal to that of the standard Brownian motion $\{\mathbb{B}(t) : 0 \le t \le 1\}$. But the former process depends on the unknown θ, which renders it useless from the testing point of view. In order to implement tests based on this process, we need to replace the parameters μ, σ by their respective estimates μ_n, σ_n. Let $\hat{U}_i := F_0((Y_i - \mu_n)/\sigma_n)$. Then, the transform of $\hat{D}_n(F_0^{-1})$ and the process of interest is $\hat{\mathcal{U}}_n(t) := \mathcal{U}_n(t, (\mu_n, \sigma_n))$ given by

$$\hat{\mathcal{U}}_n(t) := n^{-1/2} \sum_{i=1}^{n} \Big[I(\hat{U}_i \le t) - \int_0^{t \wedge \hat{U}_i} \ell_{\sigma_n}(v)' A_{\sigma_n}^{-1}(v)\, dv\, \ell_{\sigma_n}(\hat{U}_i) \Big]. \tag{9.2.12}$$

Khmaladze and Koul (2004) have shown that under the above assumptions, $\|\hat{\mathcal{U}}_n\|_\infty \to_D \|\mathbb{B}\|_\infty$, under H_{LS}. Hence, the test that rejects H_{LS} whenever $\|\hat{\mathcal{U}}_n\|_\infty > b_\alpha$ has asymptotic size α, where b_α is such that

$$P(\|\mathbb{B}\|_\infty > b_\alpha) \le \alpha. \tag{9.2.13}$$

The following formula given in Resnick (1999) can be used to obtain b_α's.

$$P(\|\mathbb{B}\|_\infty < b) = P(|\mathbb{B}(1)| < b) + 2\sum_{i=1}^{\infty} (-1)^i P((2i-1)b < \mathbb{B}(1) < (2i+1)b).$$

This formula gives $b_\alpha = 2.24241, 2.49771, 2.80705$ for $\alpha = 0.05, 0.025, 0.01$, respectively.

We remark here that the condition (9.2.8) is used only for motivating the above transformation. In fact, one only needs the estimators μ_n, σ_n to satisfy (9.2.7) in order for the above result to hold.

We shall next describe analogs of the above transformation when fitting a distribution up to an unknown location parameter only or unknown scale parameter only, i.e., for testing $H_L : F(y) = F_0(y - \mu)$, for all $y \in \mathbb{R}$ and some $\mu \in \mathbb{R}$ and testing $H_S : F(y) = F_0(y/\sigma)$, for all $y \in \mathbb{R}$ and for some $\sigma > 0$. In the former, the scale parameter is known and taken to be unity, while in the latter the location parameter is known and taken to be zero.

Consider H_L first. Recall the definition of finite Fisher information for location from Definition 2.11.1. Instead of (9.2.6) and (9.2.7), we now require the following conditions.

The density f_0 has finite Fisher information for location. (9.2.14)

The estimator $\tilde{\mu}_n$ of μ satisfies $n^{1/2}|\tilde{\mu}_n - \mu| = O_p(1)$, under H_L.

Let $\widetilde{U}_i := F_0(Y_i - \tilde{\mu}_n)$ and $\widetilde{D}_n(t) := n^{-1/2}\sum_{i=1}^n \{I(\widetilde{U}_i \le t) - t\}$. Tests of H_L are to be based on $\widetilde{D}_n$. For $0 \le t \le 1$, set

$$\tilde{\ell}(t) := \begin{pmatrix} 1 \\ -\dot{p}_0(t) \end{pmatrix}, \quad \tilde{A}(t) := E\Big(\tilde{\ell}(U)\tilde{\ell}(U) I(U \ge t)\Big) = \begin{pmatrix} 1-t & p_0(t) \\ p_0(t) & \int_t^1 (\dot{p}_0(s))^2 ds \end{pmatrix}.$$

Note that in this case there are no unknown parameters in $\tilde{\ell}$ and $\tilde{A}$.

Assume $\tilde{A}^{-1}(t)$ exists for all $0 \le t < 1$. Let

$$\Gamma(t) := \int_0^t \tilde{\ell}(s)' \tilde{A}^{-1}(s) ds = \int_{-\infty}^{F_0^{-1}(t)} \tilde{\ell}(F_0(y))' \tilde{A}^{-1}(F_0(y)) f_0(y)\, dy,$$

where the last expression is obtained by changing the variable $s = F_0(y)$ and where we used (9.2.5). Then, the transformation that yields the ADF tests for H_L is

$$\widetilde{\mathcal{U}}_n(t) = \frac{1}{\sqrt{n}} \sum_{i=1}^{n} \Big\{ I(\widetilde{U}_i \le t) - \int_0^{t \wedge \widetilde{U}_i} \tilde{\ell}(s)' \tilde{A}^{-1}(s)\, ds\, \tilde{\ell}(\widetilde{U}_i) \Big\} \qquad (9.2.15)$$
$$= \frac{1}{\sqrt{n}} \sum_{i=1}^{n} \Big\{ I(\widetilde{U}_i \le t) - \Gamma(t \wedge \widetilde{U}_i) \tilde{\ell}(\widetilde{U}_i) \Big\}.$$

Again, the ADF test that rejects H_L whenever $\|\widetilde{\mathcal{U}}_n(t)\|_\infty > b_\alpha$ has asymptotic size α.

Example 9.2.1. Fitting a normal distribution. Let Φ denote the d.f. of a $\mathcal{N}(0,1)$ r.v. Consider the problem of testing for H_L when $F_0 = \Phi$. In this case we shall give an explicit formula for computing the above transformation. Let φ denote the density of Φ and define $h(y) := \varphi(y)/[1 - \Phi(y)]$. Here $\dot{p}_0(t) = -\Phi^{-1}(t)$, $\tilde{\ell}(t) = \big(1, -\dot{p}_0(t)\big)' = \big(1, \Phi^{-1}(t)\big)'$. Note that $\tilde{\ell}(\Phi(y)) = (1, y)'$ and

$$\tilde{A}(\Phi(y)) = \begin{pmatrix} 1 - \Phi(y) & \varphi(y) \\ \varphi(y) & y\varphi(y) + 1 - \Phi(y) \end{pmatrix} = (1 - \Phi(y)) \begin{pmatrix} 1 & h(y) \\ h(y) & yh(y) + 1 \end{pmatrix},$$
$$\tilde{A}^{-1}(\Phi(y)) = \frac{1}{[1 - \Phi(y)]\,[yh(y) + 1 - h^2(y)]} \begin{pmatrix} 1 + yh(y) & -h(y) \\ -h(y) & 1 \end{pmatrix}.$$

Then,

$$\Gamma(\Phi(x)) = \int_{-\infty}^{x} (1, y) \tilde{A}^{-1}(\Phi(y)) \varphi(y)\, dy$$
$$= \int_{-\infty}^{x} \frac{1}{[yh(y) + 1 - h^2(y)]} \Big(1, y - h(y)\Big) h(y)\, dy.$$

Let $\tilde{\varepsilon}_i := Y_i - \tilde{\mu}_n, 1 \le i \le n$. After changing the variable t to $\Phi(y)$, the process of (9.2.15) here becomes $T_n(y) := \widetilde{\mathcal{U}}_n(\Phi(y))$ given by

$$T_n(y) = \frac{1}{\sqrt{n}} \sum_{i=1}^{n} \Big[I(\tilde{\varepsilon}_i \le y) - \int_0^{y \wedge \tilde{\varepsilon}_i} \frac{1 + \tilde{\varepsilon}_i (x - h(x))}{[xh(x) + 1 - h^2(x)]} h(x) dx \Big], \quad y \in \mathbb{R}. \qquad (9.2.16)$$

This representation is useful in computing $\|T_n\|_\infty$. The test that rejects H_L with $F_0 = \Phi$ whenever $\|T_n\|_\infty > b_\alpha$ has asymptotic size α.

Next, consider the hypothesis $H_S : F(y) = F_0(y/\sigma)$, for all $y \in \mathbb{R}$ and some $\sigma > 0$. Recall Definition 2.11.2 of finite Fisher information for scale. Here, we make the following assumptions.

The density f_0 has finite Fisher information for scale. (9.2.17)

The estimator $\bar{\sigma}_n$ of σ satisfies $n^{1/2}|\sigma_n - \sigma| = O_p(1)$, under H_S. (9.2.18)

Set $\bar{U}_i := F_0(Y_i/\bar{\sigma}_n)$. For $0 \le t \le 1$, let $\bar{\ell}_\sigma(t) := \big(1, -\dot{q}_0(t)/\sigma\big)'$ and define

$$\bar{A}_\sigma(t) := E\Big(\bar{\ell}_\sigma(U) \bar{\ell}_\sigma(U) I(U \ge t)\Big) = \begin{pmatrix} 1 - t & -q_0(t)/\sigma \\ -q_0(t)/\sigma & \int_t^1 \big(\dot{q}_0(s)/\sigma\big)^2 ds \end{pmatrix}.$$

Assume $\bar{A}_\sigma^{-1}(t)$ exists for all $0 \le t < 1$ and define the transformation

$$\bar{\mathcal{U}}_n(t) = \frac{1}{\sqrt{n}} \sum_{i=1}^{n} \Big\{ I(\bar{U}_i \le t) - \int_0^{t \wedge \bar{U}_i} \bar{\ell}_{\bar{\sigma}_n}(s)' \bar{A}_{\bar{\sigma}_n}^{-1}(s) ds\, \bar{\ell}_{\bar{\sigma}_n}(\bar{U}_i) \Big\}. \qquad (9.2.19)$$

The test that rejects H_S whenever $\|\bar{\mathcal{U}}_n\|_\infty > b_\alpha$ has asymptotic size α.

There are situations when $A_\sigma^{-1}(t)$ does not exist as happens when in H_L, F_0 is standard Laplace d.f. or F_0 is the d.f. of Student's t_k distribution with the degrees of freedom $k \geq 1$. Nikabadze (1987) and Tsigroshvili (1998) have shown that in the case of fitting a general parametric d.f., the Khmaladze transformation is well defined even when the inverse of the analog of $A_\sigma(t)$ does not exist. Khmaladze and Koul (2009) and Koul and Zhu (2015) adapted the general arguments of these authors to show that even when $A_\sigma^{-1}(t)$ does not exist for some $0 \leq t < 1$, the transformations defined at (9.2.10) or (9.2.15) or (9.2.19) are still well defined in the above location-scale, location only or scale only models. Roberts, Haywood and Swordson (2023) derive an explicit expression for the transformation $\mathcal{T}_t$ of (9.2.10) when F_0 is standard Laplace.

Regression models. Here we shall describe analogous goodness-of-fit tests for fitting an error d.f. F in linear and nonparametric regression models.

Let $\xi := (\xi_1, \cdots, \xi_p)'$ be a p-dimensional random vector of covariates and $X := (1, \xi')'$. Let $\beta_0, \beta_1, \cdots, \beta_p$ denote the components of a vector $\beta \in \mathbb{R}^{p+1}$, i.e., $\beta = (\beta_0, \beta_1, \cdots, \beta_p)'$. In the parametric linear regression model of interest here, the response variable Y is related to X by the relation

$$Y = \beta_0 + \beta_1\xi_1 + \cdots + \beta_p\xi_p + \sigma\varepsilon = \beta' X + \sigma\varepsilon, \quad \text{for some } \beta \in \mathbb{R}^{p+1}, \sigma > 0. \tag{9.2.20}$$

Here, the error r.v. ε is assumed to be independent of ξ. We shall also assume that the model is identified. Conditions for this are that the error d.f. F has mean zero or is symmetric about zero.

Let (ξ_i, Y_i), $i = 1, \cdots, n$ be a random sample from the model (9.2.20) and set $X_i := (1, \xi_i')'$, $i = 1, \cdots, n$. The problem of interest is to test $H_0 : F = F_0$ based on the given random sample. Let β_n, σ_n be estimators of β, σ satisfying $n^{1/2}\|\beta_n - \beta\| = O_p(1)$ and $n^{1/2}|\sigma_n - \sigma| = O_p(1)$, under H_0. Assume that $E(\|\xi\|^2)$ is finite and $\Sigma := E(XX')$ is positive definite. By Theorem 2.10.7, $S_n := n^{-1}\sum_{i=1}^n X_i X_i' \to_{a.s.} \Sigma$. Define the residual empirical process

$$\mathcal{Z}_n(y) := n^{-1/2}\sum_{i=1}^n \big[I(Y_i - \beta_n' X_i \leq y\sigma_n) - F_0(y)\big], \quad y \in \mathbb{R}.$$

The process $\mathcal{Z}_n$ provides an analog of the one sample empirical process $n^{1/2}\hat{D}_n$ suitable in linear regression. One can verify, as done in Koul (2002), that under (9.2.6), H_0, and (9.2.20),

$$\mathcal{Z}_n(y) = n^{1/2}D_n(y) + \sigma^{-1}n^{1/2}\Big[\bar{X}'(\beta_n - \beta)f_0(y) + (\sigma_n - \sigma)yf_0(y)\Big] + u_p(1). \tag{9.2.21}$$

This result is analogous to (9.2.4).

From this approximation one sees that if $E(\xi) = 0$, then $\bar{X}_n = (1, \bar{\xi}_n)' \to_p (1, 0, \cdots, 0)'$ and only the intercept parameter among the regression parameters appears in the above approximation of $\mathcal{Z}_n$. In this case, the transform $\widetilde{U}_n$ of (9.2.15) with $\widetilde{U}_i = F_0\big((Y_i - X_i'\beta_n)/\sigma_n\big), 1 \leq i \leq n$, can be used to construct ADF tests for the above H_0.

In general, the Khmaladze transform of the process $\mathcal{Z}_n$ is

$$\mathcal{K}_n(t) := n^{-1/2}\sum_{i=1}^n \Big[I(\zeta_{ni} \leq t) - \int_0^{t\wedge\zeta_{ni}} \ell_{\sigma_n}(s)A_{\sigma_n}^{-1}(s)\,ds\,\ell_{\sigma_n}(\zeta_i)\Big], \tag{9.2.22}$$

where $\zeta_{ni} := F_0((Y_i - \beta_n' X_i)/\sigma_n)$, ℓ_{σ_n} and A_{σ_n} are as in the above location-scale testing problem. This transform is similar to $\hat{\mathcal{U}}_n$ of (9.2.12) used for fitting an error d.f. up to unknown location and scale parameters. The only difference is in the residuals and the estimates of the nuisance parameters being used in these two statistics. One can show, under (9.2.6) and H_0, $\|\mathcal{K}_n\|_\infty \to_D \|\mathbb{B}\|_\infty$. Hence, the test that rejects H_0 whenever $\|\mathcal{K}_n\|_\infty > b_\alpha$ is ADF, where b_α is as in (9.2.13). The expansion (9.2.21) is crucial for proving this result.

In the linear regression model (9.2.20) with $\beta_0 = 0$ and $\sigma = 1$, the process to use for constructing goodness of fit tests is

$$\mathcal{L}_n(y) := n^{-1/2} \sum_{i=1}^{n} \big[I(Y_i - \hat{\beta}_*'\xi_i \le y) - F_0(y)\big], \qquad y \in \mathbb{R}, \tag{9.2.23}$$

where $\hat{\beta}_*$ is an estimator of the vector $\beta_* := (\beta_1, \cdots, \beta_p)'$ satisfying $n^{1/2}\|\hat{\beta}_* - \beta_*\| = O_p(1)$, under H_0. Let 0_p denote the vector of p zeros. Akin to (9.2.21), with $\bar{\xi} := n^{-1}\sum_{i=1}^n \xi_i$, we have the expansion

$$\mathcal{L}_n(y) = n^{1/2} D_n(y) + \bar{\xi}' n^{1/2}(\hat{\beta}_* - \beta_*) f_0(y) + u_p(1).$$

From this it follows that if $E(\xi) = 0_p$, then $\bar{\xi} \to_p 0_p$ and $n^{1/2}\|\mathcal{L}_n\|_\infty \to_D \|\mathbb{B}^0\|_\infty$, under H_0. Hence, there is no need for any transformation, and the test that rejects H_0 whenever

$$\|\mathcal{L}_n\|_\infty > b_\alpha^0 \tag{9.2.24}$$

will be an ADF test of the asymptotic level α, where b_α^0 is determined from the distribution of $\|\mathbb{B}^0\|_\infty$ tabulated in Owen (1962).

Linear regression model with non-random design. Let $x_{ni}, 1 \le i \le n$ be p-dimensional non-random vectors of real numbers and $\mathcal{X}_n$ denote the $n \times p$ matrix whose ith row is x_{ni}'. Let Y_{ni} denote the responses satisfying the relations

$$Y_{ni} = x_{ni}'\beta + \sigma\varepsilon_{ni}, \quad 1 \le i \le n, \tag{9.2.25}$$

for some $\sigma > 0$, $\beta \in \mathbb{R}^p$, where the error variables $\varepsilon_{n1}, \cdots, \varepsilon_{nn}$ are i.i.d. F. As before, let F_0 be a known d.f. and β_n, σ_n be estimators of β, σ, respectively. Assume the following.

(a) $\mathcal{X}_n'\mathcal{X}_n$ is non-singular and $\max_{1\le i\le n} \big\|(\mathcal{X}_n'\mathcal{X}_n)^{-1/2} x_{ni}\big\| \to 0$.

(b) $\big\|(\mathcal{X}_n'\mathcal{X}_n)^{1/2}(\beta_n - \beta)\big\| = O_p(1), \quad n^{1/2}|\sigma_n - \sigma| = O_p(1)$.

Let $\widetilde{\mathcal{K}}_n(t)$ be defined as $\mathcal{K}_n(t)$ of (9.2.22) but with ζ_{ni} replaced by $\xi_{ni} = F_0\big((Y_{ni} - x_{ni}'\beta_n)/\sigma_n\big)$, $1 \le i \le n$. Then tests of the hypothesis H_0 based on $\widetilde{\mathcal{K}}_n$ process have the same asymptotic null distribution as those based on the process $\mathcal{K}_n$ in the random design case.

Nonparametric regression model. A more general problem is that of testing a goodness-of-fit hypothesis of an error d.f. F in the nonparametric regression model

$$Y = \mu(X) + \varepsilon, \tag{9.2.26}$$

where the r.v.'s ε and X are independent, ε has mean zero and a finite positive variance, and μ is a real-valued (smooth) function defined on $\mathbb{R}$. Here ADF tests can be based on the analog of $\widetilde{\mathcal{U}}_n$ of (9.2.15), where $\widetilde{U}_i$ is replaced by nonparametric residuals. More precisely, let $\widetilde{\mu}_n$ denote a local linear smoother of Section 8.4 for a kernel K and a bandwidth $h = h_n$ and define

$$V_i := F_0\big(Y_i - \widetilde{\mu}_n(X_i)\big), \quad 1 \le i \le n,$$
$$\mathcal{V}_n(t) := n^{-\frac{1}{2}} \sum_{i=1}^{n} \Big\{ I(V_i \le t) - \int_0^{t\wedge V_i} \tilde{\ell}(s)'\tilde{A}^{-1}(s)\, ds\, \tilde{\ell}(V_i) \Big\}.$$

Because, conditionally on X, the regression model (9.2.26) is a one sample location model, the same transformation as the one used in the one sample location model to construct ADF tests yields the same for the nonparametric regression model (9.2.26) for fitting an error d.f.

An interesting fact to know here is that under some conditions an analog of (9.2.4) also holds in the current nonparametric regression setup. Suppose the following additional assumptions hold: $E(|\varepsilon|^3) < \infty$ and the error density f_0 satisfies (9.2.14); the variable X takes values in $[0,1]$ and has a density that is bounded and bounded away from zero on the interval $[0,1]$; and the regression function μ is twice continuously differentiable. Moreover, the bandwidth h_n is proportional to $n(\log(n))^{-1/4}$ and the kernel K is a three times continuously differentiable symmetric density with support $[-1,1]$. Set

$$\widetilde{F}_n(x) := n^{-1}\sum_{i=1}^{n} I(Y_i - \widetilde{\mu}_n(X_i) \le x), \qquad F_n(x) := n^{-1}\sum_{i=1}^{n} I(\varepsilon_i \le x).$$

Müller, Schick and Wefelmeyer (2007) proved that under the above conditions,

$$\widetilde{F}_n(x) = F_n(x) + \bar{\varepsilon}_n f_0(x) + u_p(n^{-1/2}).$$

A similar result with higher-dimensional X is derived in Müller, Schick and Wefelmeyer (2009). Thus, somewhat surprisingly, the first-order asymptotic behavior of the nonparametric residual empirical process $\widetilde{F}_n$ is the same as that of the one sample location model residual empirical process.

Akritas and Van Keilegom (2001) obtain similar results for nonparametric residual empirical process in a heteroscedastic regression model under additional assumptions and with different regression function estimators.

Khmaladze and Koul (2009) used the above type of uniform approximation of the nonparametric residual empirical process $\widetilde{F}_n$ to show that if f_0 satisfies (9.2.14) then, $\|\mathcal{V}_n\|_\infty \to_D \|\mathbb{B}\|_\infty$, so that the test $I(\|\mathcal{V}_n\|_\infty > b_\alpha)$ has asymptotic size α.

Now, let $\mathcal{W}_n := n^{1/2}\|\widetilde{F}_n - F_0\|_\infty$. Consider the test that rejects $H : F = F_0$ whenever $\mathcal{W}_n$ is large. One can determine the critical values of this test statistic by using the Monte Carlo method, which can be time-consuming because at every iteration one has to compute the nonparametric estimator $\widetilde{\mu}_n(x)$ of $\mu(x)$. Nevertheless, one may ask how does this test compare with the asymptotically distribution free test that rejects H whenever $\|\mathcal{V}_n\|_\infty > b_\alpha$. Khmaladze and Koul (2009) give an example where in a finite sample simulation study with $n = 200$, the empirical power of the latter test is uniformly significantly larger than that of the Monte Carlo $\mathcal{W}_n$ test at all three values of $\alpha = 0.025, 0.05, 0.10$ while at the same time, its empirical level is very close to the nominal level α, for all its three chosen values. All this is done when X is one-dimensional. At the writing of this book, the behavior of the above test when the dimension X is larger than unity is not well studied.

Linear autoregressive model. Consider the AR(p) model (2.12.6), where $\varepsilon_i, i \in \mathbb{Z}$ are i.i.d. F r.v.'s with $E(\varepsilon_0) = 0$ and $\sigma^2 := \mathrm{Var}(\varepsilon_0) = E(\varepsilon_0^2) < \infty$ and for each $i \in \mathbb{Z}$, ε_i is independent of $\boldsymbol{X}_{i-1} := (X_{i-1}, \cdots, X_{i-p})'$. We shall assume that the time series is stationary and ergodic. This assumption puts some restrictions on the domain of the parameter vector $\boldsymbol{\rho} := (\rho_1, \cdots, \rho_p)'$, c.f., see Brockwell and Davis (1991). For example, in the case of $p = 1$, $|\rho_1| < 1$ is needed for these conditions to hold.

Consider the problem of testing $\tilde{H}_0 : F = F_0$ against the alternatives that this hypothesis is not true. Let $\boldsymbol{\rho}_n$ be an estimator of $\boldsymbol{\rho}$ and define

$$\mathbb{G}_n(x) := n^{-1/2}\sum_{i=1}^{n}\{I(X_i - \boldsymbol{\rho}_n'\boldsymbol{X}_{i-1} \le x) - F_0(x)\}, \quad x \in \mathbb{R}.$$

If F_0 has a uniformly continuous and bounded density f_0 and $n^{1/2}\|\boldsymbol{\rho}_n - \boldsymbol{\rho}\| = O_p(1)$, then analogous to (9.2.21), we have the following result under $\tilde{H}_0$. With D_n as in (9.2.1) and $\bar{\boldsymbol{X}}_n := n^{-1}\sum_{i=1}^n \boldsymbol{X}_{i-1}$,

$$\mathbb{G}_n(x) = n^{1/2}D_n(x) + \bar{\boldsymbol{X}}_n' n^{1/2}(\boldsymbol{\rho}_n - \boldsymbol{\rho}) f_0(x) + u_p(1).$$

By the Ergodic Theorem, see Brockwell and Davis (1991), $\bar{\boldsymbol{X}}_n \to_p E(\boldsymbol{X}_0)$. But for the AR($p$) time series (2.12.6), $E(\boldsymbol{X}_0) = 0_p$. Hence, the second term in the above expansion vanishes and

$$\sup_{x\in\mathbb{R}} \left|\mathbb{G}_n(x) - n^{1/2}D_n(x)\right| \to_p 0. \tag{9.2.27}$$

In other words, in the above AR(p) model the asymptotic null distribution of the residual empirical process $\mathbb{G}_n$ is the same as that of the empirical process of the errors. Moreover, by (9.2.2), the test that rejects $\tilde{H}_0$ whenever $\sup_{x\in\mathbb{R}}\left|\mathbb{G}_n(x)\right| > b_\alpha^0$ is of asymptotic size α and ADF, where b_α^0 is as in (9.2.24). This is similar to what happens in the linear regression model with zero intercept when the predictor vector ξ has zero mean.

This scenario changes if the AR(p) model has non-zero mean, i.e., if for some $\mu \in \mathbb{R}$,

$$X_i = \mu + \boldsymbol{\rho}'\boldsymbol{X}_{i-1} + \varepsilon_i, \quad \forall\, i \in \mathbb{Z},$$

with $\varepsilon_i, i \in \mathbb{Z}$ as in (2.12.6). Then, with μ_n denoting a $n^{1/2}$ consistent estimator of μ, the tests based on the analog of $\widetilde{\mathcal{U}}_n$ of (9.2.15), where $\tilde{U}_i$ are replaced by $F_0\big(X_i - \mu_n - \boldsymbol{\rho}_n'\boldsymbol{X}_{i-1}\big)$, $i = 1, \cdots, n$, will be ADF here.

9.3 Bootstrap Approximation

In this section we shall discuss briefly some basic notion of the bootstrap approximation to the distributions of sequences of statistics and its application to some inference problems. In particular, the bootstrap approximation of the null distribution of some goodness-of-fit test statistics of the previous section will be described in some detail. Many results will be stated without proofs. Beginning with Efron (1979) and Singh (1981), numerous monographs have appeared on the topic, which offer a variety of details. They include Efron (1982), Hall (1992), Mammen (1992), Shao and Tu (1995) and Lahiri (2003).

Let $X_1, X_2, \cdots$ be a sequence of i.i.d. r.v.'s with d.f. F, n be a positive integer, $\boldsymbol{X}_n := (X_1, \cdots, X_n)$ and $T(\boldsymbol{X}_n, H)$ be a real-valued statistic for each H in a sufficiently large class of d.f.'s. Let $\hat{F}_n$ be the empirical d.f. of $\boldsymbol{X}_n$, i.e., $\hat{F}_n(x) := n^{-1}\sum_{i=1}^n I(X_i \le x)$, $x \in \mathbb{R}$. Now think of a population consisting of n elements $X_1, \cdots, X_n$. Let $X_1^*, \cdots, X_n^*$ denote a random sample obtained with replacement from this population, which is equivalent to saying that $X_1^*, \cdots, X_n^*$ are i.i.d. $\hat{F}_n$. Then, the conditional distribution of $T^* := T\big(\boldsymbol{X}_n^*, \hat{F}_n\big)$, given $\boldsymbol{X}_n$, is called the bootstrap distribution of $T(\boldsymbol{X}_n, F)$. It is used to approximate the distribution of $T(\boldsymbol{X}_n, F)$ or some of its characteristics like mean and variance.

In general it is hard to determine the bootstrap distribution in practice. Instead, one uses the Monte Carlo method to estimate it. Let N be a large positive integer. Take N random samples from $\hat{F}_n$ with replacement each of size n and compute T^* for each sample. Let T_j^* denote the value of T^* for the jth sample, $j = 1, \cdots, N$, and use these values to estimate the distribution of T^* or its characteristics of interest. Let P_F, E_F, Var_F denote the probability, expectation and variance, when X_i's are i.i.d. F, respectively.

The d.f. $G_n(t) := P_F\big(T(\boldsymbol{X}_n, F) \le t\big)$ is estimated by the empirical d.f. of $T_j^*, 1 \le j \le N$, i.e., by $\widetilde{G}_N(t) := N^{-1}\sum_{j=1}^N I(T_j^* \le t), t \in \mathbb{R}$. Similarly, $E_F\big(T(\boldsymbol{X}_n, F)\big)$ and $\mathrm{Var}_F\big(T(\boldsymbol{X}_n, F)\big)$ are estimated by $\bar{T}^* := N^{-1}\sum_{j=1}^N T_j^*$ and $v_{T^*}^2 := N^{-1}\sum_{j=1}^N \big(T_j^* - \bar{T}^*\big)^2$, respectively.

Denote the conditional d.f. of $T\big(\boldsymbol{X}_n^*, \hat{F}_n\big)$, given $\boldsymbol{X}_n$, by

$$G_n^*(t) := P\big(T(\boldsymbol{X}_n^*, \hat{F}_n) \le t \big| \boldsymbol{X}_n\big) = P_{\hat{F}_n}\big(T(\boldsymbol{X}_n^*, \hat{F}_n) \le t\big), \quad t \in \mathbb{R}.$$

In practice G_n^* is also estimated by $\widetilde{G}_N$. Let

$$D_n^* := \sup_{t\in\mathbb{R}} \big|G_n^*(t) - G_n(t)\big|.$$

The bootstrap distribution of $T(\boldsymbol{X}_n, F)$ is said to be *first-order accurate*, if $D_n^* \to_{a.s.} 0$. It is said to be *second-order accurate*, if $n^{1/2}D_n^* \to_{a.s.} 0$. Clearly, having second-order accuracy is more desirable compared to first-order accuracy, as it states that $D_n^* \to_{a.s.} 0$, at a rate faster than $n^{-1/2}$.

A seminal result proved by Singh (1981) is that when estimating the population mean, the bootstrap distribution of the standardized sample mean is second-order accurate under some conditions on F. To describe this result precisely, let $\mu := E_F(X)$ and $\sigma^2 := \mathrm{Var}_F(X)$. Let $\bar{X}$ and s denote the sample mean and sample standard deviation based on $\boldsymbol{X}_n$, respectively. Let $T(\boldsymbol{X}_n, F) := n^{1/2}(\bar{X} - \mu)/\sigma$. Then $T(\boldsymbol{X}_n^*, \hat{F}_n) = n^{1/2}(\bar{X}^* - \bar{X})/s_n$ and its conditional distribution, given the sample $\boldsymbol{X}_n$, is the bootstrap distribution of $n^{1/2}(\bar{X} - \mu)/\sigma$.

Let Φ denote the standard normal d.f., φ its derivative and set $\mu_3 := E_F\big((X - \mu)^3\big)$,

$$D_n := \|G_n - \Phi\|_\infty = \sup_{t\in\mathbb{R}} \big|G_n(t) - \Phi(t)\big|, \qquad \widehat{G}_n(t) := \Phi(t) + \frac{\mu_3(1 - t^2)}{6\sigma^3 n^{1/2}}\varphi(t), \quad t \in \mathbb{R}.$$

Moreover, now

$$G_n(t) := P\Big(n^{1/2}(\bar{X} - \mu) \le t\sigma\Big), \qquad G_n^*(t) := P\Big(n^{1/2}(\bar{X}^* - \bar{X}) \le ts_n \big| \boldsymbol{X}_n\Big), \quad t \in \mathbb{R}.$$

By the CLT, if σ^2 is finite and positive, then $D_n \to 0$. If one strengthens the second moment assumption to assuming that $E_F(|X|^3) < \infty$, then the Berry-Esseen Theorem, see Loève (1963), says that for some $0 < K < \infty$, $n^{1/2}D_n \le KE_F(|X|^3)/\sigma^3$, for all $n \ge 1$. This implies that D_n is of order $O\big(n^{-1/2}\big)$. A *remarkable property* of the bootstrap distribution here is that under some conditions on F, it is second-order accurate, i.e., $n^{1/2}D_n^* \to_{a.s.} 0$. In other words, the bootstrap distribution approximates the true distribution at a faster rate than the rate of $n^{-1/2}$ provided by the classical normal distribution approximation. To state this result precisely, we recall that a d.f. F is said to be non-lattice if it does not assign any positive probability to the points of the form $x + yk$, where $y > 0$, $x \in \mathbb{R}$, $k = 0, \pm 1, \pm 2, \cdots$. We are ready to state the following theorem due to Singh (1981) without proof.

Theorem 9.3.1. (a). *If* $E_F\big(X^2\big)$ *is finite, then* $D_n^* \to_{a.s.} 0$. (b). *If* $E_F\big(|X|^3\big)$ *is finite and* F *is non-lattice, then the following hold.*

$$\sup_{t\in\mathbb{R}} n^{1/2}\big|G_n^*(t) - \widehat{G}_n(t)\big| \to_{a.s.} 0, \qquad n^{1/2}D_n^* \to_{a.s.} 0.$$

As mentioned earlier, in applications, one draws a large number N of random samples, each of size n, with replacement from $\hat{F}_n$ and uses $\widetilde{G}_N$ to estimate both G_n and G_n^* and some of the characteristics of F. For example, a bootstrap confidence interval for μ is constructed as follows. Fix an $0 < \alpha < 1$. Let $\bar{X}_j^*$ denote the sample mean of the jth bootstrap sample and set $T_j^* := n^{1/2}(\bar{X}_j^* - \bar{X})/s_n, j = 1, \cdots, N$. Let z_α be as in (3.3.9) and $t_{n,\alpha}^*$ be such that $N^{-1}\sum_{j=1}^N I(|T_j^*| > t_{n,\alpha}^*) \le \alpha$. Then the bootstrap confidence interval for μ of asymptotic confidence level $(1-\alpha)$ is $[\bar{X} - s_n n^{-1/2} t_{n,\alpha}^*, \bar{X} + s_n n^{-1/2} t_{n,\alpha}^*]$. Compare this with the asymptotic confidence interval $[\bar{X} - s_n n^{-1/2} z_{\alpha/2}, \bar{X} + s_n n^{-1/2} z_{\alpha/2}]$ obtained by the CLT approximation.

Next, we shall describe the bootstrap estimator of the bias and mean squared error of an estimator of a parameter. Let $\theta \equiv \theta(F)$ be a real-valued parameter of interest and $\hat{\theta}$ be an estimator of θ based on $\boldsymbol{X}_n$, not necessarily of the form $\theta(\hat{F}_n)$. Let $T(\boldsymbol{X}_n, F) = \hat{\theta} - \theta$. Let $\hat{\theta}_j^*$ be the estimator $\hat{\theta}$ based on the jth bootstrap sample, $1 \le j \le N$. Then the bootstrap estimators of the bias $E_F(T(\boldsymbol{X}_n, F))$ and the mean squared error $E_F(|\hat{\theta} - \theta(F)|^2)$ of the estimator $\hat{\theta}$ are $N^{-1}\sum_{j=1}^N (\hat{\theta}_j^* - \hat{\theta}) = \bar{\hat{\theta}}^* - \hat{\theta}$ and $\tau^2 := N^{-1}\sum_{j=1}^N (\hat{\theta}_j^* - \hat{\theta})^2$, respectively.

Bootstrap confidence intervals for θ are obtained as follows. Let $0 < \alpha < 1$ be given. Compute $c_{n,\alpha}^*$ such that $N^{-1}\sum_{j=1}^N I(|\hat{\theta}_j^* - \hat{\theta}| > c_{n,\alpha}^*) \le \alpha$. Then, the interval $[\hat{\theta} - n^{-1/2} c_{n,\alpha}^* \tau, \hat{\theta} + n^{-1/2} c_{n,\alpha}^* \tau]$ is a bootstrap confidence interval of asymptotic confidence level $1-\alpha$, under conditions that guarantee the weak convergence of the distribution of $n^{1/2}(\hat{\theta}^* - \hat{\theta})$ to the true limiting distribution of $n^{1/2}(\hat{\theta} - \theta)$. Examples of the application of this methodology can be found in Shao and Tu (1995).

Next, we shall consider the problem of obtaining confidence bands for the d.f. F, which is assumed to be continuous. With $X_1^*, \cdots, X_n^*$ denoting the bootstrap sample, let $F_n^*(x) := n^{-1}\sum_{i=1}^n I(X_i^* \le x)$ and $\mathcal{Z}_n(x) := n^{1/2}\{F_n^*(x) - \hat{F}_n(x)\}$, $x \in \mathbb{R}$. Then, by Theorem 4.1 of Bickel and Freedman (1981), we obtain that the conditional distribution of $\|\mathcal{Z}_n\|_\infty$, given $\boldsymbol{X}_n$, converges weakly to $\sup_{x\in\mathbb{R}} |\mathbb{B}^0(F(x))|$, wp1. Because F is continuous, $\sup_{x\in\mathbb{R}} |\mathbb{B}^0(F(x))| = \sup_{0\le t\le 1} |B^0(t)| = \|B^0\|_\infty$. Hence, the conditional distribution of $\|\mathcal{Z}_n\|_\infty$, given $\boldsymbol{X}_n$, converges weakly to $\|\mathbb{B}^0\|_\infty$, wp1. These facts are useful in constructing large sample confidence bands for F as follows.

Fix an $0 < \alpha < 1$. Let b_α^0 be such that $P(\sup_{0\le t\le 1} |\mathbb{B}^0(t)| \le b_\alpha^0) = 1 - \alpha$. Choose $b_{n,\alpha}(\hat{F}_n)$ from the bootstrap distribution such that

$$P\Big(n^{1/2} \sup_{x\in\mathbb{R}} |F_n^*(x) - \hat{F}_n(x)| \le b_{n,\alpha}(\hat{F}_n) | \boldsymbol{X}_n\Big) \to_{a.s.} 1 - \alpha.$$

Then, $P(n^{1/2}\sup_{x\in\mathbb{R}} |\hat{F}_n(x) - F(x)| \le b_{n,\alpha}(\hat{F}_n)) \to 1 - \alpha$. Moreover, $b_{n,\alpha}(\hat{F}_n) \to_p b_\alpha^0$, wp1. These results follow from Corollary 4.2 in Bickel and Freedman (1981).

In practice, one uses the Monte Carlo method to estimate $b_{n,\alpha}(\hat{F}_n)$ as follows. Let F_{nj}^* denote the empirical d.f. corresponding to the jth bootstrap sample from $\hat{F}_n$ and let $\mathcal{D}_j^* := n^{1/2}\|F_{nj}^* - \hat{F}_n\|_\infty$, $1 \le j \le N$. Let $b_{n,\alpha}^*$ be such that

$$N^{-1}\sum_{j=1}^N I(\mathcal{D}_j^* > b_{n,\alpha}^*) \le \alpha. \tag{9.3.1}$$

Then, the bootstrap confidence band for F of the asymptotic confidence level $1-\alpha$ is the following set of d.f.'s H.

$$\Big\{H;\ (\hat{F}_n(x) - b_{n,\alpha}^*) \vee 0 \le H(x) \le (\hat{F}_n(x) + b_{n,\alpha}^*) \wedge 1,\ \ \forall x \in \mathbb{R}\Big\}.$$

Also, the test that rejects the hypothesis that F is the true underlying d.f. whenever $n^{1/2}\|\hat{F}_n - F\|_\infty > b^*_{n,\alpha}$ has asymptotic size α.

In general, it is possible to draw bootstrap samples from other estimators of F, other than $\hat{F}_n$. The empirical d.f. $\hat{F}_n$ assigns equal mass of $1/n$ to each observation, and the above bootstrap method which samples from this $\hat{F}_n$ is sometimes called *naive* or *ordinary* bootstrap in the literature. Moreover, $\hat{F}_n$ is a discontinuous distribution function. If F is continuous, it is desirable to estimate F by a continuous d.f. based on the data. A *smooth bootstrap* method draws random samples with replacement from an estimator of F that is continuous. One such estimator of F is given by $\widetilde{F}_n(t) := \int_{s\le t} \hat{f}_n(s)ds$, where $\hat{f}_n$ is as in (7.3.1). In the next section we see how a smooth bootstrap method is used to provide asymptotically distribution free goodness of tests pertaining to the error d.f. in the linear regression model.

9.3.1 Bootstrap Goodness-of-fit Testing in Regression

As seen in Subsection 9.2.1, the Khmaladze transform of the residual empirical process provides asymptotically distribution free tests for fitting an error d.f. in linear regression models. It is thus clearly desirable to use these tests whenever applicable. In the situations where either they are not applicable or hard to apply, one may use either Monte-Carlo or bootstrap methodology to perform a goodness-of-fit test for this testing problem. In this section we shall discuss a smooth bootstrap method to implement the Kolmogorov-Smirnov goodness-of-fit test in finite samples in the multiple linear regression model with non-random design.

More precisely, consider the linear regression model

$$Y_{ni} = x'_{ni}\beta + \varepsilon_{ni}, \quad 1 \le i \le n, \tag{9.3.2}$$

where $x_{n1}, \cdots, x_{nn}$ are p-dimensional non random vectors and the error r.v.'s $\varepsilon_{n1}, \cdots, \varepsilon_{nn}$ are i.i.d. F. Let ψ be a function from $\mathbb{R}$ to $\mathbb{R}$ such that $E\big(\psi(\varepsilon_{n1})\big) = 0$. An M estimator $\bar{\beta}_n$ of β is a solution $b \in \mathbb{R}^p$ of the equation $\sum_{i=1}^n x_{ni}\psi(Y_{ni} - x'_{ni}b) = 0_p$, where 0_p is the vector of p zeros. Let $\bar{\varepsilon}_{ni} = Y_{ni} - x'_{ni}\bar{\beta}_n$ and form the empirical process of the M residuals

$$W_n(v) := n^{-1/2}\sum_{i=1}^n \Big(I\big(\bar{\varepsilon}_{ni} \le F^{-1}(v)\big) - v\Big), \quad 0 \le v \le 1.$$

From the discussion in the above Section 9.2, e.g., see (9.2.4), it is apparent that the asymptotic distribution theory of the tests based on residual empirical processes like W_n heavily depends on the smoothness of the error d.f. For this reason we introduce a similar smoothness in the bootstrapping resampling distribution here. Suppose F has density f. Let K be a positive probability density on $\mathbb{R}$, $h > 0$ denote a window width and define density and d.f. estimates

$$f_n(x) := \frac{1}{nh}\sum_{i=1}^n K\Big(\frac{x - \bar{\varepsilon}_{ni}}{h}\Big), \quad F_n(x) := \int_{-\infty}^x f_n(y)dy, \quad x \in \mathbb{R}.$$

The smooth bootstrap method consists of the following three steps. First, draw a random sample $\varepsilon^*_{ni}, 1 \le i \le n$ from F_n with replacement and form the pseudo bootstrap observations $Y^*_{ni} = x'_{ni}\bar{\beta}_n + \varepsilon^*_{ni}, 1 \le i \le n$. The second step consists of obtaining a bootstrap estimator of β as follows. Let P_*, E_* denote the conditional probability and expectation, respectively, under this bootstrap resampling scheme and define the bootstrap version β^*_n of $\bar{\beta}_n$ as a solution b of the equations

$$\sum_{i=1}^n x_{ni}\Big[\psi(Y^*_{ni} - x'_{ni}b) - E_*\psi(\varepsilon^*_{n1})\Big] = 0_p.$$

In the third step, define the bootstrap version of W_n to be

$$W_n^*(v) := n^{-1/2}\sum_{i=1}^{n}\Big\{I(Y_{ni}^* - x_{ni}'\beta_n^* \le F_n^{-1}(v)) - v\Big\}, \quad 0 \le v \le 1.$$

We say that this bootstrap method is first-order accurate if the conditional distribution of $W_n^*(v)$, $0 \le v \le 1$, given $(x_{ni}, Y_{ni}), 1 \le i \le n$, approximates the distribution of $W_n(v)$, $0 \le v \le 1$, wp1. In order to make this statement precise, we need the following assumptions.

The density f is positive and uniformly continuous on $\mathbb{R}$. (9.3.3)

The matrix $A_n := \sum_{i=1}^n x_{ni}x_{ni}'$ is positive definite and $\max_{1\le i\le n}\|A_n^{-1/2}x_{ni}\| \to 0$. (9.3.4)

The function ψ is bounded and differentiable on $\mathbb{R}$, $E\big(\psi(\varepsilon_{n1})\big) = 0$, (9.3.5)
its derivative $\dot\psi$ is Lipschitz(δ), for some $\delta > 0$; i.e., for some $\delta > 0, 0 < C < \infty$, $|\dot\psi(x) - \dot\psi(y)| \le C|x-y|^\delta$, $\forall\, x, y \in \mathbb{R}$, and $\tau := E\big(\dot\psi(\varepsilon_{n1})\big) \ne 0$.

The kernel K has finite mean, is differentiable, and K and $\dot K$ are of bounded variation on $\mathbb{R}$. (9.3.6)

The bandwidth h satisfies $h + (n^a h)^{-1} \to 0$, for some $0 < a < 1/4$. (9.3.7)

Define $\bar x_n := n^{-1}\sum_{i=1}^n x_{ni}$, $q(v) := f\big(F^{-1}(v)\big)$ and

$$\beta_n := \tau^{-1}A_n^{-1/2}\sum_{i=1}^{n} x_{ni}\psi(\varepsilon_{ni}),$$

$$\bar W_n(v) := n^{-1/2}\sum_{i=1}^{n}\Big\{I\big(\varepsilon_{ni} \le F^{-1}(v)\big) - v\Big\} - n^{1/2}\bar x_n' A_n^{-1/2}\beta_n q(v),$$

$$\mathcal{C}_n(v,w) := \mathrm{Cov}\big(\bar W_n(v), \bar W_n(w)\big), \quad 0 \le v, w \le 1.$$

Assume that for all $0 \le v, w \le 1$, the limit

$$\mathcal{C}(v,w) := \lim_{n\to\infty} C_n(v,w) \text{ exists and is continuous.} \tag{9.3.8}$$

We are ready to state the following theorem. Its proof is available in Koul and Lahiri (1994).

Theorem 9.3.2. *Suppose the assumptions* (9.3.2) *to* (9.3.8) *hold. Then,*

$$\sup_{x\in\mathbb{R}}\Big|P_*\Big(\sup_{0\le v\le 1}|W_n^*(v)| \le x\Big) - P\Big(\sup_{0\le v\le 1}|W_n(v)| \le x\Big)\Big| \to_{a.s.} 0.$$

This result can be used in practice to either obtain confidence bands for F or to perform a goodness-of-fit test. Consider the problem of obtaining $(1-\alpha)$ level confidence band for F. Let N be a large positive integer. Construct N bootstrap versions $\|W_{nj}^*\|_\infty$ of $\|W_n^*\|_\infty$. Let $b_{n,\alpha}^*$ be the $1-\alpha$ sample quantile of the variables $\|W_{n1}^*\|_\infty, \cdots, \|W_{nN}^*\|_\infty$ and $\bar F_n(y) := n^{-1}\sum_{i=1}^n I(\bar\varepsilon_{ni} \le y)$. Then a confidence band for F of asymptotic confidence level $1-\alpha$ is the set of all d.f.'s H that satisfy

$$\bar F_n(y) - n^{-1/2}b_{n,\alpha}^* \le H(y) \le \bar F_n(y) + n^{-1/2}b_{n,\alpha}^*, \ \forall\, y \in \mathbb{R}.$$

Typically N is taken to be 1000 or more.

The above empirical distribution of $\|W_n^*\|_\infty$ can also be used to perform a goodness-of-fit test for testing $F = F_0$ as follows. Assume F_0 has a density f_0 that satisfies (9.3.3). The test that rejects the hypothesis that $F = F_0$ whenever $\|\bar W_n\|_\infty > b_{n,\alpha}^*$ is of asymptotic size α.

Theorem 9.3.2 continues to hold in the regression model $Y = \beta' X + \varepsilon$, where X is a p-dimensional random vector, independent of ε, provided the condition (9.3.4) is replaced by $E(\|X\|^2) < \infty$ and $E(XX')$ is positive definite.

There are two other bootstrap methods used in practice. We shall describe them for a bit more general setting. Let Θ be an open subset in $\mathbb{R}^q$, $q \geq 1$, $\{m(\cdot, \theta); \theta \in \Theta\}$ be a family of parametric functions from $\mathbb{R}^p$ to $\mathbb{R}$, and X_i, $1 \leq i \leq n$ be random vectors. Consider the model where one observes pairs $(X_1, Y_1), \cdots, (X_n, Y_n)$ such that, for some $\beta \in \Theta$, the errors $\varepsilon_i = Y_i - m(X_i, \beta)$, $1 \leq i \leq n$, are i.i.d. r.v.'s. Let F denote their common distribution function. If the mean of F is zero then $m(\cdot, \beta)$ is the classical regression function. If F has no finite mean but its median is zero, then $m(\cdot, \beta)$ is the conditional median function, as would be the case when fitting a Cauchy distribution.

The problem of interest is to test the null hypothesis $H_0 : F = F_0$, versus the alternatives $K : F \neq F_0$, where F_0 is a known d.f.

Let β_n be an estimator of β obtained from $(X_i, Y_i), 1 \leq i \leq n$, obeying the regression model $Y_i = m(X_i, \beta) + \varepsilon_i$, $1 \leq i \leq n$. It does not need to be an M estimator but it needs to satisfy $n^{1/2}\|\beta_n - \beta\| = O_p(1)$. Now define $\varepsilon_{ni} := Y_i - m(X_i, \beta_n)$ and

$$\hat{F}_n(y) := n^{-1} \sum_{i=1}^{n} I(\varepsilon_{ni} \leq y), \qquad \mathcal{K}_n := n^{1/2} \big\|\hat{F}_n - F_0\big\|_\infty.$$

The *naive bootstrap* method is implemented as follows. Draw a random sample $\varepsilon_{n1}^*, \cdots, \varepsilon_{nn}^*$ with replacement from the finite population consisting of n elements $\varepsilon_{n1}, \cdots, \varepsilon_{nn}$, and let

$$F_n^*(y) := n^{-1} \sum_{i=1}^{n} I(\varepsilon_{ni}^* \leq y), \quad \mathcal{K}_n^* := n^{1/2} \big\|F_n^* - \hat{F}_n\big\|_\infty.$$

Repeat this say N_1 times. The empirical distribution of $\mathcal{K}_n^*$ so obtained is called the naive bootstrap distribution of $\mathcal{K}_n$. Let $k_{n,\alpha}^*$ be $(1-\alpha)$-percentile of this distribution. The naive bootstrap $\mathcal{K}_n$-test rejects H_0 at level α whenever $\mathcal{K}_n > k_{n,\alpha}^*$. This bootstrap method is also called "wild bootstrap" in some literature, see, e.g., Hart (1997). For a given sample size, the distribution of K_n^* is the same for any sample of residuals. Thus this bootstrap method is fast and $k_{n,\alpha}^*$ can be obtained prior to sampling.

The *centered bootstrap* method is implemented as follows. Obtain ε_{ni}^* as in the naive bootstrap method. Let

$$\bar{\varepsilon}_n := n^{-1} \sum_{i=1}^{n} \varepsilon_{ni}, \qquad \tilde{\varepsilon}_n := \text{median}\{\varepsilon_{ni}, 1 \leq i \leq n\}.$$

Define the centered residuals and the associated regression model

$$\begin{aligned} \eta_{ni} &:= \varepsilon_{ni}^* - \bar{\varepsilon}_n, && \text{if mean of } F_0 \text{ is zero,} \\ &:= \varepsilon_{ni}^* - \tilde{\varepsilon}_n, && \text{if median of } F_0 \text{ is zero,} \\ \widetilde{Y}_i &:= m(X_i, \beta_n) + \eta_{ni}, && 1 \leq i \leq n. \end{aligned} \tag{9.3.9}$$

Typically in the original homoscedastic regression model, where X_i's are independent of the errors ε_i's, the error distribution has either zero mean or zero median. The centered bootstrap as described here ensures this setup conditionally, given $\{(X_i, Y_i), 1 \leq i \leq n\}$.

Now obtain an estimator $\widetilde{\beta}_n$ of β from the model (9.3.9), using the observations $\{(X_i, \widetilde{Y}_i), 1 \leq i \leq n\}$. Let $\widetilde{\varepsilon}_{ni} := \widetilde{Y}_i - m(X_i, \widetilde{\beta}_n)$ and define

$$\widetilde{F}_n(y) := n^{-1} \sum_{i=1}^{n} I(\widetilde{\varepsilon}_{ni} \leq y), \quad \widetilde{\mathcal{K}}_n := n^{1/2} \sup_{y \in \mathbb{R}} \big|\widetilde{F}_n(y) - F_n^*(y)\big|.$$

Repeat this say N_2 times. This sampling distribution of $\widetilde{\mathcal{K}}_n$ so obtained is called the centered bootstrap distribution of $\mathcal{K}_n$. Let $c^*_{n,\alpha}$ denote the $(1-\alpha)$-percentile of this distribution. Then the centered bootstrap $\mathcal{K}_n$ test rejects H_0 at level α whenever $\mathcal{K}_n > c^*_{n,\alpha}$.

In a finite sample simulation study, Koul and Sakhanenko (2005) found that the naive and centered bootstrap methods are insensitive to the significance level compared to the smooth bootstrap method.

9.4 GOF Tests based on Density Estimates

In this section we shall present some tests for fitting a known density or a parametric family of densities based on nonparametric kernel density estimators in the one sample setup.

Let f be a density and φ be a known density on $\mathbb{R}$. Consider the problem of testing the simple hypothesis $\mathcal{H}_0 : f(x) = \varphi(x)$, for all $x \in \mathbb{R}$, versus the alternatives that $\mathcal{H}_0$ is not true, based on a random sample $X_1, \cdots, X_n$ from f. Recall from (7.3.1) that a kernel density estimator corresponding to a kernel K and bandwidth h is

$$\hat{f}_n(x) := \frac{1}{nh} \sum_{i=1}^{n} K\Big(\frac{x - X_i}{h}\Big), \quad x \in \mathbb{R}.$$

Analogous to (7.3.3), under $\mathcal{H}_0$, we have

$$\nu_n(x) := E\big(\hat{f}_n(x)\big) = \int K(u)\varphi(x - hu)\, du, \quad x \in \mathbb{R}.$$

We again use the notation $I_2(\psi) := \int \psi^2(x)\, dx$, for any square integrable function ψ. Let g be a nonnegative function defined on $\mathbb{R}$ and define

$$T_n := \int \Big(\hat{f}_n(x) - \nu_n(x)\Big)^2 g(x)\, dx, \quad A := I_2(K) \int \varphi(x)g(x)\, dx, \tag{9.4.1}$$

$$S_n := n\sqrt{h}\Big(T_n - \frac{1}{nh}A\Big), \tag{9.4.2}$$

$$\mathcal{T}_n := \int \big(\hat{f}_n(x) - \varphi(x)\big)^2 g(x)\, dx, \qquad \mathcal{S}_n := n\sqrt{h}\big(\mathcal{T}_n - \frac{1}{nh}A\big). \tag{9.4.3}$$

The test statistics S_n or $\mathcal{S}_n$ can be used to test for $\mathcal{H}_0$. These tests reject $\mathcal{H}_0$ for large values of $|S_n|$ or $|\mathcal{S}_n|$. Their exact null distributions are hard to determine. Under some conditions, their asymptotic null distributions are normal, which we shall derive shortly.

Intuitively, it is more natural to use the statistic $\mathcal{T}_n$ or its standardized version $\mathcal{S}_n$ for testing $\mathcal{H}_0$. However, because of the bias in $\hat{f}_n$, the asymptotic normality of $\mathcal{S}_n$ holds under more stringent conditions on φ and h than that of S_n. As a result, tests based on S_n are more broadly applicable than those based on $\mathcal{S}_n$. Bickel and Rosenblatt (1973) proposed tests for $\mathcal{H}_0$ based on S_n.

We shall derive the asymptotic null distribution of S_n under the following conditions.

(f) The density φ is bounded and L_1-Lipschitz, see (5.6.4).

(g) The function g is nonnegative and bounded, and $I_2(\phi g) = \int \phi^2(x)g^2(x)\, dx$ is positive.

(h) The bandwidth h satisfies $h \to 0$ and $nh^2 \to \infty$.

(k) The density K is bounded and has finite mean.

Since φ and g are bounded and g is nonnegative, the integral $\int \varphi^j(y)g^k(y)\,dy$ is finite and nonnegative for each $j \geq 1$ and $k \geq 0$. As K is bounded, so is the density J defined by

$$J(t) := \int K(s)K(s+t)\,ds, \quad t \in \mathbb{R}. \tag{9.4.4}$$

Hence, K and J are square integrable. Set

$$\Gamma := I_2(\varphi g)I_2(J) = \int \varphi^2(x)g^2(x)\,dx \int J^2(t)\,dt, \quad \tau^2 := 2\Gamma.$$

The following theorem gives the asymptotic null distribution of S_n for a general g.

Theorem 9.4.1. *Suppose $\mathcal{H}_0$ and the assumptions* (f)–(k) *hold. Then,*

$$S_n \to_D \mathcal{N}(0, \tau^2). \tag{9.4.5}$$

The proof of this theorem is postponed to the end of this section. It readily follows that the test that rejects $\mathcal{H}_0$ whenever $|S_n| > \tau\, z_{\alpha/2}$ has asymptotic size α, where z_α is as in (3.3.9).

In applications, the choice of K and h are important considerations. Using some efficiency criterion, Silverman (1986) and Härdle (1991) recommend to use the Epanechnikov kernel $K(t) = (3/4)(1-t^2)I(|t| \leq 1)$. Note that this kernel satisfies the condition (k). The bandwidth h may be chosen using one of the methods discussed in Subsection 7.3.7.

The choice of $g(x) = 1, x \in \mathbb{R}$, is common in practice. For this g, $\Gamma = I_2(\varphi)I_2(J)$, $A = I_2(K)$ and the asymptotic null distribution of $S_n = n\sqrt{h}(T_n - I_2(K)/(nh))$ is $\mathcal{N}(0, 2I_2(\varphi)I_2(J))$ as long as conditions (f), (h) and (k) are satisfied.

Another interesting choice of g mentioned in Bickel and Rosenblatt (1973) is $g(x) = I(0 \leq x \leq 1)/\varphi(x)$ in the case the support of φ is $[0,1]$ and $\inf_{0\leq x\leq 1}\varphi(x) > 0$. For this g, $\int \varphi(x)g(x)\,dx = \int_0^1 dx = 1 = \int \varphi^2(x)g^2(x)\,dx$. Hence, $A = I_2(K)$, $\tau^2 = 2I_2(J)$, condition (g) holds and the asymptotic null distribution of $S_n = n\sqrt{h}(T_n - I_2(K)/(nh))$ is $\mathcal{N}(0, 2I_2(J))$, as long as conditions (f), (h) and (k) are satisfied.

The asymptotic normality of $\mathcal{S}_n$ is given in the following corollary.

Corollary 9.4.1. *Suppose the density φ is two times continuously differentiable with $\dot{\varphi}$ integrable and $\ddot{\varphi}$ square integrable. In addition to (g), (h) and (k) assume $nh^4 \to 0$ and the density kernel K has a finite second moment $\kappa_2 := \int t^2K(t)\,dt$. Then, under $\mathcal{H}_0$, $|\mathcal{S}_n - S_n| \to_p 0$.*

Consequently, the asymptotic size of the test that rejects $\mathcal{H}_0$ whenever $|\mathcal{S}_n| > \tau\, z_{\alpha/2}$ is α.

Proof. The assumption about φ having a continuous and integrable derivative implies that φ is bounded and L_1-Lipschitz with constant $\int |\dot{\varphi}(x)|\,dx$. Hence (f) holds.

Next, set $b_n(x) := \nu_n(x) - \varphi(x)$. Then, by (7.3.14),

$$\int b_n^2(x)\,dx = \frac{h^4\kappa_2^2}{4}\int \left(\ddot{\varphi}(x)\right)^2 dx + o(h^4) = O(h^4). \tag{9.4.6}$$

Now, by writing $\hat{f}_n - \varphi = \hat{f}_n - \nu_n + b_n$ and expanding the quadratic in $\mathcal{T}_n$, we obtain the decomposition $\mathcal{T}_n - T_n = \mathcal{T}_{n1} + 2\mathcal{T}_{n2}$, where

$$\mathcal{T}_{n1} := \int b_n^2(x)g(x)\,dx, \quad \mathcal{T}_{n2} := \int \left(\hat{f}_n(x) - \nu_n(x)\right)b_n(x)g(x)\,dx.$$

Therefore, $\mathcal{S}_n = S_n + n\sqrt{h}\mathcal{T}_{n1} + 2n\sqrt{h}\mathcal{T}_{n2}$.

By (9.4.6), $\mathcal{T}_{n1} \le \|g\|_\infty \int b_n^2(x)\,dx = O(h^4)$ and $n\sqrt{h}\mathcal{T}_{n1} = O(nh^{9/2}) = o(1)$. Also, by (2.4.6), $\mathcal{T}_{n2}^2 \le \mathcal{T}_{n1}T_n$. By Theorem 9.4.1,

$$T_n = O_p\Big(\frac{1}{n\sqrt{h}}\Big) + \frac{A}{nh}. \tag{9.4.7}$$

Hence, $\mathcal{T}_{n2} = O_p(h^2/\sqrt{nh})$ and $n\sqrt{h}\mathcal{T}_{n2} = O_p(\sqrt{n}\,h^2) = O_p((nh^4)^{1/2}) = o_p(1)$ and $|\mathcal{S}_n - S_n| = o_p(1)$, which, in turn, implies the last claim of the corollary. □

Next, consider the problem of fitting a parametric family of densities. More precisely, let q be a known positive integer, $\Theta \subseteq \mathbb{R}^q$ and $\{f_\theta, \theta \in \Theta\}$ be a parametric family of densities on $\mathbb{R}$. Consider the problem of testing the composite hypothesis $\mathcal{H}_C : f = f_{\theta_0}$, for some $\theta_0 \in \Theta$ versus the alternatives that $\mathcal{H}_C$ is not true, based on a random sample $X_1, \cdots, X_n$ from f.

To proceed further, for $\theta \in \Theta$, let $\nu_n(x,\theta) := \int K(u) f_\theta(x - hu)\,du$, $x \in \mathbb{R}$, and define

$$T_n(\theta) := \int \Big(\hat{f}_n(x) - \nu_n(x,\theta)\Big)^2 g(x)\,dx, \quad A(\theta) := I_2(K) \int f_\theta(x) g(x)\,dx,$$

$$S_n(\theta) := n\sqrt{h}\Big(T_n(\theta) - \frac{1}{nh}A_n(\theta)\Big), \quad \Gamma(\theta) := I_2(f_\theta g) I_2(J), \quad \tau^2(\theta) := 2\Gamma(\theta).$$

The following two theorems describe the asymptotic null distribution of $S_n(\hat\theta)$ under two different sets of conditions on the null model, where $\hat\theta$ is an estimator of θ_0, under $\mathcal{H}_C$.

Theorem 9.4.2. *Suppose $\mathcal{H}_C$ holds, (f), (g), (h) and (k) hold with $\varphi = f_{\theta_0}$, the kernel K is of the form*

$$K(y) = \int I(s \le y)\dot{K}(y)\,ds, \quad y \in \mathbb{R},$$

for some integrable function $\dot K$, and there is an estimator $\hat\theta$ satisfying

$$\Delta_k := \int \big|f_{\hat\theta}(x) - f_{\theta_0}(x)\big|^k\,dx = O_p(n^{-1}), \quad k = 1, 2. \tag{9.4.8}$$

Then, $S_n(\hat\theta) \to_D \mathcal{N}(0, \tau^2(\theta_0))$ and $\tau^2(\hat\theta) \to_p \tau^2(\theta_0)$.

Proof. We first present some preparatory results. Let $\mu_{k,\ell}(\theta) := \int f_\theta^k(x) g^\ell(x)\,dx$ for $k, \ell = 1, 2$ and $\theta \in \Theta$. By (9.4.8) and because f_{θ_0} is bounded, we have, for $\ell = 1, 2$,

$$\big|\mu_{1,\ell}(\hat\theta) - \mu_{1,\ell}(\theta_0)\big| \le \|g\|_\infty^\ell \Delta_1 = O_p(n^{-1}), \tag{9.4.9}$$

$$\big|\mu_{2,\ell}(\hat\theta) - \mu_{2,\ell}(\theta_0)\big| \le \|g\|_\infty^\ell \big[\Delta_2 + 2\|f_{\theta_0}\|_\infty \Delta_1\big] = O_p(n^{-1}). \tag{9.4.10}$$

Since K is a density, the moment inequality (2.4.7) yields, for $k \ge 1$,

$$\big|\nu_n(x,\hat\theta) - \nu_n(x,\theta_0)\big|^k \le \int \big|f_{\hat\theta}(x - hu) - f_{\theta_0}(x - hu)\big|^k K(u)\,du.$$

From this we derive with the help of the substitution $x = y + hu$ and a change of the order of integration

$$\int \big|\nu_n(x,\hat\theta) - \nu_n(x,\theta_0)\big|^k\,dx \le \Delta_k, \quad k = 1, 2. \tag{9.4.11}$$

We shall first prove $\tau^2(\hat\theta) \to_p \tau^2(\theta_0)$. By (9.4.10), this follows from

$$\big|\tau^2(\hat\theta) - \tau^2(\theta_0)\big| \le 2I_2(J)\big|\mu_{2,2}(\hat\theta) - \mu_{2,2}(\theta_0)\big| = O_p(n^{-1}) = o_p(1).$$

It remains to prove $S_n(\hat{\theta}) \to_D \mathcal{N}(0, \tau^2(\theta_0))$. Theorem 9.4.1 applied with $\varphi = f_{\theta_0}$ gives $S_n(\theta_0) \to_D \mathcal{N}(0, \tau^2(\theta_0))$. By the Slutsky Theorem 2.10.5, it thus suffices to prove $S_n(\hat{\theta}) - S_n(\theta_0) \to_p 0$. In view of the identity

$$S_n(\hat{\theta}) - S_n(\theta_0) = n\sqrt{h}\big[T_n(\hat{\theta}) - T_n(\theta_0)\big] - h^{-1/2}\big[A(\hat{\theta}) - A(\theta_0)\big],$$

this is established by verifying the following two assertions.

$$n\sqrt{h}\big[T_n(\hat{\theta}) - T_n(\theta_0)\big] = o_p(1), \qquad A(\hat{\theta}) - A(\theta_0) = o_p(h^{1/2}).$$

Since $A(\theta) = I_2(K)\mu_{1,1}(\theta)$, (9.4.9) and (h) yield the second assertion:

$$\big|A(\hat{\theta}) - A(\theta_0)\big| \le I_2(K)\|g\|_\infty \Delta_1 = O_p(n^{-1}) = o_p(h^{1/2}). \tag{9.4.12}$$

To prove the first assertion, we introduce

$$B_n := \int \big(\hat{f}_n(x) - \nu_n(x, \theta_0)\big)\big(\nu_n(x, \hat{\theta}) - \nu_n(x, \theta_0)\big) g(x)\, dx,$$
$$C_n := \int \big(\nu_n(x, \hat{\theta}) - \nu_n(x, \theta_0)\big)^2 g(x)\, dx \le \|g\|_\infty^2 \Delta_2 = O_p(n^{-1}).$$

Use the decomposition $\hat{f}_n(x) - \nu_n(x, \hat{\theta}) = \big(\hat{f}_n(x) - \nu_n(x, \theta_0)\big) - \big(\nu_n(x, \hat{\theta}) - \nu_n(x, \theta_0)\big)$ and expand the quadratic to rewrite

$$T_n(\hat{\theta}) - T_n(\theta_0) = -2B_n + C_n. \tag{9.4.13}$$

Let $F_n(x) := n^{-1}\sum_{i=1}^n I(X_i \le x)$ and F_θ denote the d.f. of the density f_θ and $\mathcal{D}_n := n^{1/2}\|F_n - F_{\theta_0}\|_\infty$. Then, arguing as in the proof of Theorem 7.3.1, we have

$$R_n := \sup_{x \in \mathbb{R}} \big|\hat{f}_n(x) - \nu_n(x, \theta_0)\big| \le \frac{\mathcal{D}_n}{n^{1/2}h} \int \big|\dot{K}(s)\big|\, ds = o_p(1).$$

This fact, (9.4.8) and (9.4.11) yield

$$|B_n| \le \|g\|_\infty R_n \int \big|\nu_n(x, \hat{\theta}) - \nu_n(x, \theta_0)\big|\, dx \le \|g\|_\infty R_n \Delta_1 = o_p(n^{-1}).$$
$$C_n \le \|g\|_\infty \int \big(\nu_n(x, \hat{\theta}) - \nu_n(x, \theta_0)\big)^2 dx \le \|g\|_\infty \Delta_2 = O_p(n^{-1}).$$

The properties of B_n and C_n, the decomposition (9.4.13) and (h) give the first assertion. □

Theorem 9.4.3. *Suppose $\mathcal{H}_C$ holds and (f), (g), (h) and (k) hold with $\varphi = f_{\theta_0}$, and there is a measurable function $\dot{f}_{\theta_0}$ from $\mathbb{R}$ to $\mathbb{R}^q$ satisfying*

$$\int \|\dot{f}_{\theta_0}(x)\|^k\, dx < \infty, \quad k = 1, 2, \tag{9.4.14}$$

$$\int \big|f_\theta(x) - f_{\theta_0}(x) - \dot{f}_{\theta_0}(x)'(\theta - \theta_0)\big|^k dx = o(\|\theta - \theta_0\|^k), \quad k = 1, 2, \tag{9.4.15}$$

as $\theta \to \theta_0$. Let $\hat{\theta}$ be an estimator of θ_0 such that $n^{1/2}\|\hat{\theta} - \theta_0\| = O_p(1)$. Then $S_n(\hat{\theta}) \to_D \mathcal{N}(0, \tau^2(\theta_0))$ and $\tau^2(\hat{\theta}) \to_p \tau^2(\theta_0)$.

Proof. Let $r_\theta(x) := f_\theta(x) - f_{\theta_0}(x) - \dot{f}_{\theta_0}(x)'(\theta - \theta_0)$, $x \in \mathbb{R}, \theta \in \Theta$. The triangle inequality gives $|f_{\hat\theta}(x) - f_{\theta_0}(x)| \le \|\dot{f}_{\theta_0}(x)\|\|\hat\theta - \theta_0\| + |r_{\hat\theta}(x)|$. From this, (9.4.15) and the integrability of $\|\dot{f}_{\theta_0}\|$, we derive

$$\int \big|f_{\hat\theta}(x) - f_{\theta_0}(x)\big|\, dx = O_p(\|\hat\theta - \theta_0\|) = O_p\big(n^{-1/2}\big).$$

Similarly, in view of the inequality $|a+b|^2 \le 2a^2+2b^2$, for all $a, b \in \mathbb{R}$, and the square integrability of $\|\dot{f}_{\theta_0}\|$, we have

$$\int |f_{\hat\theta}(x) - f_{\theta_0}(x)|^2\, dx = O_p(\|\hat\theta - \theta_0\|^2) = O_p(n^{-1}).$$

Analogous to (9.4.9) and (9.4.10), from the above two facts we obtain that

$$\big|\mu_{k,\ell}(\hat\theta) - \mu_{k,\ell}(\theta_0)\big| = O_p\big(n^{-1/2}\big), \quad k = 1, 2,\ \ell = 1, 2. \tag{9.4.16}$$

The proof of $\tau^2(\hat\theta) \to_p \tau^2(\theta_0)$ follows as in the proof of Theorem 9.4.2. Arguing as in the proof of that theorem, we derive $S_n(\hat\theta) \to_D \mathcal{N}\big(0, \tau^2(\theta_0)\big)$ by showing $\big|A(\hat\theta) - A(\theta_0)\big| = o_p\big(h^{1/2}\big)$, $n\sqrt{h}B_n = o_p(1)$ and $n\sqrt{h}C_n = o_p(1)$. The proofs of the first and the last of these statements are similar to those in Theorem 9.4.2:

$$\big|A(\hat\theta) - A(\theta_0)\big| \le I_2(K)\|g\|_\infty \int \big|f_{\hat\theta}(x) - f_{\theta_0}(x)\big|\, dx = O_p(n^{-1/2}) = o_p(h^{1/2}),$$
$$C_n \le \|g\|_\infty \int \big|f_{\hat\theta}(x) - f_{\theta_0}(x)\big|^2\, dx = O_p(n^{-1}).$$

There is a difference in the way the term B_n is analysed. Let

$$B_{n,1} := \int \big(\hat f_n(x) - \nu_n(x, \theta_0)\big) \int \dot{f}_{\theta_0}(x - hu)K(u)\, du\, g(x)\, dx,$$
$$B_{n,2} := \int \big(\hat f_n(x) - \nu_n(x, \theta_0)\big) \int r_{\hat\theta}(x - hu)K(u)\, du g(x)\, dx.$$

Then, $B_n = B'_{n,1}(\hat\theta - \theta_0) + B_{n,2}$. From (9.4.7) applied with $\varphi = f_{\theta_0}$, we obtain that $T_n(\theta_0) = O_p\big(1/(nh)\big)$. Using the Cauchy-Schwarz inequality (2.4.4) twice, we derive

$$B_{n,2}^2 \le \|g\|_\infty T_n(\theta_0) \iint r_{\hat\theta}^2(x - hu)K(u)\, du\, dx = O_p\Big(\frac{1}{nh}\Big) o_p\Big(\|\hat\theta - \theta_0\|^2\Big) = o_p\Big(\frac{1}{n^2 h}\Big).$$

Hence, $n\sqrt{h}B_{n,2} = o_p(1)$.

To deal with $B'_{n,1}(\hat\theta - \theta_0)$, for $x \in \mathbb{R}$ and $i = 1, \cdots, n$, let

$$w_n(x) := g(x) \int \dot{f}_{\theta_0}(x - hu)\, K(u)\, du,$$
$$\xi_{ni} := \int K_h(x - X_i) w_n(x)\, dx = \int K(t) w_n(X_i + ht)\, dt.$$

Then, using the definition of $\hat f_n(x)$ we rewrite

$$B_{n,1} = \frac{1}{n}\sum_{i=1}^{n} \big(\xi_{ni} - E(\xi_{ni})\big).$$

Fix an $a \in \mathbb{R}^q$. With the help of (2.4.4), we obtain

$$(a'\xi_{n1})^2 = \Big(\int a' w_n(X_1 + ht) K(t)\, dt\Big)^2 \le \int \big(a' w_n(X_1 + ht)\big)^2 K(t)\, dt.$$

By Theorem 2.3.2 and (2.4.4) used repeatedly and the fact that K is a density,

$$\begin{aligned}
\mathrm{Var}(a'\xi_{n1}) \le E\big((a'\xi_{n1})^2\big) &\le \iint \big(a' w_n(x + ht)\big)^2 K(t)\, dt\, f_{\theta_0}(x)\, dx \\
&\le \|f_{\theta_0}\|_\infty \iint \big(a' w_n(x + ht)\big)^2 K(t)\, dx\, dt \\
&= \|f_{\theta_0}\|_\infty \int \big(a' w_n(y)\big)^2 dy \\
&\le \|f_{\theta_0}\|_\infty \|g\|_\infty^2 \iint \big(a' \dot f_{\theta_0}(y - hu)\big)^2 dy K(u)\, du \\
&= \|f_{\theta_0}\|_\infty \|g\|_\infty^2 \int \big(a' \dot f_{\theta_0}(x)\big)^2 dx \le C\|a\|^2,
\end{aligned}$$

where $C = \|f_{\theta_0}\|_\infty \|g\|_\infty^2 I_2(\|\dot f_{\theta_0}\|)$. Hence $E\big((n^{1/2} a' B_{n,1})^2\big) = \mathrm{Var}\big(a'\xi_{n1}\big) \le C\|a\|^2$, for all $a \in \mathbb{R}^q$. Consequently, $\|B_{n,1}\| = O_p(n^{-1/2})$ and $n\sqrt{h} B'_{n,1}(\hat\theta - \theta_0) = O_p(\sqrt{h}) = o_p(1)$. □

From either of the above two theorems, we conclude that the asymptotic level of the test that rejects the hypothesis $\mathcal{H}_C$, whenever $|S_n(\hat\theta)|/\tau(\hat\theta) > z_{\alpha/2}$ is α.

In other words, when fitting a parametric family of densities, the asymptotic null distribution of the test statistics based on weighted L_2 distances between the nonparametric density estimates and its expectation under the estimated null parametric density is not affected by not knowing the underlying Euclidean parameter vector, and these tests are asymptotically distribution free under the regularity conditions specified in the above two theorems.

We shall now illustrate the above two theorems by some examples.

Example 9.4.1. Here, we discuss an example where Theorem 9.4.2 is applicable. Consider the problem of fitting the family of Unif$(0,\theta)$, $\theta > 0$ densities. This problem is an example of the above general framework with $q = 1$, $\Theta = (0, \infty)$ and $f_\theta(x) := \theta^{-1} I(0 < x < \theta)$. Thus we are testing the hypothesis $\mathcal{H}_C : f = f_{\theta_0}$, for some $\theta_0 > 0$, against the alternative that $\mathcal{H}_C$ is not true, based on a random sample $X_1, \cdots, X_n$. Let $X_{(n)} := \max\{X_i, 1 \le i \le n\}$ denote the largest-order statistic of the given sample. A natural estimator of θ under the given hypothesis is $\hat\theta = X_{(n)}$ and we know that $|X_{(n)} - \theta_0| = O_p(1/n)$. Direct calculations show that $\int \big|f_{X_{(n)}}(x) - f_{\theta_0}(x)\big|^j dx = O_p(|X_{(n)} - \theta_0|) = O_p(1/n)$, $j = 1, 2$. Hence, (9.4.8) is satisfied, and by Theorem 9.4.2, for any density kernel K, function g and bandwidth h satisfying the assumptions of this theorem, the asymptotic level of the test that rejects $\mathcal{H}_C$ whenever $\big|S_n(X_{(n)})\big| > \tau(X_{(n)})\, z_{\alpha/2}$ is α.

Example 9.4.2. Another model that fits Theorem 9.4.2 is the exponential shift model in which $\Theta = \mathbb{R}$ and $f_\theta(x) = e^{\theta - x} I(x > \theta)$. Each density f_θ is bounded by 1 and satisfies the L_1-Lipschitz property with constant 2. Indeed, using the substitution $y = x - \theta$ for $t > 0$ and $y = x - \theta - t$ for $t < 0$, we derive

$$\begin{aligned}
\int \big|f_{\theta+t}(x) - f_\theta(x)\big|\, dx &= \int \big|f_\theta(x - t) - f_\theta(x)\big|\, dx = \int \big|e^{|t| - y} I(y > |t|) - e^{-y} I(y > 0)\big|\, dy \\
&= \int_0^{|t|} e^{-y}\, dy + \int_{|t|}^\infty (e^{|t|} - 1) e^{-y}\, dy = (1 - e^{-|t|}) + (e^{|t|} - 1)e^{-|t|} = 2(1 - e^{-|t|}) \le 2|t|.
\end{aligned}$$

Moreover, we have

$$\int \left|f_{\theta+t}(x) - f_\theta(x)\right|^2 dx \le 2 \int \left|f_{\theta+t}(x) - f_\theta(x)\right| dx \le 2|t|.$$

An estimator of θ is given by $X_{(1)} = \min(X_1, \cdots, X_n)$. It satisfies $X_{(1)} = \theta_0 + O_p(1/n)$ under the null hypothesis, so that the condition (9.4.8) holds for this estimator. Hence, by Theorem 9.4.2, for any density kernel K, function g and bandwidth h satisfying the assumptions of this theorem, the asymptotic level of the test that rejects H_C whenever $\left|S_n\big(X_{(1)}\big)\right| > \tau\big(X_{(1)}\big)\, z_{\alpha/2}$ is α.

Example 9.4.3. Now consider the exponential rate model for which $\Theta = (0, \infty)$ and $f_\theta(x) = \theta e^{-\theta x} I(x > 0)$. This model fits Theorem 9.4.3. Clearly each density f_θ is bounded by θ and is L_1-Lipschitz with constant 2θ. Indeed, arguing as in Example 9.4.2, we obtain, upon substituting $y = \theta x$,

$$\int |f_\theta(x - t) - f_\theta(x)|\, dx = \int |e^{\theta t - y} I(y > \theta t) - e^{-y} I(y > 0)|\, dy \le 2\theta|t|.$$

Note that $f_\theta(x)$ is continuously differentiable with respect to θ with derivative $\big((1/\theta) - x\big) f_\theta(x)$, for each $x \in \mathbb{R}$. Moreover, for $\theta_0/2 < \theta < 2\theta_0$, we have the bound $|\big((1/\theta) - x\big) f_\theta(x)| \le (2/\theta_0 + x) 2\theta_0 e^{-\theta_0 x/2} I(x > 0)$. This bound is integrable and square integrable. One can then use Theorem 2.10.10 to verify (9.4.14) and (9.4.15). The estimator $\hat{\theta} := 1/\bar{X}$ satisfies $\hat{\theta} = \theta_0 + O_p(n^{-1/2})$ under the null hypothesis. Therefore, Theorem 9.4.3 can be used to provide an ADF test based on $S_n(\hat{\theta})$ to fit this model.

Example 9.4.4. Here, we shall discuss another example where Theorem 9.4.3 is applicable. Let φ be a known density and consider the problem of testing the hypothesis that the error density in the one sample location model $X = \theta + \varepsilon$ is φ, which is equivalent to saying that the density f of X is equal to φ up to an unknown location parameter. That is, we wish to test the composite hypothesis

$$\tilde{\mathcal{H}}_0 : f(x) = \varphi(x - \theta_0), \qquad \text{for all } x \in \mathbb{R} \text{ and some } \theta_0 \in \mathbb{R},$$

against the alternatives that $\tilde{\mathcal{H}}_0$ is not true. This is the special case of the above general framework where $q = 1$, $\Theta = \mathbb{R}$ and $f_\theta(x) = \varphi(x - \theta)$.

Assume the density φ is absolutely continuous with integrable and square integrable a.e. derivative $\dot{\varphi}$. Examples of φ that satisfy this condition include Cauchy, normal, double exponential and logistic densities. Actually, this condition is met if φ has finite Fisher information for location, see Definition 2.11.1. The above condition implies that φ is bounded by $\varphi(0) + \|\dot{\varphi}\|_1$ and L_1-Lipschitz with constant $\|\dot{\varphi}\|_1$, where $\|\dot{\varphi}\|_1 := \int |\dot{\varphi}(x)|\, dx$. These properties are shared by each f_θ. Moreover, we have

$$\begin{aligned} &\varphi(x - \theta) - \varphi(x - \theta_0) + \dot{\varphi}(x - \theta_0)(\theta - \theta_0) \\ &= -(\theta - \theta_0) \int_0^1 \left[\dot{\varphi}(x - \theta_0 - s(\theta - \theta_0)) - \dot{\varphi}(x - \theta_0)\right] ds, \quad \forall\, x, \theta \in \mathbb{R}. \end{aligned} \tag{9.4.17}$$

This representation is useful in verifying d (9.4.14) and (9.4.15) with $\dot{f}_{\theta_0}(x) = -\dot{\varphi}(x - \theta_0)$. Indeed, with $\delta = \theta - \theta_0$, (9.4.17), the substitution $t = x - \theta_0$, (2.4.4) and Theorem 2.3.2 yield

$$\begin{aligned} &\int \left|\varphi(x - \theta) - \varphi(x - \theta_0) + \dot{\varphi}(x - \theta_0)(\theta - \theta_0)\right|^k dx \\ &= |\delta|^k \int \left| \int_0^1 \left[\dot{\varphi}(t - s\delta) - \dot{\varphi}(t)\right] ds \right|^k dt \\ &\le |\delta|^k \int_0^1 \int \left|\dot{\varphi}(t - s\delta) - \dot{\varphi}(t)\right|^k dt\, ds, \quad k = 1, 2. \end{aligned}$$

By the integrability conditions on $\dot{\varphi}$ and Corollary 2.10.8, the last double integral tends to zero as $\delta \to 0$, for $k = 1, 2$.

Consequently, from Theorem 9.4.3, we obtain the following corollary giving the asymptotic null distribution of $S_n(\hat{\theta})$ in this case.

Corollary 9.4.2. *Suppose* $\tilde{\mathcal{H}}_0$, (h) *and* (k) hold. Let g be bounded and positive and φ be *absolutely continuous with an integrable and square integrable a.e. derivative* $\dot{\varphi}$*. Then,* $S_n(\hat{\theta})/\tau(\hat{\theta}) \to_D \mathcal{N}(0,1)$ *holds for every estimator* $\hat{\theta}$ *of the location parameter* θ_0 *that satisfies* $n^{1/2}(\hat{\theta}-\theta_0) = O_p(1)$.

Note that in this testing problem the test statistic $S_n(\hat{\theta})/\tau(\hat{\theta})$ equals

$$\frac{n\sqrt{h}\int \big(\hat{f}_n(x) - \nu_n(x-\hat{\theta})\big)^2 g(x)\,dx - h^{-1/2} I_2(K)\int \varphi(x-\hat{\theta})g(x)\,dx}{\sqrt{2I_2(J)\int \varphi^2(x-\hat{\theta})g^2(x)\,dx}}.$$

Now take $g = 1$ and substitute $x - \hat{\theta} = y$, to see that this statistic becomes

$$Z_n(\hat{\theta}) := \frac{n\sqrt{h}\int \big(\hat{f}_n(y+\hat{\theta}) - \nu_n(y)\big)^2 dy - I_2(K)/\sqrt{h}}{\sqrt{2I_2(J)I_2(\varphi)}}.$$

Note that

$$\hat{f}_n(x+\hat{\theta}) = \frac{1}{n}\sum_{i=1}^n K_h(x+\hat{\theta}-X_i) = \frac{1}{n}\sum_{i=1}^n K_h(x-\hat{\varepsilon}_i), \quad x \in \mathbb{R},$$

is the kernel density estimator of φ based on the residuals $\hat{\varepsilon}_i = X_i - \hat{\theta}$, $i = 1, \cdots, n$.

Consequently, the test that rejects $\tilde{\mathcal{H}}_0$ whenever $|Z_n(\hat{\theta})| > z_{\alpha/2}$ has asymptotic size α. In other words, this test is asymptotically distribution free (ADF) for testing the composite hypothesis $\tilde{\mathcal{H}}_0$. This phenomenon is true more generally. Lee and Na (2002), Bachmann and Dette (2005), Cheng and Sun (2008), Koul and Mimoto (2012) and Cheng, Koul, Mimoto and Balakrishna (2023) observed that the above fact continues to hold for the analogs of this test statistic for fitting an error density based on the residuals in linear and nonlinear autoregressive, generalized autoregressive conditionally heteroscedastic time series models and in linear regression models. In other words, the asymptotic null distributions of the analogs of this test statistic in all these models are not affected by not knowing the underlying nuisance parameters, and the corresponding tests are ADF. In comparison, as seen in Section 9.2, except in the linear autoregressive model with zero mean, the asymptotic null distributions of the goodness-of-fit tests based on suitably standardized residual empirical processes in these models depend on the estimators of the underlying nuisance parameters and the null error d.f. in a complicated fashion, which renders them not to be ADF. However, because of the slow consistency rate of the kernel density estimators, one needs a relatively larger sample size to justify the use of the asymptotic null distribution of the analogs of $S_n(\hat{\theta})$ statistics for implementing these tests.

We shall now focus on the proof of Theorem 9.4.1. Hall (1984) gave details of the proof of (9.4.5) for the case $g = 1$ and suggested the possibility that the same method of proof would yield the above result for a general g. The proof presented here adopts Hall's proof for this general case. We begin with a lemma that is found useful in the proof.

Lemma 9.4.1. *If* φ *is square integrable,* K *and* g *are bounded and* $h \to 0$*, then the following hold.*

$$\iint \Big\{\int K(s)K(t+s)\varphi(z-h(t+s))\,ds\Big\}^2 g(z)g(z-ht)\,dz\,dt \to \Gamma. \tag{9.4.18}$$

$$h\iint \Big(\nu_n(z-ht)\nu_n(z)\Big)^2 g(z)g(z-ht)\,dz\,dt \to \Big(\int \varphi^2(x)g(x)\,dx\Big)^2. \tag{9.4.19}$$

Proof. We first prove (9.4.18). For this, let

$$D_h(t,z) = \int K(s)K(t+s)\big[\varphi(z-h(t+s)) - \varphi(z)\big]\,ds, \quad t,z \in \mathbb{R}.$$

Then we rewrite

$$\int K(s)K(t+s)\varphi(z-h(t+s))\,ds = J(t)\varphi(z) + D_h(t,z), \quad t,z \in \mathbb{R}.$$

Use this to rewrite the left-hand side of (9.4.18) as the sum $C_1(h) + 2C_2(h) + C_3(h)$, where

$$\begin{aligned} C_1(h) &:= \iint D_h^2(t,z))g(z)g(z-ht)\,dz\,dt, \\ C_2(h) &:= \iint D_h(t,z)J(t)\varphi(z)g(z)g(z-ht)\,dz\,dt, \\ C_3(h) &:= \iint J^2(t)\varphi^2(z)g(z)g(z-ht)\,dz\,dt. \end{aligned}$$

Since $\big(C_2(h)\big)^2 \le C_1(h)C_3(h)$, it suffices to show $C_1(h) \to 0$ and $C_3(h) \to \Gamma$.

Since K is a density, the inequality (2.4.4) gives

$$D_h^2(t,z) \le \int K^2(t+s)\Big(\varphi(z-h(t+s)) - \varphi(z)\Big)^2 K(s)\,ds.$$

Using this inequality and the fact that K and g are bounded we derive

$$\begin{aligned} C_1(h) &\le \|g\|_\infty^2\|K\|_\infty \iiint K(s)K(t+s)\Big(\varphi(z-h(t+s)) - \varphi(z)\Big)^2 ds\,dz\,dt \\ &= \|g\|_\infty^2\|K\|_\infty \iint K(u)\Big(\varphi(z-hu) - \varphi(z)\Big)^2 dz\,du. \end{aligned}$$

The last identity is obtained by using the Fubini Theorem 2.3.2, $\int K(s)\,ds = 1$ and the substitution $u = t+s$ in the previous bound. Since φ is square integrable and K is a density, we obtain $C_1(h) \to 0$ from Corollary 2.10.8.

To show $C_3(h) \to \Gamma$, we write $C_3(h) = C_{31}(h) + C_{32}(h) + \Gamma$, where

$$\begin{aligned} C_{31}(h) &:= \iint J^2(t)\varphi(z)g(z)g(z-ht)\Big[\varphi(z) - \varphi(z-ht)\Big]\,dz\,dt, \\ C_{32}(h) &:= \iint J^2(t)\varphi(z)g(z)\Big[\varphi(z-ht)g(z-ht) - \varphi(z)g(z)\Big]\,dz\,dt. \end{aligned}$$

Bound $\big(C_{31}(h)\big)^2$ by $\|g\|_\infty^2\Gamma\iint J^2(t)\big(\varphi(z-ht)-\varphi(z)\big)^2\,dz\,dt$, which tends to zero, by Corollary 2.10.8 in view of the square-integrability of J and φ. Similarly, with $\psi := \varphi g$, we bound $\big(C_{32}(h)\big)^2$ by $\Gamma\iint J^2(t)\big(\psi(z-ht)-\psi(z)\big)^2\,dz\,dt$, which tends to zero as J and ψ are square integrable. This proves $C_3(h) \to \Gamma$ and completes the proof of (9.4.18).

Now we prove (9.4.19). The substitution $x = z - ht$ with $dx = hdt$ gives

$$\begin{aligned} &h\iint \Big(\nu_n(z-ht)\nu_n(z)\Big)^2 g(z)g(z-ht)\,dz\,dt \\ &= \iint \big(\nu_n(x)\nu_n(z)\big)^2 g(x)g(z)\,dz\,dx = \Big(\int \nu_n^2(x)g(x)\,dx\Big)^2. \end{aligned}$$

It thus suffices to show that $\int \nu_n^2(x)g(x)\,dx \to \int \varphi^2(x)g(x)\,dx$.

To proceed, let $b_n(x) = \nu_n(x) - \varphi(x)$ and

$$B_1 := \int \varphi^2(x)g(x)\,dx, \quad B_2 := \int \varphi(x)b_n(x)g(x)\,dx, \quad B_3 := \int b_n^2(x)g(x)\,dx.$$

Writing $\nu_n(x) = \varphi(x) + b_n(x)$ and expanding the quadratic yields

$$\int \nu_n^2(x)g(x)\,dx = \int \big[\varphi(x) + b_n(x)\big]^2 g(x)\,dx = B_1 + 2B_2 + B_3.$$

Since $B_2^2 \leq B_1 B_3$, it suffices to show $B_3 \to 0$. But the latter follows from the inequality

$$\begin{aligned} B_3 &\leq \|g\|_\infty \int \Big(\int (\varphi(x - sh) - \varphi(x))K(s)\,ds\Big)^2 dx \\ &\leq \|g\|_\infty \iint \Big(\varphi(x - hs) - \varphi(x)\Big)^2 K(s)\,dx\,ds \to 0. \end{aligned}$$

In the last step we used square-integrability of φ and Corollary 2.10.8. □

Proof. [Proof of Theorem 9.4.1]. Let $K_h(\cdot) := K(\cdot/h)/h$ and define

$$\begin{aligned} \psi_h(x, z) &:= K_h(x - z) - \nu_n(x) = K_h(x - z) - E\big(K_h(x - X)\big), \\ H_n(y, z) &:= \sqrt{h} \int \psi_h(x, y)\psi_h(x, z)g(x)\,dx, \\ T_{n1} &:= \frac{1}{n}\sum_{i=1}^n h \int \psi_h^2(x, X_i)g(x)\,dx, \qquad U_n := \sum_{1\leq i<j\leq n} H_n(X_i, X_j). \end{aligned}$$

Then $\hat{f}_n(x) - \nu_n(x) = n^{-1}\sum_{i=1}^n \psi_h(x, X_i)$. Expand the quadratic in T_n to obtain

$$\begin{aligned} n\sqrt{h}T_n &= \frac{\sqrt{h}}{n}\sum_{i=1}^n \int \psi_h^2(x, X_i)g(x)\,dx + \frac{2\sqrt{h}}{n}\sum_{1\leq i<j\leq n} \int \psi_h(x, X_i)\psi_h(x, X_j)g(x)\,dx \\ &= \frac{T_{n1}}{\sqrt{h}} + \frac{2}{n}U_n. \end{aligned}$$

Hence, the claim (9.4.5) will follow from the following two facts.

$$\text{(a)} \quad \frac{2}{n}U_n \to_D \mathcal{N}(0, \tau^2), \qquad \text{(b)} \quad T_{n1} - A = o_p\big(\sqrt{h}\big). \tag{9.4.20}$$

Proof of (9.4.20)(a). Let X be a copy of X_1 and $G_n(y, z) := E\big(H_n(X, y)H_n(X, z)\big)$, $y, z \in \mathbb{R}$. We shall apply Corollary 5.9.1. Note that $H_n(y, z) = H_n(z, y)$, for all $y, z \in \mathbb{R}$. Because X_1, X_2 are independent, we have

$$E\big(H_n(X_1, X_2)\big|X_1\big) = \sqrt{h} \int \psi_h(x, X_1)E\big(\psi_h(x, X_2)\big)g(x)\,dx = 0.$$

In addition, we need to verify the following three conditions.

$$\text{(a1)}\ E\big(H_n^2(X_1, X_2)\big) \to \Gamma, \quad \text{(a2)}\ E\big(H_n^4(X_1, X_2)\big) = o(n), \quad \text{(a3)}\ E\big(G_n^2(X_1, X_2)\big) \to 0.$$

Using the independence of X_1 and X_2, we calculate

$$\begin{aligned}E\big(H_n^2(X_1,X_2)\big) &= hE\Big(\int \psi_h(x,X_1)\psi_h(x,X_2)g(x)\,dx\Big)^2 \\ &= h\iint E\big(\psi_h(x,X_1)\psi_h(z,X_1)\big)E\big(\psi_h(x,X_2)\psi_h(z,X_2)\big)g(x)g(z)\,dx\,dz \\ &= h\iint \Big[E\big(\psi_h(x,X_1)\psi_h(z,X_1)\big)\Big]^2 g(x)g(z)\,dx\,dz.\end{aligned}$$

Let $L(z,t) = \int K(s)K(t+s)\varphi\big(z-h(t+s)\big)\,ds$. The substitution $y = x - hs$ yields

$$\begin{aligned}\Psi_h(x,z) :&= E\big(\psi_h(x,X_1)\psi_h(z,X_1)\big) = \mathrm{Cov}\big(K_h(x-X_1),K_h(z-X_1)\big) \\ &= \int \frac{1}{h^2}K\Big(\frac{x-y}{h}\Big)K\Big(\frac{z-y}{h}\Big)\varphi(y)\,dy - \nu_n(x)\nu_n(z) \\ &= \frac{1}{h}\int K(s)K\Big(\frac{z-x}{h}+s\Big)\varphi(x-sh)\,ds - \nu_n(x)\nu_n(z).\end{aligned}$$

Next, substituting $x = z - ht$ yields

$$\begin{aligned}E\big(H_n^2(X_1,X_2)\big) &= h\iint \Psi_h^2(x,z)g(x)g(z)\,dx\,dz = \iint h^2\Psi_h^2(z-ht,z)g(z-ht)g(z)\,dt\,dz \\ &= \iint \Big[L^2(z,t) - 2hL(z,t)\nu_n(z-ht)\nu_n(z) \\ &\qquad\qquad + h^2\big(\nu_n(z-ht)\nu_n(z)\big)^2\Big]g(z-ht)g(z)\,dt\,dz.\end{aligned}$$

Hence, by Lemma 9.4.1, $E\big(H_n^2(X_1,X_2)\big) \to \Gamma$. This proves (a1).

Since K_h and ν_n are densities, both bounded by $\|K_h\|_\infty = \|K\|_\infty/h$, we derive the bound

$$\big|H_n(y,z)\big| \le \sqrt{h}\|g\|_\infty 2\|K_h\|_\infty \int \big(K_h(x-z)+\nu_n(x)\big)\,dx = 4h^{-1/2}\|g\|_\infty\|K\|_\infty.$$

Hence, with $C = 4\|g\|_\infty\|K\|_\infty$, we obtain

$$E\big(H_n^4(X_1,X_2)\big) \le \frac{C^2}{h}E\big(H_n^2(X_1,X_2)\big) = O\Big(\frac{1}{h}\Big). \tag{9.4.21}$$

This yields (a2) in view of (h).

With J as in (9.4.4), let $J_h(v) = J(v/h)/h$, $v \in \mathbb{R}$. Since K_h and ν_n are densities and $\|\nu_n\|_\infty \le \|\varphi\|_\infty$, we obtain

$$\begin{aligned}|H_n(x,z)| &\le \sqrt{h}\,\|g\|_\infty \int \big[K_h(s-x)+\nu_n(s)\big]\big[K_h(s-z)+\nu_n(s)\big]\,ds \\ &\le \sqrt{h}\,\|g\|_\infty\Big(\int K_h(s-x)K_h(s-z)\,ds + 3\|\nu_n\|_\infty\Big) \\ &\le \sqrt{h}\,\|g\|_\infty\Big(\int K(t)\frac{1}{h}K\Big(\frac{x-z}{h}+t\Big)\,dt + 3\|\varphi\|_\infty\Big) \\ &= \sqrt{h}\,\|g\|_\infty\Big(J_h(x-z) + 3\|\varphi\|_\infty\Big), \quad \forall x,z \in \mathbb{R}.\end{aligned}$$

Hence, for all $x, y, z \in \mathbb{R}$,

$$\begin{aligned}&\big|H_n(x,y)H_n(x,z)\big|\\&\le h\,\|g\|_\infty^2\Big(J_h(x-y)J_h(x-z)+3\|\varphi\|_\infty\big[J_h(x-y)+J_h(x-z)\big]+9\|\varphi\|_\infty^2\Big).\end{aligned}$$

Because J is a density, we obtain

$$E\big(J_h(X-z)\big)=\int J_h(x-z)\varphi(x)\,dx=\int\varphi(z+hs)J(s)\,ds\le\|\varphi\|_\infty.$$

These bounds together yield the following bound on $G_n(y,z)$.

$$\big|G_n(y,z)\big|=\Big|E\big(H_n(X,y)H_n(X,z)\big)\Big|\le h\,\|g\|_\infty^2\Big(D(y,z)+15\|\varphi\|_\infty^2\Big), \tag{9.4.22}$$

where

$$\begin{aligned}D(y,z)&:=E\big(J_h(X-y)J_h(X-z)\big)=\int\frac{1}{h^2}J\Big(\frac{x-y}{h}\Big)J\Big(\frac{x-z}{h}\Big)\varphi(x)\,dx\\&\le\frac{\|\varphi\|_\infty}{h}\int J(s)J\Big(\frac{y-z}{h}+s\Big)\,ds,\quad y,z\in\mathbb{R}.\end{aligned}$$

The above bound is obtained by first bounding φ and then using the substitution $x=y+hs$. Using the fact that J and φ are densities and the substitution $z=y-ht$ we calculate

$$\begin{aligned}E\big(D^2(X_1,X_2)\big)&=\iint D^2(y,z)\varphi(y)\varphi(z)\,dz\,dy\\&=\iint D^2(y,y-ht))\varphi(y)\varphi(y-ht)h\,dt\,dy\\&\le\frac{\|\varphi\|_\infty^3}{h}\iint\Big(\int J(s)J(t+s)\,ds\Big)^2\varphi(y)\,dt\,dy\\&\le\frac{\|\varphi\|_\infty^3}{h}\iint J(s)J^2(t+s)\,ds\,dt=\frac{\|\varphi\|_\infty^3}{h}I_2(J).\end{aligned}$$

This bound and the inequality (9.4.22) yield (a3). This completes the verification of the conditions of Corollary 5.9.1 and also the proof of (9.4.20)(a).

Proof of (9.4.20)(b). Note the identity $T_{n1}:=n^{-1}\sum_{i=1}^n\zeta_{ni}$, where

$$\zeta_{ni}:=\int h\Big[K_h(x-X_i)-\nu_n(x)]^2g(x)\,dx,\quad i=1,\dots,n.$$

The r.v.'s ζ_{ni} are bounded. Indeed, since $\nu_n\le\|\varphi\|_\infty$ and $\int\nu_n(x)\,dx=1$, using the inequality $(a+b)^2\le2a^2+2b^2$, $a,b\in\mathbb{R}$, we calculate

$$\zeta_{ni}\le\|g\|_\infty\int\Big[2hK_h^2(x-X_i)+2h\nu_n^2(x)\Big]\,dx\le2\|g\|_\infty\Big[I_2(K)+h\|\varphi\|_\infty\Big].$$

This implies $\mathrm{Var}\big(T_{n1}\big)=O(1/n)$ and hence

$$T_{n1}=E(\zeta_{n1})+O_p\big(n^{-1/2}\big)=E(\zeta_{n1})+o_p(h^{1/2}). \tag{9.4.23}$$

Next, the substitution $y = x - hu$ gives

$$\begin{aligned}
E(\zeta_{n1}) &= \int h\mathrm{Var}(K_h(x - X_1))g(x)\,dx \\
&= \int \Big[h \int K_h^2(x-y)\varphi(y)\,dy - h\nu_n^2(x)\Big] g(x)\,dx \\
&= \iint K^2(u)\varphi(x-hu)g(x)\,du\,dx - h\int \nu_n^2(x)g(x)\,dx \\
&= A + \iint K^2(u)g(x)\big(\varphi(x-hu) - \varphi(x)\big)\,dx\,du + O(h).
\end{aligned}$$

Because φ is L_1-Lipschitz, see assumption (f), the second term in the right-hand side of the above bound is bounded by $\|g\|_\infty Lh\int |u|K^2(u)\,du$ with L a L_1-Lipschitz constant. Hence, $E(\zeta_{n1}) = A + O(h)$. Combining this with (9.4.23) yields the desired result (9.4.20)(b). This completes the proof of Theorem 9.4.1. □

9.5 Lack-of-fit Tests

Another classical problem in statistics is to assess the effect of a p-dimensional random vector X on the response variable Y. Assuming Y has finite expectation, this is often done in terms of the function $\mu(x) := E(Y|X = x)$. In practice one often is interested in fitting a given parametric model to the function μ. We assume from now on that Y has a finite second moment.

Proceeding a bit more precisely, let $\{m_b(x) : x \in \mathbb{R}^p,\ b \in \Theta\}$ be a given parametric model for the regression function with $\Theta \subset \mathbb{R}^q$. We are given a random sample $\{(X_i, Y_i) : i = 1, \cdots, n\}$ from the distribution of (X, Y) and the problem of interest here is to test the hypothesis

$$\mathcal{H}_{LF} : \mu = m_\beta, \quad \text{for some } \beta \in \Theta,$$

against the alternative that $\mathcal{H}_{LF}$ is false.

In this section we shall discuss a class of tests for this problem. These tests are based on minimized integrated square differences between a nonparametric estimator of the regression function and the null model being fitted. Moreover, assuming that the given parametric model holds, one is interested in finding the parameter in the given family that best fits the data, i.e., the corresponding minimum distance estimator.

To proceed further, let K be a density on $\mathbb{R}^p$ and $h = h_n > 0$ be a bandwidth and define $K_h(u) := h^{-p}K(u/h)$, $u \in \mathbb{R}^p$. For $x \in \mathbb{R}^p$ and $b \in \Theta$, set

$$\nu_n(x) := \frac{1}{n}\sum_{i=1}^n K_h(x - X_i)Y_i, \qquad \nu_n(x, b) := \frac{1}{n}\sum_{i=1}^n K_h(x - X_i)m_b(X_i).$$

Note that $\nu_n(x, \beta)$ is the conditional expectation of $\nu_n(x)$, given $\boldsymbol{X}_n := (X_1, \cdots, X_n)$, under $\mathcal{H}_{LF}$.

Let $\hat{w}_n(x) = w_n(x, \boldsymbol{X}_n)$, $x \in \mathbb{R}^p$, be a nonnegative weight function and define

$$T_n(b) := \int \big(\nu_n(x) - \nu_n(x, b)\big)^2 \hat{w}_n(x)\,dx, \quad b \in \Theta,$$

$$T_n^* := \inf_{b\in\Theta} T_n(b), \qquad \hat{\beta}_n := \operatorname*{argmin}_{b\in\Theta} T_n(b).$$

Then $m_{\hat{\beta}_n}$ is the member of the model $\{m_b, b \in \Theta\}$, that best fits the data with respect to the above distance.

Example 9.5.1. Let $\mathcal{C}$ be a closed rectangle of the form $[a_1, b_1] \times \cdots \times [a_p, b_p]$ in $\mathbb{R}^p$ and ϕ be a positive function on $\mathcal{C}$. A possible choice for $\hat{w}_n$ is given by

$$\hat{w}_n(x) = I(x \in \mathcal{C})\frac{\phi(x)}{\hat{f}_n^2(x)}, \quad x \in \mathbb{R}^p, \tag{9.5.1}$$

where $\hat{f}_n$ is the multivariate kernel density estimator with kernel K and bandwidth h,

$$\hat{f}_n(x) := \frac{1}{n}\sum_{i=1}^n K_h(x - X_i), \quad x \in \mathbb{R}^p.$$

For this choice,

$$T_n(b) = \int_{\mathcal{C}} \Big(\frac{\nu_n(x)}{\hat{f}_n(x)} - \frac{\nu_n(x,b)}{\hat{f}_n(x)}\Big)^2 \phi(x)\,dx. \tag{9.5.2}$$

Observe that $\hat{\mu}_n(x) := \nu_n(x)/\hat{f}_n(x)$ is a multivariate version of the Nadaraya-Watson estimator and $\nu_n(x,b)/\hat{f}_n(x)$ is the conditional expectation of $\hat{\mu}_n(x)$, given $\boldsymbol{X}_n$, when $\mu = m_b$. This choice of $\hat{w}_n$ was proposed by Koul and Ni (2004). They also allowed for the choice of a different kernel and bandwidth for estimating f than used for ν_n. See the end of this section for more on this example.

The exact null distribution of T_n^* is not easily tractable. One either uses its asymptotic null distribution or its bootstrap distribution to implement the test. The proof of the asymptotic normality for a general parametric model is long and technical. For that reason we shall derive the asymptotic null distribution only for a special parametric model.

From now on we assume

$$m_b := b'\psi, \quad b \in \Theta := \mathbb{R}^q, \tag{9.5.3}$$

where ψ is a measurable functions from $\mathbb{R}^p$ to $\mathbb{R}^q$ with $E(\|\psi(X)\|^2)$ finite and $\mathcal{H}_{LF}$ is understood to mean $\mathcal{H}_{LF}$ with m_b as in (9.5.3). Since $E(Y^2)$ and $E(\|\psi(X)\|^2)$ are finite, $\varepsilon := Y - \beta'\psi(X)$ has a finite second moment and $\sigma^2(x) := E(\varepsilon^2|X = x)$ is well defined for all x. Now set $\varepsilon_i := Y_i - \beta'\psi(X_i)$, $1 \le i \le n$, and

$$S_n(x) := \frac{1}{n}\sum_{i=1}^n K_h(x - X_i)\varepsilon_i, \quad \gamma_n(x) := \frac{1}{n}\sum_{i=1}^n K_h(x - X_i)\psi(X_i),$$

$$\hat{A}_n := \int S_n^2(x)\hat{w}_n(x)\,dx, \quad \hat{B}_n := \int S_n(x)\gamma_n(x)\hat{w}_n(x)\,dx, \quad \hat{C}_n := \int \gamma_n(x)\gamma_n(x)'\hat{w}_n(x)\,dx.$$

Under $\mathcal{H}_{LF}$,

$$\nu_n(x) - \nu_n(x,b) = \frac{1}{n}\sum_{i=1}^n K_h(x - x_i)\big(Y_i - b'\psi(X_i)\big) = S_n(x) - (b-\beta)'\gamma_n(x),$$

$$T_n(b) = \int \big(S_n(x) - (b-\beta)'\gamma_n(x)\big)^2 \hat{w}_n(x)\,dx = \hat{A}_n - 2(b-\beta)'\hat{B}_n + (b-\beta)'\hat{C}_n(b-\beta).$$

Clearly, the minimizer $\hat{\beta}_n$ of the criterion function satisfies $\hat{\beta}_n - \beta = \hat{C}_n^{-1}\hat{B}_n$ on the event where $\hat{C}_n$ is invertible and T_n^* equals $\hat{A}_n - \hat{B}_n'\hat{C}_n^{-1}\hat{B}_n$ on this event. If we show $\hat{C}_n \to_p C_0$ and $n^{1/2}\hat{B}_n \to_D \mathcal{N}_q(0, \Sigma)$ for some positive definite $q \times q$ matrices C_0 and Σ, then we can conclude

$$n^{1/2}(\hat{\beta}_n - \beta) \to_D \mathcal{N}(0, C_0^{-1}\Sigma C_0^{-1}) \quad \text{and} \quad T_n^* = \hat{A}_n + O_p(n^{-1}).$$

Furthermore, if we show $nh^{p/2}\big(\hat{A}_n - A_n\big)/\tau_n \to_D \mathcal{N}(0,1)$, for some statistics A_n and τ_n with $\tau_n > 0$, then we obtain the asymptotic normality result

$$nh^{p/2}\big(T_n^* - A_n\big)/\tau_n \to_D \mathcal{N}(0,1).$$

Consequently, the test that rejects $\mathcal{H}_{LF}$, whenever $\big|nh^{p/2}\big(T_n^* - A_n\big)/\tau_n\big| > z_{\alpha/2}$ has asymptotic size α.

To implement this plan, we impose the following conditions and introduce additional ones as needed.

(f) The random vector X has a bounded density f.

(h) The bandwidth h satisfies $h \to 0$ and $nh^{2p} \to \infty$.

(k) The density K is bounded.

(w) There is a bounded nonnegative almost everywhere continuous function w such that $\|\hat{w}_n - w\|_\infty = o_p(1)$.

Let $\hat{\varepsilon}_i := \big(Y_i - \hat{\beta}_n'\psi(X_i)\big), i = 1, \cdots, n$. The above plan will be carried out under the assumptions (f)–(w) with

$$C_0 := \int \psi(x)\psi(x)'f^2(x)w(x)\,dx, \qquad \Sigma := \int \sigma^2(x)\psi(x)\psi'(x)f^3(x)w^2(x)\,dx, \tag{9.5.4}$$

$$A_n := \frac{1}{n^2}\sum_{i=1}^n \hat{\varepsilon}_i^2 \int K_h^2(x - X_i)\hat{w}_n(x)\,dx,$$

$$\Gamma_n := \frac{2h^p}{n(n-1)}\sum_{1\le i<j\le n} \hat{\varepsilon}_i^2\hat{\varepsilon}_j^2\Big(\int K_h(x - X_i)K_h(x - X_j)\hat{w}_n(x)\,dx\Big)^2, \quad \tau_n := \big(2\Gamma_n\big)^{1/2}.$$

We are ready to carry out the first step of the above plan. For any real-valued square integrable function ψ on $\mathbb{R}^p$, let $I_2(\psi) := \int \psi^2(x)\,dx$.

Lemma 9.5.1. *Under the above conditions, $\hat{C}_n \to_p C_0$.*

Proof. Since $\hat{C}_n$ is a symmetric matrix, it suffices to show $a'\hat{C}_n a \to_p a'C_0 a$, for all $a \in \mathbb{R}^q$. Fix $a \in \mathbb{R}^q$ and set $\hat{v} := a'\gamma_n$ and $v := a'\psi f$. Note that v is square integrable as f is bounded and $E(\|\psi(X)\|^2)$ is finite. Observe the identities $a'\hat{C}_n a = \int \big(\hat{v}(x)\big)^2\hat{w}_n(x)\,dx$ and $a'C_0 a = \int v^2(x)w(x)\,dx$. Condition (w) implies $\int v^2(x)\hat{w}_n(x)\,dx \to_p \int v^2(x)w(x)\,dx$ and $\|\hat{w}_n\|_\infty = O_p(1)$. It remains to show $\int |\hat{v}^2(x) - v^2(x)|\,dx = o_p(1)$. In view of the inequality $a^2 - b^2 = (a-b)^2 + 2(a-b)b$, for all $a, b \in \mathbb{R}$, and (2.4.6), this is implied by

$$D_n := \int \big(\hat{v}(x) - v(x)\big)^2\,dx = o_p(1).$$

Let $v_2 = (a'\psi)^2 f$. Then,

$$\hat{v}(x) = \frac{1}{n}\sum_{i=1}^n K_h(x - X_j)a'\psi(X_j), \qquad \bar{v}(x) := E\big(\hat{v}(x)\big) = \int K(t)v(x - ht)\,dt,$$

$$\begin{aligned}\mathrm{Var}\big(\hat{v}(x)\big) &= E\big[\big(\hat{v}(x) - \bar{v}(x)\big)^2\big] \le \frac{1}{n}E\big[K_h^2(x - X)(a'\psi(X))^2\big]\\ &= \frac{1}{n}\int K_h^2(x-y)v_2(y)\,dy = \frac{1}{nh^p}\int K^2(s)v_2(x - hs)\,ds.\end{aligned}$$

In the last step we used the substitution $y = x - hs$ with $dy = h^p\,ds$. Hence, by (h) and (k),

$$I_{n1} := \int \mathrm{Var}(\hat{v}(x))\,dx = \int E\big[(\hat{v}(x) - \bar{v}(x))^2\big]\,dx \le \frac{1}{nh^p} I_2(K) \int v_2(x)\,dx \to 0. \quad (9.5.5)$$

Moreover, by the square integrability of v, (2.4.4) and Corollary 2.10.8,

$$I_{n2} := \int \big(\bar{v}(x) - v(x)\big)^2\,dx \le \iint \big(v(x - ht) - v(x)\big)^2\,dxK(t)\,dt \to 0.$$

Hence, $E(D_n) = \int E\big[(\hat{v}(x) - v(x))^2\big]\,dx = I_{n1} + I_{n2} \to 0$. This implies $D_n = o_p(1)$ and completes the proof. □

The next lemma proves the asymptotic normality of $n^{1/2}\hat{B}_n$, $n^{1/2}(\hat{\beta}_n - \beta)$ and the approximation of T_n^* by $\hat{A}_n$. Note that the limiting distribution of the class of standardized minimum distance estimator $n^{1/2}(\hat{\beta}_n - \beta)$, as K varies among all bounded densities on $\mathbb{R}^p$, does not depend on K.

Lemma 9.5.2. *In addition to* (f), (h), (k) *and* (w), *assume C_0 and Σ are positive definite and*

$$E\big(\sigma^4(X) + \|\psi(X)\|^4\big) < \infty. \quad (9.5.6)$$

Then, $n^{1/2}\hat{B}_n \to_D \mathcal{N}_q(0, \Sigma)$, $n^{1/2}(\hat{\beta}_n - \beta) \to_D \mathcal{N}_p(0, C_0^{-1}\Sigma C_0^{-1})$ and

$$T_n^* = \hat{A}_n + O_p(1/n). \quad (9.5.7)$$

Proof. By (9.5.6) and (2.4.4), $\sigma^2\|\psi\|^2 f$ is integrable. Thus the boundedness of f and w imply that Σ is well defined. Since the assumptions of Lemma 9.5.1 are satisfied, we have $\hat{C}_n \to_p C_0$. As C_0 and Σ are positive definite, it suffices to show $n^{1/2}\hat{B}_n \to_D \mathcal{N}_q(0, \Sigma)$, since these properties and Theorem 2.10.5 imply $n^{1/2}(\hat{\beta}_n - \beta) \to_D \mathcal{N}_p(0, C_0^{-1}\Sigma C_0^{-1})$ and (9.5.7), as explained earlier.

By the Cramér-Wold device, see Corollary 2.10.1, $n^{1/2}\hat{B}_n \to_D \mathcal{N}(0, \Sigma)$ follows if we show $n^{1/2}a'\hat{B}_n \to_D \mathcal{N}(0, a'\Sigma a)$, for every unit vector $a \in \mathbb{R}^p$. Fix such a unit vector a. Let $\hat{v}, \bar{v}, v$ be as in the previous proof. With κ equal to $\hat{v}$, $\hat{v} - \bar{v}$ or $\bar{v}$, set

$$u_n(X_i, \kappa) = \int K_h(x - X_i)\kappa(x)\hat{w}_n(x)\,dx, \quad i = 1, \cdots, n.$$

Then $n^{1/2}a'\hat{B}_n = n^{-1/2}\sum_{i=1}^n \varepsilon_i u_n(X_i, \hat{v})$. Note the identity

$$a'\Sigma a = \int \sigma^2(x)v^2(x)w^2(x)f(x)\,dx = E\big[\big(\varepsilon\, v(X)w(X)\big)^2\big].$$

By the CLT, $n^{-1/2}\sum_{i=1}^n \varepsilon_i v(X_i)w(X_i) \to_D \mathcal{N}(0, a'\Sigma a)$ and the desired result follows if we show

$$D_n := n^{-1/2}\sum_{i=1}^n \varepsilon_i\big[u_n(X_i, \hat{v}) - v(X_i)w(X_i)\big] = o_p(1).$$

In view of the identity $D_n = D_{n1} + D_{n2} + D_{n3} + D_{n4}$ with

$$D_{n1} : = n^{-1/2}\sum_{i=1}^n \varepsilon_i u_n(X_i, \hat{v} - \bar{v}),$$

$$D_{n2} : = n^{-1/2}\sum_{i=1}^n \varepsilon_i\Big(u_n(X_i, \bar{v}) - v(X_i)\int \hat{w}_n(X_i + hs)K(s)\,ds\Big),$$

$$D_{n3} : = n^{-1/2}\sum_{i=1}^n \varepsilon_i v(X_i)\int \Big(\hat{w}_n(X_i + hs) - w(X_i + hs)\Big)K(s)\,ds,$$

$$D_{n4} : = n^{-1/2}\sum_{i=1}^n \varepsilon_i v(X_i)\Big(\int w(X_i + hs)K(s)\,ds - w(X_i)\Big),$$

it suffices to verify $D_{nk} = o_p(1)$, for all $k = 1, 2, 3, 4$.

Since w is bounded, so is the function $w_h(x) := \int w(x + sh)K(s)\,ds$, $x \in \mathbb{R}^p$. Moreover, $w_h(x) \to w(x)$ for each continuity point x of w. Thus, under (w), w_h is bounded and converges to w almost everywhere. Because $E(\varepsilon|X) = 0$, we have

$$E(D_{n4}^2) = \int \sigma^2(x)v^2(x)f(x)\big(w_h(x) - w(x)\big)^2\,dx$$

and $E(D_{n4}^2) \to 0$ by Theorem 2.10.10 and the integrability of $\sigma^2 v^2 f$. Hence, $D_{n4} = o_p(1)$.

Let $\boldsymbol{X}_n := (X_1, \cdots, X_n)'$. We shall show that $E(D_{nk}^2|\boldsymbol{X}_n) = o_p(1)$, for all $k = 1, 2, 3$. Then Lemma 2.10.7 implies $D_{nk} = o_p(1)$, $k = 1, 2, 3$. We have

$$E(D_{n3}^2|\boldsymbol{X}_n) \le \frac{1}{n}\sum_{i=1}^n \sigma^2(X_i)v^2(X_i)\|\hat{w}_n - w\|_\infty^2 = o_p(1),$$

by (w) and since $E\big(\sigma^2(X)v^2(X)\big)$ is finite as $\sigma^2 v^2 f$ is integrable.

The substitution $x = X_i + hs$ and the definition of $\bar{v}$ yield

$$u_n(X_i, \bar{v}) = \iint v(X_i + h(s-t))\hat{w}_n(X_i + hs)K(s)K(t)\,ds\,dt.$$

Let $r_n(x) := \iint |v(x + h(s-t)) - v(x)|K(s)K(t)\,ds\,dt$. Then, we compute

$$E(D_{n2}^2|\boldsymbol{X}_n) \le \frac{1}{n}\sum_{i=1}^n \sigma^2(X_i)r_n^2(X_i)\|\hat{w}_n\|_\infty^2.$$

Moreover,

$$E\Big(\frac{1}{n}\sum_{i=1}^n \sigma^2(X_i)r_n^2(X_i)\Big) = \int \sigma^2(x)r_n^2(x)f(x)\,dx$$

is bounded by the square root of

$$\int \sigma^4(x)f^2(x)\,dx \iiint \big(v(x + h(s-t)) - v(x)\big)^4\,dx K(s)K(t)\,ds\,dt.$$

This bound tends to zero in view of Lemma 2.10.10, because $\sigma^4 f^2 \le \|f\|_\infty \sigma^4 f$ and $v^4 = (a'\psi f)^4 \le \|a\|^4\|f\|_\infty^3\|\psi\|^4 f$ are integrable in view of (9.5.6) and (f). This establishes $E(D_{n2}^2|\boldsymbol{X}_n) = o_p(1)$. Finally, we have the bound

$$\begin{aligned}
u_n^2(X_i, \hat{v} - \bar{v}) &= \Big(\int K_h(x - X_i)\big(\hat{v}(x) - \bar{v}(x)\big)\hat{w}_n(x)\,dx\Big)^2 \\
&\le \|\hat{w}_n\|_\infty^2 \int K_h^2(x - X_i)\,dx \int \big(\hat{v}(x) - \bar{v}(x)\big)^2\,dx, \\
&= \|\hat{w}_n\|_\infty^2 h^{-p} I_2(K) \int \big(\hat{v}(x) - \bar{v}(x)\big)^2\,dx,
\end{aligned}$$

where we used the substitution $x = X_i + hs$. By (9.5.5), we have $\int \big(\hat{v}_n(x) - \bar{v}(x)\big)^2\,dx = O_p((nh^p)^{-1})$. Indeed, we established its expected value I_{n1} is of order $O((nh^p)^{-1})$. Hence,

$$E(D_{n1}^2|\boldsymbol{X}_n) = \frac{1}{n}\sum_{i=1}^n \sigma^2(X_i)u_n^2(X_i, \hat{v} - \bar{v}) \le \frac{1}{n}\sum_{i=1}^n \sigma^2(X_i)O_p\Big(\frac{1}{nh^{2p}}\Big) = o_p(1),$$

by (h). □

Next, we analyse $\hat{A}_n$. For this, with $J(t) = \int K(s)K(t+s)\,ds$, set

$$\hat{A}_{n,1} := \frac{1}{n^2}\sum_{i=1}^{n} \varepsilon_i^2 \int K_h^2(x - X_i)\hat{w}_n(x)\,dx, \quad \Gamma := I_2(J)\int \sigma^4(x)f^2(x)w^2(x)\,dx. \quad (9.5.8)$$

Lemma 9.5.3. *Suppose* (f), (h), (k) *and* (w) *hold,* $E(\varepsilon^4)$ *and* $\int \sigma^8(x)f^2(x)\,dx$ *are finite and* Γ *is positive. Then,* $nh^{p/2}(\hat{A}_n - \hat{A}_{n,1}) \to_D \mathcal{N}(0, 2\Gamma)$.

Proof. Since $S_n(x) = \frac{1}{n}\sum_{i=1}^{n} \varepsilon_i K_h(x - X_i)$, we derive

$$S_n^2(x) = \frac{1}{n^2}\sum_{i=1}^{n} \varepsilon_i^2 K_h^2(x - X_i) + \frac{2}{n^2}\sum_{1\le i<j\le n} \varepsilon_i\varepsilon_j K_h(x - X_i)K_h(x - X_j).$$

For $\kappa = \hat{w}_n$, w or $\hat{w} - w$, we set, for $1 \le i < j \le n$,

$$\begin{aligned} W_n(X_i, X_j, \kappa) &:= h^{p/2}\int K_h(x - X_i)K_h(x - X_j)\kappa(x)\,dx \quad (9.5.9)\\ &= h^{p/2}\int K(u)K_h(X_i - X_j + hu)\kappa(X_i + hu)\,du. \end{aligned}$$

The last identity was derived using the substitution $x = X_i + hu$. Let $J_h(t) = h^{-p}J(t/h)$. Then for $1 \le i < j \le n$, we have the inequalities

$$\left|W_n(X_i, X_j, \kappa)\right| \le \|\kappa\|_\infty h^{p/2} J_h(X_i - X_j) \le \|\kappa\|_\infty h^{-p/2}\|K\|_\infty. \quad (9.5.10)$$

From the above decomposition of S_n^2 it follows that $nh^{p/2}(\hat{A}_n - \hat{A}_{n,1})$ equals $U_n(\hat{w}_n) = U_n(w) + U_n(\hat{w}_n - w)$, where for $\kappa = \hat{w}_n$, w or $\hat{w} - w$,

$$U_n(\kappa) := \frac{2}{n}\sum_{1\le i<j\le n} \varepsilon_i\varepsilon_j W_n(X_i, X_j, \kappa).$$

It thus suffices to prove that $U_n(w) \to_D \mathcal{N}(0, 2\Gamma)$ and $U_n(\hat{w}_n - w) = o_p(1)$.

To simplify notation we set

$$g(x) := \sigma^2(x)f(x), \qquad x \in \mathbb{R}^p.$$

Note that g^2 and g^4 are integrable, the latter by assumption and the former as ε has a finite fourth moment and f is bounded. We first prove $U_n(\hat{w}_n - w) = o_p(1)$. For this we set

$$R_n := \frac{4}{n^2}\sum_{1\le i<j\le n} \sigma^2(X_i)\sigma^2(X_j)h^p J_h^2(X_i - X_j).$$

Direct calculations yield

$$E\big(U_n^2(\hat{w}_n - w)\big|\boldsymbol{X}_n\big) = \frac{4}{n^2}\sum_{1\le i<j\le n} \sigma^2(X_i)\sigma^2(X_j)W_n^2(X_i, X_j, \hat{w}_n - w) \le \|\hat{w}_n - w\|_\infty^2 R_n.$$

Moreover,

$$E(R_n) = \frac{2(n-1)}{n}\iint g(x)g(y)\,h^p J_h^2(x - y)\,dx\,dy.$$

An application of (2.4.6) and the identity

$$\int h^p J_h^2(x - y)\,dy = \int J^2(t)\,dt = \int h^p J_h^2(x - y)\,dx$$

yield

$$E(R_n) \le 2\Big(\iint g^2(x)h^p J_h^2(x-y)\,dx\,dy \iint g^2(y)h^p J_h^2(x-y)\,dx\,dy\Big)^{1/2} = 2I_2(g)I_2(J).$$

This shows that $R_n = O_p(1)$ and yields $E\big(U_n^2(\hat{w}_n - w)\big|\boldsymbol{X}_n\big) = o_p(1)$, by (w). The later implies $U_n(\hat{w}_n - w) = o_p(1)$, by Lemma 2.10.7.

We shall use Corollary 5.9.1 to prove the asymptotic normality of $U_n(w)$. For this we set $\zeta_i = (\varepsilon_i, X_i)$, $i = 1, \cdots, n$ and

$$H_n(\zeta_i, \zeta_j) := \varepsilon_i\varepsilon_j W_n(X_i, X_j, w). \tag{9.5.11}$$

We calculate $G_n(\zeta_1,\zeta_2) := E\Big(H_n(\zeta_3,\zeta_1)H_n(\zeta_3,\zeta_2)\big|\zeta_1,\zeta_2\Big) = \varepsilon_1\varepsilon_2 L_n(X_1,X_2)$ with

$$L_n(X_1,X_2) = \int g(z)W_n(z,X_1,w)W_n(z,X_2,w)\,dz.$$

In view of (h), the condition of Corollary 5.9.1 are met if we establish

(a) $E\big(H_n^2(\zeta_1,\zeta_2)\big) \to \Gamma$, (b) $E\big(H_n^4(\zeta_1,\zeta_2)\big) = O\big(h^{-2p}\big)$, (c) $E\big(G_n^2(\zeta_1,\zeta_2)\big) = O\big(h^p\big)$.

Using the inequality (9.5.10) we obtain

$$E\big(H_n^4(\zeta_1,\zeta_2)\big) = E\big[\big(\varepsilon_1\varepsilon_2 W_n(X_1,X_2,w)\big)^4\big] \le E(\varepsilon_1^4)E(\varepsilon_2^4)h^{-2p}\|w\|_\infty^4\|K\|_\infty^4$$

which yields (b). Next, we compute

$$E\big(H_n^2(\zeta_1,\zeta_2)\big) = E\big\{\big(\varepsilon_1\varepsilon_2 W_n(X_1,X_2,w)\big)^2\big\} = \iint g(x)g(y)W_n^2(x,y,w)\,dx\,dy.$$

Using the substitution $y = x - ht$, this expectation equals

$$\iint g(x)g(x-ht)h^{2p}\Big(\int K(s)h^{-p}K(t+s)w(x+hs)\,ds\Big)^2\,dx\,dt$$

and converges to

$$\iint g^2(x)w^2(x)\Big(\int K(s)K(t+s)\,ds\Big)^2\,dx\,dt = I_2(J)\int \sigma^4(x)f^2(x)w^2(x)\,dx = \Gamma.$$

This convergence uses the square integrability of g, $\int K(s)K(t+s)w(x+hs)\,ds \to w(x)J(t)$, almost everywhere x, which follows from (w), the boundedness of K and w. This establishes (a).

Next, we prove (c). By (9.5.10) and the substitution $z = X_1 + ht$, we obtain

$$\begin{aligned} L_n(X_1,X_2) &\le \|w\|_\infty^2 \int g(z)h^p J_h(z-X_1)J_h(z-X_2)\,dz \\ &= \|w\|_\infty^2 \int J(t)J\Big(\frac{X_1-X_2}{h}+t\Big)g(X_1+ht)\,dt. \end{aligned}$$

Recall J is a density. The above bound, the moment inequality (2.4.7), and then the substitution $y = X_1 - hs + ht$ ($dy = h^p\,ds$) yield

$$\begin{aligned} \int L_n^2(X_1,y)\,dy &\le \|w\|_\infty^4 \iint J(t)J^2\Big(\frac{X_1-y}{h}+t\Big)g^2(X_1+ht)\,dt\,dy \\ &= \|w\|_\infty^4 h^p I_2(J)\int g^2(X_1+ht)J(t)\,dt. \end{aligned}$$

In view of the identity $L_n(X_1, X_2) = L_n(X_2, X_1)$, we derive, by using the independence of (X_1, ε_1) and (X_2, ε_2), (2.4.6) and properties of the conditional expectation (2.7.6), the bound

$$\begin{aligned} E\big(G_n^2(\zeta_1,\zeta_2)\big) &= E\big(\varepsilon_1^2 L_n(X_1,X_2)\varepsilon_2^2 L_n(X_2,X_1)\big) \\ &= \iint g(x)L_n(x,y)g(y)L_n(y,x)\,dx\,dy \\ &\le \iint g^2(x)L_n^2(x,y)\,dx\,dy \\ &\le \|w\|_\infty^4 h^p I_2(J) \iint g^2(x)g^2(x+ht)J(t)\,dt\,dx \end{aligned}$$

with the last double integral bounded by $\int g^4(x)\,dx$. This establishes (c). Since the assumptions of Corollary 5.9.1 are met, we conclude $U(w) \to_D \mathcal{N}(0, 2\Gamma)$. □

The following theorem completes the step 3 of the plan with A_n as in (9.5.4).

Theorem 9.5.1. *Suppose* (f), (h), (w), (k) *hold,* $E(\varepsilon^4)$, $E(\|\psi(X)\|^4)$ *and* $\int \sigma^8(x) f^4(x)\,dx$ *are finite,* C_0 *and* Σ *are positive definite and* Γ *is positive. Then*

$$\text{(a)}\quad nh^{p/2}(A_n - \hat{A}_{n,1}) = o_p(1), \qquad \text{(b)}\quad \Gamma_n = \Gamma + o_p(1). \tag{9.5.12}$$

Consequently, $nh^{p/2}(T_n^* - A_n)/(2\Gamma_n)^{1/2} \to_D \mathcal{N}(0,1)$.

Proof. To prove the claim about the normality of T_n^*, we start with the decomposition

$$nh^{p/2}(T_n^* - A_n) = nh^{p/2}(T_n^* - \hat{A}_n) + nh^{p/2}(\hat{A}_n - \hat{A}_{n,1}) - nh^{p/2}(A_n - \hat{A}_{n,1}).$$

By (9.5.7), Lemma 9.5.3, part (a) of the present theorem and Theorem 2.10.5, we obtain

$$nh^{p/2}(T_n^* - A_n) = O_p(h^{p/2}) + nh^{p/2}(\hat{A}_n - \hat{A}_{n,1}) - o_p(1) \to_D \mathcal{N}(0, 2\Gamma).$$

Hence, by part (b) of this theorem and Theorem 2.10.5, $nh^{p/2}(T_n^* - A_n)/(2\Gamma_n)^{1/2} \to_D \mathcal{N}(0,1)$.

We proceed to prove parts (a) and (b) of this theorem. Let $\Delta_n := n^{1/2}(\hat{\beta}_n - \beta)$. By Lemma 9.5.2, $\|\Delta_n\| = O_p(1)$. Set

$$\begin{aligned} L_n &:= \frac{h^{p/2}}{n\sqrt{n}} \sum_{i=1}^n \varepsilon_i \Delta_n' \psi(X_i) \int K_h^2(x - X_i)\hat{w}_n(x)\,dx, \\ Q_n &:= \frac{h^{p/2}}{n^2} \sum_{i=1}^n (\Delta_n' \psi(X_i))^2 \int K_h^2(x - X_i)\hat{w}_n(x)\,dx. \end{aligned}$$

Expand the quadratic $\hat{\varepsilon}_i^2 = \big(\varepsilon_i - n^{-1/2}\Delta_n'\psi(X_i)\big)^2$ to obtain $nh^{p/2}(A_n - \hat{A}_{n,1}) = -2L_n + Q_n$. Thus statement (a) follows if we show $L_n = o_p(1)$ and $Q_n = o_p(1)$. Because $E\big(|\varepsilon|\|\psi(X)\|\big)$ and $E\big(\|\psi(X)\|^2\big)$, are finite, the Markov inequality (2.4.8) gives

$$n^{-1}\sum_{i=1}^n |\varepsilon_i|\|\psi(X_i)\| + n^{-1}\sum_{i=1}^n \|\psi(X_i)\|^2 = O_p(1). \tag{9.5.13}$$

We also have the inequality

$$\int K_h^2(x - X_i)\hat{w}_n(x)\,dx \le \frac{\|\hat{w}_n\|_\infty I_2(K)}{h^p}.$$

This inequality, (9.5.13), the consequence $\|\hat{w}_n\|_\infty = O_p(1)$ of (w), and $nh^p \to \infty$, yield

$$|L_n| \le \frac{\|\hat{w}_n\|_\infty I_2(K)\|\Delta_n\|}{n(nh^p)^{1/2}} \sum_{i=1}^n |\varepsilon_i| \|\psi(X_i)\| = O_p\Big(\big(nh^p\big)^{-1/2}\Big) = o_p(1),$$

$$Q_n \le \frac{\|\hat{w}_n\|_\infty I_2(K)\|\Delta_n\|^2}{n^2 h^{p/2}} \sum_{i=1}^n \|\psi(X_i)\|^2 = O_p\Big(\big(nh^{p/2}\big)^{-1}\Big) = o_p(1).$$

This completes the proof of statement (a).

The details of the proof of statement (b) are routine but long. Let ζ_i, $W_n(X_i, X_j, \kappa)$ and $H_n(\zeta_i, \zeta_j)$ be as in the the proof of Lemma 9.5.3; see (9.5.9) and (9.5.11). Then,

$$\Gamma_n = \frac{2}{n(n-1)} \sum_{1\le i<j\le n} \hat{\varepsilon}_i^2 \hat{\varepsilon}_j^2 W_n^2(X_i, X_j, \hat{w}_n).$$

Now set

$$\Gamma_{n1} := \frac{2}{n(n-1)} \sum_{1\le i<j\le n} \varepsilon_i^2 \varepsilon_j^2 W_n^2(X_i, X_j, \hat{w}_n),$$

$$\Gamma_{n2} := \frac{2}{n(n-1)} \sum_{1\le i<j\le n} \varepsilon_i^2 \varepsilon_j^2 W_n^2(X_i, X_j, w).$$

Note that Γ_{n2} is a U-statistic with kernel H_n^2. As shown in the proof of Lemma 9.5.3, $E\big(H_n^2(\zeta_1, \zeta_2)\big) \to \Gamma$ and $E\big(H_n^4(\zeta_1, \zeta_2)\big) = o(n)$. Thus, by Problem 9 in Chapter 5, $\Gamma_{n2} \to_p \Gamma$. The desired result now follows if we show $\Gamma_n = \Gamma_{n1} + o_p(1)$ and $\Gamma_{n1} = \Gamma_{n2} + o_p(1)$. For this we rely on the following properties.

$$R_{n1} := \frac{2}{n(n-1)} \sum_{1\le i<j\le n} \varepsilon_i^2 \varepsilon_j^2 h^{-p} J^2\Big(\frac{X_i - X_j}{h}\Big) = O_p(1),$$

$$R_{n2} := \frac{2}{n(n-1)} \sum_{1\le i<j\le n} (\varepsilon_i^2 + \varepsilon_j^2) h^{-p} J^2\Big(\frac{X_i - X_j}{h}\Big) = O_p(1),$$

$$R_{n3} := \frac{2}{n(n-1)} \sum_{1\le i<j\le n} h^{-p} J^2\Big(\frac{X_i - X_j}{h}\Big) = O_p(1).$$

These are verified by showing that these nonnegative U-statistics have finite means and using the Markov inequality (2.4.8). For example, the substitution $x = y + ht$ and (2.4.6) yield

$$E(R_{n1}) = \iint g(x)g(y)h^{-p}J^2\Big(\frac{x-y}{h}\Big)\,dx\,dy = \iint g(y+ht)g(y)J^2(t)\,dt\,dy \le I_2(g)I_2(J).$$

We used $\big|\int g(y+ht)g(y)\,dy\big| \le \big(\int g^2(y+ht)\,dy \int g^2(y)\,dy\big)^{1/2} = I_2(g)$.

With the help of (9.5.10), the identity $a^2 - b^2 = (a-b)(a+b)$ and (w), we verify

$$\begin{aligned}|\Gamma_{n1} - \Gamma_{n2}| &\le \frac{2}{n(n-1)} \sum_{1\le i<j\le n} \varepsilon_i^2 \varepsilon_j^2 \big|W_n(X_i, X_j, \hat{w}_n - w)\big|\big|W_n(X_i, X_j, \hat{w}_n + w)\big| \\ &\le R_{n1}\|\hat{w}_n - w\|_\infty \|\hat{w}_n + w\|_\infty = o_p(1).\end{aligned}$$

Let $M_n = \max_{1\le i\le n} |\hat{\varepsilon}_i^2 - \varepsilon_i^2|$. Use the identity $\hat{\varepsilon}_i^2\hat{\varepsilon}_j^2 - \varepsilon_i^2\varepsilon_j^2 = (\hat{\varepsilon}_i^2 - \varepsilon_i^2)\varepsilon_j^2 + \varepsilon_i^2(\hat{\varepsilon}_j^2 - \varepsilon_j^2) + (\hat{\varepsilon}_i^2 - \varepsilon_i^2)(\hat{\varepsilon}_j^2 - \varepsilon_j^2)$ to obtain the bound $|\hat{\varepsilon}_i^2\hat{\varepsilon}_j^2 - \varepsilon_i^2\varepsilon_j^2| \le M_n(\varepsilon_j^2 + \varepsilon_i^2) + M_n^2$. This bound and (9.5.10) give the result

$$\begin{aligned}|\Gamma_n - \Gamma_{n1}| &\le \frac{2}{n(n-1)} \sum_{1\le i<j\le n} \big|\hat{\varepsilon}_i^2\hat{\varepsilon}_j^2 - \varepsilon_i^2\varepsilon_j^2\big|\, W_n^2(X_i, X_j, \hat{w}_n) \\ &\le \|\hat{w}_n\|_\infty (M_n R_{n2} + M_n^2 R_{n3}) = o_p(1),\end{aligned}$$

as $M_n = o_p(1)$ as will be shown next. The identity $\hat{\varepsilon}_i = \varepsilon_i - d_{ni}$ with $d_{ni} := n^{-1/2}\Delta_n'\psi(X_i)$ gives the bound

$$|\hat{\varepsilon}_i^2 - \varepsilon_i^2| = |-2\varepsilon_i d_{ni} + d_{ni}^2| \le 2\|\Delta_n\| n^{-1/2}|\varepsilon_i|\|\psi(X_i)\| + n^{-1}\|\Delta_n\|^2\|\psi(X_i)\|^2.$$

Since $\varepsilon\|\psi(X)\|$ and $\|\psi(X)\|^2$ have finite second moments, by Lemma 2.10.8 we obtain

$$\max_{1\le i\le n} |\varepsilon_i|\|\psi(X_i)\| = o_p(n^{1/2}), \quad \max_{1\le i\le n} \|\psi(X_i)\|^2 = o_p(n^{1/2})$$

and thus $M_n = o_p(1)$. This completes the proof of statement (b) and also of the lemma. □

We conclude this subsection by providing conditions that imply (w) with $\hat{w}_n(x) = I(x \in \mathcal{C})\phi(x)/\hat{f}_n^2(x)$ as in (9.5.1) and $w(x) = I(x \in \mathcal{C})\phi(x)/f^2(x)$, where $\hat{f}_n$ is a kernel density estimator with kernel K and bandwidth h. Assume that f and ϕ are continuous at each point in $\mathcal{C}$ and f is positive at each point in $\mathcal{C}$, then $\inf_{x\in\mathcal{C}} f(x)$ is positive and $\sup_{x\in\mathcal{C}} f(x)$ and $\sup_{x\in\mathcal{C}} \phi(x)$ are finite by the extreme value theorem, so that $\|w\|_\infty < \infty$. Thus we conclude that w is bounded and nonnegative and also continuous except possibly on the boundary points of $\mathcal{C}$. Since the boundary of a closed rectangle of the form $[a_1, b_1] \times \cdots \times [a_p, b_p]$ in $\mathbb{R}^q$ has measure zero, we conclude that w has the desired properties. Suppose $\hat{f}_n$ is uniformly consistent for f on $\mathcal{C}$, i.e., $\sup_{x\in\mathcal{C}} |\hat{f}_n(x) - f(x)| = o_p(1)$. This fact together with $\inf_{x\in\mathcal{C}} f(x) > 0$ and $\sup_{x\in\mathcal{C}} \phi(x) < \infty$ readily yield

$$\|\hat{w}_n - w\|_\infty = \sup_{x\in\mathcal{C}} \left| \frac{\phi(x)}{\hat{f}_n^2(x)} - \frac{\phi(x)}{f^2(x)} \right| = o_p(1).$$

In what follows we use the notation $s \le t$ for $s = (s_1, \cdots, s_p)'$ and $t = (t_1, \cdots, t_p)' \in \mathbb{R}^p$ to mean $s_i \le t_i$ for all $i = 1, \cdots, p$. We shall now show that the uniform consistency of $\hat{f}_n$ for f on $\mathcal{C}$ holds under the following additional assumption on the kernel K.

(k2) The support of K is the unit ball $\{t \in \mathbb{R}^p : \|t\| \le 1\}$ and there is an integrable function $\dot{K}$ on $\mathbb{R}^p$ such that

$$K(t) = \int I(s \le t)\dot{K}(s)\, ds, \quad t \in \mathbb{R}^p.$$

This assumption allows us to write

$$\begin{aligned}\hat{f}_n(x) &= \frac{1}{nh^p}\sum_{i=1}^n \int I(s \le (x - X_i)/h)\dot{K}(s)\, ds = \frac{1}{nh^p}\sum_{i=1}^n \int I(X_i \le x - hs)\dot{K}(s)\, ds \\ &= h^{-p}\int F_n(x - hs)\dot{K}(s)\, ds,\end{aligned}$$

with $F_n(t) = n^{-1}\sum_{i=1}^n I(X_i \le t)$, $t \in \mathbb{R}^p$, the joint empirical d.f. of the random vectors $X_1, \ldots, X_n$. Thus $E(\hat{f}_n(x)) = \int f(x - sh)K(s)\, ds$ equals $h^{-p}\int F(x - hs)\dot{K}(s)\, ds$ and

$$\sup_{x\in\mathbb{R}} |\hat{f}_n(x) - E(\hat{f}_n(x))| \le h^{-p}\|F_n - F\|_\infty \int |\dot{K}(s)|\, ds = O_p((nh^{2p})^{-1/2}) = o_p(1)$$

by (h) as $n^{1/2}\|F_n - F\|_\infty = O_p(1)$, by Lemma 2.12.3. Finally,

$$\sup_{x\in\mathcal{C}} \|E(\hat{f}_n(x)) - f(x)\| \le \omega(h) := \sup_{x\in\mathcal{C}, \|u\|\le h} |f(x - u) - f(x)| \to 0,$$

by the continuity of f at each point in the compact set $\mathcal{C}$.

Now, let $T_n(b)$ be as in (9.5.2), $\tilde{\beta}_n := \text{argmin}_{b\in\Theta} T_n(b)$, $\tilde{\varepsilon}_i := Y_i - \tilde{\theta}_n' \psi(X_i)$, $i = 1, \cdots, n$,

$$\tilde{A}_n := \frac{1}{n^2} \sum_{i=1}^{n} \tilde{\varepsilon}_i^2 \int_C K_h^2(x - X_i) \frac{\phi(x)}{\hat{f}_n^2(x)} \, dx,$$

$$\tilde{\Gamma}_n := \frac{2h^p}{n(n-1)} \sum_{1\le i<j\le n} \tilde{\varepsilon}_i^2 \tilde{\varepsilon}_j^2 \Big(\int_C K_h(x - X_i) K_h(x - X_j) \frac{\phi(x)}{\hat{f}_n^2(x)} \, dx \Big)^2 .$$

We summarize the asymptotic normality of this test statistic in the following corollary.

Corollary 9.5.1. *Suppose f and ϕ are continuous at each point in the compact convex set $\mathcal{C}$ and f is positive at each point in $\mathcal{C}$. In addition, suppose conditions (h), (k2) and (w) hold. Then,*

$$nh^{p/2}(T_n(\tilde{\beta}_n) - \tilde{A}_n)/(2\tilde{\Gamma}_n)^{1/2} \to_D \mathcal{N}(0, 1).$$

Consequently, the asymptotic size of the test that rejects the hypothesis that $\mu(x) = \beta'\psi(x)$, for all $x \in \mathcal{C}$ and some $\beta \in \mathbb{R}^q$, whenever $|nh^{p/2}(T_n(\tilde{\beta}_n) - \tilde{A}_n)/(2\tilde{\Gamma}_n)^{1/2}| > z_{\alpha/2}$ is α.

9.6 Comparing Two Regression Functions

In this section we consider the problem of testing the equality of two nonparametric regression curves against one-sided alternatives when the covariates are distinct and when they are common.

Distinct Covariates. First we shall focus on the distinct covariates case. Let μ_1 and μ_2 denote the two regression functions and $(X_{1,1}, Y_{1,1}), \cdots, (X_{1,n_1}, Y_{1,n_1})$ and $(X_{2,1}, Y_{2,1}), \cdots, (X_{2,n_2}, Y_{2,n_2})$ be the two independent random samples satisfying the relations

$$Y_{k,j} = \mu_k(X_{k,j}) + \varepsilon_{k,j}, \qquad j = 1, \ldots, n_k, \; k = 1, 2.$$

Here the errors $\varepsilon_{1,1}, \ldots, \varepsilon_{1,n_1}, \varepsilon_{2,1}, \cdots, \varepsilon_{2,n_2}$ are i.i.d. with density f, and the covariates $X_{1,1}, \cdots, X_{1,n_1}, X_{2,1}, \ldots, X_{2,n_2}$ are independent r.v.'s with common density g and independent of the errors. Moreover, the covariates $X_{i,j}$'s are assumed to take values in a closed bounded interval $\mathcal{I} \subset \mathbb{R}$.

We write $\mu_1 >_g \mu_2$ to denote that $\int_{\mathcal{I}} I(\mu_1(x) > \mu_2(x)) g(x)\, dx > 0$ and $\mu_1(x) \ge \mu_2(x)$ for all $x \in \mathcal{I}$. The problem of interest is to test

$$H_0 : \mu_1 = \mu_2 \quad \text{versus} \quad H_1 : \mu_1 >_g \mu_2.$$

This type of testing problem arises when one compares a new treatment with an old treatment or with a placebo over the range of a covariate and expects the new treatment to be superior over the entire range. In this setting the regression curves measure the average treatment effects over the given range of values of the covariate, and the superiority of the new treatment is expressed through the alternative hypothesis H_1.

To motivate the proposed test, let $\mathcal{H}_k(y) := P(Y_{k,1} \le y) = E\big(F(y - \mu_k(X))\big)$, $k = 1, 2$, where F and X denote the d.f. corresponding to f and a r.v. having density g, respectively. The null hypothesis implies $\mathcal{H}_1 = \mathcal{H}_2$, while under the alternative hypothesis, $\mathcal{H}_1 \le \mathcal{H}_2$, $\mathcal{H}_1 \ne \mathcal{H}_2$. Thus any distribution free two sample test of $\mathcal{H}_1 = \mathcal{H}_2$ against $\mathcal{H}_1 \le \mathcal{H}_2$ and $\mathcal{H}_1 \ne \mathcal{H}_2$ based on the data $Y_{1,1}, \cdots, Y_{1,n_1}$ and $Y_{2,1}, \cdots, Y_{2,n_2}$ can be used to test the above hypotheses. Two well known tests are the two sample t and Mann–Whitney-Wilcoxon tests based, respectively, on the statistics T and W, where

$$T := \sqrt{\frac{n_1 n_2}{n_1+n_2}}\,\frac{1}{n_1 n_2}\sum_{i=1}^{n_1}\sum_{j=1}^{n_2}(Y_{2,j}-Y_{1,i}), \qquad W := \frac{1}{n_1 n_2}\sum_{i=1}^{n_1}\sum_{j=1}^{n_2}\operatorname{sgn}(Y_{1,i}-Y_{2,j}).$$

These tests ignore the covariates. In contrast, tests that use the given covariates are expected to have greater power compared to tests that ignore the covariates like T and W do. A way of introducing the covariates is to adjust a given statistic by matching the covariates. We illustrate this by introducing the covariate matched analogs of the statistics T and W.

Accordingly, let K be a symmetric Lipschitz continuous density that is positive on $(-1,1)$ and vanishes off $(-1,1)$. Put $K_h(x) = K(x/h)/h$, $x \in \mathbb{R}$, $h > 0$. Let $G_k(\cdot, x)$ denote the conditional d.f. of $Y_{k,1}$, given $X_{k,1} = x$, i.e., $G_k(y,x) = F(y - \mu_k(x))$, $k = 1, 2$. Under H_0, $G_1 = G_2$, while under the alternative hypothesis H_1, $G_1(y,x) \le G_2(y,x)$, for all y and x with strict inequality for $(x,y) \in A$, where A is a set in $\mathbb{R}^2$ such that $\iint I\big((x,y) \in A\big)\,dx\,dy > 0$. This suggests using a locally weighted analog of W, viz,

$$\widetilde{W} = \sqrt{\frac{n_1 n_2}{n_1+n_2}}\,\frac{1}{n_1 n_2}\sum_{i=1}^{n_1}\sum_{j=1}^{n_2}\operatorname{sgn}(Y_{1,i}-Y_{2,j})K_h(X_{1,i}-X_{2,j}),$$

where h is a small positive number depending on the sample sizes. The corresponding test rejects H_0 for large values of $\widetilde{W}$. This test is called the covariate-matched Mann–Whitney–Wilcoxon test.

Similarly, the covariate matched two sample t-test rejects H_0 in favor of H_1 for large values of

$$\widetilde{T} = \sqrt{\frac{n_1 n_2}{n_1+n_2}}\,\frac{1}{n_1 n_2}\sum_{i=1}^{n_1}\sum_{j=1}^{n_2}(Y_{2,j}-Y_{1,i})\,K_h(X_{2,j}-X_{1,i}).$$

The exact null distributions of these statistics are hard to determine but their asymptotic null distributions are available.

To describe their asymptotic null distributions, let μ denote the common value of the regression functions μ_1, μ_2 under H_0 and set $e_{k,j} = Y_{k,j} - \mu(X_{k,j})$, $j = 1, \cdots, n_k$, $k = 1, 2$. Under H_0, $e_{k,j}, 1 \le j \le n_k$, $k = 1, 2$, are i.i.d. with density f. Assume the following:

$$\begin{aligned}&f \text{ has zero mean and finite variance } \sigma^2,\ \mu \text{ is continuous on } \mathcal{I},\ g \text{ is continuous}\\&\text{and bounded away from 0 on } \mathcal{I} \text{ and vanishes off } \mathcal{I},\ n_1 \wedge n_2 \to \infty,\ n_1 h \to \infty,\\&n_2 h \to \infty,\ n_1/(n_1+n_2) \text{ is bounded away from 0 and 1.}\end{aligned} \tag{9.6.1}$$

The following asymptotic normality results about $\widetilde{T}$ and $\widetilde{W}$ follow from Theorem 2.4 of Koul and Schick (1997) (KS). Under (9.6.1) and H_0, $\widetilde{T} \to_D \mathcal{N}(0, \tau^2)$, where $\tau^2 = E\big(g^2(X)\big)\sigma^2$. To give a consistent estimate of τ^2, let

$$U_{i,j} := (Y_{1,i}-Y_{2,j})K_h(X_{1,i}-X_{2,j}), \qquad \bar{U}_{i,\bullet} := \frac{1}{n_2}\sum_{j=1}^{n_2} U_{i,j},$$

$$\bar{U}_{\bullet,j} := \frac{1}{n_1}\sum_{i=1}^{n_1} U_{i,j}, \qquad \bar{U}_{\bullet,\bullet} := \frac{1}{n_1 n_2}\sum_{i=1}^{n_1}\sum_{j=1}^{n_2} U_{i,j},$$

$$\hat{\tau}^2 := \frac{1}{2}\Big(\sum_{i=1}^{n_1}(\bar{U}_{i,\bullet}-\bar{U}_{\bullet,\bullet})^2 + \frac{1}{n_2}\sum_{j=1}^{n_2}(\bar{U}_{\bullet,j}-\bar{U}_{\bullet,\bullet})^2\Big).$$

The estimator $\hat{\tau}^2$ is consistent for τ^2, under H_0. Thus the test which rejects when $\widetilde{T} > z_\alpha \hat{\tau}$ has asymptotic size α, where z_α is as in (3.3.9).

If (9.6.1) holds and f is bounded, then under H_0, $\widetilde{W} \to_D \mathcal{N}(0, \tau_W^2)$, where $\tau_W^2 = Eg^2(X)E\big[2F(\varepsilon) - 1\big]^2 = Eg^2(X)/3$. Let

$$U_{i,j} := \mathrm{sgn}(Y_{1,i} - Y_{2,j})K_h(X_{1,i} - X_{2,j}), \qquad \bar{U}_{i,\bullet} := \frac{1}{n_2}\sum_{j=1}^{n_2} U_{i,j},$$

$$\bar{U}_{\bullet,j} := \frac{1}{n_1}\sum_{i=1}^{n_1} U_{i,j}, \qquad \bar{U}_{\bullet,\bullet} = \frac{1}{n_1 n_2}\sum_{i=1}^{n_1}\sum_{j=1}^{n_2} U_{i,j}.$$

Then, a consistent estimate of τ_W^2 under H_0 is

$$\hat{\tau}_W^2 = \frac{1}{2}\bigg(\frac{1}{n_1}\sum_{j=1}^{n_1}(\bar{U}_{j,\bullet} - \bar{U}_{\bullet,\bullet})^2 + \frac{1}{n_2}\sum_{j=1}^{n_2}(\bar{U}_{\bullet,j} - \bar{U}_{\bullet,\bullet})^2\bigg).$$

The consistency of $\hat{\tau}^2$ and $\hat{\tau}_W^2$ for τ^2 and τ_W^2, respectively, is also proved in Theorem 2.4 of KS. Thus the test which rejects when $\widetilde{W} > z_\alpha \hat{\tau}_W$ has asymptotic size α.

Asymptotic power against local alternatives. The paper of KS contains several classes of tests for testing H_0 versus H_1. The $\widetilde{T}$ and $\widetilde{W}$ tests are members of the T_4 class of tests in KS. The KS paper investigated the asymptotic power of the T_4 tests against the following nonparametric local alternatives. Let $\mathcal{D}$ denote the set of all non-negative measurable functions δ defined on $\mathcal{I}$ with $0 < \int \delta^2(x)\, g(x)\, dx < \infty$. With μ denoting the common regression function under H_0, the nonparametric local alternatives considered by KS, for $\delta \in \mathcal{D}$, are

$$\mu_1 := \mu + \frac{1}{n_1}\sqrt{\frac{n_1 n_2}{n_1 + n_2}}\,\delta, \quad \mu_2 := \mu - \frac{1}{n_2}\sqrt{\frac{n_1 n_2}{n_1 + n_2}}\,\delta. \tag{9.6.2}$$

Note that $\mu_1 >_g \mu_2$ so that H_1 holds.

Suppose f has finite Fisher information for location, see Definition 2.11.1. Then, using the Hájek–Le Cam's contiguity theory, see e.g., Hájek and Šidák (1967), KS derived an upper bound on the asymptotic power of any asymptotic level α sequence of tests of H_0 against the alternatives (9.6.2). A test whose asymptotic power achieves this upper bound is said to be locally asymptotically most powerful (LAMP) test against the alternatives (9.6.2) at the given densities f, g.

For example, if $f(x) = e^{-x^2/(2\sigma^2)}/(\sqrt{2\pi}\sigma)$, $x \in \mathbb{R}$, then the $\widetilde{T}$ test is LAMP among all tests of the same asymptotic level against the alternatives (9.6.2) with $\delta = g$. If $f(x) = e^{-x}(1 + e^{-x})^{-2}$, $x \in \mathbb{R}$, then the covariate-matched Wilcoxon–Mann–Whitney test $\widetilde{W}$ is LAMP among all tests of the same asymptotic level against the alternatives (9.6.2) with $\delta = g$.

The paper of KS also contains expressions for the asymptotic power of their classes of tests against the alternatives (9.6.2). For $\delta \in \mathcal{D}$, let $\|\delta\|_1 := \int \delta(x)g(x)dx$. For any test ψ of H_0, let $\Pi_{\psi,\alpha}(\delta)$ denote its asymptotic power at the asymptotic level α against the alternatives (9.6.2). Let z_α be as in (3.3.9). From KS, we obtain that

$$\Pi_{\widetilde{T},\alpha}(\delta) = 1 - \Phi\bigg(z_\alpha - \frac{\|\delta\|_1}{(\int g^3(x)dx)^{1/2}\sigma}\bigg), \qquad \Pi_{\widetilde{W},\alpha}(\delta) = 1 - \Phi\bigg(z_\alpha - \frac{2\sqrt{3}\|\delta\|_1 \int f^2(x)dx}{(\int g^3(x)dx)^{1/2}}\bigg).$$

From these expressions we readily see that the Pitman's asymptotic relative efficiency of the $\widetilde{W}$ test, relative to the $\widetilde{T}$ test, when testing H_0 against the above alternatives, is $12\sigma^2\big(\int f^2(x)dx\big)^2$. This asymptotic relative efficiency is the same as that of the two sample Mann–Whitney–Wilcoxon test relative to the two sample t test when testing for the equality $\theta_1 = \theta_2$ of the two location parameters θ_1, θ_2 against the alternatives

$$\theta_2 = \theta_1 + \frac{1}{n_1}\sqrt{\frac{n_1 n_2}{n_1 + n_2}}\,\Delta, \quad \Delta > 0.$$

Common covariates. Now we shall discuss a test of H_0 versus H_1 in the common design case. Let $X_1, \cdots, X_n$ be i.i.d. r.v.'s with density g and d.f. G. Consider the common covariates two sample regression model $Y_{k,i} = \mu_k(X_i) + \varepsilon_{k,i}$, $i = 1, \cdots, n$, $k = 1, 2$, where $\varepsilon_{k,i}, i = 1, \cdots, n, k = 1, 2$, are i.i.d. r.v.'s with common density f and independent of the covariates $X_1, \cdots, X_n$.

To motivate a test of H_0 in this case, define

$$\Delta(x) := \mu_1(x) - \mu_2(x), \qquad \mathcal{T}(x) := \int_{-\infty}^{x} \Delta(s) g(s)\, ds, \qquad x \in \mathbb{R}.$$

Under H_0, we have $\mathcal{T}(x) = 0$ for all x, while under H_1 we have $T(x) \geq 0$ for all $x \in \mathbb{R}$ and $\mathcal{T}(x) > 0$ for all $x \geq c$ for some $c \in \mathcal{I}$. Let $D_i := Y_{1,i} - Y_{2,i}$, $1 \leq i \leq n$ and set

$$\mathcal{T}_n(x) := n^{-1/2} \sum_{i=1}^{n} D_i I(X_i \leq x), \qquad x \in \mathbb{R}.$$

Under H_0, the process $\{\mathcal{T}_n(x), x \in \mathbb{R}\}$ is centered and has covariance function

$$\mathcal{C}(x,y) = E\big(\varepsilon_{1,1} - \varepsilon_{2,1}\big)^2 G(x \wedge y) = 2\sigma^2 G(x \wedge y) = 2\sigma^2 \mathrm{Cov}\big(\mathbb{B}(G(x)), \mathbb{B}(G(y))\big),$$

where $\mathbb{B}$ is the standard Brownian motion on $[0, \infty)$. In other words, the covariance structure of $\mathcal{T}_n$ is the same as that of $\sqrt{2}\sigma\mathbb{B} \circ G$. One can show that

$$\big\|\mathcal{T}_n\big\|_\infty = \sup_{x \in \mathbb{R}} \big|\mathcal{T}_n(x)\big| \to_D \sqrt{2}\,\sigma \sup_{x \in \mathbb{R}} \big|\mathbb{B}(G(x))\big| = \sqrt{2}\,\sigma \sup_{0 \leq t \leq 1} \big|\mathbb{B}(t)\big|.$$

Note that $D_1 = \Delta(X_1) + \varepsilon_{1,1} - \varepsilon_{2,1}$ has mean $E(\Delta(X_1))$ and variance $2\sigma^2 + \mathrm{Var}\big(\Delta(X_1)\big)$. Hence, under H_0, $\mathrm{Var}(D_1) = 2\sigma^2$ and a consistent estimator of $2\sigma^2$ is given by $\tau_n^2 := (n-1)^{-1} \sum_{i=1}^{n} (D_i - \bar{D})^2$, where $\bar{D} := n^{-1} \sum_{i=1}^{n} D_i$.

For an $0 \leq \alpha \leq 1$, let b_α be as in (9.2.13). The test that rejects H_0, in favor of H_1, whenever $\tau_n^{-1} \|\mathcal{T}_n\|_\infty > b_\alpha$ will have asymptotic size α.

Let $\nu_n^2 := \sum_{i=1}^{n-1} \big(D_{i+1} - D_i\big)^2 / (2n)$, $T_n := \nu_n^{-1} \big\|\mathcal{T}_n\big\|_\infty$. Delgado (1993) used T_n to test H_0 against the two sided alternatives, $P\big(\mu_1(X) \neq \mu_2(X)\big) > 0$. This test rejects H_0, in favor of the two sided alternative, at the asymptotic level α, whenever $T_n > b_\alpha$.

Heteroscedasticity. The above results are derived under the setup where the two regression models are homoscedastic and all covariates have the same distribution. As is well known, heteroscedastic regression models are of broader interest. It is then of interest to consider the above testing problem under the heteroscedastic model, where for each $k = 1, 2$, $(X_{k,j}, Y_{k,j})$, $j = 1, \cdots, n_k$, are i.i.d. bivariate observations satisfying $Y_{k,j} = \mu_k(X_{k,j}) + \varepsilon_{k,j}, j = 1, \cdots, n_k$, with the errors so that $E\big(\varepsilon_{k,j} \,\big|\, X_{k,j}\big) = 0$, $j = 1, \cdots, n_k$, $k = 1, 2$. Further assume that density g_k of $X_{k,1}$ is bounded away from zero on $\mathcal{I}$, $k = 1, 2$, the two regression samples are independent; and the two error densities are possibly different.

Let $\mathcal{V} = \int \psi(x)\big(\mu_1(x) - \mu_2(x)\big)\, dx$, where ψ is a non-negative function that is continuous on $\mathcal{I}$ and vanishes off $\mathcal{I}$. The integral $\mathcal{V}$ is zero under H_0 and non-negative under H_1. If ψ is also positive on the interior of $\mathcal{I}$, then $\mathcal{V}$ is always positive under the alternative H_1. Let

$$\hat{\mathcal{V}} := \frac{1}{n_1 n_2} \sum_{i=1}^{n_1} \sum_{j=1}^{n_2} \hat{\nu}_1(X_{1,i})\hat{\nu}_2(X_{2,j})(Y_{1,i} - Y_{2,j})K_h\big(X_{1,i} - X_{2,j}\big),$$

where $\hat{\nu}_k, k = 1, 2$ are some statistics based on the given data. Koul and Schick (2003) propose tests based on $\hat{\mathcal{V}}$ and derive their asymptotic normality under H_0 and the local alternatives H_1 : $\mu_2 = \mu_1 + \sqrt{\frac{n_1+n_2}{n_1 n_2}}\,\delta$, where δ is a non-negative continuous function on $\mathcal{I}$. They also discuss some asymptotic optimality theory as in the KS paper.

9.7 Exercises

1. Let F_0 denote the d.f. with density $f_0(x) = \exp\{-|x|\}/2$, $x \in \mathbb{R}$. Test the hypothesis $H_0; F = F_0$ against $H_1 : F \neq F_0$, using the Kolmogorov statistic $\mathcal{D}_n$ of (9.2.1) at level $\alpha = 0.05$, based on the following random sample of size $n = 10$ from a continuous d.f. F : -0.184, -0.355, -1.184, 1.729, -2.399, 0.898, 1.508, 1.501, -0.025, 0.744.

   For $\alpha = 0.10, 0.05, 0.02, 0.01$, the respective values of d_α so that $P\big(\mathcal{D}_{10} > d_\alpha\big) \leq \alpha$ are $0.369, 0.409, 0.457, 0.489$.

2. Consider the following random sample: 0.929, 1.178, 4.768, 1.796, 2.145, 2.249, 1.652, 4.724, 1.421, 2.699, 1.507, 4.883, 2.976, 3.992, 7.264, 1.237, 2.614, 3.280, 3.333, 1.983. Let $f_0(x) := (1/4)e^{-x/4}I(x > 0)$. Use $\mathcal{D}_n$ of (9.2.1) to fit f_0 to the d.f. of this data at $\alpha = 0.05$. For your information the d_α's that satisfy $P\big(\mathcal{D}_{20} > d_\alpha\big) \leq \alpha$ are $0.265, 0.294, 0.329, 0.352$ for $\alpha = 0.10, 0.05, 0.02, 0.01$, respectively.

3. Obtain the bootstrap critical value $b^*_{n,0.05}$ as in (9.3.1) based on $N = 200$ random samples of size 20 drawn with replacement from the data given in the previous problem. Use this critical value to test the hypothesis $H : F(x) = f_0(x - \mu), \forall x \in \mathbb{R}$ and for some $\mu \in \mathbb{R}$ at $\alpha = 0.05$.

4. With $\mathcal{U}_n$ defined at (9.2.11), prove that

$$\mathrm{Cov}\big(\mathcal{U}_n(s, \theta), \mathcal{U}_n(t, \theta)\big) = s \wedge t, \ \forall 0 \leq s, t \leq 1.$$

   You may find Theorem 2.3.2 useful here.

5. Verify formula (9.2.16) used in fitting a normal distribution with unknown mean.

6. Let the r.v. X have a density f with finite Fisher information for location. Set $g(x) := \big(1, -\dot{f}(x)/f(x)\big)'$ and $B(x) := E\Big(g(X)g(X)'I(X \geq x)\Big)$, $x \in \mathbb{R}$. Show that

$$B(x) = \begin{pmatrix} 1 - F(x) & f(x) \\ f(x) & \int_x^\infty \Big(\frac{\dot{f}(y)}{f(y)}\Big)^2 f(y)dy \end{pmatrix}, \qquad x \in \mathbb{R}.$$

   (a) Show that for $f(x) = \exp(-|x|)/2$, $x \in \mathbb{R}$, $B(x)$ is singular for all $x \geq 0$.

(b) Let Φ and φ denote d.f. and density of a standard normal r.v. Show that for $f = \varphi$,

$$B(x) = \begin{pmatrix} 1 - \Phi(x) & \varphi(x) \\ \varphi(x) & x\varphi(x) + 1 - \Phi(x) \end{pmatrix}$$

and $B(x)$ is positive definite for all $x \in \mathbb{R}$.

7. Let F_0 be a d.f. having a positive and uniformly continuous density f_0 and zero mean. Let $X_1, \cdots, X_n$ be a random sample from a d.f. F and consider the problem of testing the hypothesis $H_0 : F(x) = F_0(x - \mu)$, for all $x \in \mathbb{R}$ and some $\mu \in \mathbb{R}$. Let $\bar{X}$ be the sample mean, $h = 1.5n^{-1/5}$, $\bar{\varepsilon}_i = X_i - \bar{X}$, $1 \le i \le n$, and define

$$\widetilde{f}_n(x) := (nh)^{-1}\sum_{i=1}^{n} K\Big(\frac{x - \bar{\varepsilon}_i}{h}\Big), \qquad \widetilde{F}_n(x) := \int_{y \le x} \widetilde{f}_n(y)dy, \quad x \in \mathbb{R}.$$

Let $\varepsilon_1^*, \cdots, \varepsilon_n^*$ be a random sample from $\widetilde{F}_n$, $X_i^* = \bar{X} + \varepsilon_i^*$, $\mu_n^* := \sum_{i=1}^n Y_i^*/n$ and define

$$Z_n^*(v) := n^{-1/2}\sum_{i=1}^{n}\{I(X_i^* - \mu_n^* \le \widetilde{F}_n^{-1}(v)) - v\},$$
$$Z_n(v) := n^{-1/2}\sum_{i=1}^{n}\{I(X_i - \bar{X} \le F_0^{-1}(v)) - v\}, \quad 0 \le v \le 1$$

(a) Assume K satisfies the condition (9.3.6) of Theorem 9.3.2. Verify Theorem 9.3.2 with $x_{ni} = 1$ for all $i = 1, \cdots, n$ and the rest of the entities as above is applicable so that under H_0,

$$\sup_{x \in \mathbb{R}} \Big| P_*(\|V_n^*\|_\infty \le x) - P(\|V_n\|_\infty \le x) \Big| \to 0, \quad \text{a.s.}$$

(b) Now consider the two choices for K, namely, $K_1(u) := (1/2)I(|u| \le 1)$ and $K_2(u) = 3(1 - u^2)I(|u| \le 1)/4$. Denote the corresponding $\widetilde{F}_n$ by $\tilde{F}_{n1}$ $\tilde{F}_{n2}$, respectively. Show the following:

$$\tilde{F}_{n1}(x) = \frac{1}{2n}\sum_{i=1}^{n}\Big(\frac{x - \bar{\varepsilon}_i}{h} + 1\Big)I(|x - \bar{\varepsilon}_i| \le h) + \frac{1}{n}\sum_{i=1}^{n} I(x - \bar{\varepsilon}_i > h),$$
$$\tilde{F}_{n2}(x) = \frac{3}{4n}\sum_{i=1}^{n}\Big(\frac{x - \bar{\varepsilon}_i}{h} + 1\Big)\frac{1}{3}\Big[\Big(\frac{x - \bar{\varepsilon}_i}{h}\Big)^3 + 1\Big]I(|x - \bar{\varepsilon}_i| \le h)$$
$$+ \frac{1}{n}\sum_{i=1}^{n} I(x - \varepsilon_i > h).$$

(b) Let the following data represent a random sample of size 36 from d.f. F.

-1.054, -2.641, 1.374, 2.255, -2.112, 1.133, 2.942, 3.595, -0.724, 3.621, -0.886, 1.480, 2.694, 0.476, -3.208, -0.510, 1.770, -1.015, 1.582, 2.348, 0.099, -0.015, 1.704, 1.223, 0.991, 1.401, 0.029, 0.768, 4.215, -0.322, -0.344, 0.377, -2.550, 3.840 -2.952, 3.904.

Let Φ denote the d.f. of a $\mathcal{N}(0,1)$ r.v. and consider the problem of testing the hypothesis $H : F(x) = \Phi(x - \mu), \forall x \in \mathbb{R}$ and for some $\mu \in \mathbb{R}$. Use the bootstrap test statistic $\|V_n^*\|_\infty$ with $K = K_1$ and the rest of the things as specified in the beginning to test this hypothesis. Use $N = 50$ in the bootstrap sampling.

8. Let Φ be as in the previous exercise and let φ denote its derivative. Let the following data represent a random sample of size 36 from some density f on $\mathbb{R}$: 0.928, 2.522, 0.858, 1.768, 0.920, 0.133, -0.908, -1.546, -1.020, 1.147, 1.207, 1.922, 0.310, -3.367. 0.623, 0.789, 1.556, -0.165, -0.574, 1.173, -0.010, -1.555, 0.359, -0.611, -2.716, -1.578, 1.600, 3.421, 3.592, -0.979, -1.567, 0.449, 1.031, 2.386, -1.275, 1.120. Use the statistic $S_n := nh^{1/2}\big(T_n - \int K^2(u)du/(nh)\big)$, where T_n is as in (9.4.1) with $g(y) \equiv y$, $K(y) = \big[3(1-y^2)/4\big]I\big(|y| \le 1\big)$, $h = n^{-1/5}$ and Theorem 9.4.1 to perform the large sample test of the hypothesis $H_0 : f = \varphi$ versus the alternative H_0 is not true, at $\alpha = 0.05$. Repeat this with $K(y) = (1/2)I\big(|y| \le 1\big)$ and the rest of things as in the previous part of the problem. Is your conclusion any different?

9. Suppose a density f on $\mathbb{R}$ is continuously differentiable with its derivative $\dot{f}$ satisfying $\int |\dot{f}(x)|\,dx < \infty$. Show that f is bounded and L_1-Lipschitz with constant $\int |\dot{f}(x)|\,dx$.

The following three problems use Lemma 9.5.2 and the discussion just before Corollary 9.5.1.

10. Let $\boldsymbol{X}_i = (X_{i1}, X_{i2})'$, $1 \le i \le n$ be i.i.d. uniformly distributed r.v.'s on the set $\mathcal{C}_2 := [-1,1]^2$. Consider the linear regression model $Y_i = \beta_1' X_{i1} + \beta_2 X_{i2} + \varepsilon_i$, for some $\beta = (\beta_1, \beta_2)' \in \mathbb{R}^2$, where $\varepsilon_i, 1 \le i \le n$ are i.i.d. r.v.'s having zero mean and positive and finite variance σ^2 and where ε_i is independent of $\boldsymbol{X}_i$, for all $i = 1, \cdots, n$. Let f_n be an estimator of f such that $\sup_{x \in \mathcal{C}_2} \big|(1/f_n(x)) - (1/f(x))\big| = o_p(1)$. Consider the two cases: (a) $\hat{w}_n = I_{\mathcal{C}}$ and (b) $\hat{w}_n = I_{\mathcal{C}}/f_n$. Define

$$\tilde{\beta}_n := \underset{b \in \mathbb{R}^2}{\operatorname{argmin}} \int_{\mathcal{C}_2} \Big(\frac{1}{n}\sum_{i=1}^n K_h(x - \boldsymbol{X}_i)\big(Y_i - b'\boldsymbol{X}_i\big)\Big)^2 \hat{w}_n(x)dx,$$

where the density kernel K satisfies (k) and the bandwidth h satisfies $h \to 0$ and $nh^4 \to 0$. Let $\mathcal{I}_2$ denote the 2×2 identity matrix.

(A) Show that in both cases (a) and (b), $n^{1/2}\big(\tilde{\beta}_n - \beta\big) \to_D \mathcal{N}_2\big(0, 3\sigma^2\mathcal{I}_2\big)$.

(B) Let $\hat{\beta}$ denote the least squares estimator of β. Show that $n^{1/2}(\hat{\beta} - \beta) \to_D \mathcal{N}_2\big(0, 3\sigma^2\mathcal{I}_2\big)$.

11. With $\boldsymbol{X}_i = (X_{i1}, X_{i2})'$, ε_i, $i = 1, \cdots, n$ as in the previous problem, consider the linear regression model $Y_i = \beta_0 + \beta_1' X_{i1} + \beta_2 X_{i2} + \varepsilon_i$, for some $\beta = (\beta_0, \beta_1, \beta_2)' \in \mathbb{R}^3$. Let $\hat{w}_n$, K and h be as in the previous problem and define

$$\tilde{\beta}_n := \underset{b \in \mathbb{R}^2}{\operatorname{argmin}} \int_{\mathcal{C}_2} \Big(\frac{1}{n}\sum_{i=1}^n K_h(x - \boldsymbol{X}_i)\big(Y_i - b_0 - b_1 X_{i1} - b_2 X_{i2}\big)\Big)^2 \hat{w}_n(x)dx,$$

Let

$$D := \begin{pmatrix} 1 & 0 & 0 \\ 0 & 1/3 & 0 \\ 0 & 0 & 1/3 \end{pmatrix}.$$

Show that in both cases (a) and (b), $n^{1/2}\big(\tilde{\beta}_n - \beta\big) \to_D \mathcal{N}_2\big(0, \sigma^2 D^{-1}\big)$. Is this limiting distribution the same as that of the least-squares estimator in this problem?

12. For a given positive integer p, let $\boldsymbol{X} = (X_1, \cdots, X_p)'$ be a uniformly distributed p-dimensional random vector on the unit cube $\mathcal{C}_p := [-1, 1]^p$ and consider the linear regression model $Y = \beta_1 X_1 + \cdots + \beta_p X_p + \varepsilon$, where ε is independent of $\boldsymbol{X}$, $E(\varepsilon) = 0$, and $0 < \sigma^2 := E(\varepsilon^2) < \infty$. Let $\beta = (\beta_1, \cdots, \beta_p)'$, $(\boldsymbol{X}_i, Y_i), 1 \le i \le n$ be a random sample from this model and define

$$\tilde{\beta}_n := \operatorname*{argmin}_{b \in \mathbb{R}^p} \int_{\mathcal{C}} \Big(\frac{1}{n}\sum_{i=1}^n K_h(x - \boldsymbol{X}_i)(Y_i - b'\boldsymbol{X}_i)\Big)^2 \frac{dx}{f_n^2(x)}.$$

Suppose the density kernel K satisfies (k), the bandwidth h satisfies $h \to 0$, $nh^{2p} \to \infty$ and the density estimator f_n satisfies $\sup_{x \in \mathcal{C}_p} \big|(1/f_n(x)) - (1/f(x))\big| = o_p(1)$.

Show that $n^{1/2}(\tilde{\beta}_n - \beta) \to_D \mathcal{N}_p(0, 3\sigma^2 \mathcal{I}_p)$, where $\mathcal{I}_p$ is the $p \times p$ identity matrix.

13. As in Section 9.5, let $(X_1, Y_1), \cdots, (X_n, Y_n)$ be independent copies of the pair (X, Y) which forms a regression model $Y = \mu(X) + \varepsilon$, with $E(\varepsilon|X) = 0$ and X p-dimensional. Consider testing the null hypothesis $H_0 : \mu = \mu_0$ for some known μ_0. Set

$$T_n := \int \Big(\frac{1}{n}\sum_{i=1}^n K_h(x - X_i)(Y_i - \mu_0(X_i))\Big)^2 dx$$

$$A_n := \frac{1}{n^2}\sum_{i=1}^n (Y_i - \mu_0(X_i))^2 \int K_h^2(x - X_i)\, dx.$$

(a) Show that $nh^{p/2}(T_n - A_n)$ equals

$$U_n := \frac{2\sqrt{h}}{n} \sum_{1 \le i < j \le n} (Y_i - \mu_0(X_i))(Y_j - \mu_0(X_j)) \int K_h(x - X_j) K_h(x - X_j)\, dx.$$

(b) Provide conditions on (Y, X), the density kernel K and the bandwdith h such that $U_n \to_D \mathcal{N}(0, \tau^2)$ holds under H_0 and determine τ^2.

(c) Find a test based on this result that has asymptotic level α.

14. Let $(X_1, Y_1, Z_1), \cdots, (X_n, Y_n, Z_n)$ be copies of the triple (X, Y, Z), where X, Y and Z are r.v.'s with Y and Z having finite second moments. The goal is to test the null hypothesis $H_0 : E(Y|X) = E(Z|X)$. Set $D_i = Y_i - Z_i$ and set

$$T_n := \int \Big(\frac{1}{n}\sum_{i=1}^n K_h(x - X_i) D_i\Big)^2 dx, \quad A_n := \frac{1}{n^2}\sum_{i=1}^n D_i^2 \int K_h^2(x - X_i)\, dx.$$

(a) Show that $n\sqrt{h}(T_n - A_n)$ equals

$$U_n := \frac{2\sqrt{h}}{n} \sum_{1 \le i < j \le n} D_i D_j \int K_h(x - X_j) K_h(x - X_j)\, dx.$$

(b) Provide conditions on (X, Y, Z), the density kernel K and the bandwidth h such that $U_n \to_D \mathcal{N}(0, \tau^2)$ holds under H_0 and determine τ^2.

(c) Find a test based on this result that has asymptotic level α.

Notes. The Khmaladze transformation that produces ADF tests in the one sample location-scale model for fitting an error d.f. does the same in nonlinear parametric heteroscedastic regression, parametric ARCH-GARCH, nonparametric regression and nonparametric autoregressive conditionally heteroscedastic time series models, cf., Koul (2006), Koul and Khmaladze (2009) and Koul and Zhu (2015). This transformation method has also been adapted to fitting a survival distribution under random censoring on the right, cf., Nikabadze and Stute (1997), where one does not only have a finite-dimensional nuisance parameter but also an infinite-dimensional nuisance parameter via the censoring distribution. Khmaladze (2013) develops ADF tests for fitting discrete distributions while in Khmaladze (2016, 2017), there are some new transformations that provide ADF tests for fitting a d.f. in the i.i.d. setup and for fitting conditional distributions given covariates, respectively. Kim (2020) provides an algorithm for the computation of the Khmaladze martingale transformation test statistic $\|\hat{\mathcal{U}}_n\|_\infty$ when fitting a known d.f. up to unknown location and scale parameters in the one sample setup, where $\hat{\mathcal{U}}_n$ is as in (9.2.12).

The result (9.2.27) was first proved by Boldin (1982) under the condition that F_0 has a uniformly bounded second derivative and in Koul (1992) under the weaker condition that F_0 has a uniformly continuous density f_0. See also Koul and Ossiander (1994).

Shorack (1982) proposed bootstrap M estimator β_n^*. Lahiri (1989) showed that without centering ψ by $E_*\big(\psi(\varepsilon_{n1})\big)$ in the definition of β_n^* the bootstrap distribution of a suitably standardized β_n^* does not converge to any distribution on $\mathbb{R}$, which renders the bootstrap approximation useless. The papers of Bickel and Freedman (1981) and Freedman (1981) contain many technical results useful in proving the first order accuracy of various bootstrap procedures in the regression setup. Lahiri (2003) discusses the bootstrap methodology for dependent observations where $X_1, X_2, \cdots$ is a stationary process, not necessarily independent and identically distributed.

Bickel and Rosenblatt (1973) derived the asymptotic null distribution of the statistics S_n of (9.4.1) under somewhat stronger and different conditions than those appearing in Theorem 9.4.1. As mentioned earlier the proof of Theorem 9.4.1 is adopted from Hall (1984), who discussed the case of fitting a multivariate density of a finite dimension.

The research activity on the lack-of-fit testing problem has seen an exponential growth during 1990–2010, as can be assessed by the 50 pages long review article by González-Manteiga and Crujeiras (2013). In the context of fitting a regression model with random X_i's of dimension $p \geq 1$, omnibus tests of $\mathcal{H}_{LF}$ based on certain marked residual empirical processes have been investigated by several authors. Su and Wei (1991) use such a process to fit a generalized linear regression model. Stute (1997) discusses the asymptotic distribution theory of this process in detail while bootstrap method is used to implement various tests based on this process in Stute, González–Manteiga and Presedo Quindimil (1998) when $p \geq 1,\ q \geq 1,\ m(x, \beta) = \beta'\psi(x),\ x \in \mathbb{R}^p$, with ψ a vector of q functions. Stute, Thies and Zhu (1998) discusses the martingale transform of this marked residual empirical process while its analog for fitting the generalized linear model $m(x, \beta) = \nu_\theta(\beta' x)$ is investigated in Stute and Zhu (2002), where ν_θ is a known parametric link function. Balakrishna et al. (2019) discuss an analog of this work for the linear autoregressive time series. Khmaladze (2021) discusses some ADF tests for fitting a parametric regression model when covariate vector dimension p is one or larger.

Cox, Koh, Wahba and Yandell (1988), Eubank and Spiegelman (1990), Raz (1990), Härdle and Marron (1990), Hall and Hart (1990), King, Hart and Wehrly (1991), Kulasekera (1995) and Srihera and Stute (2010) propose ad hoc tests for testing for the equality of the two regression functions against the two sided alternatives $\mu_1 \neq \mu_2$. These tests depend on various nonparametric estimators of regression functions.

Hall, Huber and Speckman (1997) and Koul and Schick (2003) discus some covariate-matched tests for testing for the equality of two regression functions against the one-sided alternatives in a heteroscedastic regression setup where neither the two design densities nor the two error densities need be the same or known. Koul and Schick (2003) also discuss some optimality theory.

Some tests of fitting a homoscedastic parametric autoregressive function of order 1 based on an analog of the marked residual process of Stute (1997) were investigated by An and Bing (1991), Diebolt and Zuber (1999), Kim and Lee (2002), Koul and Stute (1999), Stute et al. (2006) and Koul, Stute and Li (2005). Bachmann and Dette (2005) discussed an analog of the Bickel–Rosenblatt test for fitting an error density in the autoregressive time series model (2.12.6). Some other tests based on nonparametric estimators of μ appear in Hjellvik and Tjøstheim (1995), Poggi and Portier (1997) and McKeague and Zhang (1994), among others.

10

Empirical Likelihood

10.1 Introduction

In parametric models, the likelihood method plays a central role in the development of inference procedures. In a nonparametric setup one has no knowledge of an underlying parametric probability model, and the parametric likelihood approach is not feasible. But there is the notion of empirical likelihood which is used to develop model assumptions free inference procedures. One of the earliest uses of the empirical likelihood was made by Kiefer and Wolfowitz (1956, 1958) who showed that the empirical distribution function is the nonparametric maximum likelihood estimator of the population distribution function. Owen (1988, 1991, 2001) developed the empirical likelihood methodology much further. This chapter will discuss some inference procedures based on the empirical likelihood.

10.2 Nonparametric MLE of a Distribution Function

In this section we shall present the result that the empirical distribution function is the nonparametric maximum likelihood estimator (MLE) of the distribution function. We shall do so in the multivariate case.

Let q be a given positive integer. For vectors $x = (x_1, \cdots, x_q)'$ and $t = (t_1, t_2, \cdots, t_q)'$, we write $x \leq t$ to indicate $x_1 \leq t_1, x_2 \leq t_2, \cdots, x_q \leq t_q$. Let $X_1, \cdots, X_n$ be q-dimensional random vectors which are i.i.d. F. Analogous to (2.4.2), the (q-variate) empirical distribution function (e.d.f.) based on $X_1, \cdots, X_n$ is defined by

$$F_n(x) := \frac{1}{n}\sum_{i=1}^{n} I(X_i \leq x), \quad x \in \mathbb{R}^q.$$

Assume that F is completely unknown. Thus the model for F is the set $\mathcal{F}(q)$ of all d.f.'s on $\mathbb{R}^q$. For $G \in \mathcal{F}(q)$, let $J_G(x)$ denote the probability that $G \in \mathcal{F}$ assigns to $x \in \mathbb{R}^q$. In other words, $J_G(x)$ is the probability that a random vector with d.f. G takes the value x. For $q = 1$, $J_G(x)$ is the jump $G(x) - G(x_-)$ of G at x.

Given the random sample $X_1, \cdots, X_n$, the nonparametric likelihood is defined as

$$L(G) = \prod_{i=1}^{n} J_G(X_i), \quad G \in \mathcal{F}(q).$$

DOI: 10.1201/9781003585770-10

A d.f. $\hat{F}$ is said to be a nonparametric maximum likelihood estimator (NPMLE) of F if $L(G) \leq L(\hat{F})$ holds for all G in $\mathcal{F}$. It is unique if $L(G) < L(\hat{F})$ for $G \neq \hat{F}$. The following theorem shows that the e.d.f. F_n is the unique NPMLE of F.

Theorem 10.2.1. *The empirical distribution function F_n is the unique nonparametric maximum likelihood estimator of F.*

Proof. Since $X_1, \cdots, X_n$ are fixed, let $\pi_i := J_G(X_i)$. Then $L(G) = \prod_{i=1}^n \pi_i$. Note that $L(G) = 0$ if and only if $\pi_i = 0$, for some $i = 1, \cdots, n$. Hence, the maximizer has to be among those $G \in \mathcal{F}(q)$ with $\pi_i > 0$, for all $1 \leq i \leq n$. We may also assume that $\sum_{i=1}^n \pi_i = 1$. Indeed, if $\sum_{i=1}^n \pi_i = c < 1$, then the d.f. G^* with $J_{G^*}(X_i) = \pi/c$ satisfies $L(G^*) = (1/c)^n L(G) > L(G)$. Thus we need to maximize $L(G) = \prod_{i=1}^n \pi_i$ over $\mathcal{F}_1 = \{G \in \mathcal{F}(q) : \pi_i > 0, \ \forall\, i = 1, \cdots, n; \ \sum_{i=1}^n \pi_i = 1\}$. We claim that this maximization problem has a unique maximizer given by $\hat{\pi}_i = 1/n$ for $i = 1, \cdots, n$. Since $J_{F_n}(X_i) = 1/n$, the claim yields the assertion of the theorem.

To prove the claim we shall use the Lagrange multiplier method. Accordingly, let

$$h(\pi_1, \cdots, \pi_n; \lambda) = \prod_{i=1}^n \pi_i + \lambda\Big(\sum_{i=1}^n \pi_i - 1\Big), \tag{10.2.1}$$

where λ is the Lagrange multiplier. Now,

$$\frac{\partial h}{\partial \lambda} = 0 \quad \text{implies that} \quad \sum_{i=1}^n \pi_i - 1 = 0 \tag{10.2.2}$$

$$\frac{\partial h}{\partial \pi_j} = 0 \quad \text{implies that} \quad \frac{1}{\pi_j}\prod_{i=1}^n \pi_i + \lambda = 0, \ \forall\, 1 \leq j \leq n. \tag{10.2.3}$$

The last set of constraints yields that

$$\left(\frac{1}{\pi_j} - \frac{1}{\pi_k}\right)\left(\prod_{i=1}^n \pi_i\right) = 0, \quad \forall 1 \leq j \neq k \leq n.$$

We thus obtain that $\pi_j = \pi_k$, $\forall 1 \leq j \neq k \leq n$. This in turn is equivalent to saying that $\pi_1 = \pi_2 \cdots = \pi_n$. Let π denote this common value. By (10.2.2), $n\pi - 1 = 0$ or $\pi = 1/n$. Substituting this relation in (10.2.3) yields that $\pi^{n-1} + \lambda = 0$ or equivalently, $\lambda = -(\pi)^{n-1} = -(1/n)^{n-1}$. By the strict concavity of the log function and Jensen's inequality (2.4.11),

$$\frac{1}{n}\sum_{i=1}^n \log \pi_i \leq \log\Big(\frac{1}{n}\sum_{i=1}^n \pi_i\Big) = \log\Big(\frac{1}{n}\Big),$$

with equality if and only if $\pi_i = 1/n$ for all $i = 1, \cdots, n$. All the above facts yield

$$\sup_{G \in \mathcal{F}(q)} L(G) = \sup_{G \in \mathcal{F}_1} L(G) = \left(\frac{1}{n}\right)^n = L(F_n) \tag{10.2.4}$$

with unique maximizer as claimed. □

Alternate Proof. Suppose $X_1, \cdots, X_n$ takes m distinct values $z_1, \cdots, z_m$. Let $p_j := J_F(z_j)$, $n_j := \sum_{i=1}^n I(X_i = z_j)$ and $\widehat{p}_j = J_{F_n}(z_j) = n_j/n, \ j = 1, \cdots, m$. Then,

$$L(F) = \prod_{j=1}^m p_j^{n_j}, \qquad L(F_n) = \prod_{j=1}^m \widehat{p}_j^{n_j}.$$

If $p_j = 0$, for some $1 \le j \le m$, then $L(F) = 0 < L(F_n)$.

Suppose $p_j > 0$, for all $1 \le j \le m$ and $p_j \ne \widehat{p}_j$, for at least one $1 \le j \le m$. Then, by using the property that $\log(y) < y - 1$ for all $y > 0, y \ne 1$,

$$\log L(F) - \log L(F_n) = \sum_{j=1}^{m} n_j \log\Big(\frac{p_j}{\widehat{p}_j}\Big) = n \sum_{j=1}^{m} \widehat{p}_j \log\Big(\frac{p_j}{\widehat{p}_j}\Big) < n \sum_{j=1}^{m} \widehat{p}_j \Big(\frac{p_j}{\widehat{p}_j} - 1\Big) = 0.$$

Hence, $L(F) < L(F_n)$ for all $F \ne F_n$ and F_n is the NPMLE of F.

10.3 Empirical Likelihood Inference

In this section we shall discuss some inference procedures based on empirical likelihood that are broadly applicable and free of any parametric model assumptions. They mimic the classical likelihood-based inference procedures without assuming the knowledge of the underlying parametric probability model.

In the rest of this chapter, for $0 < \alpha < 1$ and a positive integer k, χ_k^2 denotes a r.v. having a chi-square distribution with k degrees of freedom, see (2.8.3), and $c_{\alpha,k}$ is defined as follows.

$$P\big(\chi_k^2 > c_{\alpha,k}\big) = \alpha. \tag{10.3.1}$$

Let q be a positive integer and $X_1, \cdots, X_n$ be q-dimensional i.i.d. random vectors. We shall first recall some parametric likelihood inference procedures. Let $X_1, \cdots, X_n$ be i.i.d. $\mathcal{N}_q(\theta, I_q)$ for some unknown $\theta \in \mathbb{R}^q$. Let θ_0 be given and consider the problem of testing $H_0 : \theta = \theta_0$ versus the alternatives $H_1 : \theta \ne \theta_0$. Let φ_θ denote the $\mathcal{N}_q(\theta, I_q)$ density and

$$L_n(\theta) := \prod_{i=1}^{n} \varphi_\theta(X_i) = (2\pi)^{-qn/2} \exp\Big(-\frac{1}{2}\sum_{i=1}^{n} \|X_i - \theta\|^2\Big),$$

$$\mathcal{L}_n(\theta_0) := \frac{L_n(\theta_0)}{\sup_{\theta \ne \theta_0} L_n(\theta)} = \frac{L_n(\theta_0)}{\sup_{\theta \in \mathbb{R}^q} L_n(\theta)}.$$

The log-likelihood ratio test rejects H_0 if $\log \mathcal{L}_n(\theta_0)$ is small. Direct algebra obtains $\log \mathcal{L}_n(\theta) = -n\|\bar{X} - \theta_0\|^2/2$ with $\bar{X} = n^{-1}\sum_{i=1}^{n} X_j$ the maximum likelihood estimate of θ. Under H_0, $n\|\bar{X} - \theta_0\|^2 \sim_D \chi_q^2$. Hence, under H_0, $-2\log \mathcal{L}_n(\theta_0) \sim_D \chi_q^2$. Thus, the test that rejects H_0 whenever $-2\log \mathcal{L}_n(\theta_0)$ exceeds $c_{\alpha,q}$ has exact level α, and the confidence set $\{\theta; -2\log \mathcal{L}_n(\theta) \le c_{\alpha,q}\} = \{\theta; n\|\bar{X} - \theta\|^2 \le c_{\alpha,q}\}$ has the exact confidence level $1 - \alpha$. These results are valid for all sample sizes.

As is well known, these types of results are true asymptotically more generally, as follows. Let $r < k$ be two positive integers, and Θ_0 be an r-dimensional hyperplane of $\mathbb{R}^k$ and Θ_0 be a fixed point in $\mathbb{R}^k$. Suppose $X_1, \cdots, X_n$ is a random sample from a density f_θ for some $\theta \in \Theta = \mathbb{R}^k$. We are interested in testing the null hypotheses $H_0 : \theta \in \Theta_0$ and $H_0 : \theta = \theta_0$. Set

$$L_n(\theta) := \prod_{i=1}^{n} f_\theta(X_i), \qquad R_n(\theta_0) := \frac{L_n(\theta_0)}{\sup_{\theta \in \Theta} L_n(\theta)}, \qquad R_n := \frac{\sup_{\theta \in \Theta_0} L_n(\theta)}{\sup_{\theta \in \Theta} L_n(\theta)}.$$

Under some regularity conditions on the given family of densities, Wilks (1938) proved that $-2\log R_n \to_D \chi_{k-r}^2$ if $\theta \in \Theta_0$ and $-2\log R_n(\theta) \to_D \chi_k^2$ if $\theta = \theta_0$. Hence, the test that rejects $H_0 : \theta \in \Theta_0$ whenever $-2\log R_n > c_{\alpha,k-r}$ has asymptotic size α and the confidence set

$\{\theta \in \mathbb{R}^k, -2\log R_n(\theta) \le c_{\alpha,k}\}$ has asymptotic confidence level $1-\alpha$. Owen (1988, 1990, 2001) has shown that these types of results also hold for the empirical likelihood.

We shall first focus on the empirical likelihood ratio test in the mean testing problem. Let $\mathcal{F}_0 := \{F \in \mathcal{F}(q); \int x\, dF(x) = 0_q\}$, where 0_q is the q-dimensional vector of zeros. Let $X_1, \cdots, X_n$ be a random sample from an $F \in \mathcal{F}(q)$. Consider the problem of testing $H_0 : F \in \mathcal{F}_0$ versus the alternatives $H_1 : F \notin \mathcal{F}_0$. Recall, with $\pi_i = J_F(X_i)$, the empirical likelihood of the sample having come from an F is $L(F) = \prod_{i=1}^n \pi_i$. Recall from (10.2.4) that $\sup_{F\in\mathcal{F}(q)} L(F) = (1/n)^n$. Let

$$\mathcal{P}_n := \Big\{p = (p_1, \cdots, p_n)' \in (0,1)^n; \sum_{i=1}^n p_i = 1\Big\}. \tag{10.3.2}$$

The analog of the parametric likelihood ratio is the empirical likelihood ratio

$$\begin{aligned}\mathcal{R}_n &:= \frac{\sup_{F\in\mathcal{F}_0} L(F)}{\sup_{F\in\mathcal{F}(q)} L(F)} = \frac{\sup\Big\{\prod_{i=1}^n \pi_i;\ \pi \in \mathcal{P}_n, \sum_{i=1}^n \pi_i X_i = 0_q\Big\}}{\left(\frac{1}{n}\right)^n} \\ &= \sup\Big\{\prod_{i=1}^n n\pi_i; \pi \in \mathcal{P}_n, \sum_{i=1}^n \pi_i X_i = 0_q\Big\}.\end{aligned}$$

Our goals are to determine $\mathcal{R}_n$ and to derive the limiting null distribution of $-2\log\mathcal{R}_n$, which in turn enables one to implement large sample empirical likelihood tests and confidence regions. Towards achieving these goals, we shall first discuss analogous problems involving non-random vectors in the next subsection. The contents of the following sections are based on the Lecture Notes of Schick (2010) and the paper of Peng and Schick (2013).

10.3.1 Some Preliminaries

For any q-dimensional vectors $y_1, \cdots, y_n$ we set

$$\bar{y} := n^{-1}\sum_{i=1}^n y_i, \qquad y^* := \max_{1\le i\le n} \|y_i\|, \qquad S_y := n^{-1}\sum_{i=1}^n y_i y_i',$$
$$\lambda_y := \inf_{\|u\|=1} (u' S_y u), \qquad \Lambda_y := \sup_{\|u\|=1} (u' S_y u).$$

Recall that λ_y and Λ_y are the smallest and largest eigenvalues of the matrix S_y, respectively.

For given vectors $x_1, x_2, \cdots, x_n$ in $\mathbb{R}^q$, define

$$\mathcal{R}(p) := \prod_{i=1}^n np_i, \quad p := (p_1, \cdots, p_n)' \in [0,1]^n, \tag{10.3.3}$$
$$\bar{\mathcal{D}} := \Big\{p = (p_1, \cdots, p_n)' \in [0,1]^n; \sum_{i=1}^n p_i = 1, \sum_{i=1}^n p_i x_i = 0_q\Big\},$$
$$\mathcal{D} := \Big\{p = (p_1, \cdots, p_n)' \in (0,1)^n; \sum_{i=1}^n p_i = 1, \sum_{i=1}^n p_i x_i = 0_q\Big\}.$$

Furthermore, define

$$\mathcal{M} := \begin{cases} \sup_{p\in\mathcal{D}} \mathcal{R}(p), & \text{if } \mathcal{D} \text{ is not empty}, \\ 0, & \text{if } \mathcal{D} \text{ is empty} \end{cases}.$$

We are interested in obtaining the maximum $\mathcal{M}$. Assume that the origin is an interior point of the convex hull generated by the vectors $x_1, \cdots, x_n$. This means that there is a small positive r such that the open ball $\{x \in \mathbb{R}^q : \|x\| < r\}$ is a subset of the set of all $y \in \mathbb{R}^q$ of the form $y = \sum_{i=1}^n p_i x_i$, for some $p_1, \cdots, p_n \in [0,1]$ with $\sum_{i=1}^n p_i = 1$, which is the convex hull generated by $x_1, x_2, \cdots, x_n$. Then $\mathcal{D}$ is not empty. Since $\mathcal{R}$ is continuous, it achieves a maximum on the compact set $\bar{\mathcal{D}}$. Since $\mathcal{D} \subset \bar{\mathcal{D}}$ and $\mathcal{R}(p) = 0$ for $p \in \bar{\mathcal{D}} \cap \mathcal{D}^c$, $\mathcal{R}$ achieves the maximum over $\mathcal{D}$. Because $\log \mathcal{R}$ is strictly concave on $(0,1)^n$, there is a unique $p^* \in \mathcal{D}$ such that $\mathcal{M} = \mathcal{R}(p^*)$. To identify p^* and $\mathcal{R}(p^*)$, we use the method of Lagrange multipliers, which amounts to solving the following $n+q+1$ equations for p_i's and the Lagrange's multipliers $\alpha \in \mathbb{R}$ and $\beta \in \mathbb{R}^q$ for which $\alpha + \beta' x_i > 0, i = 1, \cdots, n$.

$$\frac{1}{p_i} = \alpha + \beta' x_i, \quad 1 \le i \le n; \qquad \sum_{i=1}^n p_i = 1, \qquad \sum_{i=1}^n p_i x_i = 0_q.$$

Or, equivalently,

$$p_i = \frac{1}{\alpha + \beta' x_i}, \qquad 1 \le i \le n, \tag{10.3.4}$$

$$\sum_{i=1}^n \frac{1}{\alpha + \beta' x_i} = 1, \tag{10.3.5}$$

$$\sum_{i=1}^n \frac{x_i}{\alpha + \beta' x_i} = 0_q. \tag{10.3.6}$$

From these equations we obtain the identity

$$\alpha = \alpha \sum_{i=1}^n \frac{1}{\alpha + \beta' x_i} + \beta' \sum_{i=1}^n \frac{x_i}{\alpha + \beta' x_i} = \sum_{i=1}^n \frac{\alpha + \beta' x_i}{\alpha + \beta' x_i} = n.$$

Let $\xi = n^{-1}\beta$. Replace α by n in (10.3.5) and (10.3.6) to see that these equations are equivalent to

$$\frac{1}{n} \sum_{i=1}^n \frac{1}{1 + \xi' x_i} = 1, \tag{10.3.7}$$

$$\frac{1}{n} \sum_{i=1}^n \frac{x_i}{1 + \xi' x_i} = 0_q. \tag{10.3.8}$$

The following lemma gives the expression for $\mathcal{M}$ and the optimum p^*.

Lemma 10.3.1. *Suppose S_x is positive definite and the origin is an interior point of the convex hull generated by $x_1, \cdots, x_n$. Then,*

$$\mathcal{M} = \prod_{i=1}^n \frac{1}{1 + \xi' x_i}, \tag{10.3.9}$$

where ξ is the unique solution of (10.3.8) *subject to the constraints $1 + \xi' x_i > 0$, for all $i = 1, \cdots, n$.*

Proof. Because S_x is positive definite, the $q \times n$ matrix $\boldsymbol{X} := (x_1, \cdots, x_n)$ has full rank and ξ will be unique. The assumption that the origin is in the convex hull of $x_1, \cdots, x_n$ ensures the existence of p_i^*'s. Moreover, because

$$\frac{1}{n} \sum_{i=1}^n \frac{1}{1 + \xi' x_i} + \xi' \frac{1}{n} \sum_{i=1}^n \frac{x_i}{1 + \xi' x_i} = 1,$$

equation (10.3.8) implies (10.3.7) and the optimal p_i^*'s are given by

$$p_i^* = \frac{1}{n}\frac{1}{1+\xi' x_i}, \quad 1 \le i \le n, \tag{10.3.10}$$

and $\mathcal{M}$ is as in (10.3.9), where ξ is a solution of (10.3.8) subject to the constraints that $1+\xi' x_i > 0$, for all $i = 1, \cdots, n$. □

The following theorem gives a sufficient condition for the conditions of the previous lemma.

Theorem 10.3.1. *Suppose $x_1, \cdots, x_n$ in $\mathbb{R}^q$ satisfy the following condition.*

$$\lambda_x > 5x^* \|\bar{x}\|. \tag{10.3.11}$$

Then, S_x is positive definite and the origin is an interior point of the convex hull generated by $x_1, \cdots, x_n$. Hence,

$$\mathcal{M} = \prod_{i=1}^{n} \frac{1}{1+\xi' x_i}, \tag{10.3.12}$$

where ξ is the unique solution of the equation

$$\frac{1}{n}\sum_{i=1}^{n} \frac{x_i}{1+\xi' x_i} = 0_q,$$

subject to the constraints $1+\xi' x_i > 0$, for all $i = 1, \cdots, n$.

Proof. The proof follows from Lemma 10.3.1 and Lemma 10.3.3 below. □

Lemma 10.3.2. *Let $0 < \kappa < \infty$ and G be a d.f. satisfying $G(\kappa) = 1$, $G(-\kappa_-) = 0$ and having zero mean. Let $\sigma^2 := \int_{[-\kappa,\kappa]} x^2\, dG(x)$ denote the variance of G. Then,*

$$1 - G(z) \ge \frac{\sigma^2 - 2\kappa z}{2\kappa^2}, \qquad \forall\, 0 \le z < \kappa. \tag{10.3.13}$$

Proof. Fix a z in $[0, \kappa)$. Then we have the inequality

$$\begin{aligned} 1 - G(z) &\ge \int_{z<x\le\kappa} \frac{x}{\kappa}\, dG(x) = \int_{0<x\le\kappa} \frac{x}{\kappa}\, dG(x) - \int_{0<x\le z} \frac{x}{\kappa}\, dG(x) \\ &\ge \int_{0<x\le\kappa} \frac{x}{\kappa}\, dG(x) - \frac{z}{\kappa}. \end{aligned}$$

The zero mean condition implies $-\int_{-\kappa\le x\le 0} x\, dG(x) = \int_{0<x\le\kappa} x\, dG(x)$ and hence

$$2\int_{0<x\le\kappa} x\, dG(x) = \int_{-\kappa\le x\le\kappa} |x|\, dG(x) \ge \frac{\int_{-\kappa\le x\le\kappa} x^2\, dG(x)}{\kappa} = \frac{\sigma^2}{\kappa}.$$

Combine this bound with the previous one to obtain (10.3.13). □

Lemma 10.3.3. *Suppose $x_1, \cdots x_n$ in $\mathbb{R}^q$ satisfy* (10.3.11)*. Then, S_x is positive definite, and the origin is in the convex hull generated by $x_1, \cdots, x_n$.*

Proof. Clearly, the assumption (10.3.11) implies $\lambda_x > 0$, which in turn implies that S_x is positive definite. It remains to show that the origin is in the interior of the convex hull generated by $x_1, \cdots, x_n$. In the proof here u is a unit vector in $\mathbb{R}^q$. Let

$$N(u) := \sum_{i=1}^{n} I(u'x_i > 0), \qquad N := \inf_{\|u\|=1} N(u).$$

The origin is in the interior of the convex hull of $x_1, \cdots, x_n$, if, for every unit vector u, $u'x_i > 0$ holds for at least one $i = 1, \cdots, n$, i.e., if $N \geq 1$.

Now, $-u'\bar{x} \leq \|\bar{x}\|$ implies that $u'x_i - u'\bar{x} - \|\bar{x}\| \leq u'x_i$ and hence

$$N(u) \geq \sum_{i=1}^{n} I\big(u'(x_i - \bar{x}) > \|\bar{x}\|\big).$$

By the triangle inequality, $\|x_i - \bar{x}\| \leq 2x^*$. Apply (10.3.13) to the discrete uniform distribution on the points $u'(x_1 - \bar{x}), \cdots, u'(x_n - \bar{x})$ and with $z = \|\bar{x}\|$ and $\kappa = 2x^*$. This yields the inequality

$$\frac{1}{n}\sum_{i=1}^{n} I\big(u'(x_i - \bar{x}) > \|\bar{x}\|\big) \geq \frac{1}{8(x^*)^2}\Big[\frac{1}{n}\sum_{i=1}^{n}\big(u'(x_i - \bar{x})\big)^2 - 4x^*\|\bar{x}\|\Big].$$

Combining it with the inequality

$$\frac{1}{n}\sum_{i=1}^{n}\big(u'(x_i - \bar{x})\big)^2 = u'S_x u - \big(u'\bar{x}\big)^2 \geq \lambda_x - \|\bar{x}\|^2 \geq \lambda_x - x^*\|\bar{x}\|$$

yields the lower bound

$$N(u) \geq n\frac{\lambda_x - 5x^*\|\bar{x}\|}{8(x^*)^2}.$$

This bound and (10.3.11) in turn imply $N \geq 1$. Hence, the origin is an interior point of the convex hull generated by $x_1, \cdots, x_n$. □

Remark 10.3.1. In general, numerical methods are needed to obtain solutions to (10.3.8) as explicit solutions are available in only a few cases. One such case is when $q = 1$ and $x_1, \cdots, x_n$ take on only exactly two values, say a and b with $a < 0 < b$. Let $n_a := \sum_{i=1}^n I(x_i = a)$, $n_b := \sum_{i=1}^n I(x_i = b)$. Then (10.3.8) becomes

$$\frac{1}{n}\Big(\frac{n_a a}{1 + a\xi} + \frac{n_b b}{1 + b\xi}\Big) = 0.$$

Solving this equation yields the solution

$$\xi = -\frac{an_a + bn_b}{abn}.$$

Assuming $n_a \wedge n_b > 0$, for this solution we also have

$$1 + a\xi = 1 - (an_a + bn_b)/(bn) > 1 - n_b/n = n_a/n > 0,$$
$$1 + b\xi = 1 - (an_a + bn_b)/(an) > 1 - n_a/n = n_b/n > 0.$$

In terms of the proportion $f_b = n_b/n$ of b values, ξ can be written as

$$\xi = \frac{bf_b - |a|(1 - f_b)}{|a|b}. \tag{10.3.14}$$

Another case where the solution (10.3.8) is explicit is when $q = 1$ and $x_1, \cdots, x_n$ take on the values a, 0 and b with $a < 0 < b$. In this case,

$$\xi = -\frac{an_a + bn_b}{ab(n_a + n_b)}.$$

10.3.2 Some Properties of the Lagrange Multiplier ξ

Here we shall discuss some more properties of the Lagrange multiplier ξ of the previous subsection, which will be useful in analyzing empirical likelihood ratios later, where the x_i's are replaced by random vectors.

First, we recall definitions of some norms for matrices. For any matrix A, the Euclidean and operator norms are, respectively, defined to be

$$\|A\| := \big(\text{trace}(A'A)\big)^{1/2}, \qquad \|A\|_o := \sup_{\|u\|=1} \|Au\| = \sup_{\|u\|=1} \big(u'A'Au\big)^{1/2}.$$

The squared Euclidean norm $\|A\|^2$ is the sum of the eigenvalues of $A'A$ and the squared operator norm $\|A\|_0^2$ is the largest eigenvalue of $A'A$. We have the following inequalities between these norms for all Euclidean vectors x with dimension equal to the number of columns of A.

$$\|A\|_o^2 \le \|A\|^2, \qquad \|Ax\| \le \|A\|_o\|x\| \le \|A\|\,\|x\|.$$

For a symmetric matrix A, one has $\|A\|_o = \sup_{\|u\|=1} \big|u'Au\big|$ and this becomes $\|A\|_o = \sup_{\|u\|=1} \big(u'Au\big)$ if A is also nonnegative definite.

For given vectors $x_1, \cdots, x_n$ in $\mathbb{R}^q$, let

$$x^{(r)} := \sup_{\|u\|=1} \Big|n^{-1}\sum_{i=1}^{n} \big(u'x_i\big)^r\Big|, \qquad r = 2, 3, 4.$$

Then,

$$x^{(2)} = \Lambda_x, \qquad x^{(3)} \le x^*\Lambda_x, \qquad x^{(4)} \le (x^*)^2\Lambda_x. \tag{10.3.15}$$

From now on the assumption (10.3.11) is in action.

Write $\xi = \|\xi\|u$, for some unit vector u. Then,

$$0 = \frac{1}{n}\sum_{i=1}^{n} \frac{u'x_i}{1+\xi'x_i} = \frac{1}{n}\sum_{i=1}^{n} \frac{u'x_i(1+\xi'x_i - \xi'x_i)}{1+\xi'x_i} = u'\bar{x} - \|\xi\|\frac{1}{n}\sum_{i=1}^{n}\frac{\big(u'x_i\big)^2}{1+\xi'x_i},$$

which is equivalent to saying

$$u'\bar{x} = \|\xi\|\frac{1}{n}\sum_{i=1}^{n}\frac{\big(u'x_i\big)^2}{1+\xi'x_i}. \tag{10.3.16}$$

We also have

$$\begin{aligned}\lambda_x\|\xi\| \le u'S_xu\|\xi\| &= \|\xi\|\frac{1}{n}\sum_{i=1}^{n}\big(u'x_i\big)^2 \le \|\xi\|\frac{1}{n}\sum_{i=1}^{n}\frac{\big(u'x_i\big)^2(1+\|\xi\|x^*)}{1+\xi'x_i} \\ &= u'\bar{x}(1+\|\xi\|x^*) \le \|\bar{x}\|(1+\|\xi\|x^*),\end{aligned}$$

where we used (10.3.16) in the last equality. Therefore,

$$\|\xi\| \le \frac{\|\bar{x}\|}{\lambda_x - \|\bar{x}\|x^*}. \tag{10.3.17}$$

This bound together with (10.3.11) yields the following.

$$\max_{1\le j\le n} |\xi' x_j| \le \|\xi\| x^* \le \frac{\|\bar{x}\| x^*}{\lambda_x - \|\bar{x}\| x^*} < \frac{1}{4}, \tag{10.3.18}$$

$$\max_{1\le j\le n} \frac{1}{1+\xi' x_j} \le \frac{1}{1-\|\xi\| x^*} < \frac{4}{3}, \tag{10.3.19}$$

$$n^{-1}\sum_{i=1}^{n} |\xi' x_i|^r \le \|\xi\|^r x^{(r)} \le \frac{\Lambda_x (x^*)^{r-2}\|\bar{x}\|^r}{(\lambda_x - \|\bar{x}\| x^*)^r}, \quad r = 2,3,4. \tag{10.3.20}$$

The inequality (10.3.19) and the identity

$$\frac{1}{1+c} - 1 + c = \frac{c^2}{1+c}, \quad c > -1,$$

applied with $c = \xi' x_i$ yield the inequalities

$$\left|\frac{1}{1+\xi' x_i} - 1 + \xi' x_i\right| \le \frac{4}{3}(\xi' x_i)^2, \quad i = 1,\cdots,n. \tag{10.3.21}$$

Let $z_1,\cdots,z_n$ be vectors in $\mathbb{R}^m$. We now prove the inequality

$$\left\|n^{-1}\sum_{i=1}^{n}\left[\frac{z_i}{1+\xi' x_i} - z_i + z_i \xi' x_i\right]\right\| \le \frac{4}{3}\frac{\Lambda_z^{1/2}\Lambda_x^{1/2} x_* \|\bar{x}\|^2}{(\lambda_x - x^*\|\bar{x}\|)^2}. \tag{10.3.22}$$

For this we use the fact that the norm $\|v\|$ of a vector in $\mathbb{R}^m$ equals $\sup_u u'v$ where the supremum is taken over all unit vectors u in $\mathbb{R}^m$. Thus the left-hand side of (10.3.22) equals

$$\sup_u n^{-1}\sum_{i=1}^{n} u' z_i \left[\frac{1}{1+\xi' x_i} - 1 + \xi' x_i\right]$$

and is bounded by

$$\sup_u n^{-1}\sum_{i=1}^{n} |u' z_i| \frac{4}{3}(\xi' x_i)^2.$$

An application of the Cauchy-Schwarz inequality shows that the last term is bounded by $4/3$ times the square-root of the product of $n^{-1}\sum_{i=1}^{n}(\xi' x_i)^4$ and

$$\sup_u n^{-1}\sum_{i=1}^{n}(u' z_i)^2 = \sup_u u' n^{-1}\sum_{i=1}^{n} z_i z_i' u = \sup_u u' S_z u = \Lambda_z.$$

These bounds and (10.3.20) imply the inequality (10.3.22). For $z_i = S_x^{-1} x_i$, we calculate, with the help of (10.3.8) that the quantity inside the norm on the left-hand side of (10.3.22) equals

$$n^{-1}\sum_{i=1}^{n}\left[\frac{S_x^{-1} x_i}{1+\xi' x_i} - S_x^{-1} x_i + S_x^{-1} x_i x_i' \xi\right] = -S_x^{-1}\bar{x} + \xi$$

and that S_z equals S_x^{-1} yielding $\Lambda_z = \lambda_x^{-1}$. Hence, the bound (10.3.22) applied with $z_i = S_x^{-1} x_i$ yields the inequality

$$\|\xi - S_x^{-1}\bar{x}\| \le \frac{4}{3}\frac{\Lambda_x^{1/2}\|\bar{x}\|^2 x^*}{\lambda_x^{1/2}(\lambda_x - \|\bar{x}\| x^*)^2}. \tag{10.3.23}$$

This inequality will be used later in analyzing the large sample behavior of the analog of ξ when x_i's are replaced by random vectors.

The next few bounds are useful in deriving the limiting distribution of the log empirical likelihood ratio when x_i's are replaced by random vectors. Recall the inequality

$$\left|2\log(1+t) - 2t + t^2\right| \le \frac{2|t|^3}{3(1-|t|)^3}, \quad \forall\, |t| < 1. \tag{10.3.24}$$

The bounds (10.3.19), (10.3.20) and (10.3.24) with $t = \xi' x_i$, give the inequality

$$\begin{aligned}\left|2\sum_{i=1}^n \log(1+\xi' x_i) - 2n\xi'\bar{x} + n\xi' S_x \xi\right| &= \left|\sum_{i=1}^n \left[2\log(1+\xi' x_i) - 2\xi' x_i + (\xi' x_i)^2\right]\right| \\ &\le \frac{2}{3}\Big(\frac{4}{3}\Big)^3 \sum_{i=1}^n |\xi' x_i|^3 \le 2n \frac{\Lambda_x x^* \|\bar{x}\|^3}{(\lambda_x - \|\bar{x}\| x^*)^3}.\end{aligned}$$

Algebra establishes the identity

$$2\xi'\bar{x} - \xi' S_x \xi = \bar{x}' S_x^{-1}\bar{x} - \big(\xi - S_x^{-1}\bar{x}\big)' S_x \big(\xi - S_x^{-1}\bar{x}\big).$$

The inequalities (10.3.18) and (10.3.23) and the definition of Λ_x yield the bound

$$\left|\big(\xi - S_x^{-1}\bar{x}\big)' S_x \big(\xi - S_x^{-1}\bar{x}\big)\right| \le \Lambda_x \big\|\xi - S_x^{-1}\bar{x}\big\|^2 \le \frac{4\Lambda_x^2 \|\bar{x}\|^3 x^*}{9\lambda_x \big(\lambda_x - \|\bar{x}\| x^*\big)^3}.$$

Combine these bounds and the above identity to obtain

$$\left|2\sum_{i=1}^n \log(1+\xi' x_i) - n\bar{x}' S_x^{-1}\bar{x}\right| \le \frac{(2\lambda_x\Lambda_x + \Lambda_x^2) n x^* \|\bar{x}\|^3}{\lambda_x(\lambda_x - \|\bar{x}\| x^*)^3}. \tag{10.3.25}$$

The last bound is used to derive the asymptotic null distribution of $-2\log\mathcal{R}_n$ in the next subsection.

10.4 Asymptotic Distribution of $-2\log\mathcal{R}_n$

Here we shall derive the limiting null distribution of $-2\log\mathcal{R}_n$. Note that $\mathcal{R}_n$ is like the $\mathcal{M}$ of Lemma 10.3.1. For a given set of q-dimensional random vectors $X_{n1}, \cdots, X_{nn}$, let

$$X_n^* := \max_{1\le i\le n} \|X_{ni}\|, \quad \bar{X}_n := n^{-1}\sum_{i=1}^n X_{ni}, \quad S_n := n^{-1}\sum_{i=1}^n X_{ni} X_{ni}'.$$

Let λ_n, Λ_n denote the smallest and largest eigenvalues of S_n. To have consistent notation, we ought to denote them by λ_X, Λ_X, but since they also depend on n and since here we are discussing the asymptotic distribution as $n \to \infty$, it is more desirable to exhibit their dependence on n. They are indeed the analogs of λ_x, Λ_x of the previous subsection. Similarly, ξ_n denotes the analog of the Lagrange multiplier ξ of the previous subsection.

For an integer $k \ge 1$ and $\Delta \ge 0$, let $\chi_k^2(\Delta)$ denote a r.v. having the noncentral chi-square distribution with k degrees of freedom and the non-centrality parameter Δ, see (2.8.4). Note that if $X \sim_D \mathcal{N}_k(\mu, I_k)$, then $X'X \sim_D \chi_k^2(\|\mu\|^2)$. We are ready to state the main theorem, which gives the limiting null distribution of $-2\log\mathcal{R}_n$.

Theorem 10.4.1. *Suppose the random vectors $X_{n1}, \cdots, X_{nn}$ satisfy*

$$(a)\ n^{-1/2}X_n^* \to_p 0, \qquad (b)\ n^{1/2}\bar{X}_n \to_D \mathcal{N}(\nu_*, \Sigma), \qquad (c)\ S_n \to_p \Sigma.$$

for some q-dimensional vector ν_* and some positive definite symmetric $q \times q$ matrix Σ. Then the following hold.
(i) *There is a random vector ξ_n and an event B_n with $P(B_n) \to 1$, such that on the event B_n,*

$$\mathcal{R}_n = \prod_{i=1}^{n} \frac{1}{1 + \xi_n' X_{ni}}.$$

Moreover, ξ_n is uniquely defined on B_n as a solution of the equation

$$\frac{1}{n}\sum_{i=1}^{n} \frac{X_{ni}}{1 + \xi_n' X_{ni}} = 0_q$$

subject to the conditions $1 + \xi_n' X_{ni} > 0$, $i = 1, \cdots, n$.
(ii) *The stochastic expansion*

$$\left| -2\log \mathcal{R}_n - n\bar{X}_n'\Sigma^{-1}\bar{X}_n \right| \to_p 0 \tag{10.4.1}$$

holds and implies $-2\log \mathcal{R}_n \to_D \chi_q^2(\nu_*'\Sigma^{-1}\nu_*)$.
(iii) *The random vector ξ_n satisfies*

$$n^{1/2}\left\|\xi_n - \Sigma^{-1}\bar{X}_n\right\| \to_p 0. \tag{10.4.2}$$

Consequently, $n^{1/2}\xi_n \to_D \mathcal{N}_q(\Sigma^{-1}\nu_*, \Sigma^{-1})$.
(iv) *Let T be an index set and $Z_{n1}(t), \cdots, Z_{nn}(t), t \in T$ be k-dimensional random vectors. Suppose there exist $k \times q$ matrices $C(t), t \in T$ such that the following conditions hold.*

$$T_{n1} := \sup_{t\in T} \frac{1}{n}\sum_{i=1}^{n} \left\|Z_{ni}(t)\right\|^2 = O_p(1), \qquad T_{n2} := \sup_{t\in T}\left\|\frac{1}{n}\sum_{i=1}^{n} Z_{ni}(t)X_{ni}' - C(t)\right\| \to_p 0. \tag{10.4.3}$$

Then,

$$\sup_{t\in T}\left\|n^{-1/2}\sum_{i=1}^{n}\left[\frac{Z_{ni}(t)}{1 + \xi_n' X_{ni}} - Z_{ni}(t) + C(t)\Sigma^{-1}X_{ni}\right]\right\| \to_p 0. \tag{10.4.4}$$

Proof. Condition (b) implies $Y_n := n^{1/2}\Sigma^{-1/2}\bar{X}_n \to_D Y \sim_D \mathcal{N}_q(\Sigma^{-1/2}\nu_*, I_q)$. Hence, Theorem 2.10.1 implies $Y_n'Y_n \to_D Y'Y$, which in turn yields

$$n\bar{X}_n'\Sigma^{-1}\bar{X}_n \to_D \chi_q^2(\nu_*'\Sigma^{-1}\nu_*).$$

Therefore, we only need to prove (i) and equations (10.4.1), (10.4.2) and (10.4.4). The continuity of the matrix inversion on the set of invertible $q \times q$ matrices and assumption (c) imply $S_n^{-1} \to_p \Sigma^{-1}$. Let $\sigma_{\min}$ denote the smallest and $\sigma_{\max}$ the largest eigenvalue of Σ. Assumption (c) implies $\lambda_n \to_p \sigma_{\min}$ and $\Lambda_n \to_p \sigma_{\max}$. Since Σ is positive definite, we have $\sigma_{\max} \geq \sigma_{\min} > 0$. This fact and assumptions (b) and (c) yield

$$\left| n\bar{X}_n'S_n^{-1}\bar{X}_n - n\bar{X}_n'\Sigma^{-1}\bar{X}_n \right| \to_p 0. \tag{10.4.5}$$

We also have $\|\bar{X}_n\|X_n^* = n^{1/2}\|\bar{X}_n\| \cdot n^{-1/2}X_n^* \to_p 0$. Hence,

$$\lambda_n - 5\|\bar{X}_n\|X_n^* \to_p \sigma_{\min}. \tag{10.4.6}$$

Let $B_n := \{\lambda_n > 5\|\bar{X}_n\|X_n^*\}$. Since $\sigma_{\min}$ is positive, (10.4.6) implies $P(B_n) \to 1$. Finally, all the above imply the following two facts:

$$\frac{n^{1/2}X_n^*\|\bar{X}_n\|^2}{(\lambda_n - \|\bar{X}_n\|X_n^*)^2} \to_p 0, \tag{10.4.7}$$

$$\frac{nX_n^*\|\bar{X}_n\|^3}{(\lambda_n - \|\bar{X}_n\|X_n^*)^3} \to_p 0. \tag{10.4.8}$$

Since $\mathcal{R}_n$ is the same as $\mathcal{M}$ of (10.3.3) with x_i's replaced by X_{ni}'s, part (i) follows from Theorem 10.3.1. Thus, on B_n, we have the identity

$$-2\log \mathcal{R}_n = 2\sum_{i=1}^n \log\big(1 + \xi_n' X_{ni}\big).$$

The properties of λ_n and Λ_n and (10.4.8) yield

$$\frac{2\lambda_n\Lambda_n + \Lambda_n^2}{\lambda_n}\frac{nX_n^*\|\bar{X}_n\|^3}{\left(\lambda_n - X_n^*\|\bar{X}_n\|\right)^3} \to_p 0.$$

Hence, from this and (10.3.25) we obtain on B_n,

$$\Big| -2\log \mathcal{R}_n - n\bar{X}_n' S_n^{-1}\bar{X}_n\Big| \to_p 0.$$

This and (10.4.5) imply (10.4.1). Applying (10.3.23) with x_i replaced by X_{ni} and then using the properties of λ_n and Λ_n and (10.4.7) gives, on B_n,

$$n^{1/2}\|\xi_n - S_n^{-1}\bar{X}_n\| \le \frac{4}{3}\frac{\Lambda_n^{1/2}}{\lambda_n^{1/2}}\frac{n^{1/2}X_n^*\|\bar{X}_n\|^2}{\left(\lambda_n - \|\bar{X}_n\|X_n^*\right)^2} \to_p 0.$$

This proves (10.4.2). To prove (10.4.4), apply (10.3.22) with x_i replaced by X_{ni} and z_i replaced by $Z_{ni}(t)$, then use (10.4.3) and (10.4.7) to derive on B_n,

$$\begin{aligned}
&\sup_{t\in T} n^{1/2}\left\|\frac{1}{n}\sum_{i=1}^n\Big[\frac{Z_{ni}(t)}{1+\xi_n' X_{ni}} - Z_{ni}(t) + Z_{ni}(t)X_{ni}'\xi_n\Big]\right\| \\
&\le \frac{4}{3}\sup_{t\in T}\Big(\frac{1}{n}\sum_{i=1}^n\|Z_{ni}(t)\|^2\Big)^{1/2}\frac{\Lambda_n^{1/2}n^{1/2}X_n^*\|\bar{X}_n\|^2}{\left(\lambda_n - \|\bar{X}_n\|X_n^*\right)^2} \to_p 0.
\end{aligned}$$

Thus (10.4.4) follows from this and the fact that

$$\sup_{t\in T} n^{1/2}\Big|\frac{1}{n}\sum_{i=1}^n Z_{ni}(t)X_{ni}'\xi_n - C(t)\Sigma^{-1}\frac{1}{n}\sum_{i=1}^n X_i\Big| \to_p 0.$$

To obtain the latter fact, bound its left-hand side by

$$(T_{n1}\mathrm{trace}(S_n))^{1/2}n^{1/2}\|\xi_n - \Sigma^{-1}\bar{X}_n\| + T_{n2}\|n^{1/2}\Sigma^{-1}\bar{X}_n\|$$

with the help of Exercise 2, and use (10.4.2), (10.4.3) and conditions (b) and (c). □

Our next goal is to provide sufficient conditions for the assumptions (a), (b) and (c) of the previous theorem. We begin with the following lemma.

Lemma 10.4.1. *Let ν_* and Σ be as in Theorem 10.4.1. Suppose the random vectors $X_{n1}, \cdots, X_{nn}$ are independent and satisfy* (2.10.1) *and the following conditions with $\mu_{ni} = E(X_{ni})$:*

$$n^{-1/2}\sum_{i=1}^{n}\mu_{ni} \to \nu_*, \quad \frac{1}{n}\sum_{i=1}^{n}\mu_{ni}\mu_{ni}' \to 0, \quad \frac{1}{n}\sum_{i=1}^{n}E(X_{ni}X_{ni}') \to \Sigma.$$

Then, the assumptions (a), (b) and (c) of Theorem 10.4.1 hold.

Proof. The assumptions (a)–(c) are equivalent to the following conditions. $\forall\, u \in \mathbb{R}^q$, $\|u\| = 1$,

$$\max_{1\le i\le n} n^{-1/2}|u'X_{ni}| \to_p 0, \qquad n^{-1/2}\sum_{i=1}^{n}u'X_{ni} \to_D \mathcal{N}(u'\nu_*, u'\Sigma u),$$

$$u'S_n u = \frac{1}{n}\sum_{i=1}^{n}\left(u'X_{ni}\right)^2 \to u'\Sigma u.$$

This follows from the Cramér-Wold device; see Corollary 2.10.1 and the inequality $\|X_{ni}\| \le \sum_{k=1}^{q}|u_k'X_{ni}|$ valid for any orthonormal basis $u_1, \cdots, u_q$ of $\mathbb{R}^q$. Fix a unit vector u in $\mathbb{R}^q$. The random variables $\xi_{ni} := n^{-1/2}u'X_{ni}$, $i = 1, \cdots, n$ satisfy the assumptions of Corollary 2.10.7 with $\mu = u'\nu_*$ and $\sigma^2 = u'\Sigma u$ and the theorem gives the desired results. □

The conditions simplify if the random vectors $X_{n1}, \cdots, X_{nn}$ are also identically distributed. This is done in the next lemma whose proof is left as an exercise.

Lemma 10.4.2. *Let ν_* and Σ be as in Theorem 10.4.1. Suppose the random vectors $X_{n1}, \cdots, X_{nn}$ are i.i.d. and satisfy the following assumptions:*

$$n^{1/2}E(X_{n1}) \to \nu_*, \quad E(X_{n1}X_{n1}') \to \Sigma, \quad E(\|X_{n1}\|^2 I(\|X_{n1}\| > \epsilon n^{1/2})) \to 0, \quad \forall \epsilon > 0.$$

Then, the assumptions (a), (b) *and* (c) *of Theorem 10.4.1 hold.*

Because of its importance, we formulate a version of Theorem 10.4.1 for i.i.d. data and with special choices for $Z_{ni}(t)$.

Theorem 10.4.2. *Let $Z_1, \cdots, Z_n$ be i.i.d. copies of the m-dimensional random vector Z. Let g be a function from $\mathbb{R}^m$ into $\mathbb{R}^q$ such that $E(g(Z)) = 0_q$ holds and $\Sigma = E(g(Z)g'(Z))$ is positive definite. Set*

$$\mathcal{R}_n := \sup\Big\{\prod_{i=1}^{n}n\pi_i;\ \pi \in \mathcal{P}_n;\ \sum_{i=1}^{n}\pi_i g(Z_i) = 0_q\Big\}.$$

Then the following hold with $\bar{G}_n := n^{-1}\sum_{i=1}^{n}g(Z_j)$.
(i) *There is a random vector ξ_n and an event B_n with $P(B_n) \to 1$, such that on the event B_n,*

$$\mathcal{R}_n = \prod_{i=1}^{n}\frac{1}{1+\xi_n' g(Z_i)}$$

and ξ_n is uniquely defined on B_n as a solution of the equation

$$\frac{1}{n}\sum_{i=1}^{n}\frac{g(Z_i)}{1+\xi_n' g(Z_i)} = 0_q$$

subject to the conditions $1+\xi_n' g(Z_i) > 0$, $i = 1, \cdots, n$.

(ii) *The stochastic expansion*

$$\left| -2\log\mathcal{R}_n - n\bar{G}_n'\Sigma^{-1}\bar{G}_n \right| \to_p 0 \tag{10.4.9}$$

holds and implies

$$-2\log\mathcal{R}_n \to_D \chi_q^2.$$

(iii) *The random vector* ξ_n *satisfies*

$$n^{1/2}\|\xi_n - \Sigma^{-1}\bar{G}_n\| \to_p 0 \tag{10.4.10}$$

and therefore, $n^{1/2}\xi_n \to_D \mathcal{N}_q(0_q, \Sigma^{-1})$.

(iv) *For a measurable function* h *from* $\mathbb{R}^m$ *into* $\mathbb{R}$ *such that* $E(h^2(Z))$ *is finite, the following stochastic expansion holds with* $C_h = E(h(Z)g'(Z))$.

$$n^{-1/2}\sum_{i=1}^n \Big[\frac{h(Z_i)}{1+\xi_n' g(Z_i)} - h(Z_i) + C_h\Sigma^{-1}g(Z_j)\Big] \to_p 0.$$

(v) *For a measurable function* ℓ *from* $\mathbb{R}^m$ *into* $\mathbb{R}$*, the following uniform stochastic expansion holds with* $C(t) = E(I(\ell(Z) \le t)g(Z))$.

$$\sup_{t\in\mathbb{R}}\Big|n^{-1/2}\sum_{i=1}^n \Big[\frac{I(\ell(Z_i)\le t)}{1+\xi_n' g(Z_j)} - I(\ell(Z_i)\le t) + C(t)\Sigma^{-1}g(Z_i)\Big]\Big| \to_p 0.$$

Proof. We shall show that the theorem is a special case of Theorem 10.4.1 with $X_{ni} = g(Z_i)$, $i = 1, \cdots, n$. The assumptions of Lemma 10.4.2 are met with $\nu_* = 0_q$ and $\Sigma = E(g(Z)g'(Z))$. Thus we have the following versions of assumptions (a)–(c) of Theorem 10.4.1:

$$n^{-1/2}\max_{1\le i\le n}\|g(Z_j)\| \to_p 0, \tag{10.4.11}$$

$$n^{1/2}\bar{G}_n = n^{-1/2}\sum_{i=1}^n g(Z_i) \to_D \mathcal{N}_q(0_q, \Sigma), \tag{10.4.12}$$

$$\bar{S}_n := \frac{1}{n}\sum_{i=1}^n g(Z_i)g'(Z_i) \to_p \Sigma. \tag{10.4.13}$$

By Theorem 2.10.6 and Lemma 2.12.4, we readily obtain

$$\frac{1}{n}\sum_{i=1}^n h(Z_i)g(Z_i)' \to_p C_h, \tag{10.4.14}$$

$$\sup_{t\in\mathbb{R}}\Big|\frac{1}{n}\sum_{i=1}^n g(Z_i)I\big(\ell(Z_i)\le t\big) - C(t)\Big| \to_p 0. \tag{10.4.15}$$

The results are now immediate consequences of Theorem 10.4.1. For the stochastic expansion in part (iv) apply part (iv) of Theorem 10.4.1 with $Z_{ni}(t) = h(Z_j)$. For the uniform expansion in (v) apply part (iv) of Theorem 10.4.1 with $Z_{ni}(t) = I(\ell(Z_i) \le t]$. Note that $n^{-1}\sum_{i=1}^n Z_{ni}^2(t) \le 1$ holds for every real t. □

In the next theorem we allow the function g to be replaced by an estimator.

Theorem 10.4.3. *Let $Z_1, \cdots, Z_n$ be i.i.d. copies of the m-dimensional random vector Z. Let g be a function from $\mathbb{R}^m$ into $\mathbb{R}^q$ such that $E(g(Z)) = 0_q$ and $\Sigma = E(g(Z)g'(Z))$ is positive definite. Let $\hat{g}_n$ be an estimator of g based on $Z_1, \cdots, Z_n$ with the following two properties.*

$$D_{n1} := n^{-1/2} \sum_{i=1}^{n} \left(\hat{g}_n(Z_i) - g(Z_i)\right) \to_p 0_q, \tag{10.4.16}$$

$$D_{n2} := \frac{1}{n} \sum_{i=1}^{n} \left\|\hat{g}_n(Z_i) - g(Z_i)\right\|^2 \to_p 0. \tag{10.4.17}$$

Set

$$\widehat{\mathcal{R}}_n := \sup \Big\{ \prod_{i=1}^{n} n\pi_i;\ \pi \in \mathcal{P}_n;\ \sum_{i=1}^{n} \pi_i \hat{g}_n(Z_i) = 0_q \Big\}.$$

Then the following hold with $\bar{G}_n := n^{-1} \sum_{i=1}^{n} g(Z_j)$.

(i) *There is a random vector ξ_n and an event B_n with $P(B_n) \to 1$, such that on the event B_n,*

$$\widehat{\mathcal{R}}_n = \prod_{i=1}^{n} \frac{1}{1 + \xi_n' \hat{g}_n(Z_i)}$$

and the sequence ξ_n is uniquely defined on B_n as a solution of the equation

$$\frac{1}{n} \sum_{i=1}^{n} \frac{\hat{g}_n(Z_i)}{1 + \xi_n' \hat{g}_n(Z_i)} = 0_q$$

subject to the conditions $1 + \xi_n' \hat{g}_n(Z_i) > 0$, $i = 1, \cdots, n$.

(ii) *The stochastic expansion*

$$\left| -2 \log \widehat{\mathcal{R}}_n - n\bar{G}_n' \Sigma^{-1} \bar{G}_n \right| \to_p 0 \tag{10.4.18}$$

holds and implies

$$-2 \log \widehat{\mathcal{R}}_n \to_D \chi_q^2.$$

(iii) *The random vector ξ_n satisfies*

$$n^{1/2} \left\| \xi_n - \Sigma^{-1} \bar{G}_n \right\| \to_p 0 \tag{10.4.19}$$

and therefore, $n^{1/2}\xi_n \to_D \mathcal{N}_q(0_q, \Sigma^{-1})$.

(iv) *For a measurable function h from $\mathbb{R}^m$ into $\mathbb{R}$ such that $E(h^2(Z))$ is finite, the following stochastic expansion with $C_h = E(h(Z)g'(Z))$ holds.*

$$n^{-1/2} \sum_{i=1}^{n} \Big[\frac{h(Z_i)}{1 + \xi_n' \hat{g}_n(Z_i)} - h(Z_i) + C_h \Sigma^{-1} g(Z_j) \Big] \to_p 0.$$

(v) *For a measurable function l from $\mathbb{R}^m$ into $\mathbb{R}$, the following uniform stochastic expansion with $C(t) = E(I(l(Z) \le t) g(Z))$ holds.*

$$\sup_{t \in \mathbb{R}} \left| n^{-1/2} \sum_{i=1}^{n} \frac{I(l(Z_i) \le t)}{1 + \xi_n' \hat{g}_n(Z_j)} - I(l(Z_i) \le t) + C(t) \Sigma^{-1} g(Z_i) \right| \to_p 0.$$

Proof. We shall verify versions of (10.4.11) to (10.4.13) with $g(Z_i)$ replaced by $\hat{g}_n(Z_i)$, which are as follows, respectively:

$$n^{-1/2} \max_{1\le i\le n} \|\hat{g}_n(Z_i)\| \to_p 0, \tag{10.4.20}$$

$$n^{-1/2} \sum_{i=1}^{n} \hat{g}_n(Z_i) \to_D \mathcal{N}_q(0_q, \Sigma), \tag{10.4.21}$$

$$\hat{S}_n := \frac{1}{n}\sum_{i=1}^{n} \hat{g}_n(Z_i)\hat{g}_n(Z_i)' \to_p \Sigma. \tag{10.4.22}$$

These conditions represent assumptions (a), (b) and (c) of Theorem 10.4.1 with $\nu_* = 0_q$ and $X_{ni} := \hat{g}_n(Z_i)$, $i = 1, \cdots, n$. This theorem yields the present (i) to (iii) with $\bar{G}_n$ replaced by $\widehat{G}_n = n^{-1}\sum_{i=1}^{n} \hat{g}_n(Z_i)$ in (10.4.18) and (10.4.19). In view of (10.4.16) we can then replace $\hat{G}_n$ by $\bar{G}_n$ to obtain (10.4.18) and (10.4.19). In view of (10.4.16) it suffices to prove the expansions in parts (iv) and (v) with $g(Z_i)$ replaced by $\hat{g}_n(Z_i)$. These modified versions follow from (10.4.4) if we prove the following two results:

$$\frac{1}{n}\sum_{i=1}^{n} h(Z_i)\hat{g}_n(Z_i)' \to_p C_h, \tag{10.4.23}$$

$$\sup_{t\in\mathbb{R}} \left| \frac{1}{n}\sum_{i=1}^{n} I(l(Z_i) \le t)\hat{g}_n(Z_i) - C(t) \right| \to_p 0. \tag{10.4.24}$$

We are left to verify (10.4.20)–(10.4.24).

Statement (10.4.20) follows from (10.4.11) and the inequality

$$\|\hat{g}_n(Z_i)\| \le \|g(Z_i)\| + \|\hat{g}_n(Z_i) - g(Z_i)\| \le \|g(Z_i)\| + (nD_{n2})^{1/2}.$$

Statement (10.4.21) follows from (10.4.16), (10.4.12) and Theorem (2.10.5). Statement (10.4.22) follows from (10.4.13) and $\hat{S}_n - \bar{S}_n \to_p 0$. In view of Exercise 3, the latter follows if we verify $u'\hat{S}_n u - u'\bar{S}_n u \to_p 0$ for unit vectors u. Fix such a u. With $\Delta_{ni} := \hat{g}_n(Z_i) - g(Z_i)$ we calculate

$$\begin{aligned} u'\hat{S}_n u - u'\bar{S}_n u &= \frac{1}{n}\sum_{i=1}^{n} \Big[(u'\hat{g}_n(Z_i))^2 - (u'g(Z_i))^2\Big] \\ &= \frac{1}{n}\sum_{i=1}^{n} \Big[2(u'\Delta_{ni})(u'g(Z_i)) + (u'\Delta_{ni})^2\Big]. \end{aligned}$$

Applications of the Cauchy-Schwarz inequality and (10.4.17) yield

$$\begin{aligned} |u'S_n u - u'\bar{S}_n u| &\le \frac{1}{n}\sum_{i=1}^{n} 2\|\Delta_{ni}\|\|g(Z_i)\| + \frac{1}{n}\sum_{i=1}^{n} \|\Delta_{ni}\|^2 \\ &\le 2\Big(D_{n2}\frac{1}{n}\sum_{i=1}^{n} \|g(Z_i)\|^2\Big)^{1/2} + D_{n2} \to_p 0. \end{aligned}$$

Statement (10.4.23) follows from (10.4.14), (10.4.17) and the inequality

$$\frac{1}{n}\sum_{i=1}^{n} \|h(Z_i)\|\|\hat{g}_n(Z_i) - g(Z_i)\| \le \Big(\frac{1}{n}\sum_{i=1}^{n} \|h(Z_i)\|^2 D_{n2}\Big)^{1/2}.$$

Similarly, one derives (10.4.24) from (10.4.15) and (10.4.17). □

10.5 Inference about Functionals

This section contains standard nonparametric inferences about mean, quantiles and probabilities based on the empirical likelihood. Let Z be an m-dimensional random vector with unknown d.f. F and let E_F denote the expectation under F. We are interested in inference about some characteristics of F. Let $\mathcal{F}$ denote a model for F and let ν be a function from the model $\mathcal{F}$ into a subset Θ of $\mathbb{R}^k$ so that $\nu(F)$ is the characteristic of interest. Suppose there is a function g from $\mathbb{R}^m \times \Theta$ into $\mathbb{R}^q$ such that for all $F \in \mathcal{F}$,

$$E_F\big(g(Z,\nu(F))\big) = 0_q, \quad E_F\Big(\big\|g(Z,\nu(F))\big\|^2\Big) < \infty, \tag{10.5.1}$$
$$\Sigma_F := E_F\Big(g(Z,\nu(F))g(Z,\nu(F))'\Big) \text{ is positive definite.}$$

We call such a g a *criterion function* for ν.

Let $Z_1, \cdots, Z_n$ be independent copies of Z. With them we associate the empirical likelihood ratio

$$\mathcal{R}_n(\vartheta) := \sup\Big\{\prod_{i=1}^n n\pi_i;\ \pi \in \mathcal{P}_n;\ \sum_{i=1}^n \pi_i g(Z_i,\vartheta) = 0_q\Big\}, \quad \vartheta \in \Theta.$$

Theorem 10.4.2 applied with $g(Z_i) := g(Z_i, \nu(F))$, $i = 1, \cdots, n$ yields that $-2\log \mathcal{R}_n\big(\nu(F)\big)$ is asymptotically chi-square with q degrees of freedom for all F in $\mathcal{F}$:

$$-2\log \mathcal{R}_n(\nu(F)) \to_D \chi^2_q, \qquad \forall\, F \in \mathcal{F}. \tag{10.5.2}$$

This is used to construct tests and confidence regions for $\nu(F)$ as follows. As an immediate consequence of (10.5.2) we obtain that the set

$$\big\{t \in \Theta : -2\log \mathcal{R}_n(t) \le c_{\alpha,q}\big\}$$

is a confidence region for $\nu(F)$ with asymptotic confidence level $1-\alpha$, where $c_{\alpha,q}$ is as in (10.3.1).

Now, let ν_0 be known vector and consider the problem of testing the hypothesis $H_0 : \nu(F) = \nu_0$ versus the alternative $H_1 : \nu(F) \neq \nu_0$. By (10.5.2), the empirical likelihood ratio test of asymptotic size α rejects H_0 whenever $-2\log \mathcal{R}_n(\nu_0) > c_{\alpha,q}$.

Next, we discuss the asymptotic power of the above test under a sequence of local alternatives. For this assume that $\nu(F) = \nu_0$ and let h be a real-valued function satisfying $E_F(h(Z)) = 0$, $E_F(h^2(Z)) < \infty$ and $C = E_F(h(Z)g(Z)) \neq 0_q$. Define

$$h_n(z) = h(z)I\big(3|h(z)| \le n^{1/4}\big) - E_F\big(h(Z)I(3|h(Z)| \le n^{1/4})\big).$$

Then, $E_F(h_n(Z)) = 0$, $\sup_{z\in\mathbb{R}^m} n^{-1/2}|h_n(z)| < n^{-1/4}$ and

$$E_F\big((h_n(Z) - h(Z))^2\big) \le E_F\big(h^2(Z)I(3|h(Z)| > n^{1/4}\big) \to 0.$$

Let $F_n(x) := \int_{z\le x}\big(1 + n^{-1/2}h_n(z)\big)dF(z)$, $x \in \mathbb{R}^m$. We shall now derive the asymptotic distribution of $-2\log \mathcal{R}_n(\nu_0)$ when the r.v.'s $Z_1, \cdots, Z_n$ are i.i.d. F_n, for each $n \ge 1$. Towards this goal, verify, with $g(z) := g(z,\nu_0) = g(z,\nu(F))$ and using (10.5.1), the following assertions:

$$n^{1/2}E_{F_n}(g(Z)) = E_F\big(h_n(Z)g(Z)\big) \to C, \quad E_{F_n}\big(g(Z)g(Z)'\big) \to \Sigma_F,$$
$$E_{F_n}\big(\|g(Z)\|^2 I(\|g(Z)\| > \epsilon n^{1/2})\big) \to 0, \quad \forall\, \epsilon > 0.$$

The first assertion shows that $\nu(F_n)$ cannot equal ν_0 for large n. The three assertions and Lemma 10.4.2 imply that the assumptions (a), (b) and (c) of Theorem 10.4.1 are met with $X_{ni} = g(Z_i, \nu(F))$, $\nu_* = C$ and $\Sigma = \Sigma_F$. Thus Theorem 10.4.1 yields, with $\Delta = C'\Sigma^{-1}C$, that

$$-2\log \mathcal{R}_n(\nu_1(F)) \to_D \chi_q^2(\Delta).$$

This result shows that the asymptotic power of the above test under the sequence of alternatives F_n is $P(\chi_q^2(\Delta) > c_{\alpha,q})$.

Example 10.5.1. Inference about the mean. Here we shall discuss empirical likelihood inference about the population mean when the population covariance matrix is unknown but positive definite. Let $\mathcal{F}_1$ denote the set of all d.f.'s on $\mathbb{R}^q$ with a positive definite covariance matrix. Let $X_1, \cdots, X_n$ be a random sample from an $F \in \mathcal{F}_1$ and $\mu(F) = E_F(X_1)$ denote the population mean. A criterion function for μ is given by $g(x,t) = x - t$ resulting in

$$\mathcal{R}_n(\vartheta) := \sup\Big\{\prod_{i=1}^n n\pi_i;\ \pi \in \mathcal{P}_n;\ \sum_{i=1}^n \pi_i(X_i - \vartheta) = 0_q\Big\}.$$

Let μ_0 be a given mean vector in $\mathbb{R}^q$ and consider the problem of testing the hypothesis $H_0 : \mu(F) = \mu_0$ against $H_1 : \mu(F) \neq \mu_0$. The theory presented above shows that the empirical likelihood ratio test that rejects H_0 whenever $-2\log \mathcal{R}_n(\mu_0) > c_{\alpha,q}$ has asymptotic size α for every $0 < \alpha < 1$. The theory also yields that the confidence region $\{\vartheta; -2\log \mathcal{R}_n(\vartheta) \le c_{\alpha,q}\}$ for $\mu(F)$ has the asymptotic confidence level $1-\alpha$, for all $F \in \mathcal{F}_1$ and all $0 < \alpha < 1$. Perhaps we should emphasize that for the validity of both of these results we do not need to know anything more about F except that F belongs to $\mathcal{F}_1$.

Example 10.5.2. Inference about the mean under additional information. Here we shall discuss empirical likelihood inference about the population mean $\mu(F)$ when additional information about the d.f. F on $\mathbb{R}^q$ is available. More precisely, let h be a function from $\mathbb{R}^{2q}$ into $\mathbb{R}^k$ such that

$$E_F\big(h(X, \mu(F))\big) = 0_k. \tag{10.5.3}$$

The problem of interest is then to make inference about $\mu(F)$ subject to (10.5.3).

For example, if $q = 1$ and the value of F is known for some point, say $F(0) = 1/2$, then we can take $h(x,t) = 2I(x \le 0) - 1$. If $q = 2$ and the variances of the two components are known to be the same, then we can take $h(x,t) = (x_1 - t_1)^2 - (x_2 - t_2)^2$. If $q = 2$ and the components are known to be uncorrelated, then we can take $h(x,t) = x_1x_2 - t_1t_2$.

To treat such cases in general, let $\mathcal{F}_{1,h}$ denote the collection of all d.f.'s $F \in \mathcal{F}_1$ for which (10.5.3) holds, the integral $E_F\big(\|X\|^2 + \|h(X,\mu(F))\|^2\big)$ is finite, and the matrix

$$E\left\{\begin{pmatrix} X - \mu(F) \\ h(X,\mu(F)) \end{pmatrix}\begin{pmatrix} X - \mu(F) \\ h(X,\mu(F)) \end{pmatrix}'\right\}$$

is positive definite. Then a criterion function for μ is given by

$$g(x,t) = \begin{pmatrix} x - t \\ h(x,t) \end{pmatrix}.$$

Let $X_1, \cdots, X_n$ be a random sample from some $F \in \mathcal{F}_{1,h}$. We now work with the empirical likelihood ratio

$$\mathcal{R}_n(\vartheta) := \sup\left\{\prod_{i=1}^n n\pi_i;\ \pi \in \mathcal{P}_n;\ \sum_{i=1}^n \pi_i\begin{pmatrix} X_i - \vartheta \\ h(X_i,\vartheta) \end{pmatrix} = 0_{q+k}\right\}.$$

Suppose one wishes to test the hypothesis $H_0 : \mu(F) = \mu_0$, for some known μ_0, against the alternatives that H_0 is not true. Then the empirical likelihood ratio test that rejects H_0 whenever $-2\log \mathcal{R}_n(\mu_0) > c_{\alpha,q+k}$ has asymptotic size α, for all $F \in \mathcal{F}_{1,h}$ and all $0 < \alpha < 1$. The confidence region $\{\vartheta; -2\log \mathcal{R}_n(\vartheta) \le c_{\alpha,q+k}\}$ for the unknown mean $\mu(F)$ has asymptotic confidence level $1-\alpha$ for all $F \in \mathcal{F}_{1,h}$ and $0 < \alpha < 1$.

Example 10.5.3. A simultaneous confidence set for mean and median. Let $0 < \alpha < 1$ be given. Let $X_1, \cdots, X_n$ be a random sample from a continuous d.f. F having a finite second moment and a unique median. We are interested in making inference about the mean $\mu(F)$ and the median $m(F)$. As a criterion function for the pair $\nu = (\mu, m)'$ we choose $g(x, t_1, t_2) = \big(x - t_1, \text{sign}(x - t_2)\big)'$, $x, t_1, t_2 \in \mathbb{R}$, with $\text{sign}(x) = I(x > 0) - I(x < 0)$. The continuity of F guarantees that the corresponding dispersion matrix is positive definite. This leads us to the empirical likelihood ratio

$$\mathcal{R}_n(t_1, t_2) := \sup\left\{ \prod_{i=1}^n n\pi_i;\ \pi \in \mathcal{P}_n;\ \sum_{i=1}^n \pi_i \begin{pmatrix} X_i - t_1 \\ \text{sign}(X_i - t_2) \end{pmatrix} = 0_2 \right\},$$

where t_1 and t_2 are real numbers. Thus the asymptotic confidence level of the simultaneous confidence set $\{(t_1, t_2)'; -2\log \mathcal{R}_n(t_1, t_2) \le c_{\alpha,2}\}$ for the mean and median is $1 - \alpha$.

Example 10.5.4. Inference about a quantile. Let $\mathcal{F}(\beta)$ be the set of all continuous d.f.'s F with a unique β quantile $\kappa_\beta(F)$ for some fixed β, $0 < \beta < 1$. Let $Z_1, \cdots, Z_n$ be i.i.d. r.v.'s having a d.f. $F \in \mathcal{F}(\beta)$. We are interested in confidence regions and tests about the quantile $\kappa_\beta(F)$. Since the quantile satisfies $F(\kappa_\beta(F)) = \beta$, the obvious criterion function here is $g(y, t) = I(y \le t) - \beta$, and the resulting empirical likelihood ratio

$$\mathcal{R}_n(\vartheta) := \sup\left\{ \prod_{i=1}^n n\pi_i;\ \pi \in \mathcal{P}_n;\ \sum_{i=1}^n \pi_i\big(I(Z_i \le \vartheta) - \beta\big) = 0 \right\}$$

satisfies $-2\log \mathcal{R}_n(\kappa_\beta(G)) \to_D \chi_1^2$. From this it then follows that the test that rejects $H_0 : \kappa_\beta(G) = \theta_0$ against the alternative $H_1 : \kappa_\beta(F) \ne \theta_0$ whenever $-2\log \mathcal{R}_n(\theta_0) > c_{\alpha,1}$ has asymptotic size α, and the asymptotic confidence level of the confidence set $\{\vartheta \in \mathbb{R}; -2\log \mathcal{R}_n(\vartheta) \le c_{\alpha,1}\}$ is $1 - \alpha$. This example can be extended to make simultaneous inference about several quantiles. This is left as an exercise.

Remark 10.5.1. The empirical likelihood of the previous example can be explicitly expressed as

$$\mathcal{R}_n(\vartheta) = \begin{cases} 0, & F_n(\vartheta) \in \{0, 1\}, \\ \left(\dfrac{\beta}{F_n(\vartheta)}\right)^{nF_n(\vartheta)} \left(\dfrac{1-\beta}{1 - F_n(\vartheta)}\right)^{n - nF_n(\vartheta)}, & 0 < F_n(\vartheta) < 1, \end{cases}$$

with F_n the empirical distribution function. Indeed, since $I[Z_i \le \vartheta] - \beta$ takes only the values $1 - \beta$ and β, it follows from (10.3.14) that on the event $\{0 < F_n(\vartheta) < 1\}$,

$$\xi_n = \frac{F_n(\vartheta) - \beta}{\beta(1 - \beta)},$$

$$1 + \xi_n(I[Z_i \le \vartheta] - \beta) = \begin{cases} F_n(\vartheta)/\beta, & Z_i \le \vartheta, \\ (1 - F_n(\vartheta))/(1 - \beta), & Z_i > \vartheta. \end{cases}$$

If $F_n(\vartheta)$ equals 0 or 1, there are no probabilities $\pi_1, \cdots, \pi_n$ such that

$$\sum_{i=1}^{n} \pi_i(I[Z_i \le \vartheta] - \beta) = 0.$$

In this case the empirical likelihood ratio is zero.

Example 10.5.5. Inference about probabilities. Let $Y_1, \cdots, Y_n$ be independent copies of some m-dimensional random vector Y and k be a positive integer. Consider a measurable partition $\{B_0, \cdots, B_k\}$ of $\mathbb{R}^m$ and positive numbers $p_0, \cdots, p_k$ satisfying $p_0 + \cdots + p_k = 1$. To test the null hypothesis $P(Y \in B_j) = p_j$ for $j = 0, \cdots, k$, we can use the empirical likelihood ratio

$$\mathcal{R}_n := \sup\left\{ \prod_{i=1}^{n} n\pi_i;\ \pi \in \mathcal{P}_n;\ \sum_{i=1}^{n} \pi_i \begin{pmatrix} I(Y_i \in B_1) - p_1 \\ \vdots \\ I(Y_i \in B_k) - p_k \end{pmatrix} = 0_k \right\}$$

which ignores $I(Y \in B_0) - p_0$. The test that rejects the null hypothesis if $-2\log \mathcal{R}_n > c_{\alpha,q}$ has the asymptotic size α. This follows because under the null hypothesis, $-2\log \mathcal{R}_n \to_D \chi^2(k)$. Indeed, under the null hypothesis, the random vector $X = (I(Y \in B_1) - p_1, \cdots, I(Y \in B_k) - p_k)'$ has mean vector zero and positive definite covariance matrix V with entries $V_{ii} = p_i(1 - p_i)$ and $V_{ij} = -p_i p_j$ for $i, j = 1, \cdots, q$ and $i \ne j$. Note that the latter will not be true if we also include $I(Y \in B_0) - p_0$ as $\sum_{j=0}^{k} \left[I(Y \in B_j) - p_j\right] = 1 - 1 = 0$.

This empirical likelihood can be used to test whether Y has a specific distribution. We only treat the case when $m = 1$ and the null hypothesis is $H_0 : F = F_0$ for some continuous d.f. F_0. Let Q_i denote the $i/(k+1)$ quantile of F_0 and take $B_0 = (-\infty, Q_1], B_1 = (Q_1, Q_2], \cdots, B_{k-1} = (Q_{k-1}, Q_k], B_k = (Q_k, \infty)$. Under H_0, the probability that Y falls into B_i is $1/(1+k)$. For this choice of partition, the above empirical likelihood can be written as

$$\sup\left\{ \prod_{i=1}^{n} n\pi_i;\ \pi \in \mathcal{P}_n;\ \sum_{i=1}^{n} \pi_i \begin{pmatrix} I(Q_1 < Y_i \le Q_2] - p \\ \vdots \\ I(Q_k < Y_i) - p \end{pmatrix} = 0_k \right\}$$

with $p = 1/(k+1)$.

10.6 Constrained Inference

In this section we shall discuss some examples of nonparametric estimation of a univariate d.f. and its functionals, subject to linear constraints. These results will cover the case when moments or quantiles of the d.f. are known.

Let ψ be a known measurable function from $\mathbb{R}$ into $\mathbb{R}^q$. Let $\mathcal{F}_\psi$ denote the set of all univariate d.f.'s F with $E_F(\|\psi(X)\|^2) < \infty$, the matrix $\Psi_F := E_F(\psi(X)\psi(X)')$ positive definite and

$$E_F(\psi(X)) = 0_q. \tag{10.6.1}$$

An example of this is the case when F has a known mean μ and a positive finite variance σ_F^2. In this case we can take $\psi(x) = x - \mu$ and have $\Psi_F = \sigma_F^2$. Another example is the case when $F(0)$ is known, say $F(0) = 1/2$. In this case we can take $\psi(x) = I(x \le 0) - 1/2$ and have $\Psi_F = 1/4$.

Assume from now on that F belongs to $\mathcal{F}_\psi$, that X is a r.v. with d.f. F and that $X_1, \cdots, X_n$ are i.i.d. F. We shall first consider the estimation of the d.f. F, then of linear functionals of F and finally of quantiles of F.

10.6.1 Estimating the d.f. F

With the ψ of (10.6.1), we associate the empirical likelihood

$$\mathcal{R}_n := \sup\left\{ \prod_{i=1}^n n\pi_i;\ \pi \in \mathcal{P}_n;\ \sum_{i=1}^n \pi_i \psi(X_i) = 0_q \right\}.$$

We have seen that there is an event D_n whose probability tends to one and a random vector ξ_n that satisfies on D_n

$$\frac{1}{n}\sum_{i=1}^n \frac{\psi(X_i)}{1+\xi_n'\psi(X_i)} = 0_q$$

and $1+\xi_n'\psi(X_n), \cdots, 1+\xi_n'\psi(X_n)$ are positive. On the event D_n, the maximizers are of the form $\pi_i = 1/\big[n\big(1+\xi_n'\psi(X_i)\big)\big]$, $i = 1, \cdots, n$, and

$$\mathcal{R}_n = \prod_{i=1}^n \frac{1}{1+\xi_n'\psi(X_i)}.$$

The empirical d.f. $F_n(x) = n^{-1}\sum_{i=1}^n I(X_i \le x)$, $x \in \mathbb{R}$ uses the weights $\pi_1 = \cdots = \pi_n = 1/n$ that maximize $\{\prod_{i=1}^n n\pi_i;\ \pi \in \mathcal{P}_n\}$. Motivated by this we estimate the d.f. F under the constraint (10.6.1) by

$$\widehat{F}_n(x) = \frac{1}{n}\sum_{i=1}^n \frac{I(X_i \le x)}{1+\xi_n'\psi(X_i)}, \qquad x \in \mathbb{R},$$

which uses the weights that maximize $\mathcal{R}_n$. Here and below we interpret ξ_n to be 0_q off the event D_n. On the event D_n,

$$E_{\widehat{F}_n}(\psi(X)) = \frac{1}{n}\sum_{i=1}^n \frac{\psi(X_i)}{1+\xi_n'\psi(X_i)} = 0_q.$$

This shows that the empirical likelihood estimator $\widehat{F}_n$ satisfies (10.6.1).

An alternative class of estimators, one for each $b \in \mathbb{R}^q$, is given by

$$F_{n,b}(x) := F_n(x) - \frac{1}{n}\sum_{j=1}^n b'\psi(X_i) \quad x \in \mathbb{R}.$$

Under (10.6.1), these estimators are unbiased. Moreover, $\forall\, x \in \mathbb{R}^q$,

$$\mathrm{Var}\big(F_{n,b}(x)\big) = n^{-1}[F(x)(1-F(x)) + b'\Psi_F b - 2b'E_F(I(X \le x)\psi(X))].$$

As is easily checked, this variance is minimized by $b_F(x) := \Psi_F^{-1}c_F(x)$ with $c_F(x) = E_F(I(X \le x)\psi(X))$ resulting in the minimal variance $[F(x)(1-F(x)) - c_F(x)'\Psi_F^{-1}c_F(x)]/n$. As the minimizer $b_F(x)$ depends on the unknown d.f. we cannot use it and work instead with the empirical estimator

$$\hat{b}(x) = \Big(\frac{1}{n}\sum_{i=1}^n \psi(X_i)\psi(X_i)'\Big)^{-1}\frac{1}{n}\sum_{i=1}^n I(X_i \le x)\psi(X_i).$$

The estimator $\hat{b}(x)$ satisfies $\sup_{x\in\mathbb{R}} \|\hat{b}(x) - b_F(x)\| \to_p 0$ and the resulting plug-in estimator $\widetilde{F}_n(x) \equiv F_{n,\hat{b}(x)}(x)$, satisfies the uniform expansion

$$n^{1/2} \sup_{x\in\mathbb{R}} \left|\widetilde{F}_n(x) - F_n(x) + \frac{1}{n}\sum_{i=1}^{n} b_F(x)'\psi(X_i))\right| \to_p 0.$$

It follows from Theorem 10.4.2 that the estimator $\widehat{F}_n$ also satisfies

$$n^{1/2} \sup_{x\in\mathbb{R}} \left|\widehat{F}_n(x) - F_n(x) + \frac{1}{n}\sum_{i=1}^{n} b_F(x)'\psi(X_i))\right| \to_p 0. \tag{10.6.2}$$

While the estimators $\widetilde{F}_n$ and $\widehat{F}_n$ are asymptotically equivalent, the empirical likelihood estimator $\widehat{F}_n$ has the advantage of being a d.f., while this is not guaranteed for the plug-in estimator $\widetilde{F}_n$.

Fix an x. The uniform expansion shows that

$$n^{1/2}\big(\widehat{F}_n(x) - F(x)\big) = n^{-1/2}\sum_{i=1}^{n} Z_i(x) + o_p(1),$$

where $Z_i(x) := I(X_i \le x) - F(x) - c_F(x)'\Psi_F^{-1}\psi(X_i)$, $i = 1, \cdots, n$, are i.i.d. with mean zero and variance

$$\begin{aligned}\tau_F(x) &= E_F\big(I(X \le x) - F(x) - c_F(x)'\Psi^{-1}\psi(X)\big)^2 \\ &= F(x)\big(1 - F(x)\big) - c_F(x)'\Psi_F^{-1}c_F(x).\end{aligned}$$

Here we also used the identity $E_F\big([I(X \le x) - F(x)]\psi(X)\big) = c_F(x)$. The CLT and Theorem 2.10.5 imply $n^{1/2}(\widehat{F}_n(x) - F(x)) \to_D \mathcal{N}(0, \tau_F(x))$. We also have $n^{1/2}\big(F_n(x) - F(x)\big) \to_D \mathcal{N}\big(0, F(x)(1 - F(x))\big)$. Since Ψ_F is positive definite, the quantity $c_F(x)'\Psi_F^{-1}c_F(x)$ is positive unless $c_F(x) = 0_q$ and measures the amount of reduction in the asymptotic variance gained by using $\widehat{F}_n(x)$ instead of $F_n(x)$. The relative variance reduction is the fraction

$$\frac{c_F(x)'\Psi_F^{-1}c_F(x)}{F(x)(1 - F(x))}.$$

It can be substantial in some cases as will be seen latter.

10.6.2 Estimating Linear Functionals of F

Let h be a measurable real-valued function. Let $\mathcal{F}_{\psi,h}$ be the set of all F in $\mathcal{F}_\psi$ such that $E_F(h^2(X))$ is finite. Consider the problem of estimating $\mu_h(F) = E_F(h(X))$ when F belongs to $\mathcal{F}_{\psi,h}$. The classical plug-in estimator of this functional is $\mu_h(F_n) := n^{-1}\sum_{i=1}^{n} h(X_i)$. It satisfies $n^{1/2}\big(\mu_h(F_n) - \mu_h(F)\big) \to_D \mathcal{N}(0, \sigma_h^2)$ with $\sigma_h^2 := \mathrm{Var}_F(h(X))$. This estimator does not take into consideration the constraint (10.6.1). A better option is the plug-in estimator

$$\mu_h(\widehat{F}_n) = \frac{1}{n}\sum_{i=1}^{n} \frac{h(X_i)}{1 + \xi_n'\psi(X_i)}.$$

By Theorem 10.4.2, with $C_h = E_F(h(X)\psi(X))$,

$$n^{1/2}\big(\mu_h(\widehat{F}_n) - \mu_h(F)\big) = n^{-1/2}\sum_{i=1}^{n} Y_i + o_p(1),$$

where $Y_i := h(X_i) - \mu_h(F) - C_h'\Psi_F^{-1}\psi(X_i)$, $i = 1,\cdots,n$, are i.i.d. with mean zero and variance $\tau_h := \sigma_h^2 - C_h'\Psi_F^{-1}C_h$. Hence, by Theorem 2.10.5 and the CLT, $n^{1/2}(\mu_h(\widehat{F}_n) - \mu_h(F)) \to_D \mathcal{N}(0,\tau_h)$, Thus the empirical likelihood estimator $\mu_h(\widehat{F}_n)$ provides an asymptotic variance reduction by the amount $C_h'\Psi_F^{-1}C_h$, which is positive unless C_h is the zero vector. The relative variance reduction factor is $\left(C_h'\Psi_F^{-1}C_h\right)/\sigma_h^2$.

10.6.3 Estimating Quantiles of F

Of interest is also the problem of estimating quantiles of F subject to (10.6.1). Fix $0 < \gamma < 1$. We are working with the model $F_{\psi,\gamma}$ consisting of all d.f. in $\mathcal{F}_\psi$ that have a unique γ quantile and a positive derivative at the quantile. Fix F in $\mathcal{F}_{\psi,\gamma}$. Let $\zeta := \kappa_\gamma(F)$ denote its unique γ quantile. Let $\hat{\zeta}_n$ be a γ quantile of $\widehat{F}_n$, i.e., $\widehat{F}_n(\hat{\zeta}_n-) \le \gamma \le \widehat{F}_n(\hat{\zeta}_n)$. Then, $\sqrt{n}(\hat{\zeta}_n - \zeta) \to_D \mathcal{N}(0, \tau/\dot{F}^2(\zeta))$, where $\tau = \gamma(1-\gamma) - c_\zeta'\Psi_F^{-1}c_\zeta$ and $c_\zeta := E_F\left(I(X \le \zeta)\psi(X)\right)$. Note that this variance is smaller than the asymptotic variance of the sample quantile $\gamma(1-\gamma)/\dot{F}^2(\zeta)$. The relative variance reduction fraction is

$$\frac{c_\zeta'\Psi_F^{-1}c_\zeta}{\gamma(1-\gamma)}.$$

We shall now sketch a proof of this asymptotic normality result. For this fix a real t and set $\zeta_{n,t} = \zeta + n^{-1/2}t$. We need to show

$$G_n(t) := P_F(n^{1/2}(\hat{\zeta}_n - \zeta) \le t) \to P(\sqrt{\tau}Z/\dot{F}(\zeta) \le t),$$

where $Z \sim_D \mathcal{N}(0,1)$. Since the event $\{n^{1/2}(\hat{\zeta}_n - \zeta) \le t\} = \{\hat{\zeta}_n \le \zeta_{n,t}\}$ equals $\{\widehat{F}_n(\zeta_{n,t}) \ge \gamma\}$, we calculate

$$G_n(t) = P_F\left(n^{1/2}(\widehat{F}_n(\zeta_{n,t}) - F(\zeta_{n,t})) \ge -n^{1/2}\left(F(\zeta_{n,t}) - F(\zeta)\right)\right).$$

Here we used the identity $F(\zeta) = \gamma$. By the differentiability assumption on F, $n^{1/2}(F(\zeta_{n,t}) - F(\zeta)) \to t\dot{F}(\zeta)$. It follows from the uniform stochastic expansion (10.6.2), the continuity of F at ζ and an application of Theorem 2.10.9 to the random variables

$$Y_{ni} = I(X_i \le \zeta_{n,t}) - F(\zeta_{n,t}) - c_{\zeta_{n,t}}'\Psi_F^{-1}\psi(X_j), \quad j = 1,\cdots,n,$$

that $n^{1/2}\left(\widehat{F}_n(\zeta_{n,t}) - F(\zeta_{n,t})\right)$ is asymptotically normal with mean zero and variance τ. This shows that

$$G_n(t) \to P(\sqrt{\tau}Z \ge -t\dot{F}(\zeta)) = P(-\sqrt{\tau}Z/\dot{F}(\zeta) \le t)$$

with Z standard normal. This is the desired result as $-Z \sim_D \mathcal{N}(0,1)$.

Example 10.6.1. Estimation when the mean is known. We shall now specialize the previous results to the case when the mean of F is known and F has a positive finite variance $\sigma^2(F)$. For simplicity in notation we assume that the mean is zero. Then we can apply the results of the previous subsections with $\psi(x) = x$ and $\Psi_F = \sigma^2(F)$. The empirical likelihood estimator of F is then

$$\widehat{F}_n(x) = \frac{1}{n}\sum_{i=1}^{n}\frac{I(X_i \le x)}{1 + \xi_n X_i}.$$

Here ξ_n is a solution to the equation

$$\frac{1}{n}\sum_{i=1}^{n}\frac{X_i}{1 + \xi_n X_i} = 0$$

subject to the constraints $1 + \xi_n X_i > 0$ for $i = 1, \cdots, n$ on the event $D_n = \{\min(X_1, \cdots, X_n) < 0 < \max(X_1, \cdots, X_n)\}$ and is zero otherwise. In this setting, $c_F(x) = E_F\big(XI(X \le x)\big)$. The function $x \mapsto c_F(x)$ is nonincreasing on the interval $(-\infty, 0]$, nondecreasing on the interval $[0, \infty)$, has limit 0 at minus and plus infinity, and is thus negative for x satisfying $0 < F(x) < 1$. Thus the variance reduction $c_F(x)^2/\sigma^2(F)$ is positive for such x. It is maximal at $x = 0$ and can be substantial. For the standard normal distribution and $x = 0$, the variance reduction is $1/(2\pi)$. For the standard double exponential distribution (with density $f(y) = \exp(-|y|)/2$) and $x = 0$, the variance reduction is $1/8$. In both cases the relative variance reduction is at least fifty percent.

Next, consider the problem of estimating the variance $\sigma^2(F)$ of F. Under the assumption of mean zero, an often used estimator of $\sigma^2(F)$ is

$$\hat\sigma_1^2 = \frac{1}{n}\sum_{i=1}^n X_i^2,$$

while the empirical likelihood estimator is

$$\hat\sigma_2^2 = \frac{1}{n}\sum_{i=1}^n \frac{X_i^2}{1+\xi_n X_i}.$$

Assume that F has a finite fourth moment. Then $n^{1/2}\big(\hat\sigma_j^2 - \sigma^2(F)\big)$ is asymptotically normal with mean zero and variance τ_j, $j = 1, 2$, where $\tau_1 = \mathrm{Var}_F(X^2) = E_F\big(X^2 - \sigma^2(F)\big)$ and $\tau_2 = \tau_1 - \big(E_F(X^3)\big)^2/\sigma^2(F)$. Here we used the classical central limit theorem with the variables $X_i^2 - \sigma^2(F)$ for the first estimator and the above theory with $h(x) = x^2$ and $\psi(x) = x$ for the empirical likelihood estimator. Thus the empirical likelihood estimator $\hat\sigma_2^2$ is asymptotically superior unless the third moment of F is zero. There is no improvement for distributions that are symmetric about zero.

Next, consider the problem of estimating the median $m(F)$ of F. For this assume that F has a positive derivative $\dot F(m(F))$ at its (unique) median $m(F)$. We estimate the median by a median $\hat m_n$ of the empirical likelihood estimator $\widehat F_n$ of F, which satisfies

$$\widehat F_n(\hat m_n-) \le 1/2 \le \widehat F_n(\hat m_n).$$

Under the given assumptions, $n^{1/2}\big(\hat m_n - m(F)\big) \to_D \mathcal{N}(0, \tau_F^2)$, where

$$\tau_F^2 := \frac{1}{4\big(\dot F(m(F)\big)^2} - \frac{(E_F(XI(X \le m(F))))^2}{\big(\dot F(m(F))\big)^2\sigma^2(F)}.$$

By Corollary 3.3.1, $n^{1/2}\big(\widetilde X_n - m(F)\big) \to_D \mathcal{N}(0, 1/(2\dot F(m(F))^2)$, where $\widetilde X_n$ is the sample median. Thus, the relative variance reduction by the estimator $\hat m_n$ over $\widetilde X_n$ is the fraction

$$\frac{4\big(E_F(XI(X \le m(F)))\big)^2}{\sigma^2(F)}.$$

For the standard normal and double exponential distributions, this fraction is $2/\pi \approx .6366$ and $1/2$, respectively.

Example 10.6.2. Estimation when the median is known. In this example we treat estimation of F and of its mean when F has a known median. We restrict ourselves to the case when the median is zero and F is continuous at 0. Then we have $F(0) = 1/2$ and can take $\psi(x) = I(x \le 0) - 1/2$ and have $\Psi_F = 1/4$. The empirical likelihood estimator of F is then

$$\widehat{F}_n(x) = \frac{1}{n}\sum_{i=1}^{n} \frac{I(X_i \le x)}{1 + \xi_n[I(X_i \le 0) - 1/2]}, \quad x \in \mathbb{R}.$$

As shown in Remark 10.5.1, on the event $D_n := \{0 < F_n(0) < 1\}$, $\xi_n = 4(F_n(0) - 1/2)$ and $1 + \xi_n(I(X_i \le 0) - 1/2) = 2F_n(0)I(X_i \le 0) + 2(1 - F_n(0))I(X_i > 0)$, $i = 1, \cdots, n$. Using this we derive on D_n the identity

$$\widehat{F}_n(x) = \frac{1}{2}\frac{F_n(x \wedge 0)}{F_n(0)} + \frac{1}{2}\frac{F_n(x \vee 0) - F_n(0)}{(1 - F_n(0))}, \quad x \in \mathbb{R}.$$

Thus, on D_n, we have $\widehat{F}_n(0) = 1/2$, i.e., 0 is also a median of $\widehat{F}_n$. We compute

$$c_F(x) = E_F\big(I(X \le x)\big[I(X \le 0) - \frac{1}{2}\big]\big) = F(x \wedge 0) - F(x)F(0).$$

Using $F(0) = 1/2$, the variance reduction for an x for which $c_F(x) \ne 0$ is

$$4\big(c_F(x)\big)^2 = \begin{cases} F^2(x), & x \le 0, \\ \big(1 - F(x)\big)^2, & x > 0 \end{cases}.$$

This variance reduction is $1/4$ for $x = 0$.

Next, consider the problem of estimating the mean $\mu(F)$ of F. For this assume also that the variance $\sigma^2(F)$ of F is finite. The plug-in empirical likelihood estimator $\mu(\widehat{F}_n)$ is given by

$$\frac{1}{2F_n(0)}\frac{1}{n}\sum_{i=1}^{n} X_i I(X_i \le 0) + \frac{1}{2(1 - F_n(0))}\frac{1}{n}\sum_{i=1}^{n} X_i I(X_i > 0).$$

It satisfies

$$\mu(\widehat{F}_n) = \frac{1}{n}\sum_{i=1}^{n}\Big[X_i - 4c_0\big(I(X_i \le 0) - \frac{1}{2}\big)\Big] + o_p(n^{-1/2})$$

with $c_0 = E_F\big(X(I(X \le 0) - \frac{1}{2})\big)$. The relative variance reduction is

$$4\big\{E_F\big[X\big(I(X \le 0) - \frac{1}{2}\big)\big]\big\}^2/\sigma^2(F).$$

This fraction becomes $2/\pi$ for the standard normal distribution and $1/2$ when X has density $f(x) = e^{-|x|}/2$, $x \in \mathbb{R}$.

10.7 Estimated Criterion Function

Let Z be an m-dimensional random vector with d.f. F belonging to some model $\mathcal{F}$. Let k, r and q be positive integers and ν and γ be functions from $\mathcal{F}$ into $\mathbb{R}^k$ and $\mathbb{R}^r$, respectively. Let g be an $\mathbb{R}^q$-valued criterion function for the pair $(\nu', \gamma')'$ satisfying the following conditions:

$$E_F\big(\|g(Z, \nu(F), \gamma(F))\|^2\big) < \infty, \quad E_F\big(g(Z, \nu(F), \gamma(F))\big) = 0_q,$$
$$\Sigma_F := E_F\Big(g(Z, \nu(F), \gamma(F))g'(Z, \nu(F), \gamma(F))\Big) \text{ is positive definite.}$$

Let $Z_1, \cdots, Z_n$ be independent copies of Z.

Suppose we are interested in inference about $\nu(F)$ only, in which case $\gamma(F)$ plays the role of a nuisance parameter. Here we estimate $\gamma(F)$ by say $\hat{\gamma}_n$ and then work with the estimated criterion function $\hat{g}_n(z,t) = g(z,t,\hat{\gamma}_n)$. The resulting empirical likelihood ratio is

$$\hat{\mathcal{R}}_n(\vartheta) := \sup\Big\{\prod_{i=1}^n n\pi_i;\ \pi \in \mathcal{P}_n;\ \sum_{i=1}^n \pi_i \hat{g}_n(Z_j,\vartheta) = 0_q\Big\}, \quad \vartheta \in \Theta,$$

where $\Theta \subseteq \mathbb{R}^k$ is the range of ν. In order to apply Theorem 10.4.3, we require

$$n^{-1/2}\sum_{i=1}^n \Big(g(Z_j,\nu(F),\hat{\gamma}_n) - g(Z_j,\nu(F),\gamma(F))\Big) \to_p 0_q, \tag{10.7.1}$$

$$\frac{1}{n}\sum_{i=1}^n \Big\|g(Z_j,\nu(F),\hat{\gamma}_n) - g(Z_i,\nu(F),\gamma(F))\Big\|^2 \to_p 0. \tag{10.7.2}$$

Then Theorem 10.4.3 yields the conclusion

$$-2\log \mathcal{R}_n(\nu(F)) \to_D \chi^2_q.$$

Using this result, it is straightforward to construct confidence regions and tests for the functional ν.

Now consider the case when $\nu(F)$ is known, say $\nu(F) = \nu_0$, and we want to estimate $\mu_h(F) = E_F(h(X))$ for some function h with $E_F(h^2(X))$ finite. Then we work with the submodel $\mathcal{F}_{0,h} = \{F \in \mathcal{F} : \nu(F) = \nu_0, E_F(h^2(X)) < \infty\}$. As estimator we take

$$\hat{\mu}_h := \frac{1}{n}\sum_{i=1}^n \frac{h(Z_i)}{1 + \xi_n' \hat{g}_n(Z_i,\nu_0)}$$

with ξ_n the Lagrange multiplier for the r.v.'s $\hat{g}_n(Z_1,\nu_0),\cdots,\hat{g}_n(Z_n,\nu_0)$. In view of Theorem 10.4.3 we need the following two conditions:

$$n^{-1/2}\sum_{i=1}^n \Big(g(Z_j,\nu_0,\hat{\gamma}_n) - g(Z_j,\nu_0,\gamma(F))\Big) \to_p 0_q, \tag{10.7.3}$$

$$\frac{1}{n}\sum_{i=1}^n \Big\|g(Z_j,\nu_0,\hat{\gamma}_n) - g(Z_i,\nu_0,\gamma(F))\Big\|^2 \to_p 0, \tag{10.7.4}$$

in order to conclude

$$\hat{\mu}_h = \frac{1}{n}\sum_{i=1}^n [h(Z_i) - C_h \Sigma_F^{-1} g(Z_j,\nu_0,\gamma(F))] + o_p(n^{-1/2})$$

with $C_h = E_F(h(Z)g(Z,\nu_0,\gamma(F)))$.

Next, we consider again the case when $\nu(F)$ is known, say $\nu(F) = \nu_0$, but now we wish to estimate the distribution function G_F of a random variable $\ell(Z)$ for some measurable function ℓ. The model is then simply $\mathcal{F}_0 = \{F \in \mathcal{F} : \nu(F) = \nu_0\}$. The corresponding empirical likelihood estimator is

$$\widehat{G}_n(t) = \frac{1}{n}\sum_{i=1}^n \frac{I(\ell(Z_i) \le t)}{1 + \xi_n \hat{g}_n(Z_j,\nu_0)}.$$

Under (10.7.3) and (10.7.4) we have the uniform stochastic expansion

$$n^{1/2}\sup_{t\in\mathbb{R}}\Big|\widehat{G}_n(t) - \frac{1}{n}\sum_{i=1}^n \Big[I(\ell(Z_i) \le t] - C_t \Sigma_F^{-1} g(Z_i,\nu_0,\gamma(F))\Big]\Big| \to_p 0$$

with $C_t = E\{I(\ell(Z) \le t) g(Z, \nu_0, \gamma(F)\}$.

We should point out that the requirements (10.7.1) and (10.7.3) are rather strong. Under mild assumptions one expects

$$\begin{aligned} D_{n1} &:= n^{-1/2} \sum_{i=1}^{n} [g(Z_i, \nu(F), \hat{\gamma}_n) - g(Z_i, \nu(F), \gamma(F))] \\ &= n^{-1/2} \sum_{i=1}^{n} \dot{g}(Z_i, \nu(F), \gamma(F))(\hat{\gamma}_n - \gamma(F)) + o_p(n^{-1/2}) \\ &= \Delta n^{1/2} (\hat{\gamma}_n - \gamma(F)) + o_p(n^{-1/2}) \end{aligned}$$

with $\Delta = E\{\dot{g}(Z, \nu(F), \gamma(F))\}$ and $\dot{g}(z, \nu(F), \gamma(F))$ the derivative of the map $t \mapsto g(z, \nu(F), t)$ at $t = \gamma(F)$. Thus (10.7.2) typically requires Δ to be zero.

We shall now illustrate the above theory with some examples.

Example 10.7.1. Confidence region for a variance. Let the r.v.'s $Y_1, \cdots, Y_n$ be i.i.d. F. Assume that F has a finite fourth moment, a positive variance $\sigma^2(F)$ and a finite central fourth moment $\nu_4(F)$ greater than $\sigma^4(F)$. The problem of interest is to construct a confidence region for the variance $\sigma^2(F)$. If the mean $\mu(F)$ of F were known, we could use the criterion function $g(y,t) = (y - \mu(F))^2 - t$, $y \in \mathbb{R}, t > 0$. Indeed, the r.v. $g(Y_1, \sigma^2(F))$ has mean zero and finite positive variance $\nu_4(F) - \sigma^4(F)$. As $\mu(F)$ is unknown, we replace it by the sample mean $\bar{Y}_n$ and work with the estimated criterion function

$$\hat{g}_n(y,t) = (y - \bar{Y}_n)^2 - t, \qquad x \in \mathbb{R}, t > 0.$$

The corresponding empirical likelihood is then

$$\mathcal{R}_n(t) := \sup\Big\{ \prod_{i=1}^{n} n\pi_i;\ \pi \in \mathcal{P}_n;\ \sum_{i=1}^{n} \pi_i \big[(Y_i - \bar{Y}_n)^2 - t\big] = 0 \Big\}, \quad t > 0.$$

Use the identity $Y_i - \bar{Y}_n = (Y_i - \mu(F)) + (\mu(F) - \bar{Y}_n)$ to obtain the following.

$$\begin{aligned} D_{n1} :&= \frac{1}{n} \sum_{i=1}^{n} \Big[\hat{g}_n(Y_i, \sigma^2(F)) - g(Y_i, \sigma^2(F)) \Big] \\ &= \frac{1}{n} \sum_{i=1}^{n} \Big[(Y_i - \bar{Y}_n)^2 - (Y_i - \mu(F))^2 \Big] = -(\bar{Y}_n - \mu(F))^2 = O_p(n^{-1}), \\ D_{n2} :&= \frac{1}{n} \sum_{i=1}^{n} \Big| \hat{g}_n(Y_i, \sigma^2(F)) - g(Y_i, \sigma^2(F)) \Big|^2 \\ &= \frac{1}{n} \sum_{i=1}^{n} \Big| (Y_i - \bar{Y}_n)^2 - (Y_i - \mu(F))^2 \Big|^2 \\ &\le \frac{1}{n} \sum_{i=1}^{n} \Big[2|Y_i - \mu(F)|\,|\mu(F) - \bar{Y}_n| + |\mu(F) - \bar{Y}_n)|^2 \Big]^2 \to_p 0. \end{aligned}$$

Hence, (10.7.1) and (10.7.2) are satisfied for the present problem with $\nu(F) = \sigma^2(F)$ and $\gamma(F) = \mu(F)$. Therefore, $-2 \log \mathcal{R}_n(\sigma^2(F)) \to_D \chi_1^2$ and the set $\{t > 0; \hat{-}2 \log \mathcal{R}_n(t) \le c_{\alpha,1}\}$ is an asymptotic $1 - \alpha$ confidence set for the variance $\sigma^2(F)$.

Example 10.7.2. Estimating F when the variance is known. Assume that F has a finite fourth moment, known positive variance σ_0^2 and a central fourth moment $\nu_4(F)$ greater than σ_0^4. We are interested in estimating F based on a random sample $Y_1, \cdots, Y_n$ from F. Since the variance is known, we work with the estimator

$$\widehat{F}_n(t) = \frac{1}{n}\sum_{i=1}^{n} \frac{I(Y_i \le t)}{1 + \xi_n[(Y_i - \bar{Y}_n)^2 - \sigma_0^2]}, \qquad t \in \mathbb{R}.$$

It follows from the results in the previous example that (10.7.3) and (10.7.4) are satisfied. Thus we have the uniform stochastic expansion

$$n^{1/2}\sup_{t\in\mathbb{R}}\left|\widehat{F}_n(t) - \Big(\mathbb{F}_n(t) - \frac{C(t)}{\nu_4(F) - \sigma_0^4}\frac{1}{n}\sum_{i=1}^{n}\big[(Y_i - \mu(F))^2 - \sigma_0^2\big]\Big)\right| \to_p 0$$

with $C(t) = E_F(I(Y \le t)[(Y - \mu(F))^2 - \sigma_0^2])$.

Example 10.7.3. Confidence region for a covariance. Let $\mathcal{F}$ denote the set of all bivariate d.f.'s F with the following properties. If $(X, Y) \sim_D F$, then X, Y and $XY - E_F(Y)E_F(Y)$ have finite positive variances. For this model we take $\nu(F) := \mathrm{Cov}_F(X, Y)$ to be the covariance and $\gamma(F) = (E_F(X), E_F(Y))'$ to be the vector of marginal means. Then a criterion function is given by

$$g((x, y)', s, (t_1, t_2)') = (x - t_1)(y - t_2) - s.$$

Indeed, by the choice of $\mathcal{F}$, $g((X, Y)', \nu_F, \gamma(F))$ has mean zero and a positive finite variance. Let $(X_1, Y_1)', \cdots, (X_n, Y_n)'$ be a random sample from a d.f. F in $\mathcal{F}$. Take $\hat{\gamma}_n := (\bar{X}_n, \bar{Y}_n)'$ the vector of the marginal sample means $\bar{X}_n = n^{-1}\sum_{i=1}^{n} X_i$ and $\bar{Y}_n = n^{-1}\sum_{i=1}^{n} Y_i$. Then the conditions (10.7.1) and (10.7.2) are met. We leave their verification as an exercise. By the provided theory, the confidence region $\{\vartheta; -2\log\hat{\mathcal{R}}_n(\vartheta) \le c(\alpha, 1)\}$ has asymptotic confidence level $1 - \alpha$, where

$$\hat{\mathcal{R}}_n(\vartheta) := \sup\Big\{\prod_{i=1}^{n} n\pi_i;\ \pi \in \mathcal{P}_n;\ \sum_{i=1}^{n}\pi_i[(X_i - \bar{X}_n)(Y_i - \bar{Y}_n) - \vartheta] = 0\Big\}, \qquad \vartheta \in \mathbb{R}.$$

Example 10.7.4. Estimation with uncorrelated bivariate observations. Let $\mathcal{F}$ be as in the previous problem and take $\mathcal{F}_0 = \{F \in \mathcal{F} : \nu(F) = 0\}$. Let $(X_1, Y_1)', \cdots, (X_n, Y_n)'$ be a sample from an F in $\mathcal{F}_0$. Then X_i and Y_i are uncorrelated for each $i = 1, \cdots, n$. Abbreviate X_1 and Y_1 to X and Y. Our goal is to estimate the d.f. G_F of $\max(X, Y)$. The empirical likelihood estimator in the present case is

$$\widehat{G}_n(t) = \frac{1}{n}\sum_{j=1}^{n}\frac{I\big(\max(X_i, Y_i) \le t\big)}{1 + \xi_n(X_i - \bar{X}_n)(Y_i - \bar{Y}_n)}, \qquad t \in \mathbb{R}.$$

with $\bar{X}_n$ and $\bar{Y}_n$ as in the previous example. It follows from the previous example that the requirements (10.7.3) and (10.7.4) are met. Thus we have the uniform expansion

$$n^{1/2}\sup_{t\in\mathbb{R}}\left|\widehat{G}_n(t) - \frac{1}{n}\sum_{i=1}^{n} I[\max(X_i, Y_i) \le t] - \frac{C(t)}{\tau_F} g_F(X_i, Y_i)\right| \to_p 0$$

with $C(t) = E_F\{I(\max(X, Y) \le t) g_F(X, Y)), \tau_F := E_F(g_F^2(X, Y))$ and $g_F(X, Y) = (X - E_F(X))(Y - E_F(Y))$. For a fixed $t \in \mathbb{R}$, the estimator $\widehat{G}_n(t)$ provides a relative asymptotic variance reduction of

$$\frac{C^2(t)}{\tau_F G_F(t)(1 - G_F(t))}$$

over the classical estimator $\mathbb{G}_n(t) = n^{-1}\sum_{i=1}^{n} I\big(\max(X_i, Y_i) \le t\big)$.

10.8 Regression

In the regression model of interest here, we observe a response variable Y and a p-dimensional covariate vector X such that for some $\theta \in \mathbb{R}^p$, $\varepsilon = Y - \theta' X$ and X are independent. To identify the p-dimensional parameter θ assume that the $p \times p$ matrix $E(XX')$ is well defined and has rank p. In what follows we address several approaches to the inference about θ. These are motivated by assumptions about $\varepsilon = Y - \theta' X$. The first example treats the case when ε has mean zero and finite positive variance σ^2. In the second example, ε has median 0 and a d.f. that is continuous at 0. The third example treats the case when ε is symmetric about zero. In all examples, $(X_1', Y_1)', \cdots, (X_n', Y_n)'$ denote independent copies of $(X', Y)'$.

Example 10.8.1. Here we treat the case when $\varepsilon = Y - \theta' X$ has mean zero and finite positive variance σ^2. Then $X\varepsilon$ has mean zero and a positive definite dispersion matrix $E(XX')\sigma^2$. This suggests the empirical likelihood ratio

$$\mathcal{R}_n(\vartheta) := \sup\Big\{ \prod_{i=1}^n n\pi_i;\ \pi \in \mathcal{P}_n;\ \sum_{i=1}^n \pi_i X_i (Y_i - \vartheta' X_i) = 0 \Big\}.$$

Example 10.8.2. Here we treat the case when $\varepsilon = Y - \theta' X$ has median zero and a d.f. that is continuous at 0. Then $X\mathrm{sign}(\varepsilon)$ has mean zero and a positive definite dispersion matrix $E(XX')$. This suggests the empirical likelihood ratio

$$\mathcal{R}_n(\vartheta) := \sup\Big\{ \prod_{i=1}^n n\pi_i;\ \pi \in \mathcal{P}_n;\ \sum_{i=1}^n \pi_i X_i \mathrm{sign}(Y_i - \vartheta' X_i) = 0 \Big\}.$$

Example 10.8.3. Here we treat the case when $\varepsilon = Y - \theta' X$ has a distribution that is symmetric about zero, i.e., ε has the same distribution as $-\varepsilon$. Let h be a bounded odd increasing function. In this case $Xh(\varepsilon)$ has mean zero and positive dispersion matrix $E(XX')E(h^2(\varepsilon))$. This suggests the empirical likelihood ratio

$$\mathcal{R}_n(\vartheta) := \sup\Big\{ \prod_{i=1}^n n\pi_i;\ \pi \in \mathcal{P}_n;\ \sum_{i=1}^n \pi_i X_i h(Y_i - \vartheta' X_i) = 0 \Big\}.$$

In all of the above examples, under the given assumptions, $-2\log \mathcal{R}_n(\theta) \to_D \chi_p^2$, by Theorem 10.4.2. Use this fact and a routine argument to construct the empirical likelihood confidence region for θ of asymptotic confidence level $1 - \alpha$ and the empirical likelihood tests about θ of asymptotic level α in all these examples.

10.9 Exercises

1. Show that $\|uv'\| = \|u\|\|v\|$ for any vectors $u \in \mathbb{R}^m$ and $v \in \mathbb{R}^k$.
2. Let $u_1, \cdots, u_n$ be vectors in $\mathbb{R}^m$ and $v_1, \cdots, v_n$ be vectors in $\mathbb{R}^k$. Verify the inequality

$$\Big\| \frac{1}{n} \sum_{i=1}^n u_i v_i' \Big\|^2 \le \frac{1}{n} \sum_{i=1}^n \|u_i\|^2 \frac{1}{n} \sum_{i=1}^n \|v_i\|^2.$$

Hint: Use Exercise 1 and then apply the inequality (2.4.4).

3. Let $A, A_1, A_2, \cdots$ be symmetric $q \times q$ matrices. Then the following are equivalent: (a) $A_n \to A$; (b) $v'A_n v \to v'Av$ for all $v \in \mathbb{R}^q$; (c) $u'A_n u \to u'Au$ for all unit vectors $u \in \mathbb{R}^q$.

4. Let $a < 0 < b$. Suppose $q = 1$ and $x_1, \cdots, x_n$ take on the three values $a, 0, b$. Let $n_a := \sum_{i=1}^n I(x_i = a)$, $n_b := \sum_{i=1}^n I(x_i = b)$. Show that the solution ξ of the equation (10.3.8) is given by
$$\xi = -\frac{an_a + bn_b}{ab(n_a + n_b)}.$$

5. Prove Lemma 10.4.2.

6. Show that Theorem 10.4.3 remains valid if $\hat{g}_n(Z_1), \cdots, \hat{g}_n(Z_n)$ are replaced by random vectors $X_{n1}, \cdots, X_{nn}$.

7. Let $r_F := 4\{E_F[X(I(X \le 0) - \frac{1}{2})]\}^2/\mathrm{Var}_F(X)$. Show that the reduction factor r_F equals $2/\pi$ when $X \sim_D \mathcal{N}(0,1)$ and equals $1/2$ when X has density $f(x) := e^{-|x|}/2$, $x \in \mathbb{R}$.

8. Compute r_F when $F(x) = 1/(1 + e^{-x})$, $x \in \mathbb{R}$. For this distribution, $V_F(X) = \pi^2/3$.

9. *Inference about several quantiles.* Generalize Example 10.5.4 to simultaneous inference about several quantiles. For this fix a positive integer r and $0 < \beta_1 < \cdots < \beta_r < 1$ and let $\mathcal{F}(\beta_1, \cdots, \beta_r)$ denote the set of all continuous d.f.'s F with unique β_i^{th} quantile $\kappa_{\beta_i}(F)$, $i = 1, \cdots, r$. Construct an asymptotic level $1 - \alpha$ simultaneous confidence region based on a random sample $Z_1, \cdots, Z_n$ from some F in $\mathcal{F}(\beta_1, \cdots, \beta_m)$. What assumptions are needed?

10. *Estimation of F when a quantile is known.* Let $0 < \gamma < 1$. Assume that F has a known γ-quantile ζ and is continuous at ζ. Then (10.6.1) is satisfied with $\psi(x) = I(x \le \zeta) - \gamma$ and have $\Psi_F = \gamma(1 - \gamma)$.

(a) Show that the empirical likelihood estimator of F subject to (10.6.1) with the above ψ is
$$\widehat{F}_n(x) = \frac{1}{n}\sum_{i=1}^n \frac{I(X_i \le x)}{1 + \xi_n[I(X_i \le \zeta) - \gamma]}, \quad x \in \mathbb{R},$$
where $\xi_n = (F_n(\zeta) - \gamma)/(\gamma(1 - \gamma))$ on the event $D_n := \{0 < F_n(\zeta) < 1\}$.

(b) Show that on D_n, the estimator in part (a) satisfies
$$\widehat{F}_n(x) = \gamma\frac{F_n(x \wedge \zeta)}{F_n(\zeta)} + (1 - \gamma)\frac{F_n(x \vee \zeta) - F_n(\zeta)}{(1 - F_n(\zeta))}, \quad x \in \mathbb{R}.$$

(c) Show that $\widehat{F}_n(\zeta) = \gamma$ on D_n and
$$c_F(x) := E(I(X \le x)(I(X \le \zeta) - \gamma) = F(x \wedge \zeta) - F(x)F(\zeta)$$
which is non-zero for $0 < F(x) < 1$. For such an x, show that the asymptotic variance reduction by this estimator over the classical estimator $F_n(x)$ is
$$\frac{c_F^2(x)}{\gamma(1 - \gamma)} = \begin{cases} F^2(x)(1 - \gamma)/\gamma, & x \le \zeta, \\ \gamma(1 - F(x))^2/(1 - \gamma), & x > \zeta. \end{cases}$$

11. *Estimate the mean when the variance is known.* What assumptions are needed? What is the asymptotic behavior of the empirical likelihood estimator? Find a formula for the relative asymptotic variance reduction over the sample mean. Is there a variance reduction if the distribution is symmetric about the mean?
12. Verify the conditions (10.7.1) and (10.7.1) as needed in Example 10.7.3.
13. Let $(X,Y)'$ be a bivariate random vector with d.f. F. Let $(X_1,Y_1)',\cdots,(X_n,Y_n)'$ be n independent copies of (X,Y). We are interested in obtaining empirical likelihood confidence regions for $P(X > Y)$ based on the given sample in each of the following cases. Formulate the proper model $\mathcal{F}$ for each case.

    (a) No additional information is available.
    (b) X and Y have the same mean.
    (c) X and Y have the same variance.
    (d) X and Y have the same mean and variance.
    (e) X and Y are uncorrelated.

14. Let $(X,Y)'$ be a bivariate random vector. We are interested in estimating the distribution function H of $\min(X,Y)$ based on independent copies $(X_1,Y_1)',\cdots,(X_n,Y_n)'$ of $(X,Y)'$ in each of the following cases. In each case state the assumptions needed and the asymptotic behavior of the estimator.

    (a) The mean of X is zero. (b) The median of Y is zero.
    (c) $P(X > Y) = 1/2$. (d) X and Y have the same mean.
    (e) X and Y have the same variance. (f) X and Y have the same mean and variance.
    (g) X and Y are uncorrelated.

Notes: Ever since the publication of the papers of Owen (1988, 1990) there has been an exponential growth in the research pertaining to empirical likelihood. Owen (1991) discusses some empirical likelihood inference for linear models. Owen (2001) is a nice monograph to learn more about empirical likelihood methodology and numerous applications.

Hjort, McKeague and Van Keilegom (2009) extend the scope of the empirical likelihood methodology; they allow estimated constraints and also treat the case when the number q of constraints grows to infinity with the sample size. Peng and Schick (2013) generalize the latter to also allow for estimated constraints. Qin and Lawless (1994) treat maximum empirical likelihood estimation for smooth criterion functions. Peng and Schick (2018) generalize this theory further to allow for irregular and estimated criterion functions.

11

Survival Analysis

11.1 Introduction

Survival analysis is concerned with analysing the times of the occurrences of some events. For example, in reliability studies, the time to failure of a machine is of interest. Or in medical science, the time of the occurrence of a disease or the time it takes to cure a patient of a disease is of interest. All these times are referred to as lifetimes. A lifetime is the time of the occurrence of an event. A life test is a testing procedure that tests for the lifetime of an object. For example, in an industry setting a unit is put on test till its first failure or till a certain time has elapsed. In health-related studies, a patient is put on a treatment till s/he recovers or for a certain period of time. In some situations one does not observe the true lifetime but a censored lifetime. As an example consider a clinical experiment where patients are put on a treatment for a certain time period and different patients enter into the study at different times. If a patient is cured within the period of study, then we have a complete observation. But if a patient is not cured during the time period or leaves the study for some reason, then we do not have a complete observation. In this case we have a randomly right-censored observation. This is very common in clinical studies.

In survival analysis, lifetime is the basic random variable and the main concern is the analysis of a lifetime distribution or related entities like the survival function, cumulative hazard function and failure rate function. There are parametric models used to carry out such analysis, see, e.g., the monographs by Kalbfleisch and Prentice (1982) and Lawless (1982). Examples include Weibull and gamma distributions. But often it is not clear which parametric model best fits the given data. Also the inference based on parametric model assumptions are restrictive in scope. For these reasons it is desirable to have available nonparametric inference procedures, which are more broadly applicable. This chapter contains some nonparametric survival analysis inference procedures.

The focus of the next section is the definitions of survival, cumulative hazard and failure rate functions and their relations to each other. The nonparametric estimation of the cumulative hazard and survival functions when complete data is observed and when data is randomly censored on the right are the contents of the subsequent sections. Nonparametric maximum likelihood and nonparametric Bayes estimates of the lifetime d.f. and survival function and some of their properties like consistency are discussed in this chapter. It also contains a brief discussion of randomly right-censored linear regression and Cox's proportional hazards regression models.

DOI: 10.1201/9781003585770-11

11.2 Survival and Cumulative Hazard Functions

This section contains the definitions of survival, cumulative hazard and failure rate functions and the description of their relations to each other.

Let X be a r.v. having d.f. F, set $\bar{F} := 1 - F$ and recall $P(X < x) = F(x_-)$. In survival analysis, X represents a lifetime, which is a nonnegative r.v. But the notions defined below are valid for any r.v. taking values on $\mathbb{R}$. For that reason we define them over $\mathbb{R}$.

The survival function of X is defined to be

$$S(x) := P(X > x) = 1 - F(x) = \bar{F}(x), \quad x \in \mathbb{R}.$$

Assume first that F is continuous and has a right-hand derivative at each x. Then the corresponding hazard or failure rate function is defined to be

$$\lambda(x) := \lim_{\epsilon \downarrow 0} \frac{P(x \leq X < x + \epsilon | X \geq x)}{\epsilon}, \qquad x \in \mathbb{R}.$$

The hazard function $\lambda(x)$ is traditionally interpreted to be the instantaneous rate of failure of an item at time x, given that the item survived till just before x. The cumulative hazard function is defined to be

$$\Lambda(x) := \int_{s \leq x} \lambda(s)\, ds, \quad x \in \mathbb{R}.$$

Suppose F is differentiable with f denoting its derivative. Then, for any x with $S(x) > 0$,

$$\lambda(x) = \lim_{\epsilon \downarrow 0} \frac{P(x \leq X < x + \epsilon | X \geq x)}{\epsilon} = \frac{f(x)}{1 - F(x)} = -\frac{d \ln S(x)}{dx},$$
$$\Lambda(x) = \int_{s \leq x} \lambda(s) ds = -\ln S(x) = -\ln \big(1 - F(x)\big).$$

Or, equivalently,

$$S(x) = \exp\Big(-\int_{s \leq x} \lambda(s)\, ds\Big) = e^{-\Lambda(x)}, \qquad x \in \mathbb{R},\ S(x) > 0.$$

This gives a one-to-one correspondence between the survival and cumulative hazard functions. From these relations we also obtain

$$f(x) = \lambda(x) S(x) = \lambda(x) e^{-\Lambda(x)}, \qquad x \in \mathbb{R},\ S(x) > 0. \tag{11.2.1}$$

Suppose X is discrete taking k values $-\infty < x_1 < x_2 < \cdots < x_k < \infty$. Let $p(x_j) := P(X = x_j)$. Set $x_0 = -\infty$, $x_{k+1} := \infty$, $S(x_0) = 1, S(x_{k+1}) = 0$. Then, for all $j = 1, \cdots, k$,

$$p(x_j) = S(x_{j-1}) - S(x_j),$$
$$\lambda(x_j) := P\big(X = x_j \big| X \geq x_j\big) = \frac{p(x_j)}{S(x_{j-1})} = 1 - \frac{S(x_j)}{S(x_{j-1})}.$$

Moreover, for any $x \in \mathbb{R}$,

$$S(x) = P(X > x) = \sum_{j: x_j > x} p(x_j) = \sum_{j: x_j > x} S(x_{j-1}) \lambda(x_j).$$

We claim the identity

$$S(x) = \prod_{j:x_j \le x} [1 - \lambda(x_j)], \qquad \forall x \in \mathbb{R}, \tag{11.2.2}$$

where the product over an empty set is interpreted as 1. Thus for $x < x_1$, the right-hand side of the above identity is 1 while by definition, $S(x) = 1$. For $x \ge x_1$, there exists a positive integer $1 < i < k$ such that $x_i \le x < x_{i+1}$ and

$$\prod_{j:x_j \le x} [1 - \lambda(x_j)] = \prod_{j=1}^{i} \frac{S(x_j)}{S(x_{j-1})} = \frac{S(x_1)}{S(x_0)} \frac{S(x_2)}{S(x_1)} \cdots \frac{S(x_i)}{S(x_{i-1})} = S(x_i) = P(X > x_i) = S(x).$$

Therefore, an alternative representation for the survival function is

$$S(x) = \prod_{j:x_j \le x} \left(1 - \frac{p(x_j)}{S(x_{j-1})}\right) = \prod_{j:x_j \le x} \left(1 - \frac{\Delta F(x_j)}{1 - F(x_{j-})}\right),$$

where $\Delta F(y) := F(y) - F(y_-)$, for any $y \in \mathbb{R}$.

In general, the cumulative hazard function corresponding to any d.f. F is defined to be

$$\Lambda(x) = \int_{s \le x} \frac{dF(s)}{1 - F(s_-)}, \quad x \in \mathbb{R}. \tag{11.2.3}$$

This in turn yields that

$$S(x) = 1 - \int_{s \le x} S(s_-)\, d\Lambda(s), \quad x \in \mathbb{R}. \tag{11.2.4}$$

11.3 Estimation of Λ and S

In this section we shall discuss the estimation of the cumulative hazard function Λ and the survival function S based on a random sample $X_1, \cdots, X_n$ from F. Before proceeding to describe an estimator of S, we need to introduce the notion of a continuous time martingale and discuss the Doob–Meyer decomposition of the empirical d.f.

Let $\{M(x), x \in \mathbb{R}\}$ be a stochastic process and $\{\mathcal{F}_x, x \in \mathbb{R}\}$ be a family of σ-fields such that $M(x)$ is $\mathcal{F}_x$-measurable for every $x \in \mathbb{R}$ and $\mathcal{F}_x \subseteq \mathcal{F}_y$ for $x \le y$. We say that $\{M(x), x \in \mathbb{R}\}$ is a martingale with respect to the σ-fields $\{\mathcal{F}_x, x \in \mathbb{R}\}$ if $E\big(|M(x)|\big) < \infty$, for all $x \in \mathbb{R}$, the map $x \mapsto M(x)$ is right continuous and has left limits and for all $x < y$,

$$E\big(M(y)\big|\mathcal{F}_x\big) = M(x), \quad \text{a.s.} \tag{11.3.1}$$

Now, let X be a r.v. having d.f. F and define, with Λ as in (11.2.3),

$$N(x) = I(X \le x), \qquad M(x) := N(x) - \int_{u \le x} I(X \ge u)\, d\Lambda(u), \tag{11.3.2}$$
$$\mathcal{F}_x := \sigma\text{-field}\{N(u), -\infty < u \le x\}, \qquad x \in \mathbb{R}.$$

Here and in the sequel, for any stochastic process $\{\mathcal{Y}(t), t \in \mathbb{R}\}$, the notation σ-field$\{\mathcal{Y}(t), -\infty < t \le x\}$ denotes the sigma field generated by the stochastic process $\{\mathcal{Y}(t), -\infty < t \le x\}$.

The following theorem proves the martingale property of $\{M(x), x \in \mathbb{R}\}$.

Theorem 11.3.1. *The process* $\{M(x), x \in \mathbb{R}\}$ *of* (11.3.2) *is a mean zero martingale with respect to* $\{\mathcal{F}_x, x \in \mathbb{R}\}$.

Proof. Clearly, $x \mapsto M(x)$ is right continuous and has left limits, $M(x)$ is $\mathcal{F}_x$-measurable for all $x \in \mathbb{R}$, and $\mathcal{F}_x \subseteq \mathcal{F}_y$ for all $x \leq y$. Since $[1 - F(u_-)]\, d\Lambda(u) = dF(u)$ by (11.2.3), $\forall x \in \mathbb{R}$,

$$E\big|M(x)\big| \leq 1 + \int_{u \leq x} [1 - F(u_-)]\, d\Lambda(u) = 1 + \int_{u \leq x} dF(u) = 1 + F(x) \leq 2. \tag{11.3.3}$$

By Theorem 2.3.2, for every $x \in \mathbb{R}$, we have

$$E\Big(\int_{u \leq x} I(X \geq u)\, d\Lambda(u)\Big) = \int_{u \leq x} \bar{F}(u_-) \frac{dF(u)}{\bar{F}(u_-)} = F(x).$$

Hence,

$$E\big(M(x)\big) = F(x) - E\Big(\int_{u \leq x} I(X \geq u)\, d\Lambda(u)\Big) = F(x) - F(x) = 0, \quad \forall x \in \mathbb{R}.$$

Let $s < x$. Then we have

$$\begin{aligned} M(x) - M(s) &= N(x) - N(s) - \int_{(s,\,x]} I(X \geq u)\, d\Lambda(u) \\ &= I(s < X \leq x) - \int_{(s,\,x]} I(X \geq u)\, d\Lambda(u) \\ &= I(X > s)\Big\{I(s < X \leq x) - \int_{(s,\,x]} I(X \geq u)\, d\Lambda(u)\Big\}. \end{aligned}$$

If $F(s) = 1$, then the right hand is zero almost surely, which in turn yields $E\big(M(x) - M(s)|\mathcal{F}_s\big) = 0$, almost surely. Now assume $F(s) < 1$. The event $\{X > s\}$ belongs to $\mathcal{F}_s$ because it equals $\{N(v) = 0, v \leq s\}$. Next, one verifies the identity $\mathcal{F}_s = \big\{A \in \mathcal{F}_x : A \cap \{X > s\} \in \{\emptyset, \{X > s\}\}\big\}$. Using the results of Example 2.7.3 we derive

$$E\big(M(x) - M(s)|\mathcal{F}_s\big) = I(X > s)\frac{P(s < X \leq x) - \int_{(s,x]} S(u_-)\, d\Lambda(u)}{1 - F(s)} = 0,$$

in view of the identity $P(s < X \leq x) = F(x) - F(s) = \int_{(s,x]} dF(u) = \int_{(s,x]} S(u_-)\, d\Lambda(u)$. □

The above result shows that for every $x \in \mathbb{R}$, $N(x)$ can be decomposed into two parts. One is the nondecreasing right continuous function $A(x) := \int_{u \leq x} I(X \geq u)\, d\Lambda(u)$ and the other is the martingale part $M(x)$, i.e.,

$$N(x) = \int_{u \leq x} I(X \geq u)\, d\Lambda(u) + M(x) = A(x) + M(x), \quad x \in \mathbb{R}.$$

This is called the Doob-Meyer decomposition of the counting process N, and A is called its compensator. This type of result holds for a large class of counting processes. See Loève (1963).

Because of the linearity property of the expectation and sum operators, for r.v.'s $X_1, \cdots, X_n$ that are i.i.d. F,

$$M_n(x) = \sum_{i=1}^{n}\Big(I(X_i \leq x) - \int_{u \leq x} I(X_i \geq u)\, d\Lambda(u)\Big) = nF_n(x) - n\int_{u \leq x} (1 - F_n(u_-))\, d\Lambda(u)$$

is a mean zero martingale with respect to $\mathcal{F}_{x,n} = \sigma\{I(X_i \le s), s \le x, i = 1, \cdots, n\}$, for each $n \ge 1$. Thus the Doob–Meyer decomposition of the empirical d.f. is

$$nF_n(x) = n \int_{u \le x} (1 - F_n(u_-))\, d\Lambda(u) + M_n(x).$$

If F is continuous, then $\Lambda = -\ln \bar{F}$ and

$$M_n(x) = nF_n(x) + \sum_{i=1}^{n} \ln \bar{F}(X_i \wedge x)$$

is a mean zero martingale in x with respect to $\mathcal{F}_{x,n}$, for each $n \ge 1$.

The above martingale type property continues to hold even when X_i's have different distributions. Let F_i and Λ_i denote the d.f. and the cumulative hazard function of X_i, $1 \le i \le n$, respectively. Then,

$$\widetilde{M}_n(x) = \sum_{i=1}^{n} \Big\{ I(X_i \le x) - \int_{u \le x} I(X_i \ge u)\, d\Lambda_i(u) \Big\},$$

is a means zero martingale with respect to the σ-field $\mathcal{F}_{x,n}$, for each $n \ge 1$.

Estimation of Λ. Let $X_1, \cdots, X_n$ be a random sample from F. Assume F to be continuous. Then, $P(X_i \ne X_j, 1 \le i \ne j \le n) = 1$. In the rest of this section, all X_i's are assumed to be distinct. To derive an estimate of Λ based on this random sample, define

$$N_\bullet(x) := \sum_{i=1}^{n} I(X_i \le x), \qquad Y_\bullet(x) := \sum_{i=1}^{n} I(X_i \ge x),$$

$$\widehat{\Lambda}(x) := \int_{s \le x} \frac{I(Y_\bullet(s) > 0)}{Y_\bullet(s)}\, dN_\bullet(s), \qquad \Lambda^*(x) := \int_{s \le x} I(Y_\bullet(s) > 0)\, d\Lambda(s).$$

Let $X_{(1)} < \cdots < X_{(n)}$ denote the order statistics obtained from the given sample. Because both $Y_\bullet$ and $N_\bullet$ jump at the same points $X_{(1)}, \cdots, X_{(n)}$,

$$\widehat{\Lambda}(x) = \int_{s \le x} \frac{dN_\bullet(s)}{Y_\bullet(s)} = \sum_{i=1}^{n} \frac{I(X_{(i)} \le x)}{Y_\bullet(X_{(i)})} = \sum_{i=1}^{n} \frac{I(X_{(i)} \le x)}{n - i + 1}.$$

Let

$$M(x) := N_\bullet(x) - \int_{s \le x} Y_\bullet(s)\, d\Lambda(s), \qquad \mathcal{M}(x) := \int_{s \le x} \frac{I(Y_\bullet(s) > 0)}{Y_\bullet(s)}\, dM(s).$$

Using $dM(x) = dN_\bullet(x) - Y_\bullet(x)\, d\Lambda(x)$, we derive

$$\mathcal{M}(x) = \int_{s \le x} \frac{I(Y_\bullet(s) > 0)}{Y_\bullet(s)}\, dN_\bullet(s) - \int_{s \le x} I(Y_\bullet(s) > 0)\, d\Lambda(s) = \widehat{\Lambda}(x) - \Lambda^*(x).$$

Because $E\big(\mathcal{M}(x)\big) = 0$, for all x, we obtain

$$E(\widehat{\Lambda}(x)) = E(\Lambda^*(x)), \quad \forall x \in \mathbb{R}.$$

Moreover, because $Y_\bullet(x) \sim_D \text{Bin}(n,\, 1 - F(x_-))$,

$$\begin{aligned} E(\Lambda^*(x)) &= \int_{s\le x} P(Y_\bullet(s) > 0)\, d\Lambda(s) = \int_{s\le x} (1 - P(Y_\bullet(s) = 0))\, d\Lambda(s) \\ &= \Lambda(x) - \int_{s\le x} P(Y_\bullet(s) = 0)\, d\Lambda(s) = \Lambda(x) - \int_{s\le x} F^n(s_-)\, d\Lambda(s), \quad \forall x \in \mathbb{R}. \end{aligned}$$

Hence $\widehat{\Lambda}$ is a negatively-biased estimator of Λ. However, the bias tends to zero, typically, exponentially fast, as $n \to \infty$. In fact,

$$\sup_{x\le b} \left|E(\widehat{\Lambda}(x)) - \Lambda(x)\right| = \sup_{x\le b} \left|E(\Lambda^*(x)) - \Lambda(x)\right| \le \Lambda(b) F^n(b_-) \to 0$$

exponentially fast, for every $b \in \mathbb{R}$ with $F(b_-) < 1$ and $\Lambda(b) < \infty$.

Estimation of S. The estimation of S is based on the above estimator $\widehat{\Lambda}$ of Λ. Upon replacing Λ by $\widehat{\Lambda}$ in (11.2.4) we are led to an estimator $\widehat{S}$ of S defined as a solution of the equation

$$\widehat{S}(x) = 1 - \int_{s\le x} \widehat{S}(s_-)\, d\widehat{\Lambda}(s) = 1 - \int_{s\le x} \widehat{S}(s_-)\, \frac{dN_\bullet(s)}{Y_\bullet(s)}.$$

The relation

$$\widehat{S}(x_-) - \widehat{S}(x) = -\Delta\widehat{S}(x) = \frac{\widehat{S}(x_-)\Delta N_\bullet(x)}{Y_\bullet(x)},$$

yields

$$\widehat{S}(x) = \widehat{S}(x_-)\left(1 - \frac{\Delta N_\bullet(x)}{Y_\bullet(x)}\right), \quad \forall x \in \mathbb{R}. \tag{11.3.4}$$

Recall that $Y_\bullet$ and $N_\bullet$ jump at the same points, namely, $X_{(1)}, \cdots, X_{(n)}$.

We shall now give a computational formula for $\widehat{S}$. Clearly, for $x < X_{(1)}$, $\Delta N_\bullet(x) = 0$, $Y_\bullet(x) = n$. Also, since $S(-\infty) = 1$, define

$$\widehat{S}(x) = 1, \quad \forall x < X_{(1)}.$$

By (11.3.4),

$$\widehat{S}(X_{(1)}) = \widehat{S}(X_{(1)_-})\Big(1 - \frac{1}{n}\Big) = 1 - \frac{1}{n}.$$

Because $\widehat{S}$ jumps only at $X_{(1)}, \cdots, X_{(n)}$, $\widehat{S}(x) = \widehat{S}(x_-)$, for all $x \ne X_{(i)}$, $1 \le i \le n$. Hence, we define

$$\widehat{S}(x) = \widehat{S}(X_{(1)}), \qquad X_{(1)} \le x < X_{(2)}.$$

Continuing like this, define

$$\begin{aligned} \widehat{S}(X_{(2)}) &= \widehat{S}(X_{(2)_-})\Big(1 - \frac{1}{n-1}\Big) = \Big(1 - \frac{1}{n}\Big)\Big(1 - \frac{1}{n-1}\Big) = \frac{n-1}{n}\frac{n-2}{n-1} = 1 - \frac{2}{n}, \\ \widehat{S}(x) &= \widehat{S}(X_{(2)}), \qquad X_{(2)} \le x < X_{(3)}, \end{aligned}$$

and, by induction,

$$\widehat{S}(X_{(i)}) = \prod_{j=1}^{i}\Big(1 - \frac{1}{n-j+1}\Big) = \prod_{j=1}^{i}\frac{n-j}{n-j+1} = \frac{n-1}{n}\frac{n-2}{n-1}\cdots\frac{n-i}{n-i+1} = 1 - \frac{i}{n},$$

$$\widehat{S}(x) = 1 - \frac{i}{n}, \qquad X_{(i)} \le x < X_{(i+1)},\ i = 0, 1, \cdots, n,$$

where $X_{(0)} = -\infty$ and $X_{(n+1)} = \tau_F := \sup\{x : F(x) < 1\}$. We thus have an estimator $\widehat{S}$ of S, which in turn gives an estimator $\widehat{F}$ of F given by

$$\widehat{F}(x) = 1 - \widehat{S}(x) = \frac{i}{n}, \qquad X_{(i)} \le x < X_{(i+1)}, i = 0, 1, \cdots, n.$$

Observe that this $\widehat{F}$ is the same as the empirical d.f. F_n of (2.4.2).

11.4 Some Censoring Notions

In this section we shall discuss several notions of censored data. In later sections we shall develop inference procedures only for the randomly censored data on the right.

Type I Censoring. In Type I censoring a unit is put on a life test and the test is terminated if either it fails or until a fixed time t_0. Suppose n units are put on life test. Let X_i denote the lifetime of the ith unit, and set $\widetilde{X}_i := X_i \wedge t_0$ and $\delta_i := I(X_i \le t_0)$, $i = 1, \cdots, n$. Then the data consists of $(\widetilde{X}_1, \delta_1), \cdots, (\widetilde{X}_n, \delta_n)$.

Type II Censoring. Here n units are put on a life test, and one stops observing the units as soon as the rth failure occurs, where r is a fixed positive integer. With X_i denoting the ith unit's lifetime, the data consists of $\{(\widetilde{X}_i, \delta_i); i = 1, \cdots, n\}$, where $\widetilde{X}_i = X_i \wedge X_{(r)}$ and $\delta_i = \{X_i \le X_{(r)}\}$, for $i = 1, \cdots, n$. Or equivalently, the data consists of $X_{(1)}, \cdots, X_{(r)}$, where $X_{(j)}$ denotes jth order statistics of $X_1, \cdots, X_n$.

Random Censoring on the Right. Suppose that $X_1, \cdots, X_n$ are true lifetimes and $U_1, \cdots, U_n$ are another set of r.v.'s. In randomly censored data on the right, the observed data consists of the pairs $\{(\widetilde{X}_1, \delta_1), \cdots, (\widetilde{X}_n, \delta_n)\}$, where $\widetilde{X}_i := X_i \wedge U_i$ and $\delta_i := I\big(X_i \le U_i\big)$, for $i = 1, \cdots, n$. The r.v.'s U_i's are called censoring r.v.'s. Observe that if all U_i's equal a constant t_0, then randomly right-censored data reduces to Type I censored data.

A natural example of randomly right-censored data is given by a series system consisting of k components, where one is interested in assessing the lifetime distribution of the 1st component based on the observations about the lifetime of the whole system, and whether the system failed because of component 1 or not.

Let T_j, $1 \le j \le k$ denote the lifetimes of the k components of the series system. Let $X = T_1$ and $U := \min\{T_2, T_3, \cdots, T_k\}$. Then the lifetime of the series system is

$$\widetilde{X} := \min\{T_1, T_2, T_3, \cdots, T_k\} = \min\{T_1, U\} = X \wedge U.$$

Thus here the data based on one unit consists of $\widetilde{X}$ and $\delta = I(X \le U)$. In general, put n items on a test. Let T_{ij} be the lifetime of component j for the ith item, $j = 1, \cdots, k$; $i = 1, \cdots, n$, and $U_i = \min_{2 \le j \le k} T_{ij}$. Then the data consists of $\widetilde{X}_i = T_{i1} \wedge U_i$ and $I\big(T_{i1} \le U_i\big)$, for $i = 1, \cdots, n$.

Another example is when n patients are put on a treatment for a certain period of time. The ith patient "may leave" during the time of study. Then $\widetilde{X}_i = X_i$ if $\delta_i = 1$ and $X_i > \widetilde{X}_i$ if $\delta_i = 0$, $i = 1, \cdots, n$.

Interval Censoring Case I. This is an example about the time of the onset of an event which is usually hard to observe. Imagine a patient visits a doctor and upon examination, the doctor says that the patient is having a certain disease, but no one knows the time X of the onset of the disease. In this case one observes the inspection time T and $\delta := I(X \leq T)$. This is called interval-censored data type I. It is also called current status data. Here, one does not observe X and the problem of interest is to make inference about the d.f. F of X based on n observations $\{(T_1, \delta_1), \cdots, (T_n, \delta_n)\}$, where $\delta_i = I(X_i \leq T_i)$, for $i = 1, \cdots, n$.

Interval Censoring Case II. Now imagine the situation where a doctor inspects a patient at time U and finds that a disease is not present. And then the same doctor examines the same patient some time later at time V, $V > U$, and finds that the disease is present. If X denotes the time of the onset of the disease, then here the observation consists of U, V, δ, γ, where $\delta := I(X \leq U) = 0$, $\gamma := I(U < X \leq V) = 1$. This is called the interval censored data of case 2.

Now if n such patients are in the experiment then the data consists of $\{(U_i, V_i, \delta_i, \gamma_i); i = 1, \cdots, n\}$, where $\delta_i = I(X_i \leq U_i)$ and $\gamma_i = I(U_i < X_i \leq V_i)$. Groeneboom and Wellner (1992) discuss the nonparametric MLE of F for the interval censoring cases I and II, where X is never observed.

Mixed Case Interval Censoring. Suppose a unit is put on test and inspected at a random number of inspection times. Let X denote the lifetime of the unit, K denote a positive random integer and $\boldsymbol{Y} = \{Y_{k,j} : 1 \leq j \leq k\}$ a triangular array of random variables satisfying $Y_{k,1} < \cdots < Y_{k,k}$. Assume that $(K, \boldsymbol{Y})$ and X are independent. On the event $\{K = k\}$, one then observes the pair (L, R) where L and R denote the left and right endpoints of the interval among $(-\infty, Y_{k,1}], (Y_{k,1}, Y_{k,2}], \ldots, (Y_{k,k}, \infty)$ that contains the lifetime. This formulation allows for a different number of inspection times if n units are put on test. For n units, one then obtains the data $(L_1, R_1), \ldots, (L_n, R_n)$. This is the mixed case interval censoring model introduced by Schick and Yu (2000). These authors studied the consistency of the nonparametric MLE of the d.f. F of X.

11.5 Randomly Right-Censored Data

In this section we shall discuss the likelihood of randomly censored data on the right, the Nelson-Aalen estimator of the cumulative hazard function and the Kaplan-Meier estimator of the survival function.

Likelihood function. Let $X_1, \cdots, X_n$ and $U_1, \cdots, U_n$ be two independent sets of independent r.v.'s. Let F_i, G_i denote the d.f.'s of X_i, U_i, respectively, $1 \leq i \leq n$. Suppose F_i has a density f_i for each $i = 1, \cdots, n$. We observe $\{(\widetilde{X}_1, \delta_1), \cdots, (\widetilde{X}_n, \delta_n)\}$, where $\widetilde{X}_i = X_i \wedge U_i$ and $\delta_i = I(X_i \leq U_i)$, $i = 1, \cdots, n$. Then,

$$P(\widetilde{X}_i \leq x_i, \delta_i = 1) = P(X_i \leq x_i, X_i \leq U_i) = \int_{x \leq x_i} f_i(x)\,[1 - G_i(x_-)]\,dx,$$

$$P(\widetilde{X}_i \leq x_i, \delta_i = 0) = P(U_i \leq x_i, X_i > U_i) = \int_{x \leq x_i} [1 - F_i(x)]\,dG_i(x).$$

It follows that for all $d_i \in \{0,1\}, 1 \le i \le n$,

$$dP(\widetilde{X}_i \le x_i, \delta_i = d_i) = \big[f_i(x_i)\big(1 - G_i(x_{i\_})\big)dx_i\big]^{d_i} \big[(1 - F_i(x_i))dG_i(x_i)\big]^{1-d_i}.$$

Let $(x,d) := \{(x_i, d_i), 1 \le i \le n\}$. Then, the likelihood function is

$$\begin{aligned} L_n(x,d) &= dP\Big(\widetilde{X}_i \le x_i, \delta_i = d_i, i = 1, \cdots, n\Big) \\ &= \prod_{i=1}^{n} \Big(f_i(x_i)\, dx_i\Big)^{d_i} \Big[1 - F_i(x_i)\Big]^{1-d_i} \Big[1 - G_i(x_{i-})\Big]^{d_i} \Big[dG_i(x_i)\Big]^{1-d_i}. \end{aligned}$$

Ignoring the factor involving G_i's, the relevant likelihood of the data $(\widetilde{X}_i, \delta_i), 1 \le i \le n$ is

$$L_n((\widetilde{X}, \delta)) \propto \prod_{i=1}^{n} f_i^{\delta_i}(\widetilde{X}_i)\big(1 - F_i(\widetilde{X}_i)\big)^{1-\delta_i}.$$

If $F_i = F$, for all $i = 1, \cdots, n$, and F has density f, then the relevant part of the likelihood is

$$L_n((\widetilde{X}, \delta)) \propto \prod_{i=1}^{n} f^{\delta_i}(\widetilde{X}_i)\big(S(\widetilde{X}_i)\big)^{1-\delta_i}. \tag{11.5.1}$$

In the parametric setup this can be used to arrive at maximum likelihood estimators of the given parameters.

11.5.1 Estimating Λ under Random Censoring

Just as in the uncensored case we shall first prove the martingale property of basic counting processes arising under right random censoring. Let X, U be independent r.v.'s with $X \sim_D F$, $U \sim_D G$, and $\widetilde{X} = X \wedge U$, $\delta = I(X \le U)$. For any d.f. L, let Λ_L denotes the cumulative hazard function associated with L. Define, for $x \in \mathbb{R}$,

$$\begin{aligned} &N_u(x) := I\big(\widetilde{X} \le x, \delta = 1\big), \qquad N_c(x) := I\big(\widetilde{X} \le x, \delta = 0\big), \\ &M_u(x) := N_u(x) - \int_{s \le x} I\big(\widetilde{X} \ge s\big)\, d\Lambda_F(s), \quad M_c(x) := N_c(x) - \int_{s \le x} I\big(\widetilde{X} \ge s\big)\, d\Lambda_G(s), \\ &\mathcal{F}_x := \sigma\text{-field}\big\{N_u(s), N_c(s), s \le x\big\}. \end{aligned}$$

We are ready to state an analog of Theorem 11.3.1 for the N_u, N_c processes.

Theorem 11.5.1. *The processes $\{M_u(x), x \in \mathbb{R}\}$ and $\{M_c(x), x \in \mathbb{R}\}$ are mean zero martingales with respect to the σ-fields $\{\mathcal{F}_x, x \in \mathbb{R}\}$.*

Proof. We shall give details only for the M_u process. For the M_c process the proof is similar. Fix an $x \in \mathbb{R}$ and $s < x$. Argue as for (11.3.3) to obtain that $E(|M_u(x)|) \le 2$. Let

$$D := E\Big(M_u(x) - M_u(s)\big|\mathcal{F}_s\Big) = E\Big(N_u(x) - N_u(s)\big|\mathcal{F}_s\Big) - E\Big(\int_{(s,x]} \{\widetilde{X} \ge u\} d\Lambda_F(u)\big|\mathcal{F}_s\Big).$$

We need to show that $D = 0$, a.s. If $P(\widetilde{X} > s) = 0$, then $I(\widetilde{X} > s) = 0$ a.s. and $D = 0$, a.s., trivially. Assume $P(\widetilde{X} > s) > 0$. Consider

$$\begin{aligned} E\Big(N_u(s) - N_u(x)\big|\mathcal{F}_s\Big) &= E\Big(I\big(s < \widetilde{X} \le x, \delta = 1\big)\big|\mathcal{F}_s\Big) \\ &= E\Big(I\big(s < \widetilde{X} \le x, \delta = 1\big)\big|N_u(z), N_c(z),\, z \le s\Big). \end{aligned}$$

To compute the above conditional expectation, it suffices to consider the two cases: (i) $N_u(z) = 1$ or $N_c(z) = 1$ for some $z \le s$ and (ii) $N_u(z) = 0 = N_c(z)$, for all $z \le s$. In case (i), $\widetilde{X} \le s$ and the above expression is 0. The event $\{N_u(z) = 0 = N_c(z)$, for all $z \le s\}$ of case (ii) is equal to the event $\{\widetilde{X} > s\}$. Therefore, in case (ii),

$$E\Big(N_u(x) - N_u(x)|\mathcal{F}_s\Big) = P(s < \widetilde{X} \le x, \delta = 1|\widetilde{X} > s) = \frac{P(s < X \le x, X \le U)}{P(\widetilde{X} > s)}$$
$$= \frac{\int_{(s,x]} (1 - G(z_-))\, dF(z)}{P(\widetilde{X} > s)}.$$

Similarly, by Theorem 2.3.2,

$$E\Big(\int_{(s,x]} I(\widetilde{X} \ge z)\, d\Lambda_F(z)|\mathcal{F}_s\Big) = \int_{(s,x]} P(\widetilde{X} \ge z|\mathcal{F}_s)\, d\Lambda_F(z)$$
$$= \frac{\int_{(s,x]}(1 - G(z_-))(1 - F(z_-))\, \frac{dF(z)}{(1 - F(z_-))}}{P(\widetilde{X} > s)}.$$

Therefore, $D = 0$. □

Let $X_1, \cdots, X_n$ be i.i.d. F r.v.'s, and $U_1, \cdots, U_n$ be another set of i.i.d. r.v.'s with d.f. G. Assume that X_i's are independent of U_j's. Randomly right-censored data consists of $\widetilde{X}_i := X_i \wedge U_i$ and $\delta_i = I(X_i \le U_i)$, $1 \le i \le n$. For $x \in \mathbb{R}$, define

$$N_{iu}(x) := I(\widetilde{X}_i \le x,\, \delta_i = 1), \quad N_{ic}(x) := (\widetilde{X}_i \le x,\, \delta_i = 0), \quad \widetilde{Y}_\bullet(x) := \sum_{i=1}^n I(\widetilde{X}_i \ge x),$$
$$\mathcal{F}_{x,n} := \sigma\text{-field}\{N_{iu}(s),\, N_{ic}(s);\, s \le x,\, i = 1, \cdots, n\},$$
$$M_{iu}(x) := N_{iu}(x) - \int_{s \le x} I(\widetilde{X}_i \ge s)\, d\Lambda_F(s),$$
$$M_{\bullet u}(x) := \sum_{i=1}^n M_{iu}(x) = N_{\bullet u}(x) - \int_{s \le x} \widetilde{Y}_\bullet(s)\, d\Lambda_F(s).$$

Again, because of the linearity property of the expectation, $\{M_{\bullet u}(x),\, x \in \mathbb{R}\}$ is a mean zero martingale with respect to $\{\mathcal{F}_{x,n}, x \in \mathbb{R}\}$. Let

$$\widetilde{\Lambda}(x) := \int_{s \le x} \frac{dN_{\bullet u}(s)}{\widetilde{Y}_\bullet(s)}, \qquad \widetilde{\Lambda}^*(x) := \int_{s \le x} I(\widetilde{Y}_\bullet(s) > 0)\, d\Lambda_F(s).$$

Then,

$$\mathcal{M}_u(x) := \int_{s \le x} \frac{I(\widetilde{Y}_\bullet(s) > 0)}{\widetilde{Y}_\bullet(s)}\, dM_{\bullet u}(s)$$
$$= \int_{s \le x} \frac{I(\widetilde{Y}_\bullet(s) > 0)}{\widetilde{Y}_\bullet(s)}\, dN_{\bullet u}(s) - \int_{s \le x} I(\widetilde{Y}_\bullet(s) > 0)\, d\Lambda_F(s)$$
$$= \int_{s \le x} \frac{dN_{\bullet u}(s)}{\widetilde{Y}_\bullet(s)} - \int_{s \le x} I(\widetilde{Y}_\bullet(s) > 0)\, d\Lambda_F(s)$$
$$= \widetilde{\Lambda}(x) - \widetilde{\Lambda}^*(x),$$

where the third equality holds because the two processes $N_{\bullet u}(s)$ and $\widetilde{Y}_{\bullet}(s)$ have common discontinuity points.

The estimator $\widetilde{\Lambda}(x)$ is called Nelson-Aalen estimator of the cumulative hazard function Λ_F corresponding to the d.f. F. It is a negatively biased estimator. To see this, let $H(x) = P(\widetilde{X}_i \le x)$, $x \in \mathbb{R}$. Note that $\widetilde{Y}_{\bullet}(s) \sim_D \text{Bin}(n, 1 - H(s_-))$. Because $E(\mathcal{M}_u(x)) = 0$,

$$E(\widetilde{\Lambda}(x)) = E(\widetilde{\Lambda}^*(x)) = \int_{s \le x} \Big(1 - P(\widetilde{Y}_{\bullet}(s) > 0)\Big)\, d\Lambda_F(s) = \Lambda_F(x) - \int_{s \le x} H^n(s_-)\, d\Lambda_F(s).$$

Let $b \in \mathbb{R}$ be such that $H(b_-) < 1$ and $\Lambda_F(b) < \infty$. Then,

$$\sup_{x \le b} \left| E(\widetilde{\Lambda}(x)) - \Lambda_F(x) \right| \le H^n(b_-)\Lambda_F(b) \to 0,$$

exponentially fast. Thus, $\widetilde{\Lambda}(x)$ is negatively biased estimator of $\Lambda_F(x)$ and its bias tends to zero exponentially fast, uniformly in $x \le b$.

Furthermore, suppose there are no ties in the given data and let $\widetilde{X}_{(1)} < \widetilde{X}_{(2)} < \cdots < \widetilde{X}_{(n)}$ denote the ordered $\widetilde{X}_i$'s and $\delta_{(i)}$'s denote the corresponding δ's. Then,

$$\widetilde{\Lambda}(x) = \sum_{i=1}^{n} \frac{\delta_{(i)} I(\widetilde{X}_{(i)} \le x)}{\widetilde{Y}_{\bullet}(\widetilde{X}_{(i)})} = \sum_{i=1}^{n} \frac{\delta_{(i)} I(\widetilde{X}_{(i)} \le x)}{n - i + 1}.$$

11.5.2 Kaplan-Meier Estimator $\widetilde{F}$ of F

We continue to assume that there are no ties among $\widetilde{X}_i$'s. Analogous to the uncensored case (11.3.4), define $\widetilde{S}$ as a solution of the equation

$$\widetilde{S}(x) = 1 - \int_{s \le x} \widetilde{S}(s_-)\, d\widetilde{\Lambda}(s) = 1 - \int_{s \le x} \widetilde{S}(s_-) \frac{dN_{\bullet u}(s)}{\widetilde{Y}_{\bullet}(s)}.$$

Then the jump size of $\widetilde{S}$ at x is

$$-\Delta\widetilde{S}(x) = \widetilde{S}(x_-) - \widetilde{S}(x) = \widetilde{S}(x_-) \frac{\Delta N_{\bullet u}(x)}{\widetilde{Y}_{\bullet}(x)}.$$

Since $N_{\bullet u}$ jumps only at those $\widetilde{X}_{(i)}$'s with $\delta_{(i)} = 1$, i.e., only at uncensored observations, we see that $\widetilde{S}$ will also jump only at such observations. Thus $\widetilde{S}(x) = 1$ for x smaller than all uncensored observations and

$$\widetilde{S}(x) = \widetilde{S}(x-)\Big(1 - \frac{\Delta N_{\bullet u}(x)}{\widetilde{Y}_{\bullet}(x)}\Big), \quad x \in \mathbb{R}.$$

Suppose there are r uncensored observations, $\widetilde{X}_{(i_1)} < \widetilde{X}_{(i_2)} < \cdots < \widetilde{X}_{(i_r)}$. Then

$$1 - \frac{\Delta N_{\bullet u}(X_{(i_k)})}{\widetilde{Y}_{\bullet}(X_{(i_k)})} = 1 - \frac{1}{n - i_k + 1} = \frac{n - i_k}{n - i_k + 1}, \quad k = 1, \cdots, r.$$

This in turn leads to

$$\widetilde{S}(X_{(i_k)}) = \prod_{j=1}^{k} \frac{n - i_j}{n - i_j + 1}, \qquad k = 1, \cdots, r.$$

From this and the piecewise constant nature of $\widetilde{S}$ we conclude

$$\widetilde{S}(x) = \prod_{i:\widetilde{X}_{(i)} \le x} \left(\frac{n-i}{n-i+1}\right)^{\delta_{(i)}}. \tag{11.5.2}$$

Accordingly, the corresponding estimator of F is

$$\widetilde{F}(x) := 1 - \prod_{i;\widetilde{X}_{(i)} \le x} \left(\frac{n-i}{n-i+1}\right)^{\delta_{(i)}}. \tag{11.5.3}$$

The estimators $\widetilde{S}$ and $\widetilde{F}$ are called the Kaplan-Meier (KM) estimators of S and F, respectively. If there is no censoring, $\delta_i = 1$ for all $i = 1, \cdots, n$ and $\widetilde{F}(x)$ reduces to the empirical d.f. $F_n(x)$ of (2.4.2).

We shall now give the following three examples of computing $\widetilde{S}$. In each example $n = 5$.
(a) $\delta_{(1)} = 1,\ \delta_{(2)} = 1,\ \delta_{(3)} = 0,\ \delta_{(4)} = 1,\ \delta_{(5)} = 1.$
(b) $\delta_{(1)} = 1,\ \delta_{(2)} = 0,\ \delta_{(3)} = 0,\ \delta_{(4)} = 1,\ \delta_{(5)} = 1.$
(c) $\delta_{(1)} = 1,\ \delta_{(2)} = 1,\ \delta_{(3)} = 0,\ \delta_{(4)} = 1,\ \delta_{(5)} = 0.$
In the cases (a), (b) and (c), $\widetilde{S}$ is given as follows, respectively.

$$\begin{aligned}
\widetilde{S}(x) &= 1, & x &< \widetilde{X}_{(1)},\\
&= \left(\tfrac{4}{5}\right), & \widetilde{X}_{(1)} \le x &< \widetilde{X}_{(2)},\\
&= \left(\tfrac{4}{5}\right)\left(\tfrac{3}{4}\right), & \widetilde{X}_{(2)} \le x &< \widetilde{X}_{(3)},\\
&= \left(\tfrac{4}{5}\right)\left(\tfrac{3}{4}\right), & \widetilde{X}_{(3)} \le x &< \widetilde{X}_{(4)},\\
&= \left(\tfrac{4}{5}\right)\left(\tfrac{3}{4}\right)\left(\tfrac{1}{2}\right), & \widetilde{X}_{(4)} \le x &< \widetilde{X}_{(5)},\\
&= 0, & x &\ge \widetilde{X}_{(5)}.\\
\widetilde{S}(x) &= 1, & x &< \widetilde{X}_{(1)},\\
&= \left(\tfrac{4}{5}\right), & \widetilde{X}_{(1)} \le x &< \widetilde{X}_{(2)},\\
&= \left(\tfrac{4}{5}\right), & \widetilde{X}_{(2)} \le x &< \widetilde{X}_{(4)},\\
&= \left(\tfrac{4}{5}\right)\left(\tfrac{1}{2}\right), & \widetilde{X}_{(4)} \le x &< \widetilde{X}_{(5)},\\
&= 0, & x &\ge \widetilde{X}_{(5)}.\\
\widetilde{S}(x) &= 1, & x &< \widetilde{X}_{(1)},\\
&= \left(\tfrac{4}{5}\right), & \widetilde{X}_{(1)} \le x &< \widetilde{X}_{(2)},\\
&= \left(\tfrac{4}{5}\right)\left(\tfrac{3}{4}\right), & \widetilde{X}_{(2)} \le x &< \widetilde{X}_{(3)},\\
&= \left(\tfrac{4}{5}\right)\left(\tfrac{3}{4}\right), & \widetilde{X}_{(3)} \le x &< \widetilde{X}_{(4)},\\
&= \left(\tfrac{4}{5}\right)\left(\tfrac{3}{4}\right)\left(\tfrac{1}{2}\right), & x &\ge \widetilde{X}_{(4)}.
\end{aligned}$$

A few comments are in order. Let $\tau_F := \sup\{x; F(x) < 1\}$. If the support of F is $\mathbb{R}$, then $\tau_F = \infty$. First, observe that we did not need to know the actual values of $\widetilde{X}_{(i)}$'s to compute this estimator.

Secondly, the estimator jumps at only uncensored observations. The third point and perhaps the most important point is that it matters whether the largest observation $\widetilde{X}_{(n)}$ is censored or not. In the first two examples, it was uncensored and the estimator $\widetilde{S}(x) = 0$, for all $x \geq X_{(n)}$. This is consistent with the fact that $S(\tau_F) = 0$. In the last example the largest observation is censored and $\widetilde{S}(x) = 3/10$ for $x \geq \widetilde{X}_{(5)}$. In other words, here $\widetilde{F}$ puts the mass of $3/10$ at τ_F.

This phenomenon is true for any sample size. If $\widetilde{X}_{(n)}$ is censored, then $\widetilde{F}$ puts a positive mass at τ_F. This is the source of many technical difficulties when deriving various theoretical results about $\widetilde{F}$ or $\widetilde{S}$. It also creates the difficulty in applications. Often in practice to circumvent these technical difficulties one assumes that the last observation is uncensored.

Recall $H(x) = P(\widetilde{X} \leq x) = 1 - P(X > x, U > x) = 1 - (1 - F(x))(1 - G(x))$. Gill (1980) showed that $-F(x)H^n(x_-) \leq E\big(\widetilde{F}(x) - F(x)\big) \leq 0, \forall\, x,\, n \geq 2$. Thus, if $F(x_-) \vee G(x_-) < 1$, then $H(x_-) < 1$ and $\widetilde{F}(x)$ is a negatively biased estimator of $F(x)$ with the bias tending to zero exponentially fast.

Let $J(s) := I\big(\widetilde{Y}_\bullet(s) > 0\big)$. Gill (1980) derived the following expression for the variance of $\widetilde{F}$:

$$\text{Var}\big(\widetilde{F}(x)\big) = S^2(x)E\Big(\int_{s\leq x} \frac{\widetilde{S}^2(s)}{S(s)}\frac{J(s)}{\widetilde{Y}_\bullet(s)}\frac{1}{S(s)^2}dF(s)\Big).$$

To proceed further we need to describe the KM estimator of the censoring d.f. G. By symmetry, where one interchanges the role of X's and U's, the KM estimator of G is given by

$$\widetilde{G}(x) := 1 - \prod_{i:\widetilde{X}_{(i)}\leq x} \Big(\frac{n-i}{n-i+1}\Big)^{1-\delta_{(i)}}. \tag{11.5.4}$$

Gill (1983) also proposed the following estimator of the above variance:

$$\begin{aligned}
\widetilde{V}(x) :&= \widetilde{S}^2(x)\int_{s\leq x} \frac{J(s)}{\widetilde{Y}_\bullet(s)}\frac{1}{\widetilde{S}(s)}\,d\widetilde{F}(s) \\
&= \widetilde{S}^2(x)\sum_{j=1}^{n} \frac{J(\widetilde{X}_{(j)})}{\widetilde{Y}_\bullet(\widetilde{X}_{(j)})\widetilde{S}(\widetilde{X}_{(j)})}\frac{\delta_{(j)}}{n(1-\widetilde{G}(\widetilde{X}_{(j)-}))} \\
&= \widetilde{S}^2(x)\sum_{j=1}^{n} \frac{1}{n-j+1}\frac{1}{\prod_{i=1}^{j}\Big(\frac{n-i}{n-i+1}\Big)^{\delta_{(i)}}}\frac{\delta_{(j)}}{n\prod_{i=1}^{j}\Big(\frac{n-i}{n-i+1}\Big)^{1-\delta_{(i)}}}.
\end{aligned}$$

Consistency. Gill (1980) proved the following results about the consistency of $\widetilde{F}$ and $\widetilde{\Lambda}$.

(a) $\sup_{s\leq x}\big|\widetilde{\Lambda}(s) - \Lambda_F(s)\big| = o_p(1)$, (b) $\sup_{s\leq x}\big|\widetilde{F}(s) - F(s)\big| = o_p(1)$, $\forall\, x < \tau_H$.

(c) If $F(\tau_H) = F(\tau_{H_-})$ and $G(\tau_H) < 1$, then $\sup_{s\leq \tau_H}\big|\widetilde{F}(s) - F(s)\big| = o_p(1)$.

In the next section we discuss the estimation of the mean of F under random right censoring.

11.5.3 Susarla-Van Ryzin Estimator of the Mean

Considering the problem of estimating the mean $\mu := \int x dF(x)$ in the presence of right random censoring. Susarla and Van Ryzin (1980) were the first to propose an estimator of μ as follows.

Assuming $G(x_-) < 1$, for all $x \in \mathbb{R}$, they observed that

$$E\Big(\frac{\delta \widetilde{X}}{1 - G(\widetilde{X}_-)}\Big) = \iint \frac{x I(x \le u)}{1 - G(x_-)} \, dG(u)\, dF(x) = \mu. \tag{11.5.5}$$

Hence, if G is known then an estimator of μ is the average

$$\frac{1}{n} \sum_{i=1}^{n} \frac{\delta_i \widetilde{X}_i}{1 - G(\widetilde{X}_{i-})}.$$

But G is rarely known. Susarla and Van Ryzin replaced G by its KM estimator $\widetilde{G}$ of (11.5.4) in this formula, which leads to the estimator

$$\widetilde{\mu} := \frac{1}{n} \sum_{i=1}^{n} \frac{\delta_i \widetilde{X}_i}{1 - \widetilde{G}(\widetilde{X}_{i-})}. \tag{11.5.6}$$

On the other hand, a natural estimator of μ is the plug in estimator, where one replaces F by its KM estimator in the definition of μ, i.e., one estimates μ by $\widehat{\mu} := \int x \, d\widetilde{F}(x)$. We shall show that this estimator is the same as the Susarla–Van Ryzin estimator $\widehat{\mu}$. To show this, we need to know the jumps of $\widetilde{F}$. The jump of $\widetilde{F}$ at $\widetilde{X}_{(k)}$ for $1 \le k \le n$ is

$$\begin{aligned}
J_k &= \widetilde{F}(\widetilde{X}_{(k)}) - \widetilde{F}(\widetilde{X}_{(k)_-}) = \widetilde{S}(\widetilde{X}_{(k)_-}) - \widetilde{S}(\widetilde{X}_{(k)}) \qquad (11.5.7)\\
&= \prod_{i=1}^{k-1} \Big(\frac{n-i}{n-i+1}\Big)^{\delta_{(i)}} - \prod_{i=1}^{k} \Big(\frac{n-i}{n-i+1}\Big)^{\delta_{(i)}} \\
&= \prod_{i=1}^{k-1} \Big(\frac{n-i}{n-i+1}\Big)^{\delta_{(i)}} \Big[1 - \Big(\frac{n-k}{n-k+1}\Big)^{\delta_{(k)}}\Big] \\
&= \prod_{i=1}^{k-1} \Big[\Big(\frac{n-i+1}{n-i}\Big)^{1-\delta_{(i)}} \Big(\frac{n-i}{n-i+1}\Big)\Big] \frac{\delta_{(k)}}{n-k+1} \\
&= \prod_{i=1}^{k-1} \Big(\frac{n-i}{n-i+1}\Big) \frac{1}{n-k+1} \prod_{i=1}^{k-1} \Big(\frac{n-i+1}{n-i}\Big)^{1-\delta_{(i)}} \delta_{(k)} \\
&= \frac{1}{n} \frac{\delta_{(k)}}{\prod_{i=1}^{k-1} \Big(\frac{n-i}{n-i+1}\Big)^{1-\delta_{(i)}}} = \frac{1}{n} \frac{\delta_{(k)}}{1 - \widetilde{G}(\widetilde{X}_{(k)-})}.
\end{aligned}$$

Hence,

$$\widehat{\mu} = \frac{1}{n} \sum_{k=1}^{n} \frac{\delta_{(k)}}{1 - \widetilde{G}(\widetilde{X}_{(k)-})} \widetilde{X}_{(k)} = \frac{1}{n} \sum_{i=1}^{n} \frac{\delta_i \widetilde{X}_i}{1 - \widetilde{G}(\widetilde{X}_{i-})} = \widetilde{\mu}.$$

An alternate estimator of μ. Another estimator of μ is obtained by replacing X_i by its least squares predictor $X_i^* := E\big(X_i \big| \widetilde{X}_i, \delta_i\big)$, given the data $(\widetilde{X}_i, \delta_i), i = 1, \cdots, n$. The calculation of X_i^* is a special case of the following more general result whose proof is left as an exercise.

Lemma 11.5.1. *Let X and U be independent r.v.'s with respective d.f.'s F and G and set $\widetilde{X} = X \wedge U$ and $\delta = I(X \le U)$. Let h be a function such that $h(X, U)$ has finite expectation. Then*

$$E\big(h(X, U) | \tilde{X}, \delta\big) = \delta \, \frac{\int_{s \ge \widetilde{X}} h(\widetilde{X}, s)\, dG(s)}{1 - G(\widetilde{X}_-)} + (1 - \delta) \, \frac{\int_{s > \widetilde{X}} h(s, \widetilde{X})\, dF(s)}{1 - F(\widetilde{X})}.$$

Assume F to be continuous. Lemma 11.5.1 applied with $h(x,u) = x$ yields the identity

$$X_i^* = \widetilde{X}_i\delta_i + (1-\delta_i)\frac{\int_{x>\widetilde{X}_i} x\,dF(x)}{1-F(\widetilde{X}_i)} = \widetilde{X}_i\delta_i + (1-\delta_i)\frac{\int_{x>\widetilde{X}_i} x\,dF(x)}{1-F(\widetilde{X}_{i-})}.$$

Then $n^{-1}\sum_{i=1}^n X_i^*$ would be an unbiased estimator of μ, provided F is known. Since F is unknown, we plug in the KM estimator $\widetilde{F}$ for F in X_i^* and denote the new variables by $\widehat{X}_i^*$. The alternate estimator of μ is

$$\widehat{\mu}_* = \frac{1}{n}\sum_{i=1}^n \widehat{X}_i^*.$$

We shall show that this estimator is the same as $\widetilde{\mu}$. By interchanging the role of X_i's and U_i's in J_k of (11.5.7), the jumps $J_{ci} := \widetilde{G}(\widetilde{X}_{(i)}) - G(\widetilde{X}_{(i)-})$ of the KM estimator $\widetilde{G}$ of G are

$$J_{ci} := \frac{1}{n}\frac{1-\delta_{(i)}}{1-\widetilde{F}(\widetilde{X}_{(i)-})}, \quad 1 \le i \le n. \tag{11.5.8}$$

Hence, for all $y \in \mathbb{R}$,

$$\widetilde{G}(y_-) = \int_{x<y} d\widetilde{G}(x) = \frac{1}{n}\sum_{i=1}^n \frac{1-\delta_{(i)}}{1-\widetilde{F}(\widetilde{X}_{(i)-})} I(\widetilde{X}_{(i)} < y). \tag{11.5.9}$$

Now consider

$$\begin{aligned}
\widehat{\mu}_* &= \frac{1}{n}\sum_{i=1}^n \widehat{X}_i^* = \frac{1}{n}\sum_{i=1}^n \widetilde{X}_{(i)}\delta_{(i)} + \frac{1}{n}\sum_{i=1}^n (1-\delta_{(i)})\frac{\int_{x>\widetilde{X}_{(i)}} x\,d\widetilde{F}(x)}{1-\widetilde{F}(\widetilde{X}_{(i)-})} \\
&= \int x(1-\widetilde{G}(x_-))\,d\widetilde{F}(x) + \frac{1}{n}\sum_{i=1}^n \frac{(1-\delta_{(i)})}{1-\widetilde{F}(\widetilde{X}_{(i)-})}\frac{1}{n}\sum_{j=1}^n \frac{I(\widetilde{X}_{(j)} > \widetilde{X}_{(i)})\widetilde{X}_{(j)}\delta_{(j)}}{1-\widetilde{G}(\widetilde{X}_{(j)-})} \\
&= \int x(1-\widetilde{G}(x_-))\,d\widetilde{F}(x) + \frac{1}{n}\sum_{j=1}^n \frac{\widetilde{X}_{(j)}\delta_{(j)}}{1-\widetilde{G}(\widetilde{X}_{(j)-})}\sum_{i=1}^n \frac{1}{n}\frac{(1-\delta_{(i)})}{1-\widetilde{F}(\widetilde{X}_{(i)-})} I(\widetilde{X}_{(i)} < \widetilde{X}_{(j)}) \\
&= \int x(1-\widetilde{G}(x-))\,d\widetilde{F}(x) + \frac{1}{n}\sum_{j=1}^n \frac{\widetilde{X}_{(j)}\delta_{(j)}}{1-\widetilde{G}(\widetilde{X}_{(j)-})}\widetilde{G}(\widetilde{X}_{(j)-}) \\
&= \int x(1-\widetilde{G}(x_-))\,d\widetilde{F}(x) + \int x\widetilde{G}(x_-)\,d\widetilde{F}(x) = \int x\,d\widetilde{F}(x) = \widehat{\mu} = \widetilde{\mu}.
\end{aligned}$$

In the above, the last but second equality follows from (11.5.9).

The Susarla-Van Ryzin estimator $\widetilde{\mu}$ of μ was initially obtained by arguing intuitively using the identity (11.5.5). At the time the formula for the jumps of $\widetilde{F}$ was not readily available. As we have seen above it is equivalent to the plug-in estimator of $\mu = \int x\,dF(x)$ where F is replace by the KM estimators $\widetilde{F}$. Later on we will show that $\widetilde{F}$ is the nonparametric MLE of F and hence the Susarla-Van Ryzin estimator $\widetilde{\mu}$ is a nonparametric MLE of μ.

In general, if $J(F)$ is a real-valued functional of F then the nonparametric MLE of $J(F)$ is $J(\widetilde{F})$. For example, the median $\widetilde{m}_n$ of $\widetilde{F}_n$ would be a nonparametric MLE of the median of F.

Asymptotic distribution of $\widetilde{\mu}$. Assume $U \wedge X \geq 0$ so that F and G are d.f.'s satisfying $F(t) = G(t) = 0$ for $t < 0$, $\mu := \int_0^\infty (1 - F(x))\,dx$ and $\tau_F = \infty = \tau_G = \tau_H$. Let

$$\mu(x) = \int_{0 \leq s \leq x} (1 - F(s))\,ds, \qquad \bar{\mu}(x) := \int_{s > x} (1 - F(s))\,ds,$$
$$\mu_n(x) := \int_{0 \leq s \leq x} (1 - \widetilde{F}(s))\,ds, \qquad x \geq 0.$$

We have

$$n^{1/2}(\mu_n - \mu)(x) = -\int_{0 \leq s \leq x} n^{1/2}(\widetilde{F}(s) - F(s))\,ds, \quad x \geq 0.$$

The following result is Corollary 3.2 of Gill (1983), where $T := \widetilde{X}_{(n)}$. Let

$$C(x) := \int_{s \leq x} \frac{dF(s)}{(1 - F(s_-))^2 (1 - G(s_-))}, \qquad x \in \mathbb{R}.$$

Theorem 11.5.2. *If $\sigma^2 := \int_0^\infty \bar{\mu}^2(s)\,dC(s)$ is finite and $n^{1/2}\bar{\mu}(T) \to_p 0$, then*

$$n^{1/2}(\mu_n(T) - \mu) \to_D \mathcal{N}(0, \sigma^2). \tag{11.5.10}$$

Because

$$\mu_n(T) = \int_{s \leq T} s d\widetilde{F}(s) = \frac{1}{n} \sum_{i=1}^{n} \frac{\delta_i \widetilde{X}_i}{1 - G(\widetilde{X}_{i_-})} I(\widetilde{X}_i \leq T) = \widetilde{\mu},$$

the result (11.5.10) is equivalent to saying that $n^{1/2}(\widetilde{\mu} - \mu) \to_D \mathcal{N}(0, \sigma^2)$.

Suppose F is continuous, $M := \sup_{x \geq 0} \bar{\mu}(x)/\bar{F}(x)$ is finite and the following condition holds.

$$\int_0^\infty \frac{dF(s)}{1 - G(s_-)} < \infty. \tag{11.5.11}$$

Then,

$$\sigma^2 = \int_0^\infty \bar{\mu}^2(s)\,dC(s) = \int_0^\infty \frac{\bar{\mu}^2(s)\,dF(s)}{(1 - F(s))^2 (1 - G(s_-))} \leq M^2 \int_0^\infty \frac{dF(s)}{1 - G(s_-)} < \infty.$$

Also, $\bar{\mu}(T) \leq M\bar{F}(T)$ and hence $n^{1/2}\bar{\mu}(T) \to_p 0$ is implied by $n^{1/2}\bar{F}(T) \to_p 0$.

Alternately, because $\bar{\mu}(x) = \int I(s > x)\bar{F}(s)\,ds$, by Theorem 2.3.2,

$$E(\bar{\mu}(T)) = \int E(I(s > T)\bar{F}(s)\,ds = \int_0^\infty H^n(s_-)\bar{F}(s)\,ds,$$
$$n^{1/2}E(\bar{\mu}(T)) = \int_0^\infty n^{1/2}H^n(s_-)\bar{F}(s)\,ds.$$

By (2.4.8), $n^{1/2}\bar{\mu}(T) \to_p 0$ will be implied by

$$\int_0^\infty n^{1/2}H^n(s_-)\bar{F}(s)\,ds \to 0. \tag{11.5.12}$$

Example 11.5.1. Suppose $\bar{F}(x) = e^{-\theta x}$, $\bar{G}(x) = e^{-\gamma x}$, $x > 0$, for some positive θ and γ. Then, $\tau_H = \infty$ and

$$\int_0^\infty \frac{dF(s)}{1 - G(s_-)} = \theta \int_0^\infty e^{-(\theta-\gamma)s}\, ds = \frac{\theta}{\theta - \gamma} I(\theta > \gamma) + \infty I(\theta \le \gamma).$$

Hence, here (11.5.11) and the asymptotic normality result of (11.5.10) holds only when $\theta > \gamma$. Moreover, in this case

$$\begin{aligned}
\bar{\mu}(x) &= \int_{s>x} e^{-\theta s}\, ds = \frac{e^{-\theta s}}{\theta}, \\
C(x) &= \int_{s \le x} \frac{dF(s)}{\big(1 - F(s)\big)^2 \big(1 - G(s_-)\big)} = \theta \int_{0<s\le x} e^{(\theta+\gamma)s}\, ds = \theta \frac{e^{(\theta+\gamma)x} - 1}{\theta + \gamma}, \\
\sigma^2 &= \int_0^\infty \bar{\mu}^2(s) dC(s) = \frac{1}{\theta} \int e^{-(\theta-\gamma)s} ds = \frac{1}{\theta(\theta - \gamma)} < \infty.
\end{aligned}$$

Example 11.5.2. Suppose F is continuous, has positive and finite variance, and there is no censoring, i.e., G puts all its mass at ∞. Then, $\widetilde{\mu} = \bar{X}_n = n^{-1} \sum_{i=1}^n X_i$. Moreover,

$$\begin{aligned}
\sigma^2 &= \int_0^\infty \bar{\mu}^2(s)\, dC(s) = \int \Big(\int_{y>s} \bar{F}(y) dy \Big)^2 \frac{dF(s)}{\big(1 - F(s)\big)^2} \\
&= 2 \int_0^\infty \iint_{s<y<z<\infty} \bar{F}(y) \bar{F}(z)\, dy\, dz \frac{dF(s)}{\big(\bar{F}(s)\big)^2} \\
&= 2 \iint_{0<y<z<\infty} \bar{F}(y) \bar{F}(z)\, dy\, dz \int_0^y d\big(- \big(\bar{F}(s)\big)^{-1}\big) \\
&= 2 \iint_{0<y<z<\infty} F(y) \bar{F}(z)\, dy\, dz = \mathrm{Var}(X),
\end{aligned}$$

where $X \sim_D F$. See Exercise 1 at the end of this chapter for the proof of the last equality.

11.6 Self-consistency of $\widetilde{F}$

Recall from (2.4.2), $nF_n(x) = \sum_{i=1}^n I(X_i \le x)$. A d.f. F_0 is said to be a self-consistent estimator of F, if and only if,

$$F_0(x) = E_{F_0}\Big(F_n(x) \big| (\widetilde{X}_i, \delta_i), 1 \le i \le n\Big) = \frac{1}{n} \sum_{i=1}^n E_{F_0}\Big(I(X_i \le x) \big| \widetilde{X}_i, \delta_i\Big), \quad \forall x \in \mathbb{R}.$$

The last equality follows because $(\widetilde{X}_i, \delta_i), 1 \le i \le n$ are independent. Lemma 11.5.1 applied with $h(X, U) = I(X \le x)$ gives the identity

$$\begin{aligned}
E_{F_0}\Big(I(X \le x) \big| \widetilde{X}, \delta\Big) &= \delta I(\widetilde{X} \le x) + (1 - \delta)\Big(1 - \frac{S_0(x)}{S_0(\widetilde{X})}\Big) I(\widetilde{X} \le x) \\
&= I(\widetilde{X} \le x) - (1 - \delta) \frac{S_0(x)}{S_0(\widetilde{X})} I(\widetilde{X} \le x).
\end{aligned}$$

Therefore, F_0 is a solution of the self-consistency equation if the corresponding survival function S_0 satisfies the following equation, where $\widetilde{N}(x) := \sum_{i=1}^{n} I(\widetilde{X}_{(i)} > x)$.

$$\begin{aligned} S_0(x) &= \frac{1}{n}\sum_{i=1}^{n} I(\widetilde{X}_i > x) + \frac{1}{n}\sum_{i=1}^{n}(1-\delta_i)\frac{S_0(x)}{S_0(\widetilde{X}_i)} I(\widetilde{X}_i \le x) \\ &= \frac{1}{n}\widetilde{N}(x) + \frac{1}{n}\sum_{i=1}^{n}(1-\delta_{(i)})\frac{S_0(x)}{S_0(\widetilde{X}_{(i)})} I(\widetilde{X}_{(i)} \le x), \quad \forall x \in \mathbb{R}. \end{aligned} \tag{11.6.1}$$

Theorem 11.6.1. *Suppose $\widetilde{X}_{(1)} < \widetilde{X}_{(2)} < \cdots < \widetilde{X}_{(n)}$ holds. Then the KM Estimator $\widetilde{S}(x)$ of* (11.5.2) *is the unique solution of the self-consistency equation* (11.6.1) *for all $x < \widetilde{X}_{(n)}$.*

Proof. Fix an $x \in \mathbb{R}$. The equation (11.6.1) is equivalent to

$$S_0(x)\left(1 - \frac{1}{n}\sum_{i=1}^{n}\frac{1-\delta_{(i)}}{S_0(\widetilde{X}_{(i)})} I(\widetilde{X}_{(i)} \le x)\right) = \frac{\widetilde{N}(x)}{n}. \tag{11.6.2}$$

Thus, if $x < \widetilde{X}_{(1)}$, then $S_0(x) = 1$. To proceed further, define

$$d_0 := n = \frac{\widetilde{N}(\widetilde{X}_{(1)-})}{S_0(\widetilde{X}_{(1)-})}, \quad d_k := n - \sum_{i=1}^{k}\frac{1-\delta_{(i)}}{S_0(\widetilde{X}_{(i)})}, \quad 1 \le k \le n-1.$$

By (11.6.2), for each $k = 1, \cdots, n-1$ and $\widetilde{X}_{(k)} \le x < \widetilde{X}_{(k+1)}$, we obtain

$$S_0(x) d_k = \widetilde{N}(x). \tag{11.6.3}$$

Hence, for each $k = 1, \cdots, n-1$,

$$d_k = d_{k-1} - \frac{1-\delta_{(k)}}{S_0(\widetilde{X}_{(k)})} = \frac{\widetilde{N}(\widetilde{X}_{(k)-})}{S_0(\widetilde{X}_{(k)-})} - \frac{1-\delta_{(k)}}{S_0(\widetilde{X}_{(k)})}. \tag{11.6.4}$$

Thus, if $\delta_{(k)} = 0$, then, by first using (11.6.3) and then (11.6.4), we derive

$$\widetilde{N}(\widetilde{X}_{(k)_-}) - 1 = \widetilde{N}(\widetilde{X}_{(k)}) = S_0(\widetilde{X}_{(k)}) d_k = S_0(\widetilde{X}_{(k)})\frac{\widetilde{N}(\widetilde{X}_{(k)-})}{S_0(\widetilde{X}_{(k)_-})} - 1,$$

which is equivalent to $S_0(\widetilde{X}_{(k)}) = S_0(\widetilde{X}_{(k)_-})$. Similarly, if $\delta_{(k)} = 1$, then

$$S_0(\widetilde{X}_{(k)}) = \frac{\widetilde{N}(\widetilde{X}_{(k)})}{d_{k-1}} = \frac{\widetilde{N}(\widetilde{X}_{(k)})}{\widetilde{N}(\widetilde{X}_{(k)-})}\frac{\widetilde{N}(\widetilde{X}_{(k)-})}{d_{k-1}} = \frac{\widetilde{N}(\widetilde{X}_{(k)})}{\widetilde{N}(\widetilde{X}_{(k)-})} S_0(\widetilde{X}_{(k)-}).$$

The above give the identity

$$\frac{S_0(\widetilde{X}_{(k)})}{S_0(\widetilde{X}_{(k)-})} = \left(\frac{n-k}{n-k+1}\right)^{\delta_{(k)}}, \qquad k = 1, \cdots, n-1.$$

This relation is the same as given at (11.5.2), as was to be proved. □

11.7 Nonparametric MLE under Random Censoring

We shall now show that $\widetilde{F}$ is also the nonparametric MLE of F in a more general setting where ties in the observations are allowed. In the previous sections we had assumed there are no ties among $\widetilde{X}_i$'s. The details below are as in Kaplan and Meier (1958).

Suppose n units are put on a life test. Let there be k distinct times of death (non-censoring times) denoted by $Z_1 < Z_2 < \cdots < Z_k$. Let d_j denote the number of deaths at Z_j, n_j the number of units alive just before Z_j (and uncensored) and L_i the censoring time of the ith unit, whose lifetime is not observed. If the censoring time L_i and the uncensoring time Z_j coincide, always place L_i on the immediate right of the Z_j. Then, the KM estimator of $S(x)$ is

$$\widehat{S}(x) = \prod_{j:Z_j \le x}^{k} \frac{n_j - d_j}{n_j}. \tag{11.7.1}$$

If $d_j = 1$, for all j, then $k = n$, $Z_j = \widetilde{X}_{(j)}$, $n_j \equiv n - j + 1$, $L_j = U_j$ and

$$\widehat{S}(x) = \prod_{j;\widetilde{X}_j \le x}^{n} \left(\frac{n-j}{n-j+1}\right)^{\delta_{(j)}} = \widetilde{S}(x), \qquad \forall x \in \mathbb{R},$$

where $\widetilde{S}$ is as in the equation (11.5.2).

Now, let $u_{1j}, u_{2j}, \cdots, u_{r_j j}$ denote the censoring times of r_j units in the interval $(Z_{j-1}, Z_j]$. Then the likelihood is given by

$$L(S) = \prod_{j=1}^{k} \left(\prod_{i=1}^{r_j} S(u_{ij\_})(S(Z_j-) - S(Z_j))^{d_j} \right) \prod_{i=1}^{r_{k+1}} S(u_{i,k+1}-).$$

We have the inequalities $S(Z_j-) \le S(u_{ij}-) \le S(Z_{j-1})$, $i = 1, \cdots, r_j$, for $j = 1, \cdots, k$ and $S(u_{i,k+1}-) \le S(Z_j)$ for $i = 1, \cdots, r_{k+1}$. Thus, if we set $s_0 = 1$ and $s_j = S(Z_j)$, $j = 1, \ldots, k$, then we have the inequality

$$L(S) \le \prod_{j=1}^{k} s_{j-1}^{r_j} (s_{j-1} - s_j)^{d_j} s_k^{r_{k+1}} = \prod_{j=1}^{k} s_j^{r_j+1} (s_{j-1} - s_j)^{d_j}.$$

Now set $p_j = s_j / s_{j-1}$ and $q_j = 1 - p_j$. Then $s_j = \prod_{i=1}^{j} p_i$ and the above inequality can be written as

$$L(S) \le \prod_{j=1}^{k} (p_1 p_2 \cdots p_{j-1} q_j)^{d_j} (p_1 \cdots p_j)^{r_j+1} = \prod_{j=1}^{k} q_j^{d_j} p_j^{r_j+1} M$$

with

$$M = \prod_{j=1}^{k} \prod_{i=1}^{j-1} p_i^{d_j + r_{j+1}} = \prod_{i=1}^{k} p_i^{\sum_{j=i+1}^{k} (d_j + r_{j+1})} = \prod_{i=1}^{k} p_i^{n_i - d_i - r_{i+1}}.$$

as $n_i = \sum_{j=i}^{k} (d_j + r_{j+1})$. This establishes the inequality

$$L(S) \le \prod_{j=1}^{k} q_j^{d_j} p_j^{n_j - d_j}.$$

It is now easily checked that the above bound is maximized by $\widehat{p}_j = (n_j - d_j)/n_j$, $j = 1, \cdots, k$. Since $s_j = p_j s_{j-1}$, we obtain that the nonparametric MLE of S is the $\widehat{S}$ of (11.7.1).

11.8 Nonparametric Bayes Estimation of F

In this section we shall derive nonparametric Bayes estimators of the survival function when there is no censoring and when there is random censoring on the right. Let $\mathcal{F}$ denote the set of all d.f.'s on $\mathbb{R}$ and W be a d.f. on $\mathbb{R}$. We shall work with the loss function

$$L(F,G) := \int \left|F(t) - G(t)\right|^2 dW(t), \quad F,G \in \mathcal{F}.$$

For a Bayesian analysis we need a prior distribution on $\mathcal{F}$ resulting in a random d.f. $\mathbb{F}$, and a random sample $\xi_1, \cdots, \xi_n$ from a distribution Q_F, given $\mathbb{F} = F$. We then have to find a random d.f. $\mathbb{G}$ based on the observations $\xi_1, \cdots, \xi_n$, i.e., $\mathbb{G}(t) = g(t, \xi_1, \cdots, \xi_n)$ for some measurable function g, which minimizes the risk $E\big(L(\mathbb{F}, \mathbb{G})\big)$. Using the Fubini Theorem 2.3.2 and Example 5.4.1, we obtain the inequality

$$E\big(L(\mathbb{F},\mathbb{G})\big) = \int E\Big(\left|\mathbb{F}(t) - \mathbb{G}(t)\right|^2\Big)\, dW(t) \geq \int E\Big(\left|\mathbb{F}(t) - E\big(\mathbb{F}(t)|\xi_1, \cdots, \xi_n\big)\right|^2\Big)\, dW(t).$$

This shows that a minimizer is given by a random d.f. $\mathbb{G}$ satisfying $\mathbb{G}(t) = E\big(\mathbb{F}(t)|\xi_1, \cdots, \xi_n\big)$, a.s. Thus, to implement this we need a prior distribution and derive the corresponding posterior distribution. For this we follow Ferguson (1973) who introduced the Dirichlet prior, which he showed to be a conjugate prior. A prerequisite for all this is the understanding of Dirichlet distributions and Dirichlet processes. This will be provided next following the work of Ferguson (1973).

11.8.1 Dirichlet Distributions

Recall that for $\alpha > 0$, a Gamma$(\alpha, 1)$ r.v. has moment generating function $M_\alpha(t) = (1-t)^{-\alpha}$, $t < 1$, see Example 2.9.5. This function has pointwise limit $M_0(t) = 1$, $t < 1$, as α tends to zero from the right. As M_0 is the moment generating function of the constant 0, we conclude that if $X_\alpha \sim$ Gamma$(\alpha, 1)$, then $X_\alpha \to_D 0$ as α tends to zero from the right. In view of this, in the following we interpret Gamma$(0,1)$ as the distribution degenerate at 0 so that the family of distributions $\{\text{Gamma}(\alpha, 0), \alpha \geq 0\}$ is well defined.

Also recall that if $X_\alpha \sim_D$ Gamma$(\alpha, 1)$, $Y \sim_D$ Gamma$(\beta, 1)$, X_α and Y are independent, α and β are positive, then $X_\alpha/(X_\alpha + Y)$ has a Beta(α, β) distribution, see Example 2.9.5. If we let α tend to zero, then, by the above, $X_\alpha \to_D 0$ and this implies $X_\alpha/(X_\alpha + Y) \to_D 0$. In view of this, in the following we interpret Beta$(0, \beta)$ as the distribution degenerate at 0. For a later use, recall that if $Y \sim_D$ Beta(a, b), $a \geq 0, b > 0$, then $E(Y) = a/(a+b)$ and $E\big(Y^2\big) = a(a+1)/[(a+b)(a+b+1)]$.

Definition 11.8.1. $\big[$**Dirichlet distribution**$\big]$. Let $Z_1, \cdots, Z_k$ be independent r.v.'s with $Z_i \sim_D$ Gamma$(\alpha_i, 1)$, where $\alpha_1, \cdots, \alpha_k$ are nonnegative and $\alpha_\bullet = \alpha_1 + \cdots + \alpha_k$ is positive. Let

$$Y_i = \frac{Z_i}{Z_1 + \cdots + Z_k}, \quad 1 \leq i \leq k.$$

Then the joint distribution of $(Y_1, \cdots, Y_k)$, denoted by $\mathcal{D}(\alpha_1, \cdots, \alpha_k)$, is known as the Dirichlet distribution.

Lemma 11.8.1. *Let $(Y_1, \cdots, Y_k)$ be distributed according to $\mathcal{D}(\alpha_1, \cdots, \alpha_k)$. Then the following hold with $\alpha_\bullet$ as in the definition.*

(a) *If $\alpha_1, \cdots, \alpha_k$ are positive, then $(Y_1, \cdots, Y_{k-1})$ has density*

$$f(y_1, \cdots, y_{k-1}) := \frac{\Gamma(\alpha_\bullet)}{\prod_{i=1}^k \Gamma(\alpha_i)} \prod_{i=1}^{k-1} y_i^{\alpha_i - 1} \Big(1 - \sum_{j=1}^{k-1} y_j\Big)^{\alpha_k - 1} I\big((y_1, \cdots, y_{k-1}) \in \mathcal{S}\big).$$

where $\mathcal{S} := \big\{(y_1, \cdots, y_{k-1}) : y_1 > 0, \cdots, y_{k-1} > 0, \sum_{i=1}^{k-1} y_i < 1\big\}$.
(b) *For each $j = 1, \cdots, k$, $Y_j \sim_D \text{Beta}(\alpha_j, \alpha_\bullet - \alpha_j)$ and thus*

$$E(Y_j) = \frac{\alpha_i}{\alpha_\bullet}, \qquad E(Y_i^2) = \frac{\alpha_i(\alpha_i + 1)}{\alpha_\bullet(\alpha_\bullet + 1)},$$

(c) *Let $A_1, \cdots, A_m$ be a partition of $\{1, \cdots, k\}$ with each A_i nonempty. Then,*

$$\Big(\sum_{i \in A_1} Y_i, \cdots, \sum_{i \in A_m} Y_i\Big) \sim_D \mathcal{D}\Big(\sum_{i \in A_1} \alpha_i, \cdots, \sum_{i \in A_m} \alpha_i\Big).$$

(d) *For $1 \leq i < j \leq k$, we have*

$$E(Y_i Y_j) = \frac{\alpha_i \alpha_j}{\alpha_\bullet(\alpha_\bullet + 1)}.$$

We leave the proof of this lemma as an exercise.

11.8.2 Dirichlet Process

Let $\mathcal{X}$ be a nonempty set, $\mathscr{A}$ be a sigma field on $\mathcal{X}$, and α be a positive multiple of a probability measure on $\mathscr{A}$.

Definition 11.8.2. A process $\Pi := \{\Pi(A) : A \in \mathscr{A}\}$ indexed by the σ-field $\mathscr{A}$ is called a Dirichlet(α) process, and we write $\Pi \sim_D \mathcal{D}(\alpha)$, if for each finite $\mathscr{A}$-measurable partition $A_1, \cdots, A_k$ of $\mathcal{X}$

$$(\Pi(A_1), \cdots, \Pi(A_k)) \sim_D \mathcal{D}\big(\alpha(A_1), \cdots, \alpha(A_k)\big).$$

Ferguson (1973) showed that a Dirichlet process defines a probability measure on the set of all probability measures on $\mathscr{A}$ and then calculated the posterior distribution given a random sample from this random probability measure. Here is his result. Let η_x denote the degenerate distribution at x.

Theorem 11.8.1. *Let $\Pi \sim \mathcal{D}(\alpha)$, and given $\Pi = Q$, let $(X_1, \cdots, X_n)$ be a random sample from Q. Then, the conditional distribution of Π, given $X_1, \cdots, X_n$, is $\mathcal{D}\big(\alpha + \sum_{i=1}^n \eta_{X_i}\big)$.*

Let us now specialize this to the case when $(\mathcal{X}, \mathscr{A})$ is $(\mathbb{R}, \mathscr{B})$. Suppose that $\Pi \sim \mathcal{D}(\alpha)$ and given $\Pi = Q$, $(X_1 \cdots, X_n)$ is a random sample from F, the d.f. of the distribution Q. By Theorem 11.8.1, the posterior distribution of Π, given X, is $\mathcal{D}\big(\alpha + \sum_{i=1}^n \eta_{X_i}\big)$. In particular, with $\mathbb{F}$ defined by $\mathbb{F}(t) = \Pi((-\infty, t])$, $t \in \mathbb{R}$, this implies

$$\big(\mathbb{F}(t) \big| X_1, \cdots, X_n\big) \sim_D \text{Beta}\Big(\alpha(-\infty, t] + nF_n(t),\ \alpha(\mathbb{R}) + n - \alpha(-\infty, t] - nF_n(t)\Big),$$

with $F_n(t) = n^{-1} \sum_{i=1}^n I(X_i \leq t)$. Hence, the nonparametric Bayes estimator of F is

$$\widehat{F}(t) = \frac{\alpha(-\infty, t] + nF_n(t)}{\alpha(\mathbb{R}) + n} = p_n F_0(t) + (1 - p_n) F_n(t), \quad t \in \mathbb{R},$$

where $p_n = \alpha(\mathbb{R}) / \big(\alpha(\mathbb{R}) + n\big)$ and $F_0(t) = \alpha(-\infty, t] / \alpha(\mathbb{R})$.

As $n \to \infty$, $\|\widehat{F} - F_n\|_\infty \to 0$. Hence, by Lemma 2.12.2, $\widehat{F}$ is a strongly consistent estimator of F in the sup norm, i.e., $\|\widehat{F} - F\|_\infty \to_{a.s.} 0$. If $\alpha(\mathbb{R}) \to \infty$, then $\|\widehat{F} - F_0\|_\infty \to 0$.

The estimator $\widehat{F}$ is the Bayes estimator of F with respect to the Dirichlet prior $\mathcal{D}(\alpha)$ and the loss function L when there is no censoring. The next sub-section describes its analog for randomly right-censored data.

11.8.3 Susarla-Van Ryzin Estimator of F

We shall now describe the Bayes estimator of F under the right random censorship with respect to the loss function L and the prior $\mathcal{D}(\alpha)$. Susarla and Van Ryzin (1976) were the first to obtain this estimator.

Let Π be a Dirichlet(α) process and $\mathbb{F}(x) := \Pi(-\infty, x]$, $x \in \mathbb{R}$. Given $\Pi = Q$, let $X_1, \cdots, X_n$ be a random sample from F, the d.f. of Q. Furthermore, let $U_1, \cdots, U_n$ be independent r.v.'s that are independent of $X_1, \cdots, X_n$ and Π. We observe $(\delta_i, \widetilde{X}_i)$, $i = 1, \cdots, n$, where $\delta_i = I(X_i \le U_i)$ and $\widetilde{X}_i = X_i \wedge U_i$. The Bayes estimator of F with respect to the Dirichlet(α) prior and the loss function L is

$$\widehat{F}_{SV}(x) = E\Big\{\mathbb{F}(x)\big|(\delta_1, \widetilde{X}_1), \cdots, (\delta_n, \widetilde{X}_n)\Big\}, \quad x \in \mathbb{R}.$$

Susarla and Van Ryzin (1976) derived an expression for this estimator. To describe it here, we need some more notation. Suppose there are k uncensored observations. Arrange the $\widetilde{X}_i$'s so that $\widetilde{X}_1, \cdots \widetilde{X}_k$ are the uncensored observations and $\widetilde{X}_{k+1}, \cdots \widetilde{X}_n$ are the censored observations. Let $\widetilde{X}'_{k+1} < \widetilde{X}'_{k+2} < \cdots < \widetilde{X}'_m$ denote the distinct observations among $\widetilde{X}_{k+1}, \cdots, \widetilde{X}_n$. Let λ_j denote the number of censored observations that are equal to $\widetilde{X}'_j$, $j = k+1, \cdots, m$. Clearly, $\lambda_j \ge 1$, $\forall j = k+1, \cdots, m$. Let

$$\widetilde{Y}_\bullet(x) := \sum_{i=1}^n I(\widetilde{X}_i \ge x), \quad \widetilde{Y}_\bullet^+(x) := \sum_{i=1}^n I(\widetilde{X}_i > x), \quad x \in \mathbb{R}.$$

Then, the Susarla-Van Ryzin (SV) Bayes estimator $\widehat{F}_{SV}$ of F under right random censoring is $1 - \widehat{S}_{SV}$, where

$$\widehat{S}_{SV}(x) = \frac{\alpha[x, \infty) + \widetilde{Y}_\bullet^+(x)}{\alpha(\mathbb{R}) + n} \prod_{j=k+1}^{\ell} \frac{\alpha[\widetilde{X}'_j, \infty) + \widetilde{Y}_\bullet(\widetilde{X}'_j)}{\alpha[\widetilde{X}'_j, \infty) + \widetilde{Y}_\bullet(\widetilde{X}'_j) - \lambda_j},$$

for $\widetilde{X}'_\ell \le x < \widetilde{X}'_{\ell+1}$, $\ell = k, \cdots, m$, where $\widetilde{X}'_k := -\infty$, $\widetilde{X}'_{m+1} := \infty$.

Suppose there are no ties and all $\widetilde{X}_i$'s are distinct. Then $\lambda_j = 1$ for all $j = k+1, \cdots, m$ and for every $x \in \mathbb{R}$, the estimator $\widehat{S}_{SV}(x)$ converges to the Kaplan-Meyer estimator $\widetilde{S}(x)$ of (11.5.2), as $\alpha(\mathbb{R}) \to 0$.

Unlike the Kaplan-Meier estimator, the computation of the Bayes estimators, needs the knowledge of the $\widetilde{X}_i$'s and also the prior measure α. Both KM and SV Bayes estimators are discontinuous at uncensored observations. However, the SV Bayes estimator is smoother compared to the KM estimator at the censored observations, provided $\alpha[x, \infty)$ is continuous in x. To overcome not knowing α, Susarla and Van Ryzin (1978) discuss an empirical Bayes approach which provides a method of estimating α.

11.9 Randomly Right-Censored Linear Regression

In this section we shall consider the linear regression model when responses are censored at random on the right. The responses here could be log of the survival times.

Let $\boldsymbol{z}_i$ be $p \times 1$ non-random real vectors and suppose the responses Y_i's follow the linear regression model

$$Y_i = \boldsymbol{z}_i'\boldsymbol{\beta} + \varepsilon_i, \quad i = 1, \cdots, n, \tag{11.9.1}$$

for some $\boldsymbol{\beta} \in \mathbb{R}^p$. Here, the error variables $\varepsilon_1, \cdots, \varepsilon_n$ are i.i.d. F with mean zero and finite variance. Let $U_1, \cdots, U_n$ be independent r.v.'s and independent of $Y_1, \cdots, Y_n$. In the right randomly censored linear regression model one observes $(\delta_1, \widetilde{Y}_1), \cdots, (\delta_n, \widetilde{Y}_n)$, where $\widetilde{Y}_i := Y_i \wedge U_i$ and $\delta_i := I(Y_i \le U_i), 1 = 1, \cdots, n$. In what follows we always assume that the $n \times p$ matrix $\boldsymbol{\mathcal{Z}}$, whose ith row is $\boldsymbol{z}_i'$, has rank p. This implies that $\boldsymbol{\mathcal{Z}}'\boldsymbol{\mathcal{Z}}$ is positive definite.

We shall discuss two estimators of $\boldsymbol{\beta}$. One is by Buckley and James (1979) (BJ), and the other is by Koul, Susarla and Van Ryzin (1981).

We begin with describing the BJ estimator. Let $\boldsymbol{Y} := (Y_1, \cdots, Y_n)'$. Recall that in the uncensored case the least squares estimator is obtained by solving for $\boldsymbol{b}$ the normal equations $\boldsymbol{\mathcal{Z}}'\boldsymbol{\mathcal{Z}}\boldsymbol{b} = \boldsymbol{\mathcal{Z}}'\boldsymbol{Y}$. In these equations the only thing that is not known when data is randomly right censored is the vector $\boldsymbol{Y}$. A natural predictor of this vector is $\boldsymbol{Y}^* := (Y_1^*, \cdots, Y_n^*)'$, where $Y_i^* = E(Y_i|\delta_i, \widetilde{Y}_i)$. Thus one can define an estimator of $\boldsymbol{\beta}$ to be a solution of the equations $\boldsymbol{\mathcal{Z}}'\boldsymbol{\mathcal{Z}}\boldsymbol{b} = \boldsymbol{\mathcal{Z}}'\boldsymbol{Y}^*$. In order to implement this procedure we need to get a more accurate description of $\boldsymbol{Y}^*$. For that purpose, let

$$Y_i(\boldsymbol{b}) := \frac{\int_{\{y > \widetilde{Y}_i - \boldsymbol{z}_i'\boldsymbol{b}\}} y\, dF(y)}{1 - F(\widetilde{Y}_i - \boldsymbol{z}_i'\boldsymbol{b})}, \qquad e_i(\boldsymbol{b}) := \delta_i(\widetilde{Y}_i - \boldsymbol{z}_i'\boldsymbol{b}) + (1 - \delta_i)Y_i(\boldsymbol{b}).$$

Then, for all $1 \le i \le n$, we obtain, with the help of Lemma 11.5.1, the identity

$$\begin{aligned} Y_i^*(\boldsymbol{\beta}) :&= E(Y_i \mid \delta_i, \widetilde{Y}_i) = \boldsymbol{z}_i'\boldsymbol{\beta} + E(Y_i - \boldsymbol{z}_i'\boldsymbol{\beta} \mid \delta_i, \widetilde{Y}_i - \boldsymbol{z}_i'\boldsymbol{\beta}) \\ &= \boldsymbol{z}_i'\boldsymbol{\beta} + \delta_i(\widetilde{Y}_i - \boldsymbol{z}_i'\boldsymbol{\beta}) + (1 - \delta_i)Y_i(\boldsymbol{\beta}). \end{aligned}$$

Therefore, we have the model

$$Y_i^*(\boldsymbol{\beta}) = \boldsymbol{z}_i'\boldsymbol{\beta} + e_i(\boldsymbol{\beta}), \qquad 1 \le i \le n,$$

where $E(e_i(\boldsymbol{\beta})) = 0$, because $E(Y_i^*(\boldsymbol{\beta})) = E(Y_i) = \boldsymbol{z}_i'\boldsymbol{\beta}$. In this model, the normal equations are $\boldsymbol{\mathcal{Z}}'\boldsymbol{\mathcal{Z}}\boldsymbol{b} = \boldsymbol{\mathcal{Z}}'\boldsymbol{Y}^*(\boldsymbol{b})$. Since $\boldsymbol{\mathcal{Z}}'\boldsymbol{\mathcal{Z}}$ is assumed to be nonsingular, the least squares estimator is a solution $\boldsymbol{b}$ of the normal equations $\boldsymbol{b} = (\boldsymbol{\mathcal{Z}}'\boldsymbol{\mathcal{Z}})^{-1}\boldsymbol{\mathcal{Z}}'\boldsymbol{Y}^*(\boldsymbol{b})$. But, because F is unknown, $\boldsymbol{Y}^*(\boldsymbol{b})$ is unknown, for every $\boldsymbol{b}$. For each $\boldsymbol{b}$, replace F in $\boldsymbol{Y}^*(\boldsymbol{b})$ by the KM estimator $\widehat{F}_{\boldsymbol{b}}$ based on $\{(\widetilde{Y}_i - \boldsymbol{z}_i'\boldsymbol{b}, \delta_i)_{i=1}^n\}$ to obtain an estimate $\widehat{\boldsymbol{Y}}(\boldsymbol{b})$ of $\boldsymbol{Y}^*(\boldsymbol{b})$. The Buckley-James estimator $\widehat{\boldsymbol{\beta}}_{BJ}$ is then a solution of the equations $\boldsymbol{b} = (\boldsymbol{\mathcal{Z}}'\boldsymbol{\mathcal{Z}})^{-1}\boldsymbol{\mathcal{Z}}'\widehat{\boldsymbol{Y}}(\boldsymbol{b})$.

Buckley and James (1979) reported that in the simple linear regression case, where $p = 2$ and the first column of $\boldsymbol{\mathcal{Z}}$ is $(1, \cdots, 1)'$, solutions of these normal equations may oscillate between two values, in which case the estimator is taken to be the average of the two values. They also adopt the common practice of treating the last observation as uncensored. Although we have described this estimator in the multiple linear regression setup, its computation in the case $p \ge 3$ is difficult.

In comparison the following estimator of Koul, Susarla and Van Ryzin (1981) (KSV) is explicitly defined and easy to compute. Let G_i denote the d.f. of U_i, $i = 1, \cdots, n$ such that $G_i(y_-) < 1$, for all $i = 1, \cdots, n$. Arguing as for (11.5.5), we obtain that

$$E\Big(\frac{\delta_i \widetilde{Y}_i}{1 - G_i(\widetilde{Y}_{i-})}\Big) = E(Y_i) = \boldsymbol{z}_i'\boldsymbol{\beta}, \quad i = 1, \cdots, n.$$

We thus have the new regression model

$$\mathcal{W}_i := \frac{\delta_i \widetilde{Y}_i}{1 - G_i(\widetilde{Y}_{i-})} = \boldsymbol{z}_i'\boldsymbol{\beta} + \eta_i, \quad i = 1, \cdots, n,$$

where η_i are independent errors with mean zero. But G_i's are not known and hence $\mathcal{W}_i$'s are not observable.

If for some d.f. G, $G_1 = G_2 = \cdots = G_n = G$, we estimate G by the KM estimator

$$\widetilde{G}(y) = 1 - \prod_{i=1, \widetilde{Y}_{(i)} \le y}^{n} \Big(\frac{n-i}{n-i+1}\Big)^{1-\delta_{(i)}}$$

and $\mathcal{W}_i$ by $\widetilde{\mathcal{W}}_i = \delta_i \widetilde{Y}_i / \big(1 - \widetilde{G}(\widetilde{Y}_{i-})\big)$, for $i = 1, \cdots, n$. Let $\widetilde{\boldsymbol{\mathcal{W}}}$ denote the vector $\big(\widetilde{\mathcal{W}}_1, \cdots, \widetilde{\mathcal{W}}_n\big)'$. Then, the KSV estimator of $\boldsymbol{\beta}$ is

$$\widehat{\boldsymbol{\beta}}_{KSV} = \big(\boldsymbol{\mathcal{Z}}'\boldsymbol{\mathcal{Z}}\big)^{-1}\boldsymbol{\mathcal{Z}}'\widetilde{\boldsymbol{\mathcal{W}}}.$$

A clear advantage of this estimator over the Buckley-James (BJ) estimator is that it is explicitly defined and easy to compute. One just uses the classical least square computation method using the data $\big(\boldsymbol{\mathcal{Z}}, \widetilde{\boldsymbol{\mathcal{W}}}\big)$. This definition is valid also when $\boldsymbol{z}_i$'s are random vectors, as long as for each i, Y_i is conditionally independent of U_i, given $\boldsymbol{z}_i$.

Asymptotic distribution of $\widehat{\boldsymbol{\beta}}_{KSV}$. The asymptotic normality of the KSV estimator has been proved under different conditions by several authors. In KSV, the estimator is defined slightly differently. Let M_n be a sequence of positive numbers tending to infinity and let $V_i := \widetilde{\mathcal{W}}_i I(\widetilde{Y}_i \le M_n), i = 1, \cdots, n$ and $\boldsymbol{V} := (V_1, \cdots, V_n)'$. KSV defined the estimator to be $\widetilde{\boldsymbol{\beta}}_{KSV} := \big(\boldsymbol{\mathcal{Z}}'\boldsymbol{\mathcal{Z}}\big)^{-1}\boldsymbol{\mathcal{Z}}'\boldsymbol{V}$ and proved its mean square consistency under fairly general conditions. They also proved the asymptotic normality of $n^{1/2}(\widetilde{\boldsymbol{\beta}}_{KSV} - \boldsymbol{\beta}_n)$ under some conditions, where $\boldsymbol{\beta}_n := (\boldsymbol{\mathcal{Z}}'\boldsymbol{\mathcal{Z}})^{-1}\boldsymbol{\mathcal{Z}}'\nu_n$, with $\nu_n := E\big(Y_1 I(Y_1 \le M_n), \cdots, Y_n I(Y_n \le M_n)\big)'$. This result is similar to that of Theorem 11.5.2 pertaining to the SV estimator of the mean.

Srinivasan and Zhou (1994) used the counting process and martingale approach to prove the asymptotic normality of $\widehat{\boldsymbol{\beta}}_{KSV}$. For the sake of an interested reader, we shall state the assumptions and the main result only, without proofs. These authors allowed the independent errors $\varepsilon_1, \ldots, \varepsilon_n$ to have different d.f.'s $F_1, \cdots, F_n$ and densities $f_1, \cdots, f_n$, respectively. For $i = 1, \cdots, n$, let H_i denote the d.f. of $\tilde{Y}_i$ and u_{ji} denote the (j, i)th element of the matrix $(\boldsymbol{\mathcal{Z}}'\boldsymbol{\mathcal{Z}})^{-1}$, $j = 1, \cdots, p$. Let $\bar{F}_n := n^{-1}\sum_{i=1}^n F_i$ and

$$C_n(x) := \int_{s \le x} \frac{dG(s)}{\big(1 - \bar{F}_n(s)\big)\big(1 - G(s)\big)^2}, \quad s \in \mathbb{R}.$$

Now consider the following assumptions, where $\mathbb{B}$ is the standard Brownian motion on $[0, \infty)$.

(A1) For some $0 < \tau < \infty$,

$$\max_{1\le j\le p} \sup_{n\ge 1} \int_\tau^\infty \Big[\sum_{i=1}^n u_{ji} x f_i(x)\Big]^2 dC_n(x) < \infty.$$

(A2) The limit $\bar{h}_j(x) := \lim_{n\to\infty} \sum_{i=1}^n u_{ji} x f_i(x)$ exists for all $x \in \mathbb{R}$ and, for some $0 < \tau < \infty$,

$$\max_{1\le j\le p} \sup_{n\ge p} E\Big(\int_\tau^\infty \big|\mathbb{B}(C_n(x))\bar{h}_j(x)\big|\, dx\Big) < \infty.$$

(A3)

$$\lim_{t\to\infty} \lim_{n\to\infty} n \sum_{i=1}^n u_{ji}^2 \frac{\Big[\int_t^\infty s f_i(s) ds\Big]^2}{1 - H_i(t)} = 0, \quad \forall j = 1, \cdots, p.$$

(A4) For each $\gamma > 0$, the functions $f_i(x)/[1 - H_i(x)]$ are of bounded variation on $[-\gamma, \gamma]$ uniformly in $1 \le i \le n$.

(A5) For all $\gamma > 0$

$$\max_{1\le j\le p} \limsup_n \int_\gamma^\infty \Big(n \sum_{i=1}^n u_{ji}^2 \frac{x^2 f_i^2(x)}{1 - H_i(x)}\Big)^{1/2} dx < \infty.$$

(A6) The limit $\lim_n n^{-1} \boldsymbol{\mathcal{Z}}'\boldsymbol{\mathcal{Z}}$ exists and is positive definite, and $n^{-1} \sum_{i=1}^n \|\boldsymbol{z}_i\|^4 = O(1)$.

Let

$$\gamma_j(s, y) := \frac{\sum_{r=1}^n u_{jr} \int_s^y x\, dF_i(x)}{\sum_{r=1}^n \bar{H}_r(s)} - \frac{u_{ji} \int_s^y x\, dF_i(x)}{\bar{H}_i(s)},$$

$$\sigma_{jk}(y) := \lim_{n\to\infty} n \sum_{i=1}^n u_{ji} u_{ki} \int_{-\infty}^y \Big[\frac{1}{\bar{G}(s)} - \frac{\int_s^y x dF_i(x)}{\bar{H}_i(s)}\Big]^2 \bar{G}(s)\, dF_i(s)$$
$$+ \lim_{n\to\infty} n \sum_{i=1}^n \int_{-\infty}^y \gamma_j(s, y)\gamma_k(s, y)\bar{F}_i(s)\, dG(s), \quad 1 \le j, k \le p,$$

where these limits are assumed to exist and be finite. Let $\Sigma(y)$ denote the matrix whose (j, k)th entry is $\sigma_{jk}(y)$. Also, let $T_n := \max_{1\le i\le n} \widetilde{Y}_i$ and

$$\beta_j^* := \sum_{i=1}^n u_{ji} \int_{x\le T_n} s\, dF_i(s), \quad 1 \le j \le p, \quad \boldsymbol{\beta}^* := (\beta_1^*, \cdots, \beta_p^*)'.$$

We are now ready to state the following theorem due to Srinivasan and Zhou (1994).

Theorem 11.9.1. *Consider the above setup with (A1)–(A6) satisfied. Assume that $\Sigma(y)$ is well defined for all $y \in \mathbb{R}$ and $\Sigma(\infty) := \lim_{y\to\infty} \Sigma(y)$ exists and is positive definite. In addition, assume that*

$$\sup_{t>0, 1\le i\le n} E\big(\varepsilon_i - t \big| \varepsilon_i > t\big) < \infty.$$

Then, as $n \to \infty$, $n^{1/2}\big(\widehat{\boldsymbol{\beta}}_{KSV} - \boldsymbol{\beta}^*\big) \to_D \mathcal{N}_p\big(0, \Sigma(\infty)\big)$.

As in the case of Theorem 11.5.2, if also $n^{1/2}(\boldsymbol{\beta}^* - \boldsymbol{\beta}) \to_p 0$, then we obtain $n^{1/2}(\widehat{\boldsymbol{\beta}}_{KSV} - \boldsymbol{\beta}) \to_D \mathcal{N}_p(0, \Sigma(\infty))$.

Lai, Ying and Zheng (1995) (LYZ) proved perhaps the most useful asymptotic normality result about a class of estimators of $\boldsymbol{\beta}$, which includes the KSV estimator. We shall now describe their result about the KSV estimator. These authors considered the right consored regression model with random covariates. More precisely, their observations are $(Z_1', \tilde{Y}_1, \delta_1)', \cdots, (Z_n', \tilde{Y}_n, \delta_n)'$, with $\tilde{Y}_i = Y_i \wedge U_i$ and $\delta = I(Y_i \leq U_i)$, based on independent random vectors $(U_1, Y_1, Z_1')', \cdots, (U_n, Y_n, Z_n')'$ where, for each $i = 1, \cdots, n$, the r.v. U_i is independent of (Y_i, Z_i') and has d.f. G, and the real-valued response Y_i and the p-dimensional covariate vector Z_i satisfy $Y_i = Z_i'\beta + \varepsilon_i$ for some $\boldsymbol{\beta} \in \mathbb{R}^p$ and some unobservale error variable ε_i with d.f. F. Let L_i denote the d.f. of Y_i. Recall for any d.f. L, $\tau_L := \sup\{s; L(s) < 1\}$. For a d.f. Ψ, Lai et al. define $\Psi^{-1}(1) := \inf\{x : \Psi(x) = 1\}$. Note that $\Psi^{-1}(1) = \tau_\Psi$. For a Borel set B in $\mathbb{R}^p$ and $y \in \mathbb{R}$, let

$$\mu_n(B, y) := \frac{1}{n}\sum_{i=1}^{n} P(Z_i \in B, Y_i \leq y), \qquad \bar{L}_n(y) := \frac{1}{n}\sum_{i=1}^{n} L_i(y), \qquad \tau_n := \tau_{\bar{L}_n}.$$

Consider the following conditions.

(B0) For all $i = 1, \cdots, n$, $\tau_G \geq \tau_{L_i}$.

(B1) The following limits exist for every $y \in \mathbb{R}$:

$$\Gamma_0(y) := \lim_{n\to\infty} n^{-1}\sum_{i=1}^{n}(1 - L_i(y)), \quad \Gamma_1(y) := \lim_{n\to\infty} n^{-1}\sum_{i=1}^{n} E\big(Z_i I(Y_i > y)\big).$$

(B2) For all $y \leq \tau := \inf\{y : \Gamma_0(y) = 0\}$, $\Delta G(y)\Delta\Gamma_j(y) = 0$, $j = 0, 1$, and

$$\int_{-\infty}^{\tau} \frac{\left\| \int_{t>y} t\, d\Gamma_1(t)\right\|^2}{\big[(1 - G(y))\Gamma_0(y_-)\big]^2}\Gamma_0(y)\, dG(y) < \infty.$$

(B3) There is $y_\epsilon < \tau_n$ for every $\epsilon > 0$ such that for all large n,

$$\int_{z\in\mathbb{R}^p}\int_{y_\epsilon}^{\tau_n} \frac{\|z\||y|\, d\mu_n(z, y)}{(1 - G(y))\big(1 - \bar{L}_n(y)\big)^{1/2}} < \infty.$$

(B4) The random vectors $\xi_i := Z_i\Big(\{\delta_i\widetilde{Y}_i/1 - G(\widetilde{Y}_{i-})\} - Z_i'\boldsymbol{\beta}\Big), 1 \leq i \leq n$ satisfy the Lindeberg-Feller conditions: $\forall\, \epsilon > 0$

$$n^{-1}\sum_{i=1}^{n} E\Big(\|\xi_i\|^2 I(\|\xi_i\| > \epsilon n^{1/2})\Big) \to 0.$$

(B5) The limit $\Sigma_1 := \lim_{n\to\infty} n^{-1}\sum_{i=1}^{n} E\Big(\Big\{\frac{\delta_i\widetilde{Y}_i}{1 - G(\widetilde{Y}_{i-})} - Z_i'\boldsymbol{\beta}\Big\}^2 Z_i Z_i'\Big)$ exists and is a positive definite matrix.

(B6) With $\boldsymbol{Z}_n$ denoting the $n \times p$ matrix whose ith row is Z_i', $1 \leq i \leq n$, $n^{-1}\boldsymbol{Z}_n'\boldsymbol{Z}_n \to_p A$, a positive definite symmetric $p \times p$ matrix.

Let

$$\Sigma := \Sigma_1 - \int_{-\infty}^{\tau} \frac{\left(\int_{t>y} t\, d\Gamma_1(t)\right)\left(\int_{s>y} s\, d\Gamma_1(s)\right)'}{(1-G(y))(1-G(y_-))\Gamma_0^2(y_-)} \Gamma_0(y)\, dG(y).$$

We are now ready to state the following theorem, which is part (ii) of Theorem 2 of LYZ applied with the function $\alpha = 0$. The above conditions (B1) to (B6) are equivalent to (C1) to (C6) of LYZ applied with $\alpha = 0$, respectively. Similarly, the above Σ_1 and Σ are equivalent to V_1 and V of Theorem 2 of LYZ with $\alpha = 0$.

Theorem 11.9.2. *Under the above setup and conditions (B0)–(B6),*

$$n^{1/2}(\widehat{\boldsymbol{\beta}}_{KSV} - \boldsymbol{\beta}) \to_D \mathcal{N}_p(0, A^{-1}\Sigma A^{-1}).$$

LYZ also proposed a consistent estimator of Σ. To describe this estimator, let $\widetilde{Y}_{(1)} \le \widetilde{Y}_{(2)} \le \cdots \widetilde{Y}_{(n)}$ denote the ordered $\widetilde{Y}_i$'s and $\delta_{(i)}, Z_{(i)}$ denote the corresponding δ_j, Z_j's and let

$$B_{nj} := \frac{1}{n-j+\delta_{(j)}} \sum_{i=j}^{n} \frac{\delta_{(i)} Z_{(i)} \widetilde{Y}_{(i)}}{(1-\widetilde{G}(\widetilde{Y}_{(i)-}))}.$$

Then a consistent estimator of Σ proposed by LYZ is

$$\widehat{\Sigma}_n = n^{-1}\sum_{i=1}^{n}\left(\frac{\delta_{(i)}\widetilde{Y}_{(i)}}{(1-\widetilde{G}(\widetilde{Y}_{(i)-}))} - Z'_{(i)}\widehat{\boldsymbol{\beta}}_{KSV}\right)^2 Z_{(i)}Z'_{(i)} + n^{-1}\sum_{j=1}^{n} B_{nj}B'_{nj}\frac{(n-j)(1-\delta_{(j)})}{n-j+1}.$$

11.10 The Proportional Hazards Model

The data available in regression problems with right-censored failure times are independent observations on the triple $(\widetilde{Y}, \delta, Z)$, where $\widetilde{Y} = Y \wedge U$, $\delta = I(Y \le U)$ and $Z = (z_1, \cdots, z_p)'$ is a p-dimensional column vector of covariates. The inference questions in this setting are most often about the conditional distribution of failure, given the covariates.

Suppose the conditional survival function is of the form $S(x|Z) = e^{-h(Z)x}$, for some function h of Z. Then the conditional hazard rate function $\lambda(x|Z) = h(Z)$ is constant in x and the conditional cumulative hazard function $\Lambda(x|Z) = -\ln S(x|Z) = h(Z)x$. Thus, for any two different values Z_1 and Z_2 of the covariate vector Z, the ratio of the two conditional hazard rate functions $\lambda(x|Z_1)/\lambda(x|Z_2) = h(Z_1)/h(Z_2)$ is constant in x. Whenever a failure rate function at two different covariate levels are proportional to each other, we say that the model obeys a proportional hazards model.

A useful proportional hazards model is obtained by having

$$\lambda(x|Z) = \lambda_0(x)e^{Z'\boldsymbol{\beta}}, \quad x \in \mathbb{R}, \tag{11.10.1}$$

for some $\boldsymbol{\beta} := (\beta_1, \cdots, \beta_p)' \in \mathbb{R}^p$ and some failure rate function λ_0 not depending on the covariates. This model is called the Cox's proportional hazard model (PHM) and λ_0 is called the baseline hazard function. This model was proposed in Cox (1972) for the regression analysis of the lifetimes. It has been used widely since then in applications.

The survival function of the Cox model is

$$S(x|Z) = \exp\Big(- \int_{s \le x} \lambda_0(s)\, ds\, g(Z)\Big) = S_0(x)^{g(Z)}, \tag{11.10.2}$$

where $g(Z) = e^{Z'\boldsymbol{\beta}}$ and $S_0(x) := \exp\big(- \int_{s \le x} \lambda_0(s)\, ds\big)$. This fact shows that in this PHM, $S(x|Z_1) \ge S(x|Z_2)$, for all x, if, and only if, $g(Z_1) \le g(Z_2)$. Here, S_0 is called the baseline survival function and covariates Z act multiplicatively on the hazard rate function of a lifetime.

Sometimes one uses the so-called accelerated life testing model, where $T > 0$ denotes the survival or lifetime, $Y = \ln(T)$, $\mu(z) = E\big(Y|Z = z\big)$ and

$$Y = \ln T = \mu(Z) + \sigma\varepsilon, \quad \text{for some } \sigma > 0,$$

where ε is a random error with $E\big(\varepsilon\big|Z\big) = 0$, a.s.

This model and the PHM coincide only when the survival distribution is Weibull. To see this, let F_ε denote the d.f. of ε, $S_\varepsilon = 1 - F_\varepsilon$, $\alpha(z) := e^{\mu(z)}$, $S_*(\cdot) := S_\varepsilon\big(\ln(\cdot)^{1/\sigma}\big)$ and let P_z denote the conditional probability measure, given $Z = z$. Then,

$$\begin{aligned} S_T\big(t|Z = z\big) &= P_z\big(T > t\big) = P_z(\ln T > \ln t) \\ &= S_\varepsilon\Big(\frac{\ln t - \mu(z)}{\sigma}\Big) = S_\varepsilon\Big(\ln\big(te^{-\mu(z)}\big)^{1/\sigma}\Big) \\ &= S_*\Big(\frac{t}{\alpha(z)}\Big). \end{aligned}$$

Hence,

$$\lambda_T\big(t|Z = z\big) = \frac{1}{\alpha(z)}\lambda_*\Big(\frac{t}{\alpha(z)}\Big) = e^{-\mu(z)}\lambda_*\big(te^{-\mu(z)}\big).$$

Suppose T has a Weibull density with parameters $\gamma > 0$ and $\theta > 0$,

$$f_T(t) = \gamma\theta(\gamma t)^{\theta-1}e^{-(\gamma t)^\theta} I(t > 0).$$

Then,

$$S_T(t) = e^{-(\alpha t)^\theta}, \quad \lambda_T(t) = \gamma\theta(\gamma t)^{\theta-1}, \quad t \ge 0, \lambda > 0, \theta > 0.$$

The above representations of the two hazard rates are used to show that the proportional hazards model and the log-linear model coincide only when the survival distribution is Weibull. See Kalbfleisch and Prentice (1980) or Doksum (1982) for a proof of this statement.

11.10.1 Cox's Partial Likelihood

From (11.2.1) and (11.10.2), the density of the ith unit of the Cox's proportional hazards model, given Z_i, is

$$f_i(x|Z_i) = \theta_i\lambda_0(x)\big(S_0(x)\big)^{\theta_i}, \quad \theta_i := \exp\big(\boldsymbol{\beta}'Z_i\big), \ 1 \le i \le n. \tag{11.10.3}$$

Under random right censorship, one observes $(\widetilde{X}_i, \delta_i) := (T_i \wedge U_i, I(T_i \le U_i)), i = 1, \cdots, n$, where $(T_1, U_1), \cdots, (T_n, U_n)$ are independent random vectors, T_i and U_i are conditionally independent given Z_i, for each $1 \le i \le n$, and T_i has the above conditional density given Z_i. Suppose that there are exactly k distinct uncensored observations $\widetilde{X}_{\pi_1} < \widetilde{X}_{\pi_2} < \cdots < \widetilde{X}_{\pi_k}$, where π_j is the index of the observation whose rank is j among the observations. Let $Z_{(1)}, Z_{(2)}, \cdots, Z_{(k)}$ denote the corresponding covariate vectors. Let R_i denote the set of individuals alive and uncensored just prior to $\widetilde{X}_{\pi_i}$ and $D := \{i, \delta_i = 1, i = 1, \cdots, n\}$.

Let $G_1, \cdots, G_n$ be d.f.'s on $\mathbb{R}$. Suppose $U_i \sim_D G_i$, $i = 1, \cdots, n$. Let

$$\Lambda_n := \prod_{i \in D} \Big(\lambda_0(\widetilde{X}_i) \sum_{j \in R_i} \theta_j\Big) \prod_{i=1}^{n} \Big(S_0(\widetilde{X}_i)\Big)^{\theta_i}.$$

Then, the likelihood for $\boldsymbol{\beta}$ based on $(\widetilde{X}_i, \delta_i)$, $i = 1, \cdots, n$ is

$$\begin{aligned}
L_n(\boldsymbol{\beta}) &= \prod_{i=1}^{n} f_i^{\delta_i}(\widetilde{X}_i|Z_i)\Big(1 - F_i(\widetilde{X}_i|Z_i)\Big)^{1-\delta_i} \\
&= \prod_{i=1}^{n} \Big(\theta_i \lambda_0(\widetilde{X}_i)(S_0(\widetilde{X}_i))^{\theta_i}\Big)^{\delta_i} \Big(S_0(\widetilde{X}_i)\Big)^{\theta_i(1-\delta_i)} \\
&= \prod_{i=1}^{n} \Big(\theta_i \lambda_0(\widetilde{X}_i)\Big)^{\delta_i} \Big(S_0(\widetilde{X}_i)\Big)^{\theta_i} \\
&= \prod_{i \in D} \Big(\theta_i \lambda_0(\widetilde{X}_i)\Big) \prod_{i=1}^{n} \Big(S_0(\widetilde{X}_i)\Big)^{\theta_i} \\
&= \prod_{i \in D} \frac{\theta_i}{\sum_{j \in R_i} \theta_j} \prod_{i \in D} \Big(\lambda_0(\widetilde{X}_i) \sum_{j \in R_i} \theta_j\Big) \prod_{i=1}^{n} \Big(S_0(\widetilde{X}_i)\Big)^{\theta_i} \\
&= \prod_{i \in D} \frac{\exp\big(\boldsymbol{\beta}' Z_{(i)}\big)}{\sum_{j \in R_i} \exp\big(\boldsymbol{\beta}' Z_{(j)}\big)} \Lambda_n.
\end{aligned}$$

Cox (1972, 1975) proposed the ratio

$$\mathcal{L}_n(\boldsymbol{\beta}) := = \prod_{i \in D} \frac{\exp\big(\boldsymbol{\beta}' Z_{(i)}\big)}{\sum_{j \in R_i} \exp\big(\boldsymbol{\beta}' Z_{(j)}\big)}$$

as a likelihood for the purpose of making inference about $\boldsymbol{\beta}$. It is also known as partial likelihood. The log likelihood and the $(j, \ell)^{\text{th}}$ entry of the Fisher information matrix, respectively, are as follows, for $1 \le j, \ell \le p$.

$$\log \mathcal{L}_n(\boldsymbol{\beta}) = \sum_{i=1}^{k} Z'_{(i)}\boldsymbol{\beta} - \sum_{i=1}^{k} \log\Big(\sum_{j \in R_i} e^{Z'_{(j)}\boldsymbol{\beta}}\Big),$$

$$\begin{aligned}
\mathcal{I}_{j\ell}(\boldsymbol{\beta}) := -\frac{\partial^2 \mathcal{L}_n(\boldsymbol{\beta})}{\partial \beta_j \partial \beta_\ell} = \sum_{i=1}^{k} \Bigg[& \frac{\sum_{r \in R_i} z_{rj} z_{r\ell} e^{Z'_{(i)}\boldsymbol{\beta}}}{\sum_{r \in R_i} e^{Z'_{(r)}\boldsymbol{\beta}}} \\
& - \frac{1}{\Big(\sum_{r \in R_i} e^{Z'_{(r)}\boldsymbol{\beta}}\Big)^2} \Big(\sum_{r \in R_i} z_{rj} e^{Z'_{(i)}\boldsymbol{\beta}}\Big)\Big(\sum_{r \in R_i} z_{r\ell} e^{Z'_{(i)}\boldsymbol{\beta}}\Big)\Bigg].
\end{aligned}$$

Define $\widehat{\boldsymbol{\beta}}$ to be a solution of the likelihood equation $\partial \mathcal{L}_n(\boldsymbol{\beta})/\partial \boldsymbol{\beta} = 0$. Assume that $\mathcal{I}(\beta)^{-1}$ exists. In practice people essentially work under the premise that $n^{1/2}(\widehat{\boldsymbol{\beta}} - \boldsymbol{\beta}) \to_D \mathcal{N}_p(0, \mathcal{I}(\boldsymbol{\beta})^{-1})$ and use $\mathcal{I}(\widehat{\boldsymbol{\beta}})^{-1}$ for an estimate of the Fisher information matrix $\mathcal{I}(\boldsymbol{\beta})$. The proof of this result is beyond the scope of this book.

Interested readers may refer to the monographs by Kalbfleisch and Prentice (1980), Fleming and Harrington (1991), Tsiatis (2006) and the references therein, for more on Cox's likelihood and the proofs of the consistency and asymptotic normality of $\widehat{\beta}$.

11.11 Exercises

1. Let F be a continuous d.f. on $[0, \infty)$ with finite variance τ^2. Let X, X_1, X_2 be i.i.d. F r.v.'s. Show that $\tau^2 = \text{Var}(X) = E\big((X_1 - X_2)^2/2\big) = 2\iint_{0<s<t<\infty} F(s)\bar{F}(t)\, ds\, dt$.
2. Find the survival function S, the hazard function λ and the cumulative hazard function Λ for each of the following densities.
$$\text{(a) } f(x) = \exp(-x)I(x > 0), \quad \text{(b) } f(x) = \frac{I(x > 0)}{(1 + x)^2},$$
$$\text{(c) } f(x) = 2x\exp(-x^2)I(x > 0).$$
3. Let X have probability mass function $p(x) = \theta(1 - \theta)^x$, $x = 0, 1, 2, \cdots$, for some $\theta \in (0, 1)$. Find the survival function S, the hazard rate function λ and the cumulative hazard function Λ.
4. Prove the following: A function S from $\mathbb{R}$ to $\mathbb{R}$ is a survival function if and only if S is nonincreasing, right-continuous and $S(-\infty) := \lim_{t\to-\infty} S(t) = 1$ and $S(\infty) = \lim_{t\to\infty} S(t) = 0$.
5. Let S be a survival function. Show that S^r is a survival function for each $r > 0$.
6. Let S be a survival function. Is $2S/(1 + S)$ a survival function? Justify your answer.
7. Let S_1 and S_2 be two survival functions.
(a) Is their product S_1S_2 a survival function?
(b) Is their average $(S_1 + S_2)/2$ a survival function?
(c) Is $\max(S_1, S_2)$ a survival function?
(d) Is $\min(S_1, S_2)$ a survival function?
You must justify your answers.
8. Let F_i and Λ_i denote the d.f. and the cumulative hazard function of X_i, $1 \le i \le n$, respectively, and let
$$\widetilde{M}_n(x) = \sum_{i=1}^{n}\Big\{I(X_i \le x) - \int_{u\le x} I(X_i \ge u)d\Lambda_i(u)\Big\}.$$
Show that for each $n \ge 1$, $\widetilde{M}_n(x), x \in \mathbb{R}$ is a means zero martingale with respect to the σ-field $\mathcal{F}_{x,n} = \sigma\{I(X_i \le s), s \le x, i = 1, \cdots, n\}$.
9. Prove Lemma 11.5.1.
10. Suppose the d.f. F of X has density f.
(a) Obtain the likelihood functions of the Type I and II censored data.
(b) Let $S_0(x) = e^{-x}$, $x \ge 0$ and $S_\theta(y) := S_0(y - \theta)$, $y, \theta \in \mathbb{R}$. Obtain the MLE of θ for the Type I and II censored data.

11. In a randomly right-censored setup suppose F has density $f_\theta(x) = \theta e^{-\theta x} I(x \ge 0)$, for some $\theta > 0$. Show that the maximizer of the likelihood (11.5.1) with respect to θ is $\hat{\theta} = \sum_{i=1}^n \delta_i / \sum_{i=1}^n \widetilde{X}_i$.

12. Derive the equation (11.5.8) that gives the jumps of the KM estimator $\widetilde{G}$ of G directly from the definition of $\widetilde{G}$ given at (11.5.4).

13. Let $(U_1, X_1), \cdots, (U_n, X_n)$ be independent copies of (U, X), where U and X are independent with respective d.f.'s G and F. For $i = 1, \cdots, n$, set $\widetilde{X}_i := \min(X_i, U_i)$ and $\delta_i := I(X_i \le U_i)$. Assume that F is continuous and that

$$\bar{G}(x) = \{\bar{F}(x)\}^\beta, \quad x \in \mathbb{R},$$

for some $\beta > 0$. This is called the Koziol-Green model, see Koziol and Green (1976). Prove the following:

(a) The r.v.'s $\delta_1, \cdots, \delta_n$ are i.i.d. Bernoulli$(1/(1+\beta))$ and independent of $\widetilde{X}_1, \cdots, \widetilde{X}_n$.

(b) The r.v.s $\widetilde{X}_1, \cdots, \widetilde{X}_n$ are i.i.d. with d.f. $1 - \bar{F}^{\beta+1}$.

(c) Show that if the observations $(\widetilde{X}_i, \delta_1), \cdots, (\widetilde{X}_n, \delta_n)$ satisfy the above (a) and (b) and $0 < P(X \le U) < 1$, then the Koziol-Green model must be satisfied for some $\beta \ge 0$.

14. Let

$$c_{ni} = \frac{1}{n-i+1}, \quad a_{ni} = \sum_{j=1}^{i} c_{nj}, \quad \tau_n^2 = \sum_{j=1}^{i} c_{nj}^2, 1 \le i \le n,$$

$$\widehat{\Lambda}_1(x) = \sum_{i=1}^{n} c_{ni} I(\widetilde{X}_{(i)} \le x), \qquad \widehat{\Lambda}_2(x) = \sum_{i=1}^{n} \delta_{(i)} c_{ni} I(\widetilde{X}_{(i)} \le x), \quad x \in \mathbb{R},$$

$$f(i, n, p) = \binom{n}{i} p^i (1-p)^{n-i}, \quad 1 \le i \le n, 0 \le p \le 1,$$

$$H(x) = P(\widetilde{X} \le t), \quad t \in \mathbb{R}, \ \gamma = \frac{1}{1+\beta}.$$

Assume that F is continuous and that the Koziol-Green model holds. Then, prove the following, for every $x \in \mathbb{R}$.

$$\text{(a) } E(\widehat{\Lambda}_1(x)) = \sum_{i=1}^{n} f(i, n, H(x)) a_{ni}, \quad \text{(b) Bias}(\widehat{\Lambda}_2(x)) = \gamma \text{Bias}(\widehat{\Lambda}_1),$$

$$\text{(c) Var}(\widehat{\Lambda}_1(x)) = \sum_{i=0}^{n} f(i, n, H(x)) a_{ni}^2 - \Big(\sum_{i=0}^{n} f(i, n, H(x)) a_{ni}\Big)^2,$$

$$\text{(d) Var}(\widehat{\Lambda}_2(x)) = \gamma^2 \text{Var}(\widehat{\Lambda}_1(x)) + \gamma(1-\gamma) \sum_{i=0}^{n} f(i, n, H(x)) \tau_{ni}^2.$$

15. **MD estimation of a scale parameter.** The following exercise is on some minimum distance estimators of the scale parameter of the two sample randomly censored scale model. Let F, G_1, G_2 be d.f.'s on $[0, \infty)$. Consider the following two sample randomly

censored setup, where $\{X_{1j}/\theta; 1 \le j \le n_1\}$ are i.i.d. F for some $\theta > 0$, $\{X_{2j}; 1 \le j \le n_2\}$ are i.i.d. F, $\{U_{ij} : 1 \le j \le n_i\}$ are i.i.d. G_i, $i = 1, 2$. Define, for $i = 1, 2$,

$$\widetilde{X}_{ij} = X_{ij} \wedge U_{ij}, \quad \delta_{ij} = \{X_{ij} \le U_{ij}\}, \quad j = 1, \cdots, n_i,$$

$$T_i := \max_{1 \le j \le n_i} \widetilde{X}_{ij}, \quad Y_{i\bullet}(x) := \sum_{j=1}^{n_i} I(\widetilde{X}_{ij} \ge x),$$

$$\widehat{F}_i(x) = \begin{cases} 1 - \prod_{\widetilde{X}_{ij} \le x, j=1}^{n_i} \left(1 - \frac{\delta_{ij}}{Y_{i\bullet}(\widetilde{X}_{ij})}\right), & x < T_i; \\ 1 & x \ge T_i; \end{cases},$$

$$K(t) = \int \left(\widehat{F}_1(xt) - \widehat{F}_2(x)\right)^2 L(dx), \quad t \ge 0,$$

where L is a nondecreasing right continuous function on $[0, \infty)$.

(a) Show that

$$\begin{aligned} K(t) = & \frac{1}{n_1 n_2} \sum_{j=1}^{n_1} \sum_{k=1}^{n_2} a_{1j} a_{2k} \left| L(\frac{\widetilde{X}_{1j}}{t}) - L(\widetilde{X}_{2k}-) \right| \\ & - \frac{1}{n_1^2} \sum_{j=1}^{n_1} \sum_{k=1}^{n_1} a_{1j} a_{2k} \left| L(\frac{\widetilde{X}_{1j}}{t}-) - L(\frac{\widetilde{X}_{1k}}{t}-) \right| \\ & - \frac{1}{n_2^2} \sum_{j=1}^{n_2} \sum_{k=1}^{n_2} a_{2j} a_{2k} \left| L(\widetilde{X}_{2j}-) - L(\widetilde{X}_{2k}-) \right|, \end{aligned}$$

where a_{ij}/n_i are the jumps of $\widehat{F}_i$ at the observations $\widetilde{X}_{ij}$, $j = 1, \cdots, n_i$; $i = 1, 2$.

(b) Compute the estimator $\widehat{\theta} := \text{argmin}_t K(t)$ in the case $L(x) = \ln(x)$. Let $\widetilde{\theta}$ be defined as a solution t of the equation

$$\int \widehat{F}_1(xt) d\widehat{F}_2(x) = \frac{1}{2}, \quad t \ge 0. \tag{11.11.4}$$

Are the two estimators $\widehat{\theta}$ and $\widetilde{\theta}$ the same?

Now let $\widetilde{F}_i$ denote the general KM estimator of F_i that may allow some mass to escape to infinity. Let $\bar{\theta}$ denote the analog of $\widetilde{\theta}$ when $\widehat{F}_i$'s are replaced by $\widetilde{F}_i$'s in (11.11.4). Is there any difference between these two estimators.

(a) Compute the estimators $\widetilde{\theta}, \bar{\theta}$ and $\widehat{\theta}$ with $L(x) = \ln(x)$, and $L(x) = 1 - e^{-x}, x > 0$, for the following data:

$$X_{1j} : 1741, 2689, 388, 3913, 750, 3850, 3823, 3086, \qquad \delta_{1j} : 1, 1, 1, 0, 1, 0, 0, 1,$$
$$X_{2j} : 460, 130, 611, 3820, 3388, 3445, \qquad \delta_{2j} : 1, 1, 1, 0, 0, 1.$$

Repeat the above computation for $\widehat{\theta}$ with $L = \ln$ and $\bar{\theta}$, after you modify the data by making the last observation censored in each sample. Now are the two estimators the same?

16. Prove (11.5.12) if $\int_1^\infty \left(-\log(H(s_-))\right)^{1/2} \bar{F}(s)\, ds < \infty$.

17. Prove Lemma 11.8.1.

18. Take $p = 1$ in the linear regression model (11.9.1) and let U_i's be as there. Let $\widetilde{Y}_i := Y_i \wedge U_i$. Let $\widetilde{F}_b$ and $\widetilde{G}_b$, respectively, denote the Kaplan-Meier estimators of the error d.f. F and the censoring d.f. G based on the residuals $X_i(b) := \widetilde{Y}_i - z_i b$ and $\delta_i := I(Y_i \leq U_i), 1 \leq i \leq n$. Let b be such that none of the residuals are tied. Let $X_{(1)}(b) < X_{(2)}(b) < \cdots < X_{(n)}(b)$ denote the ordered residuals and $\delta_{(i)}(b)$ denote the corresponding δ_i's. Then,

$$\widetilde{F}_b(x) := 1 - \prod_{i=1; X_{(i)}(b) \leq x}^{n} \left(\frac{n-i}{n-i+1}\right)^{\delta_{(i)}(b)},$$

$$\widetilde{G}_b(x) := 1 - \prod_{i=1, \widetilde{X}_{(i)}(b) \leq x} \left(\frac{n-i}{n-i+1}\right)^{1-\delta_{(i)}(b)}.$$

Show that the jump $J_k(b)$ of $\widetilde{F}_b$ at $X_{(k)}(b)$ for $1 \leq k \leq n$ is

$$J_k(b) = \frac{1}{n} \frac{\delta_{(k)}(b)}{1 - \widetilde{G}_b(\widetilde{X}_{(k)}(b)_-)}.$$

Notes: Aalen (1978) laid the foundation of using counting processes and their Doob-Meyer decomposition for the development of various inference procedures in survival analysis. In particular, the martingale result of Theorem 11.3.1 and the derivation of $\widehat{\Lambda}$ and $\widehat{F}$ discussed in Section 11.3 are from this work of Aalen. Nelson (1969, 1972) introduced Nelson-Aalen estimator in connection with some engineering problems. See also Nelson (1982) for some other applications of this estimator. The results under random censoring discussed in Section 11.5 are based on the work of Gill (1980, 1983) and Yang (1991). Gill (1980) also discusses several other inference procedures under right random censoring. The monographs of Fleming and Harrington (1991) and Andersen, Borgan, Gill and Keiding (1993) discuss many more aspects of counting processes and martingales in survival analysis including the proof of the asymptotic normality of the Cox's maximum partial likelihood estimator and some optimal properties of Nelson-Aalen estimator. See also Efron (1977) for efficiency considerations in the Cox's model.

The KM estimator being nonparametric MLE was proved in Kaplan and Meier (1958) while the nonparametric Bayes estimation of the survival function is based on the papers by Ferguson (1973) and Susarla and Van Ryzin (1976). Ferguson and Phadia (1979) derive a Bayes estimate of the d.f. F when the prior distribution of F is a process neutral to the right. Doksum (1974) discusses the posterior distribution of random probability measure when the prior is neutral to the right. The minimum distance estimators in the randomly censored two sample scale model were developed in Yang (1995) and Yang and Koul (1995).

Susarla, Tsai and Van Ryzin (1984) were the first to show the equivalence of $\widehat{\mu}_*$ to $\widetilde{\mu}$. Zheng (1984), Leurgans (1987) and Zhou (1992a, b) propose some other estimators of the regression parameter vector β in (11.9.1). Fygenson and Zhou (1994) and Zhou (1999) discuss some more properties of the KSV estimator. Lai, Ying and Zheng (1995) prove the asymptotic normality of a class of estimators that includes the KSV estimator. The asymptotic normality of the Buckley-James estimator is derived in Lai and Ying (1991). The papers of Stute and Wang (1993) and Stute (1995) contain some results about analogs of the strong law of large numbers and central limit theorem under random censorship, respectively.

Cox (1972, 1975) introduced the proportional hazards model and partial likelihood. Ever since, these notions have been widely used in survival analysis.

12

Bibliography

Aalen, O. (1978). Nonparametric inference for a family of counting processes. *Ann. Statist.*, **6**, 701–726.

Akritas, M. G. and Van Keilegom, I. (2001). Nonparametric estimation of the residual distribution. *Scand. J. Statist.*, **28**, 549–567.

An, H.-S. and Bing, C. (1991). A Kolmogorov–Smirnov type statistic with application to test for nonlinearity in time series. *Internat. Statist. Rev.*, **59**, 287–307.

Andersen, P. K., Borgan, Ø., Gill, R. D. and Keiding, N. (1993). *Statistical models based on counting processes.* Springer Series in Statistics. Springer-Verlag, New York.

Athreya, K. B. and Lahiri, S. N. (2006). *Measure theory and probability theory*. Springer Texts in Statistics. Springer, New York.

Bachmann, D. and Dette, H. (2005). A note on the Bickel–Rosenblatt test in autoregressive time series. *Stat. Probab. Lett.*, **74**, 221–234.

Balakrishna, N., Koul, H. L., Ossiander, M. and Sakhanenko, L. (2019). Fitting a p-th order parametric generalized linear autoregressive multiplicative error model. *Sankhyā: The Indian J. Statist.*, **81-B**, Supplement 1, pp S103–S122.

Bennett, G. (1962). Probability inequalities for the sum of independent r.v.'s. *J. Amer. Statist. Assoc.*, **57**, 33–45.

Beran, R. J. (1977). Minimum Helinger distance estimates for parametric models. *Ann. Statist.*, **5**, 445-463.

Beran, R. J. (1978). An efficient and robust adaptive estimator of location. *Ann. Statist.*, **6**, 292-313.

Bhapkar, V. P. (1961). A nonparametric test for the problem of several samples, *Ann. Math. Statist.*, **32**, 1108–1117.

Bhapkar, V. P. and Deshpande, J. V. (1968). Some nonparametric tests for multisample problems. *Technometrics*, **10(3)**, 575–578.

Bickel, P. J. and Doksum, K. (1977). *Mathematical Statistics: Basic Ideas and Selected Topics.* Vol. 1, Second Edition. Prenctice Hall, New Jersey.

Bickel, P. J. and Freedman, D. (1981). Some asymptotic theory for the bootstrap. *Ann. Statist.*, **9**, 1196–1217.

Bickel, P. J. and Ritov, J. (1988). Estimating integrated squared density derivatives: Sharp best order of convergence estimates. *Sankhya: The Indian Journal of Statist., Series A*, **50**, 381–393.

Bickel P. J. and Rosenblatt, M. (1973). On some global measures of the deviations of density function estimates. *Ann. Statist.*, **6**, 1071–1095.

Billingsley, P. (1968). *Convergence of Probability Measures.* Wiley Series in Probability and Mathematical Statistics. John Wiley & Sons, Inc., New York.

DOI: 10.1201/9781003585770-12

Boldin, M. V. (1982). Estimation of the distribution of noise in autoregression scheme. *Theor. Probab. Appl.*, **27**, 866–871.

Bosq, D. (1998). *Nonparametric statistics for stochastic processes*. 2nd Edition, Springer.

Bouzebda, S. and Nemouchi, B. (2020). Uniform consistency and uniform in bandwidth consistency for nonparametric regression estimates and conditional U-statistics involving functional data. *J. Nonparametr. Stat.*, **32**, no. 2, 452–509.

Bowman, A. W. (1984). An alternative method of cross-validation for smoothing if density estimates. *Biometrika*, **71**, 353–360.

Bowman, A. W. (1985). A comparative study of some kernel-based nonparametric density estimators. *J. Statist. Comput. Simulation*, **21**, 313-327.

Brockwell, P. J. and Davis, R. A. (1991). *Time Series: Theory and Methods*. Springer Science & Business Media. New York.

Brown, B. M. (1971). Martingale central limit theorems. *Ann. Math. Statist.*, **42**, 59–66.

Buckley, J. and James, I. (1979). Linear regression with censored data. *Biometrika*, **66 (3)**, 429–436.

Chaudhuri, P. (1996). On a geometric notion of quantiles for multivariate data. *J. Amer. Statist. Assoc.*, **91(434)**, 862–872.

Cheng, F. and Sun, S. (2008). Agoodness-of-fit test of the errors in nonlinear autoregressive time series models. *Statist. Probab. Lett.*, **78**, 50–59.

Cheng, F., Koul, H. L., Mimoto, N. and Balakrishna, N. (2023). An analog of the Bickel Rosenblatt test for error density in the linear regression model. *Research Papers in Statistical Inference for Time Series and Related Models*, pp 291–323. Springer, Singapore.

Cheng, K. F. and Serfling, R. J. (1981). On estimation of a class of efficiency - related parameters. *Scand. Act. J.*, 83–92.

Chernoff, H. and Savage, I. R. (1958). Asymptotic normality and efficiency of certain nonparametric tests. *Ann. Math. Statist.*, **29**, 972–994.

Chernozhukov V., Galichon. A., Hallin, M. and Henry, M. (2017). Monge-Kantorovich depth, quantiles, ranks, and signs. *Ann. Stat.*, **45**, 223–256.

Cox, D. R. (1972). Regression models and life tables (with discussion). *J. Roy. Stat. Soc. B*, **34**,187–220.

Cox, D. R. (1975). Partial likelihood. *Biometrika*, **62**, 269–276.

Cox, D., Koh, E., Wahba, G. and Yandell, B. S. (1988). Testing the (parametric) null model hypothesis in (semiparametric) partial and generalized spline models. *Ann. Statist.* **16**, 113–119.

Delgado, M. A. (1993). Testing the equality of nonparametric curves. *Statist. Probab. Lett.*, **17**, 199–204.

Devroye, L. and Györfi, L. (1985). *Nonparametric density estimation. The L_1 view*. Wiley Series in Probability and Mathematical Statistics: Tracts on Probability and Statistics. John Wiley & Sons, Inc., New York.

Dhar, S. K. (1992). Computation of certain minimum L_2-distance type estimators under the linear model. *Comm. Statist. Simulation Comput.*, **21(1)**, 203–220.

Dhar, S. K. (1993). Computation of certain minimum distance estimators in AR[k] model. *J. Amer. Statist. Assoc.*, **88 (421)**, 278–283.

Diebolt, J. and Zuber, J. (1999). Goodness-of-fit tests for nonlinear heteroscedastic regression models. *Statist. Probab. Lett.*, **42**, 53–60.

Doksum, K. A. (1974). Tailfree and neutral random probabilities and their posterior distributions. *Ann. Probab.*, **2**, 183–201.

Doksum, K. A. (1982). On the performance of estimates in proportional hazard and log-linear models *IMS Lecture Notes Monogr. Ser.*, **2**, 74–84. Institute of Mathematical Statistics, Hayward, CA.

Donoho, D.L. and Liu, R. C. (1988a). The "automatic" robustness of minimum distance functionals. *Ann. Statist.*, **16**, 552–586.

Donoho, D.L. and Liu, R. C. (1988b). Pathologies of some minimum distance estimators. *Ann. Statist.*, **16**, 587–608.

Durbin, J. (1973). *Distribution theory for test based on the sample d.f.* SIAM, Philadelphia.

Durbin, J. (1976). *Kolomogorov-Smirnov tests when parameters are estimated. Empirical distributions and Processes.* Springer-Verlag, Lecture Notes in Math., #**566**.

Dvoretzky, A., Kiefer, J. and Wolfowitz, J. (1956). Asymptotic minimax character of the sample distribution function and of the classical multinomial estimator. *Ann. Math. Statist.*, **27**, 642–669.

Efron, B. (1977). Efficiency of Cox's likelihood function for censored data. *J. Amer. Statist. Assoc.*, **72**, 557–565.

Efron, B. (1979). Bootstrap methods: another look at the jackknife *Ann. Statist.*, **7(1)**, 1–26.

Efron, B. (1982). The jackknife, the bootstrap and other resampling plans. *CBMS-NSF Regional Conf. Ser. in Appl. Math.*, **38**. *Society for Industrial and Applied Mathematics, Philadelphia, PA.*

Einmahl, U. and Mason, D. (2000). An empirical process approach to the uniform consistency of kernel-type function estimators. *J. Theoret. Probab.*, **13(1)**, 1–137.

Einmahl, U. and Mason, D. (2005). Uniform in bandwidth consistency of kernel-type function estimators. *Ann. Statist.*, **33(3)**, 1380–1403.

Eubank, R. L. and Spiegelman, C. H. (1990). Testing the goodness-of-fit of a linear model via nonparametric regression techniques. *J. Amer. Statist. Assoc.*, **85**, 387-392.

Eubank, R. L. and Speckman, P. L. (1993). Confidence bands in nonparametric regression. *J. Amer. Statist. Assoc.*, **88**, 1287–1301.

Fan, J. and Gijbels, I. (1996). *Local Polynomial Modelling and Its Applications.* Springer Science+Business Media, Dordrecht.

Fan, J. and Yao, Q. (2003). *Nonlinear Time Series: Nonparametric and Parametric Methods.* Springer Series in Statisics, Springer, New York.

Ferguson, T. S. (1973). A Bayesian analysis of some nonparametric problems. *Ann. Statist.*, **1**, 209–230.

Ferguson, T. S. and Phadia, E. G. (1979). Bayesian nonparametric estimation based on censored data. *Ann. Statist.*, **7**, 163–186.

Fine, T. (1966). On the Hodges and Lehmann shift estimator in the two sample problem. *Ann. Math. Statist.*, **37**, 1814–1818.

Fleming, T. R. and Harrington, D. P. (1991). *Counting Processes and Survival Analysis.* Wiley & Sons. New York.

Freedman, D. A. (1981). Bootstrapping regression models. *Ann. Statist.*, **9**, 1218–1228.

Frees, E. W. (1994). Estimating densities of functions of observations. *J. Amer. Statisti. Assoc.*, **89**, 517–525.

Fygenson, M. and Zhou, M. (1994). On using stratification in th analysis of linear regression models with right censoring. *Ann. Statist.*, **22**, 747–762.

Gastwirth, J. L. (1966). On robust procedures. *J. Amer. Statist. Assoc.*, **61**, 929–948.

Gastwirth, J. L. and Wolff, S. S. (1968). An elementary method for obtaining lower bounds on the asymptotic power of rank tests. *Ann. Math. Statist.*, **39**, 2128–2130.

Ghosal, P. and Sen, B. (2022). Multivariate ranks and quantiles using optimal transport: consistency, rates and nonparametric testing. *Ann. Statist.*, **50**, 1012–1037.

Gill, R. D. (1980). *Censoring and stochastic integrals*. Mathematical Centre Tracts, **124**. Mathematisch Centrum, Amsterdam.

Gill, R. D. (1983). Large sample behaviour of the product-limit estimator on the whole line. *Ann. Statist.*, **11(1)**, 49–58.

Giné E. and Mason, D. M. (2007). On local U-statistics processes and the estimation of densities of functions of several sample variables. *Ann. Statist.*, **35**, 1105–1145.

Giné, E. and Nickl, R. (2008). A simple adaptive estimator of the Integrated sqaure of a density. *Bernoulli*, **14**, 47–61. https://doi.org/10.3150/07-BEJ110.

Giné, E. and Guillou, A. (2002). Rates of strong uniform consistency for multivariate kernel density estimators. *Ann. Inst. H. Poincaré Probab. Statist.*, **38(#6)**, 907–921.

Godambe, V. P. (1997). *Estimating functions: a synthesis of least squares and maximum likelihood methods. IMS Lecture Notes Monogr. Ser.*, **32**, 5–15. Institute of Mathematical Statistics, Hayward, CA.

Godambe, V. P. and Thompson, M. E. (1984). Robust estimation through estimating equations. *Biometrika*, **71(1)**, 115–125.

González–Manteiga, W. and Crujeiras, R. M. (2013). An updated review of Goodness-of-fit tests for regression models. *Test.*, **22**, 361–411.

Green, P. J. and Silverman, B. W. (1994). *Nonparametric regression and generalized linear models. A roughness penalty approach*. Monographs on Statistics and Applied Probability, **58**. Chapman & Hall, London.

Groeneboom, P. and Wellner, J. A. (1992). *Information bounds and nonparametric maximum likelihood estimation. DMV Seminar, 19*. Birkhäuser Verlag, Basel.

Gutenbrunner, C. and Jurečková, J. (1992). Regression rank scores and regression quantiles. *Ann. Statist.* **20(1)**, 305–330.

Hájek, J. (1968). Asymptotic normality of simple linear rank statistics under alternatives. *Ann. Math. Statist.*, **39**, 325–346.

Hájek, J. (1969). *A Course in Nonparametric Statistics*. Holden-Day, San Francisco, U.S.A.

Hájek, J. and Šidák, Z. (1967). *Theory of Rank Tests*. Academic Press, New York.

Hall, P. (1984). Central limit theorem for integrated square error of multivariate nonparametric density estimators. *J. Mult. Analysis*, **14**, 1–16.

Hall, P. (1992). *The bootstrap and Edgeworth expansion*. Springer Ser. Statist. Springer-Verlag, New York.

Hall, P. and Hart, J. D. (1990). Bootstrap test for difference between means in nonparametric regression. *J. Amer. Statist. Assoc.*, **85**, 1039-1049.

Hall, P. and Heyde, C. C. (1980). Martingale limit theory and its application. *Probab. Math. Statist.*, Academic Press, Inc. [Harcourt Brace Jovanovich, Publishers], New York-London.

Hall, P. and Johnstone, I. (1992). Empirical Functionals and Efficient Smoothing Parameter Selection *J. R. Statist. Soc. B*, **54(2)**, 475–530.

Hall, P., Huber, C. and Speckman, P. L. (1997). Covariate-matched one-sided tests for the di4erence between functional means. *J. Amer. Statist. Assoc.*, **92**, 1074– 1083.

Hallin, M. (2022). Measure transportation and statistical decision theory. *Annu. Rev. Stat. Appl.*, **9**, 401–424.

Hallin, M., Hlubinka, D. and Hudecová, S. (2023). Efficient fully distribution-free center-outward rank tests for multiple-output regression & MANOVA. *J. Amer. Statist. Assoc.*, **118(543)**, 1923–1939.

Hampel, F. R. (1968). Contributions to the Theory of Robust Estimation. Ph.D. Thesis, Uinversity of California, Berkeley. ProQuest LLC, Ann Arbor, MI, 1968, 109 pp.

Hampel, F. R. (1971). A general qualitative definition of robustness. *Ann. Math. Statist.*, **42**, 1887–1896.

Hansen, B. E. (2008). Uniform convergence rates for kernel estimation with dependent data. *Econometric Theory*, **24**, 726–748.

Härdle, W. (1991). *Smoothing techniques: with implementation in S.* Springer, New York.

Härdle, W. and Marron, J. S. (1990). Semiparametric comparisons of regression curves. *Ann. Statist.* **18**, 63–89.

Härdle, W. and Mammen, E. (1991). Bootstrap methods in nonparametric regression. *NATO Adv. Sci. Inst. Ser. C: Math. Phys. Sci., 335.* Kluwer Academic Publishers Group, Dordrecht, 111–123.

Härdle, W., Hall, P. and Marron, J. S. (1988). How far are automatically chosen regression smoothing parameters from their optimum? (with discussion). *J. Amer. Statist. Assoc.* **83**, 86–99.

Hart, J. D. (1997). *Nonparametric smoothing and lack-of-fit tests.* Springer Series in Statistics. Springer-Verlag, New York.

Helland, I. S. (1982). Central limit theorems for martingales with discrete or continuous time. *Scand. J. Statist.*, **9**, 79–94.

Hernández, E. and Weiss, G. (1996). A first course on wavelets. *Stud. Adv. Math.* CRC Press, Boca Raton, FL.

Hettmansperger, T. P. (1984). *Statistical inference based on ranks.* Wiley Series in Probability and Mathematical Statistics: Probability and Mathematical Statistics. John Wiley & Sons, Inc., New York.

Hettmansperger, T. P. and McKean, J. W. (1978). Statistical inference based on ranks. *Psychometrika*, **43(1)**, 69–79.

Hjellvik, V. and Tjostheim, D. (1995). Nonparametric test of linearity for time series. *Biometrika* **82**, 351–368.

Hjellvik, V. and Tjøstheim, D. (1996). Nonparametric statistics for testing of linearity and serial independence. *J. Nonparametric Statist.* **6**, 223–251.

Hjort, N. L., McKeague, I. W. and Van Keilegom, I. (2009). Extending the scope of empirical likelihood. *Ann. Statist.*, **37**, 1079–1111.

Hodges, J. and Lehmann, E. (1956). The efficiency of some nonparametric competitors of the t-test. *Ann. Math. Statist.*, **27**, 324–355.

Hodges, J. L. and Lehmann, E. L. (1960). Comparison of the normal scores and Wilcoxon tests. *Proc. of the 4th Berkeley Symposium on Math. Statist. & Probab.* **1**, 307–317. University of California Press, Berkeley.

Hodges Jr., J. L. and Lehmann, E. L. (1963). Estimates of location based on rank tests. *Ann. Math. Stat.* **34**, 598–611.

Hoeffding, W. (1948). A class of statistics with asymptotically normal distribution. *Ann. Math. Statis.*, **19**, 293–325.

Hsing, T. and Wu, W. (2004).On weighted U-statistics for stationary processes. *Ann. Probab.*, **32**, 1600–1631.

Huber, P. J. (1964). Robust estimation of a location parameter. *Ann. Math. Statist.* **35**, 73-–101.

Huber, P. J. (1967). The behavior of maximum likelihood estimates under nonstandard conditions. *Proceeding of 5th Berkeley Symposium in Probability and Mathematical Statist.*, **1**, 221–233. Univ. Calif. Press, Berkeley, CA.

Huber, P. J. (1973). Robust regression: Asymptotics, Conjectures and Monet Carlo. *Ann. Math. Statist.* **1**, 799–821.

Huber, P. J. (1981). *Robust statistics.* Wiley Series in Probability and Mathematical Statistics. John Wiley & Sons, Inc., New York.

Jaeckel, L. A. (1972). Estimating regression coefficients by minimizing the dispersion of the residuals. *Ann. Math. Statist.*, **43**, 1449–1458.

Jensen, J. L. W. V. (1906). "Sur les fonctions convexes et les inégalités entre les valeurs moyennes". *Acta Mathematica.* **30(1)**, 175–193.

Jurečková, J. (1969). Asymptotic linearity of a rank statistic in regression parameter. *Ann. Math. Statist.*, **40**, 1889–1900.

Jurečková, J. (1971). Nonparametric estimate of regression coefficients. *Ann. Math. Statist.*, **42**, 1328–1338.

Kalbfleisch, J. D. and Prentice, R. L. (1980). *The statistical analysis of failure time data.* Wiley Series in Probability and Mathematical Statistics. John Wiley & Sons, New York-Chichester-Brisbane.

Kaplan, E. L. and Meier, P. (1958). Nonparametric estimation from incomplete observation. *J. Amer. Statist. Assoc.*, **53**, 457–481.

Khmaladze, E. V. (1981). A martingale approach in the theory of good -ness-of-fit tests. (Russian) *Teor. Veroyatnost. i Primenen.*, **26(2)**, 246–265.

Khmaladze, E. V. (2013). Note on distribution free testing for discrete distributions. *Ann. Statist.*, **41(6)**, 2979–2993.

Khmaladze, E. V. (2016). Unitary transformations, empirical processes and distribution free testing. *Bernoulli*, **22(1)**, 563–588.

Khmaladze, E. V. (2017). Distribution free testing for conditional distributions given covariates. *Statist. Probab. Lett.*, **129**, 348–354.

Khmaladze, E. V. (2021). Distribution-free testing in linear and parametric regression. *Ann. Inst. Statist. Math.*, **73(6)**, 1063–1087.

Khmaladze, E. V. and Koul, H. L. (2004). Martingale transforms goodness-of-fit tests in regression models. *Ann. Statist.*, **32**, 995–1034.

Khmaladze, E. V. and Koul, H. L. (2009). Goodness-of-fit problem for errors in nonparametric regression: distribution free approach. *Ann. Statist.*, **37**, 3165–3185.

Kiefer, J. and Wolfowitz, J. (1958). On the deviations of the empiric distribution function of vector chance variables. *Trans. Amer. Math. Soc.*, **87**, 173–186.

Kim, J. (2018). A fast algorithm for the coordinate-wise minimum distance estimation. *J. Statist. Computation and Simulation*, **88(3)**, 482–497, DOI: 10.1080/00949655.2017.1392527.

Kim, J. (2020). Implementation of a goodness-of-fit test through Khmaladze martingale transformation. *Comput. Statist.*, **35**, 1993–2017. https://doi.org/10.1007/s00180-020-00971-7.

Kim, Y. and Lee, S. (2002). On the Kolmogorov-Smirnov type test for testing nonlinearity in time series. *Comm. Statist. Theory Methods*, **31(2)**, 299–309.

King, E., Hart, J. D. and Wehrly, T. E. (1991). Testing the equality of two regression curves using linear smoothers. *Statist. & Probab. Letters*, **12**, 239–247.

Kloke, J. D. and McKean, J. W. (2012). Rfit: Rank-based estimation for linear models. The R Journal, 4, 57–64.

Kloke, J. D. and McKean, J. W. (2014). *Nonparametric statistical methods using R.* Boca Raton, FL: Chapman-Hall.

Koenker, R. (1997). ℓ_1 computation: An interior monologue. *L_1-Statistical Procedures and Related Topics. IMS Lecture Notes - Monograph Ser., 31.* Inst. Math. Statist., Hayward, CA.

Koenker, R. and Bassett, G. Jr. (1978). Regression quantiles. *Econometrica*, **46**, 33–50.

Koenker, R. and d'Orey, V. (1987). Algorithm AS 229: Computing regression quantiles. *J. Roy. Statist. Soc. Ser. C*, **36**, 383–393.

Koenker, R. and d'Orey, V. (1993). Remark on AS R92. A remark on Algorithm AS 229; Computing dual regression quantiles and regression rank scores. *J. Roy. Statist. Soc. Ser. C*, **43**, 410–414.

Kolmogorov A. N. (1933). Sulla determinazione empirica di una legge di distribuzione. *G. Ist. Ital. Attuari.*, **4**, 83–91.

Koroljuk, V. S. and Borovskich, Yu. V. (1994). *Theory of U-statistics*. Mathematics and its Applications. Vol. 273 (Translated by P. V. Malyshev and D. V. Malyshev from the 1989 Russian original ed.). Dordrecht: Kluwer Academic Publishers Group.

Koul, H. L. (1992). *Weighted empiricals and linear models*. Institute of Mathematical Statistics Lecture Notes—Monograph Series, **21**, Hayward, CA.

Koul, H. L. (1996). Asymptotics of some estimators and sequential residual empiricals in nonlinear time series. *Ann. Statist.*, **24**, 380–404.

Koul, H. L. (2002). *Weighted empirical processes in dynamic nonlinear models. 2nd edition.* In: **Lecture Notes Series in Statistics, 166**. Springer, New York, N.Y, USA.

Koul, H. L. (2006). Model Diagnostics via Martingale Transforms: A Brief Review. In *Frontiers in Statistics*, pp 183–206. Imperial College Press, London, UK. Edts: Jianqing Fan and Hira L. Koul.

Koul, H. L. (2011). Minimum distance lack-of-fit tests for fixed design. *J. Statist. Plann. Inference*, **141**, 65–79.

Koul, H. L. and Lahiri, S. N. (1994). On bootstrapping M–estimated residual porcesses in multiple linear regression. *J. Multivariate Anal.*, **49**, 255–265.

Koul, H. L. and Mimoto, N. (2012). A goodness-of-fit test for GARCH innovation density. *Metrika*, **75**, 127–149.

Koul, H. L. and Ni, P. (2004). Minimum distance regression model checking. *J. Statist. Plann. Inference*, **119(1)**, 109–141.

Koul, H. L. and Sakhanenko, L. (2005). Goodness-of-fit testing in regression: A finite sample comparison of bootstrap methodology and Khmaladze transformation. *Statist. Prob. Lett.* **74**, no. 3, 290-302.

Koul, H. L. and Schick, A. (1997). Testing for the equality of two nonparametric regression curves. *J. Statist. Plann. Inference*, **65(2)**, 293–314.

Koul, H. L. and Schick, A. (2003). Testing for superiority among two regression curves. *J. Statist. Plann. Inference*, **117(1)**, 15–33.

Koul, H. L. and Stute, W. (1999). Nonparametric model checks in time series. *Ann. Statist.* **27**, 204-237.

Koul, H. L. and Zhu, X. (2015). Goodness-of-fit of error distribution in nonparametric ARCH(1) model. *J. Mult. Analysis*, **137**, 141–160.

Koul, H. L., Sievers, G. L. and McKean, J. W. (1987). An estimator of the scale parameter for the rank analysis of linear models under general score functions. *Scand. J. Statist.*, **14**, 131–141.

Koul, H. L., Stute, W. and Li, F. (2005). Model diagnosis for SETAR time series. *Statistca Sinica.* **15**, no. 3, 795-817.

Koul, H. L., Susarla, V. and Van Ryzin, J. (1981). Regression analysis with randomly right-censored data. *Ann. Statist.*, **9**, 1276–1288.

Koziol, J. A. and Green, S. B. (1976). A Cramér–von Mises statistic for randomly censored data. *Biometrika*, **63**, 465–474.

Kruskal, W. H. (1952). A nonparametric test for the several sample problem *Ann. Math. Statist.*, **23**, 525–540.

Kruskal, W. H. and Wallis, W. A. (1952). Use of ranks in one-criterion variance analysis. *J. Amer. Statist. Assoc.*, **47**, 583–621.

Kulasekera, K. B. (1995). Comparison of regression curves using quasi - residuals. *J. Amer. Statist. Assoc.*, **90**, 1085-1093.

Lahiri, S. N. (1989). Bootstrap approximation to the distributions of M estimators. Ph.D. Thesis, Department of Statistics and Probability, Michigan State University.

Lahiri, S. N. (2003). *Resampling methods for dependent data.* Springer Ser. Statist. Springer-Verlag, New York.

Lai, T. L. and Ying, Z. (1991). Large sample theory of a modified Buckley–James estimator for regression analysis with censored data. *Ann. Statist.*, **19**, 1370–1402.

Lai, T., Ying, Z. and Zheng, Z. (1995). Asymptotic normality of a class of adaptive statistics with applications to synthetic data methods for censored regression. *J. Mult. Analysis*, **52**, 259–279.

Lawless, J. F. (2003). *Statistical Models and Methods for Lifetime Data.* John Wiley & Sons, Inc. Hoboken, New Jersey.

Lee, A. J. (1990). *U-Statistics: Theory and Practice.* Marcel Dekker, New York.

Lee, S. and Na, S. (2002). On the Bickel–Rosenblatt test for first-order autoregressive models. *Statist. Probab. Lett.* **56**, no. 1, 23–35.

Lehmann, E. L. (1959). *Testing statistical hypotheses.* John Wiley & Sons, Inc., New York; Chapman & Hall, Ltd., London.

Lehmann, E. L. (1963). Nonparametric confidence intervals for a shift parameter. *Ann. Math. Statist.*, **34**, 1507–1512.

Lehmann, E. L. (1975). *Nonparametrics.* Holden Day, San Francisco.

Lehmann, E. L. (1983). *Theory of point estimation.* Wiley Series in Probability and Mathematical Statistics. John Wiley & Sons, Inc., New York.

Lehmann, E. L. and Romano, J. P. (2005). *Testing Statistical Hypotheses.* Springer Texts Statist. Springer, New York.

Leurgans, S. (1987). Linear models, random censoring and synthetic data. *Biometrika*, **63**, 449–464.

Liu, R. Y. (1992). Data depth and multivariate rank tests. In L_1 *Statistics and Related Methods*, ed. Y Dodge, pp. 279–294. Amsterdam: North-Holland

Liu, R. Y. and Singh, K. (1993). A quality index based on data depth and multivariate rank tests. *J. Am. Stat. Assoc.*, **88**, 257–260.

Liu, R. Y. and McKean, J. W. (2016). *Robust Rank-Based and Nonparametric Methods.* Springer Proceedings in Mathematics & Statistics, **168**. Springer Nature, Switzerland.

Loève, M. (1963). *Probability Theory. Third edition.* D. Van Nostrand Co., Inc., Princeton, N. J.

Mack, Y. P. and Silverman, B. W. (1982). Weak and strong uniform consistency of kernel regression estimates. *Z. Wahrsch. Gebiete*, **61**, 405–415.

Major, P. (1994). Asymptotic distribution of weighted U-statistics. *Ann. Probab.*, **22**, 1514–1535.

Mammen, E. (1989). Asymptotics with increasing dimension for robust regression with application to bootstrap. *Ann. Statist.* **17**, 382–400.

Mammen, E. (1992). Bootstrap, wild bootstrap, and asymptotic normality. *Probab. Theory Related Fields*, **93(4)**, 439–455.

Mann, H. and Whitney, D. R. (1947). On a test whether one of two random variables is stochastically larger than the other. *Ann. Math. Statist.*, **18**, 50–60.

Marden, J. (1999). Multivariate rank tests. In *Design of Experiments and Survey Sampling*, ed. S. Ghosh, pp. 401–32. Marcel Dekker, New York.

Massart, P. (1990). The tight constant in the Dvoretzky-Kiefer-Wolfowitz inequality. *Ann. Probab.*, **18 (3)**, 1269–1283.

McKeague, I. W. and Zhang, M. J. (1994). Identification of nonlinear time series from the first order cumulative characteristics. *Ann. Statist.*, **22**, 495–514.

McKean, J. W. and Hettmansperger, T. P. (2016). Rank-based analysis of linear models and beyond: a review. *Robust rank-based and nonparametric methods*, 1–24, Springer Proc. Math. Stat., **168.**

Müller, U. U., Schick, A. and Wefelmeyer, W. (2007). Estimating the error distribution function in semi-parametric regression. *Statist. Decisions*, **25**, 1–18.

Müller, U. U., Schick, A. and Wefelmeyer, W. (2009). Estimating the error distribution function in nonparametric regression with multivariate covariates. *Statist. Probab. Lett.*, **79**, 957–964.

Nadaraya, E. A. (1964). On Estimating Regression. *Theor. of Probab. & Its Appl.*, **9(1)**, 141–142. doi:10.1137/1109020.

Naaman, M. (2021). On the tight constant in the multivariate Dvoretzky-Kiefer-Wolfowitz inequality. *Statist. Probab. Lett.*, **173**, Paper No. 109088, 8 pp.

Nelson, W. (1969). Hazard plotting for incomplete failure data. *J. Qual. Technol.* **1**, 27–52.

Nelson, W. (1972). Theory and applications of hazard plotting for censored failure data. *Technometrics*, **14**, 945–965.

Nelson, W. (1982). *Applied life data analysis*. Wiley, New York.

Nikabadze, A. M. (1987). A method for constructing likelihood tests for parametric hypotheses in $\mathbb{R}^m$. *Teor. Veroyatn. Primen.*, **32**, 594–598.

Nikabadze, A. M. and Stute, W. (1997). Model checks under random censorship. *Statist. Probab. Lett.*, **32(3)**, 249–259.

Oja, H. (1999). Affine invariant multivariate sign and rank tests and corresponding estimates: a review. *Scand. J. Stat.*, **26**, 319–343.

Oja, H. (2010). *Multivariate Nonparametric Methods with R: An Approach Based on Spatial Signs and Ranks*. New York: Springer.

Oja, H. and Paindaveine, D. (2005). Optimal signed-rank tests based on hyperplanes. *J. Stat. Plann. Inference*, **135**, 300–323.

Oja, H. and Randles, R. H. (2004). Multivariate nonparametric tests. *Stat. Sci.*, **19**, 598–605.

O'Neil, K. A. and Redner, R. A. (1993). Asymptotic distribution of weighted U-statistics of degree two. *Ann. Probab.*, **21**, 1159–1169.

Owen, A. B. (1988). Empirical likelihood ratio confidence intervals for a single functional. *Biometrika*, **75, no. 2**, 237–249.

Owen, A. B. (1990). Empirical likelihood ratio confidence regions. *Ann. Statist.* **18**, 90–120.

Owen, A. B. (1995). Nonparametric likelihood confidence bands for a distribution function. *J. Amer. Statist. Assoc.*, **90(430)**, 516–521.

Owen, A. B. (2001). *Empirical Likelihood, 1st Edition*. Chapman & Hall/ CRC, New York.

Owen, D. B. (1962). *Handbook of statistical tables*. Addison-Wesley Publishing Co., Inc., Reading, Mass.-London.

Parzen, E. (1962). On estimation of a probability density function and mode. *Ann. Math. Statist.*, **33**, 1065–1076.

Peng, H. and Schick, A. (2013). Empirical likelihood approach to goodness of fit testing. *Bernoulli*, **19(3)**, 954–981.

Peng, H. and Schick, A. (2018). Maximum empirical likelihood estimation and related topics. *Electron. J. Stat.*, **12(2)**, 2962–2994.

Pfanzagl, J. (1994). *Parametric statistical theory. With the assistance of R. Hamböker.* De Gruyter Textbook. Walter de Gruyter & Co., Berlin.

Poggi, J. M. and Portier, B. (1997). A test of linearity for functional autoregressive models. *J. Time Ser. Anal.* **18(6)**, 615–639.

Politis, D. N. (2014). Bootstrap confidence intervals in nonparametric regression without an additive model. *Springer Proc. Math. Stat.*, **74**, 271–282. Springer, New York.

Pollard, D. (1984). *Convergence of stochastic processes. Springer Ser. Statist.*, Springer-Verlag, New York.

Portnoy, S. (1984). Asymptotic behavior of M estimators of p regression parameters when p^2/n is large; I. Consistency. *Ann. Statist.*, **12**, 1298–1309.

Portnoy, S. (1985). Asymptotic behavior of M estimators of p regression parameters when p^2/n is large; II. Normal approximation. *Ann. Statist.*, **13(4)**, 1403–1417.

Portnoy, S. (1997). On Computation of Regression Quantiles: Making the Laplacian Tortoise Faster. *Lecture Notes-Monograph Series, 31* on L_1*-Statistical Procedures and Related Topics*,187–200. Institute of Mathematical Statistics.

Portnoy, S. and Koenker, R. (1997). The Gaussian hare and the Laplacian tortoise: computability of squared-error versus absolute-error estimators. *Statist. Sci.*, **12(4)**, 279—300.

Prakasa Rao, B. L. S. (1983). *Nonparametric functional estimation*. Probability and Mathematical Statistics. Academic Press, Inc. [Harcourt Brace Jovanovich, Publishers], New York.

Puri, M. L. and Sen, P. K. (1971). *Nonparametric Methods in Multivariate Analysis*. John Wiley & Sons, Inc., New York-London-Sydney.

Qin, J. and Lawless, J. F. (1994). Empirical likelihood and general estimating equations *Ann. Statist.*, **22**, 300–325.

Rao, C. R. (1973). *Linear Statistical Inference and its Applications: Second Edition*. John Wiley & Sons, Inc. New York.

Rao, K. C. (1972). The Kolmogoroff, Cramer-von Mises chi squares statistics for goodness-of-fit tests in the parametric case. (abstract). *Bull. Inst. Math. Statist.*, **1**, 87.

Raz, J. (1990). Testing for no effect when estimating a smooth function by nonparametric regression: a randomization approach. *J. Amer. Statist. Assoc.*, **85**, 132-138.

Resnick, S. I. (1999). *A Probability Path*. Birkhäuser Boston, Inc., Boston, MA.

Rifi, M. and Utzet, F. (2000). On the asymptotic behavior of weighted U-statistics. *J. Theoret. Probab.*, **13(1)**, 141–167.

Roberts, B., Haywood, J. and Swordson, E. (2023). Distribution free testing for the family of Laplace distributions. *Comm. Statist.- Simulation and Computation*,
DOI: 10.1080/03610918.2023.2258560.

Rosenblatt, M. (1956). Remarks on some nonparametric estimates of a density function. *Ann. Math. Statist.*, **27**, 832–837.

Rudemo, M. (1982). Empirical choice of histogram and kernel density estimators. *Scand. J. Statist.*, **9**, 65–78.

Rudin, W. (1974). *Real and Complex Analysis (2nd edition)*. New York: McGraw - Hill.

Rüschendorf, L. (1977). Consistency of estimators for multivariate density functions and for the mode. *Sankhyā: The Indian Journal of Statistics*, **39(3)**, Series A, 243–250.

Scheffé, H. (1947). A useful convergence theorem for probability distributions. *Ann. Math. Statist.*, **18**, 434–438.

Schick, A. (2010). *Empirical Likelihood Lecture Notes*. Michigan State University.

Schick, A. and Wefelmeyer, W. (2004). Root-n consistent density estimators for sums of independent random variables. *J. Nonparametr. Statist.*, **16**, 925–935.

Schick, A. and Wefelmeyer, W. (2007). Root-n consistent density estimators of convolutions in weighted L_1-norms. *J. Statist. Plann. Inference*, **137**, 1765–1774.

Schick, A. and Yu, Q. (2000). Consistency of the GMLE with mixed case interval-censored data. *Scand. J. Statist.*, **27**, 45–55.

Schuster, E. and Yakowitz. (1979). Contributions to the theory of nonparametric regression, with application to system identification. *Ann. Statist.*, **7**, 139–149.

Schweder, T. (1975). Window estimation of the asymptotic variance of rank estimators of location. *Scand. J. Statist.*, **2**, 113–126.

Scott, D. W. (1992). *Multivariate Density Estimation: Theory, Practice, and Visualization*. John Wiley & Sons, Inc. New York.

Sen, P. K. (1963a). On the estimation of relative potency in dilusion (-direct) assays by distribution-free method. *Biometrics*, **19**, 532–552.

Sen, P.K. (1963b). On the properties of U-statistics when the observations are not independent. I. Estimation of non-serial parameters in some stationary stochastic process. *Calcutta Statist. Assoc. Bull.*, **12**, 69–92.

Sen, P. K. (1966). On a distribution–free method of estimating asymptotic efficiency of a class of nonparametric tests. *Ann. Math. Statist.*, **37**, 1759–1770.

Sen, P.K. (1992). Introduction to Hoeffding (1948) A Class of Statistics with Asymptotically Normal Distribution. In: Kotz, S., Johnson, N. L. *Breakthroughs in Statistics*, Vol I, pp 299–307. Springer-Verlag.

Serfling, R. J. (1980). *Approximation theorems of mathematical statistics*. John Wiley & Sons. New York.

Serfling, R. J. (2002a). Generalized quantile processes based on multivariate depth functions, with applications in nonparametric multivariate analysis. *J. Multivariate Anal.*, **83(1)**, 232–247.

Serfling, R. J. (2002b). Quantile functions for multivariate analysis: approaches and applications. *Statist. Neerlandica*, **56(2)**, 214–232.

Shao, J. and Tu, D. (1995). *The Jackknife and Bootstrap*. Springer-Verlag New York, Inc. N.Y.

Shorack, G. (1982). Bootstrapping robust regression. *Comm. Statist., A*, **11**, 961–972.

Silverman, B. W. (1986). *Density Estimation for Statistics and Data Analysis*. Chapman & Hall.

Simonoff, J. S. (1996). *Smoothing methods in Statistics*. Springer–Verlag New York, Inc.

Singh, K. (1981). On the asymptotic accuracy of Efron's bootstrap. *Ann. Statist.*, **9(6)**, 1187–1195.

Srihera, R. and Stute, W. (2010). Nonparametric comparison of regression functions. *J. Mult. Analysis*, **101**, 2039–2059.

Srinivasan, C. and Zhou, M. (1994). Linear regression with censoring. *J. Mult. Analysis*, **49**, 179–201.

Stigler, S.M. (1969). Linear functions of order statistics. *Ann. Math. Statist.*, **40**, 770–788.

Stigler, S. M. (1974). Linear functions of order statistics with smooth weight functions. *Ann. Statist.*, **2**, 676–693.

Stute, W. (1995). The central limit theorem under random censorship. *Ann. Statist.*, **23**, 422–439.

Stute, W. (1997). Nonparametric model checks for regression. *Ann. Statist.*, **25**, 613–641.

Stute, W. and Wang, J.-L. (1993). The strong law of under random censorship. *Ann. Statist.*, **21**, 1591–1607.

Stute, W. and Zhu, L. (2002). Model checks for generalized linear models. *Scand. J. Statist.* **29**, 535–545.

Stute, W., González-Manteiga, W. and Presedo Quindimil, M. (1998). Bootstrap approximations in model checks for regression. *J. Amer. Statist. Assoc.* **93**, no. 441, 141–149.

Stute, W., Thies, S. and Zhu, Li-Xing. (1998). Model checks for regression: an innovation process approach. *Ann. Statist.*, **26**, 1916–1934.

Stute, W., Presedo Quindimil, M., González Manteiga, W. and Koul, H. L. (2006). Model checks of higher order time series. *Statist. Probab. Lett.*, **76 (13)**, 1385–1396.

Su, J. Q. and Wei, L. J. (1991). A lack-of-fit test for the mean function in a generalized linear model. *J. Amer. Statist. Assoc.* **86**, 420–426.

Susarla, V. and van Ryzin, J. (1976). Nonparametric Bayesian estimation of survival curves from incomplete observations. *J. Amer. Satisti. Assoc.*, **71**, 897–902.

Susarla, V. and van Ryzin, J. (1978). Empirical Bayes estimation of distribution (survival) function from right censored observations. *Ann. Statist.*, **6(4)**, 740–754.

Susarla, V. and Van Ryzin, J. (1980). Large sample theory for an estimator of the mean survival time from censored samples. *Ann. Statist.*, **8(5)**, 1002–1016.

Susarla, V., Tsai, W.-Y. and Van Ryzin, J. (1984). A Buckley-James-type estimator for the mean with censored data. *Biometrika*, **71(3)**, 624–625.

Thewarapperuma, P. S. (1987). On estimation of some density functionals under regression and one sample model. Ph.D. Thesis, Michigan State University.

Tsiatis, A. A. (2006). *Semiparametric theory and missing data.* Springer Series in Statistics. Springer, New York.

Tsigroshvili, Z. (1998). Some notes on goodness-of-fit tests and innovation martingales. (English. English, Georgian summary) *Proc. A. Razmadze Math. Inst.* **117**, 89–102.

van Eeden, C. (1972). An analogue of signed ranks statistics of Jrečková's asymptotic linearity theorem for rank statistics. *Ann. Math. Statist.* **43**, 791–802.

van der Vaart, A. W. and Wellner, J. (1996). *Weak Convergence and Empirical Processes with Applications to Statistics.* Springer-Verlag, New York.

Vellaisamy, P. and Punnen, A. P. (2001). On the nature of the binomial distribution. *J. Appl. Probab.*, **38**, 36–44.

Vorličková, D. (1969). Asymptotic properties of rank tests under discrete distributions. *Z. Wahrscheinlichkeitstheorie und Verw. Gebiete*, **14**, 275–289.

Vorličková, D. (1972). Asymptotic properties of rank tests of symmetry under discrete distributions. *Ann. Math. Statist.*, **43**, 2013–2018.

Wallis, W. A. (1952). Rough-and-ready statistical tests. *Industrial Quality Control*, **8**, 35–40.

Wand, M. P. and Jones, M. C. (1995). *Kernel Smoothing*. Chapmann & Hall / CRC. U.S.A.

Wang, J. and Yang, L. (2009). Polynomial spline confidence bands for regression curves *Statist. Sinica.*, **19(1)**, 325–342.

Watson, G. S. (1964). Smooth regression analysis. *Sankhya: The Indian Journal of Statistics, Series A. 26 (4)*, 359–372.

Wilcoxon, F. (1945). Individual comparison by ranking methods. *Biometrics Bull.*, **1**, 80–83.

Wilcoxon, F. (1947). Probability tables for individual comparison by ranking methods. *Biometrics*, **3**, 119–122.

Wilks, S. S. (1938). The large sample distribution of the likelihood ratio for testing composite hypotheses. *Ann. Math. Statist.*, **9**, 60–62.

Wolfowitz, J. (1953). Estimation by minimum distance method. *Ann. Inst. Stat. Math.*, **5**, 9-23.

Wolfowitz, J. (1954). Estimation of the components of stochastic structures. *Proc. Nat. Acad. Sci.*, **40**, 602-606.

Wolfowitz, J. (1957). Minimum distance estimation method. *Ann. Math. Statist.* **28**, 75-88.

Yang, S. (1991). Hellinger distance estimation of parameter in the random censorship model. *Ann. Statist.*, **19(2)**, 579–602.

Yang, S. (1995). Koul-de Wet-type estimation in censored regression using the conditional empirical process. *J. Statist. Plann. Inference*, **46(3)**, 351–363.

Yang, S. and Koul, H. L. (1995). Minimum distance estimation of the center of symmetry with randomly censored data. *Metrika*, **42(2)**, 79–97.

Zheng, Z. (1984). Regression analysis with censored data. Ph.D. Thesis, Columbia University, New York.

Zhou, M. (1992a). Asymptotic normality of the 'synthetic data' regression estimators for censored data. *Ann. Statist.*, **20(2)**, 1002–1021.

Zhou, M. (1992b). M estimation in censored linear models. *Biometrika*, **79(4)**, 837–841.

Zhou, M. (1999). Regression analysis with censored data: Extensions of Koul - Susarla - Van Ryzin approach. *Statist. & Probab. Letters*, **41(3)**, 229–236.

Zuo, Y. and He, X. (2006). On the limiting distributions of multivariate depth-based rank sum statistics and related tests. *Ann. Stat.*, **34**, 2879–2896.

Zuo, Y. and Serfling, R. J. (2000). General notions of statistical depth function. *Ann. Statist.*, **28**, 461–482.

Subject Index

Author Index